MOLECULAR ELECTRONICS

Biosensors and Biocomputers

MOLECULAR ELECTRONICS

Biosensors and Biocomputers

Edited by
Felix T. Hong

Wayne State University
Detroit, Michigan

Plenum Press • New York and London

Library of Congress Cataloging-in-Publication Data

Symposium on Molecular Electronics: Biosensors and Biocomputers (1987
Santa Clara, Calif.)
Molecular electronics : biosensors and biocomputers / edited by
Felix T. Hong.
p. cm.
"Proceedings of the Office of Naval Research Symposium on
Molecular Electronics: Biosensors and Biocomputers, held in
conjunction with the Nineteenth Annual Meeting of the Fine Particle
Society, held July 19-22, 1988, in Santa Clara, California"--T.p.
verso.
Includes bibliographical references.
ISBN 978-1-4615-7484-2 ISBN 978-1-4615-7482-8 (eBook)
DOI 10.1007/978-1-4615-7482-8
1. Molecular electronics--Congresses. 2. Biosensors--Congresses.
3. Conscious automata--Congresses. I. Hong, Felix T. II. Fine
Particle Society. Meeting (19th : 1988 : Santa Clara, Calif.)
III. Title.
TK7874.S938 1987
629.8'92--dc20 89-26645
CIP

Proceedings of the Office of Naval Research
and the National Science Foundation Symposium on
Molecular Electronics: Biosensors and Biocomputers,
held in conjunction with the Ninteenth Annual Meeting of
the Fine Particle Society, held July 19-22, 1988,
in Santa Clara, California

Softcover reprint of the hardcover 1st edition 1989

A Division of Plenum Publishing Corporation
233 Spring Street, New York, N.Y. 10013

IN MEMORIAM

DR. FORREST L. CARTER

1930–1987

A Pioneer in Molecular Electronics

PREFACE

The dream of developing a biocomputer should not be dismissed as a sheer fantasy. Although there is naturally some doubt as to whether it is possible to design a computer using carbon-based components as in living organisms, instead of silicon-based components as in existing computers, the fact that an average brain often outperforms the most sophisticated computer in terms of the complexity of tasks, if not in terms of speed, is a living testimony to this possibility. The remaining question is to what extent a biocomputer can mimic a living organism and whether it is possible to design and fabricate such a biocomputer within the foreseeable future. This volume does not attempt to provide immediate and exact answers to these questions but instead attempts to provide a vision and a progress report of the initial efforts. This volume is mainly a collection of papers presented at the Symposium on Molecular Electronics - Biosensors and Biocomputers, sponsored by the Division of Biotechnology, Health and Environment of the Fine Particle Society, held from July 19-22, 1989 at the Society's 19th Annual Meeting in Santa Clara, California. Also included are articles contributed by those who planned to attend the conference but were unable to do so.

The emergence of the field of molecular electronics is largely the consequence of one person's crusade, that of Forrest L. Carter. As described by the leading article written by Hank Wohltjen, a former colleague of Forrest L. Carter, he was a staunch believer of the possibility of molecular electronic devices and biocomputers. He was a pioneer of molecular electronics and often appeared to be a lone prophet before his time. In recent years, he succeeded in gathering an ever increasing number of followers. Regrettably, he passed away during the planning of this Symposium. Thus, the Symposium which was originally planned to honor Forrest L. Carter became a Memorial Symposium, instead. In 1988 we also had the sad news of the death of Richard Feynman, who was hailed as the Godfather of Molecular Electronics (NanoTechnology) by the contributors of an article in this volume, and the death of Yonosuke Kobatake, who was a contributor to this volume. This volume thus also serves as a tribute to these two scientists.

Molecular electronics is the ideal meeting ground of a diverse group of investigators - physicists, chemists, biologists, computer scientists, engineers, etc. It is a truly interdisciplinary and multidisciplinary endeavor. Therefore, the grouping of articles is rather arbitrary, merely a matter of convenience and by no means a suggestion of the recommended reading sequence.

The keynote lecture was given by Hans Kuhn, a pioneer in the study of organized supramolecular assemblies, using the celebrated Langmuir-Blodgett (LB) technique. The LB technique was originally developed by Irving Langmuir and Katharine Burr Blodgett, and has gained an increasing popularity in recent years, especially in the field of molecular electronics. Although the section titles imply major emphases by investigators in molecular electronics, the absence of some key words does not imply the opposite. For example, synthetic conducting polymers have become a *tour de force* in molecular electronics. As another example, the advent of the technique of site-directed mutagenesis has made protein engineering a highly emphasized discipline in its own right. The readers will find these examples in this volume. On the other hands, I believe that there is really no fundamental difference between synthetic polymers and naturally occurring polymers as far as molecular electronics is concerned; many techniques for fabricating assemblies and for characterization of assemblies may be shared. Speaking of fabricating molecular assemblies, self-assembly, which is the hallmark of a living system, is a goal still largely eluding the investigators of molecular electronics. However, significant results have already appeared in the limited utilization of the self-assembly property of certain synthetic and biological molecules, as exemplified by the use of the LB technique and the technique of forming artificial bilayer lipid membranes (BLM).

Finally, there have been some debates about the appropriateness of using the term "molecular electronics". For me, the term "electronics" is generic, and is not restricted to the reference of electron movements alone but should also include movements of protons, nuclei, and even photons. The titles of the Symposium bring biosensors under the coverage of molecular electronics and place equal emphasis on the (short term) goal of biosensor development and the (long term) goal of biocomputer development because they share the common infrastructure.

The organizer of the Symposium would like to acknowledge the financial support of the Office of Naval Research and the National Science Foundation. The Forrest L. Carter Lecture was supported by Mitsubishi Electric Company, Tokyo, and Wayne State University, Detroit, Michigan. The unfailing support of Yosry A. Attia, Chairman of the Division of Biotechnology, Health and Environment of the Fine Particle Society, is also deeply appreciated. Significant editorial assistance provided by Filbert H. Hong during the preparation of this volume is also acknowledged.

Felix T. Hong

CONTENTS

MOLECULAR DEVICE AND BIOSENSOR TECHNOLOGY

GENERAL AND THEORETICAL ASPECTS OF MOLECULAR ELECTRONICS

FORREST L. CARTER

A PIONEER IN MOLECULAR ELECTRONICS

Hank Wohltjen

Microsensor Systems, Inc.
Springfield, VA 22151

Prior to his untimely death in 1987, Forrest L. Carter was employed as a chemist by the U.S. Naval Research Laboratory in Washington, DC. In 1978, he spent a sabbatical year at the CNRS X-Ray Laboratory in Grenoble, France. From 1957 to 1964 he worked at the Westinghouse Research Laboratory in Pittsburgh, Pennsylvania. He has published over 80 scientific papers and edited several books on the topic of Molecular Electronic Devices.

His staunch belief in the molecular electronic device concept led him to organize three seminal international symposia on the subject.[1-3] In the MED field, he is perhaps best known for his theoretical investigations of soliton transport and molecular means for switching and storing information using solitons. Experimentally, Dr. Carter had a long-term interest in inorganic material preparation and characterization primarily by x-ray crystallography and electron spectroscopy. His theoretical interests included the development of bidirectional orbitals for valence bonding for high coordination compounds, interrelating chemical bond information to the shape on the Fermi surface in reciprocal and k-space, a mathematical expression for the coordination number and a calculation of the polyhedral cell's atomic volume.

He was a member of the American Chemical Society, American Crystallographic Association, and Washington Crystal Colloquium. Dr. Carter received the A.B. degree (1951) in chemistry and physics from Harvard University and Ph.D. degree (1955) in chemistry from the California Institute of Technology.

Among his friends he was noted for his outstanding dancing, fencing, and diving abilities, his keen wit, and his charming personality.

REFERENCES

1. F. L. Carter, ed., "Molecular Electronic Devices," 386 p., Marcel Dekker, New York (1982).
2. F. L. Carter, ed., "Molecular Electronic Devices II," 825 p., Marcel Dekker, New York (1987)
3. F. L. Carter, R. E. Siatkowski, and H. Wohltjen, eds., "Molecular Electronic Devices," 685 p., North-Holland, Amsterdam (1988).

FORREST L. CARTER LECTURE: ORGANIZED MONOLAYERS - BUILDING BLOCKS IN CONSTRUCTING SUPRAMOLECULAR DEVICES

Hans Kuhn

Max-Planck Institute for Biophysical Chemistry
Göttingen, FRG

PROLOGUE

I first met Forrest Carter at his second meeting on Molecular Electronic Devices.[1] Dr. Carter had brought together a broadly interdisciplinary group of people - computer scientists, biologists, physicists, and chemists - to contribute to a very stimulating meeting. He gave a most enthusiastic view on the future possibilities of molecular switches, molecular computers, and cellular automata. Dr. Carter's enthusiasm fascinated many scientists. By his ideas on soliton switching and its implications to molecular electronics, preparative chemists were stimulated to new synthetic activities. Dr. Carter's excitement and his devotion to molecular electronics generated interest in many parts of the world.

I observed this development not without skepticism. On the one hand, I felt very similar to Dr. Carter. I felt that chemistry is in the beginning of a fascinating transition process, interest changing from pure substances, from isolated molecular species, to complex supramolecular systems. I felt that chemists will approach a new goal of synthesizing molecules that self-organize to form complex assemblies where molecules interlock and interact in a purpose-oriented manner, like parts in a machine. I felt that every effort should be made in supporting such attempts towards a supramolecular engineering which would revolutionize technology in many fields.

On the other hand, I feared that the problems in molecular engineering are so difficult that there would be no hope for fast success in challenging technological approaches, that too many would start with much ambition and expectation, become disappointed, and give up again.

Today, as I see it, the problems are not easier, but the scientific community begins to realize more and more the enormous difficulty of supramolecular engineering, the necessity of many small steps by many researchers, of a healthy optimism, driving fruitful work but expecting no more than a slow and steady development.

So I think Forrest Carter's immense effort gave an impetus in the right direction. He has made an important seminal contribution to the very exciting and challenging adventure of molecular and supramolecular engineering.

INTRODUCTION

My objective in this presentation is not so much to consider the great goals in the far future viewed by Forrest Carter but to focus more on some immediate problems in supramolecular engineering, the fabrication of some planned arrangements of molecules of various kinds.

As I have already mentioned, constructing molecules so that they self-organize, forming an assembly where each molecule has its individual place, is a most exciting new goal in chemistry, and should have a great potential interest in any field where materials with basically new properties depending on the exact localization of each molecular component is required, e.g. in the developments of microelectronics, integrated optics, memories with high storage capacity, micro-sensors, pyro-/piezo-/electroluminescent devices, catalysts for complex reactions, arrangements for economic solar energy conversion, and molecular computers. I think that a world of new chemically achievable possibilities is opening up and that we are in the beginning of a most creative phase of preparative chemistry.

However, today such clever molecules have not yet become available. Techniques to fabricate simple supramolecular machines with less intricate molecules are of interest, on the one hand to set a paradigm for molecular and supramolecular engineering, and to demonstrate that simple molecular machinery can actually be made, while on the other hand to learn more about design principles, and about the self-assembly of complex molecular systems.

A simple paradigm, which is an easy way to achieve organized arrangements of molecules acting as molecular functional units, proposed in the early 1960s, is to incorporate functional component molecules into monolayers of appropriate matrix molecules, and to assemble the monolayers in a planned manner. In this way individual molecules are picked up and manipulated, organized assemblies are constructed, and simple supramolecular functional units are realized.[2] With this paradigm, the Langmuir-Blodgett technique[3] for constructing multilayers of fatty acids, which had been developed in the early 1930s, is used for a new purpose.[4]

The monolayers are produced at the water surface and deposited on a glass slide in a planned sequence. In the present case dye molecules with hydrocarbon substituents are incorporated and tightly packed in a fatty acid matrix layer. Dye D is in the first layer, dye E in the second layer (Fig. 1). The hydrocarbon chains keep the chromophores D and E at a fixed distance. D is excited with UV light.

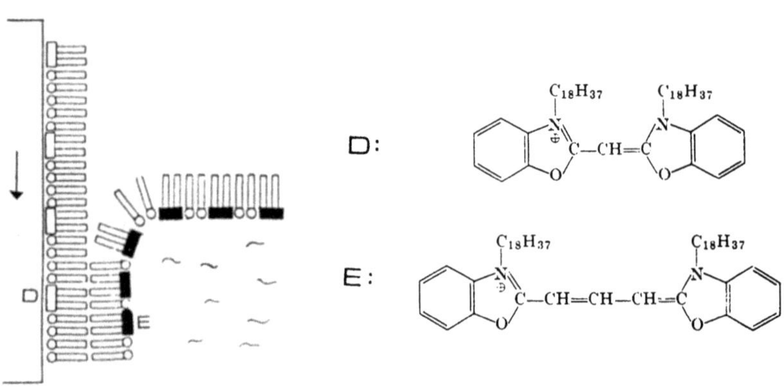

Fig. 1. Supramolecular structure formed by assembling dyes in fatty acid salt monolayers.

This early example of an artificial supramolecular functional unit is a good case to focus on the principle. The solid parts of the machinery are dye molecules: the blue fluorescing dye incorporated in the first layer acting as the energy donor D, and the yellow fluorescing dye acting as the acceptor E. The moving part of the machinery is the quantum of excitation energy transferred from donor to acceptor. The blue fluorescence of D is quenched and the sensitized yellow fluorescence of E appears. Energy transfer (seen by the dramatic change in fluorescence color when the distance between the donor and the acceptor molecule is changed) is a simple measure of distances in the molecular range.

It has proved to be very useful in the development of new techniques to manipulate molecular assemblies at molecular dimensions with molecular accuracy (contacting and separating individual monolayers in an assembly, Fig. 2).[5]

The monolayer technique as a method to arrange molecules in a planned manner is very useful for obtaining simple organized arrangements of cooperating molecules. It should be of interest to discuss some of these simple cases which have no practical use but may stimulate future developments by making chemists become aware of the importance of constructing supramolecular machinery. The investigation of these simple cases is helpful in developing new techniques and more complex systems.

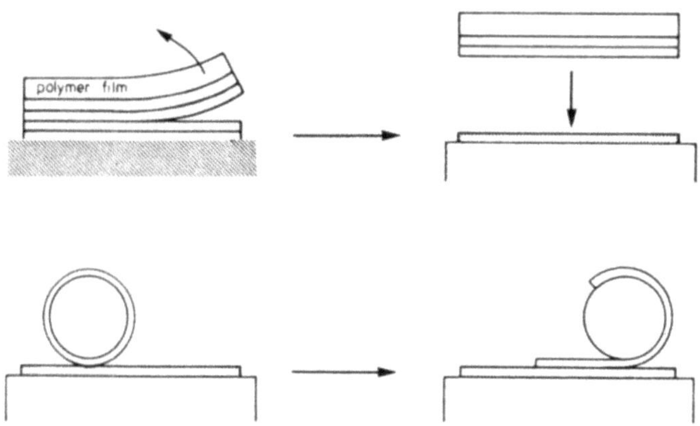

Fig. 2. Various techniques to manipulate monolayer assemblies: Cleavage, contacting, and printing monolayers.

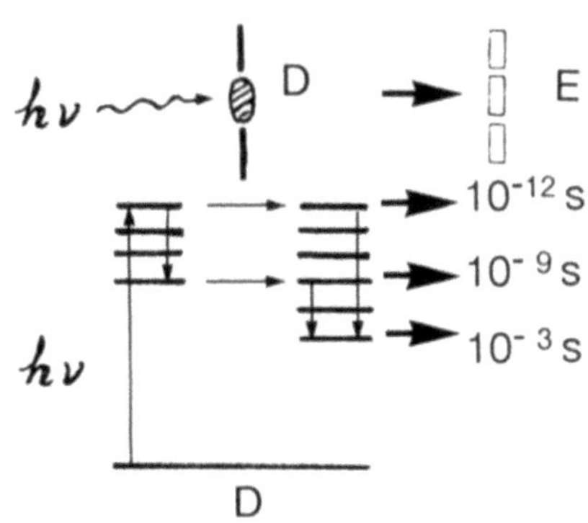

Fig. 3. Path of de-excitation of the dye D by appropriately arranging the acceptor E.

Path of De-Excitation

The arrangement in Fig. 1 (photoexcited dye D and energy acceptor E) is useful for studying decay processes after photoexcitation. Depending on the exact nature of D and E and on their distance, the excitation energy can be transferred from D to E before or after vibronic relaxation, or after transition to the triplet state (Fig. 3). Thus the path of de-excitation can be investigated.[6]

Echo of Radiation Field of Excited Molecules Stimulating Life Time of the Excited State

A striking effect can be investigated by fixing a luminescent molecule at a small distance from a silver surface, using monolayers as exact spacers. The life time of the luminescence depends on this distance. The oscillator representing the luminescent molecule is affected by the echo of the radiation field arriving at the molecule with a retardation of 10^{-15} s. The molecule is interacting with its own radiation field reflected at the metal surface, arriving at the excited molecule long before the molecule finally emits a light quantum.[7]

Dipole Moment and Polarizability of the Excited State from the Shift of the Absorption Band in an Electric Field

Dipole moments and polarizabilities of excited states can be measured in monolayer assemblies sandwiched between two semitransparent metal electrodes. The shift of the absorption band in an electric field is measured. The evaluation is easy, due to the uniform orientation of the chromophores in the monolayer.[8]

Surface Plasmon Dye Interaction

Surface plasmon dye interactions can be studied in organized assemblies.[9] The maximum energy transfer from an excited dye to a surface plasmon is achieved at a distance of 200 Å between the dye and the silver surface.

Tunneling in Monolayers

A fatty acid monolayer is sandwiched between two metal electrodes. A voltage is applied and the current, which is due to electron tunneling, is measured. The tunneling effect can be easily investigated in this arrangement (Fig. 4). The thickness of the insulator is changed by varying the number of carbon atoms in the fatty acid chain. The current decreases exponentially with increasing thickness, typical for quantum mechanical tunneling.[10,11]

Quantum mechanical tunneling in various superstructures can be measured. In a fatty acid multilayer an exponential decrease of the current with increasing thickness per monolayer was measured indicating tunneling between interlayer sites.[12] A single monolayer of the C20 fatty acid sandwiched between C16 monolayers determines the resistance of the current flow as expected from the tunneling theory.[13]

It is important to avoid large defects that would lead to dominance of leakage currents over tunneling currents. The present results and effects of skeletonizing monolayers[11] indicate the absence of such defects, in contrast to cases where discrepancies were observed.[14]

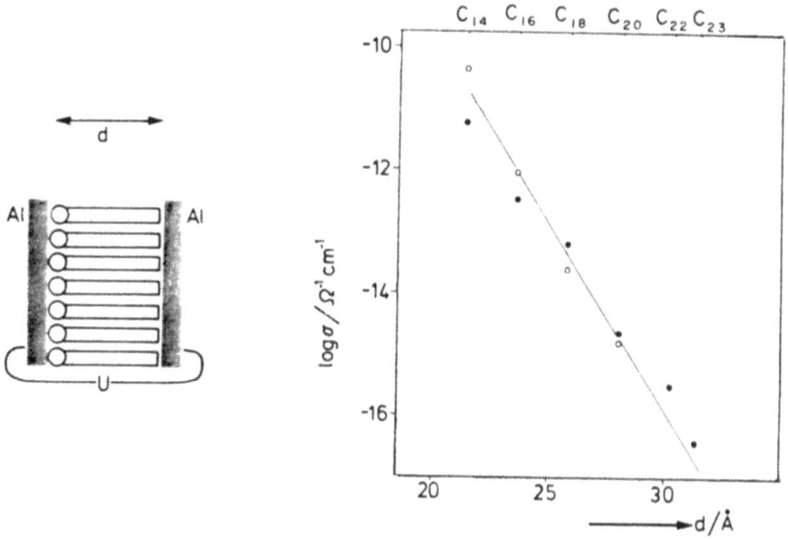

Fig. 4. Exponential decrease of conductivity with increasing monolayer thickness d as expected from the tunneling theory. • = LB; o = Adsorption.

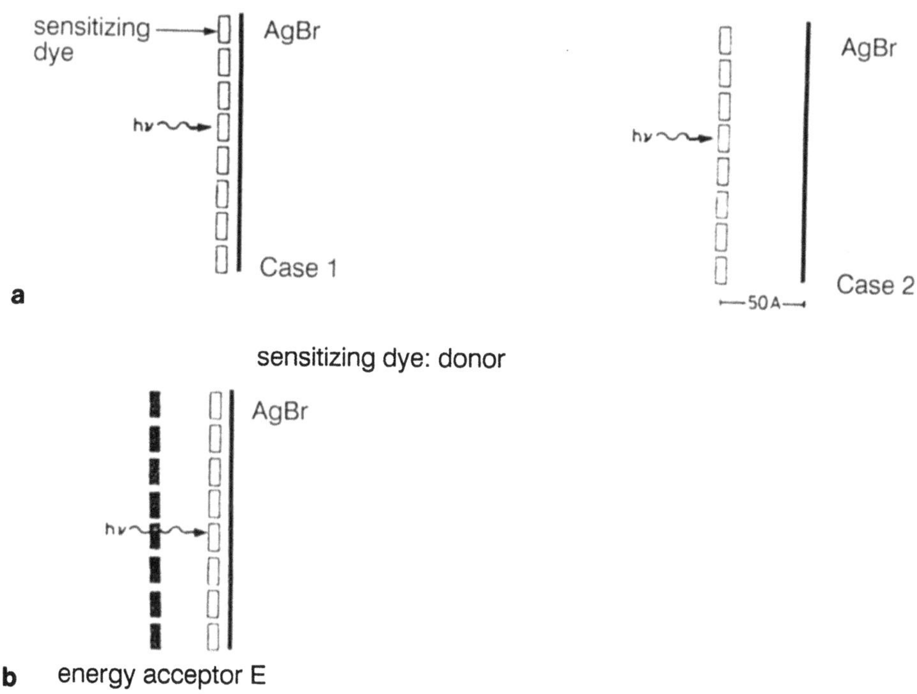

Fig. 5. AgBr film coated with organized monolayer assemblies. (a) Sensitizing dye in direct contact and at a distance of 50 Å from the surface. (b) AgBr, the sensitizing dye and the energy acceptor E in planned positions. The acceptor E acts as a competitor of AgBr.

Mechanism of Photographic Sensitization

Energy and electron transfer processes in the sensitization of photographic systems can be investigated with planned molecular arrangements of a sensitizing dye in direct contact (Fig. 5a, case 1) and at a distance of 50 Å from the silver bromide surface (case 2). The action of the dye in sensitizing the photographic process is compared. The result allows us to discriminate between sensitization mechanisms, energy or electron transfer. An additional acceptor can be put in a distinct position to act as a competitor of silver bromide (Fig. 5b). This type of experiment was very useful in elucidating the nature of the photographic sensitization process.[15]

ORGANIZED MONOLAYERS - STRUCTURAL ELEMENTS OF MOLECULAR DEVICES

For fabricating more elaborate systems than those described above many different special monolayers have to be used as building blocks in the molecular architecture. Knowledge on possible constructional components is important in planning the organized assemblies.

Monolayers of Molecules with Differently Positioned Functional Groups

The Langmuir-Blodgett layers obtained by spreading fatty acids at the water surface are of interest as neutral spacers. Amphiphilic molecules can be incorporated, and the layers can be functionalized. The functional group can be at the water interface, at the air interface, or in between.[2] A molecule with all three functional components was recently synthesized and incorporated in an intended manner in a fatty acid matrix layer.[16]

Self-organizing Monolayers

Molecules of different kinds can be tailored to match and to self-organize by interlocking at the water surface (Fig. 6).[17] These layers are prototypes demonstrating the planned construction of molecules with the purpose of matching one another in an intended manner. The development of more and more complex and sophisticated self-organizing molecular assemblies is a great challenge in the future development. Molecular self-organization should be seen as a key feature of supramolecular engineering.

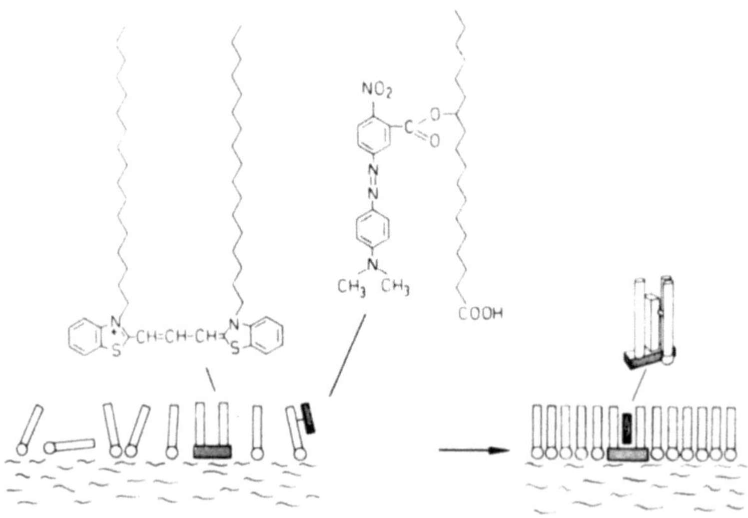

Fig. 6. Example of a self-organizing monolayer.

Compound Monolayer by Adsorption at a Preformed Monolayer

A compound monolayer can be formed by the adsorption of an appropriate species at a preformed monolayer or by co-spreading the two components. The matrix layer can be a mixture of a neutral and a charged species which acts as the counterion of the adsorbant. It rearranges in a pattern governed by the charges of the adsorbant. In the optimal situation the surface density of the positively charged head groups of the matrix layer matches the surface density of the densely packed adsorbant (Fig. 7a).[18] Its reflection spectrum is independent of dilution up to 10^{-7} M, showing a strong binding. If the ratio of the ammonium groups in the lipid is higher (Fig. 7b) the monolayer is expanded to reach neutrality. If it is lower (Fig. 7c), the reflection band is smaller since less dye is adsorbed. Polarized reflection spectra at an inclined incidence show that the transition moment is in the layer plane. Such compound monolayers can be transferred to a solid support if the interaction of the components in the complex is sufficiently strong, i.e. if the complex is compact and charges are compensated.

The matrix monolayer can be a polymer. Strong monolayers and stable assemblies can be obtained. An example is the structure shown in Fig. 8,[19] obtained by spreading poly-uridylic acid with a surface active counter-ion adogen (obtained by ion exchange) at the surface of water containing a cyanine dye in high dilution (10^{-5} M). The reflection spectrum shows an aggregate band. The assembly can be transferred to a glass slide. By slightly changing the architecture of the matrix layer the aggregate (narrow band) becomes no longer observed, showing the sensitivity on the molecular architecture.

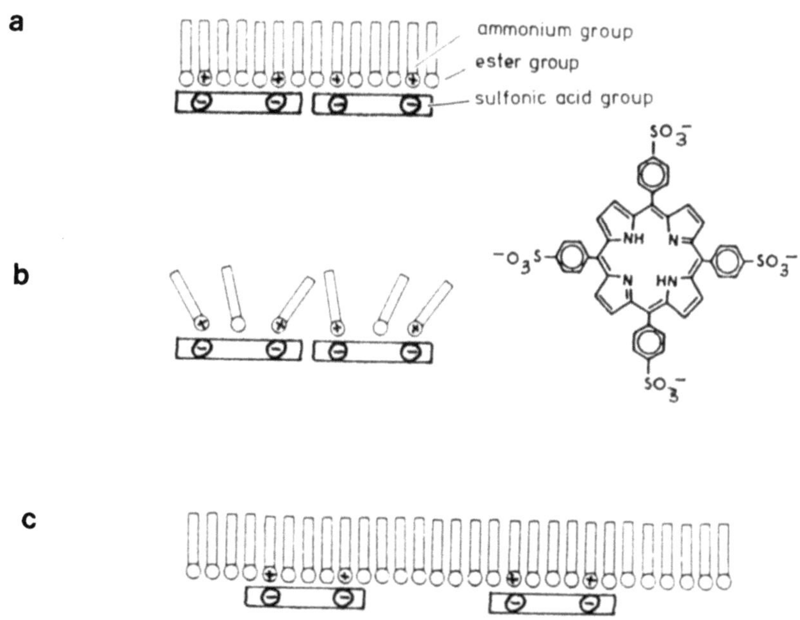

Fig. 7. Example of a compact compound monolayer obtained by adsorption of a dye with negative charges to a monolayer with an appropriate density of positive charges (a). If the density of the charges in the matrix layer is higher this layer is expanded (b). If it is lower less dye is adsorbed (c).

Two-Dimensional Crystalline Domains in Monolayers

Another type of self-organization in monolayers is the formation of domains of dyes whose size, morphology and spectroscopic properties can be controlled by the surface layer and by an inhomogeneous electric field.[20] A great variety of domain structures can thus be obtained, which are of potential interest for constructing molecular devices.

Monolayer Assemblies Obtained by Controlled Adsorption and Chemisorption, *in situ* Modification and Reaction

Monolayers of well defined architecture can be formed by controlled adsorption and chemisorption of appropriately interlocking component molecules from a solution to the solid surface.[21] The surfactant molecules can have reactive end groups which can be chemically modified *in situ*. They react with molecules in solution and more complex layer assemblies can be built in this manner.

Monolayers with Planned Holes of Molecular dimensions

Using a solution of a chemisorbing surfactant with a small amount of a non-chemisorbing surfactant, a mixed monolayer is obtained. By dissolving the non-chemisorbing molecules, holes of molecular dimensions can be produced. Each hole can be filled with a molecule matching the hole.[22] Recently, this method has been used to produce holes by adsorbing molecules with interconnected parts. The holes act as templates for pairs of molecules, as indicated for bacteriochlorophyll by surface enhanced resonance Raman spectroscopy.[23]

ORGANIZED MONOLAYERS - FUNCTIONAL COMPONENTS OF MOLECULAR DEVICES

Conducting Monolayers

A simple LB monolayer can be used as an insulator. A monolayer, depending on the nature of the functional groups and their arrangement, can act as a conductor, in-plane[24] or across constituting molecular wires.[25] Such layers are of potential interest in microelectronics.

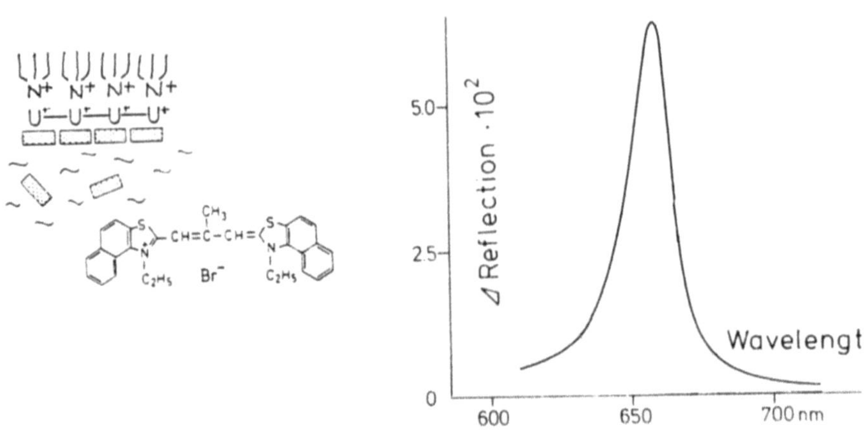

Fig. 8. Adsorption of a dye to a polymer. The dye molecules are well organized forming an aggregate, provided that each molecular component has an appropriate architecture.

Exciton Motion in Monolayers

A monolayer can be constructed to conduct excitons over distances of up to 1,000 Å and lead them to a trap with a dilution ratio of one trap molecule per 10,000 molecules of the host. The fluorescence of the host is quenched and the sensitized fluorescence of the trap is observed. It is important for this dramatic effect to have the dye chromophore of the host tightly packed to form an aggregate in order to conduct excitons efficiently.[26]

Photoinduced Intramolecular Charge Separation in Monolayers and Second Order Harmonic Generation

Monolayers with incorporated chromophores can be made to exhibit a strong photoinduced intramolecular charge separation and are useful as nonlinear optical materials and electrochromic materials. An example is the pair of dyes shown in Fig. 9a, which exhibits a strong second order harmonic generation.[27] The dye in the first layer carries the hydrocarbon chain at the bottom, the dye in the second layer on the top. Thus, both dyes are polarized in the same direction. The system is further fixed by hydrogen bonding. An interesting case showing second order harmonic generation is quinquethienyl in an amphiphilic acid or amine monolayer (Fig. 9b).[28] The chromophore is *a priori* apolar, but the required polarity is produced by the electric field of the environment acting on the strongly polarizable molecule. The sign of the polarization depends on the sign of the field which is different for acid and amine. The second order harmonic generation is then used to investigate the field effects.

Monolayers Acting as Sensors, Receptors, or Switches

Monolayers of dipolar molecules were made to function as piezosensors.[29] Monolayers of molecules with photochromic, thermochromic electroluminescent functional groups were made to function as sensors, effecters, or switches.[30,31] Monolayers were made with recognition regions for small molecules or for groups serving as receptors. An example is β-cyclodextrin with long chain hydrocarbon substituents.[32] It acts as a receptor for azobenzene derivatives, and *cis-trans* isomerization of the dye can be achieved in the enclosure.

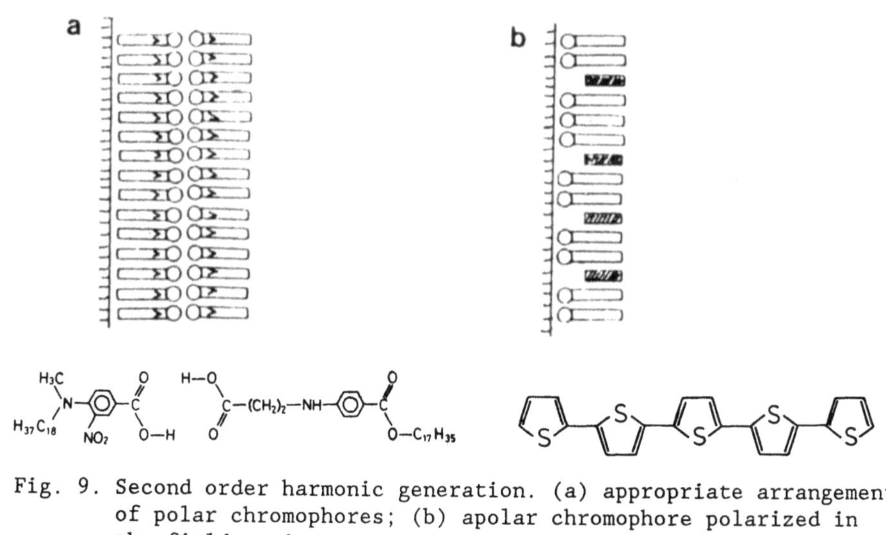

Fig. 9. Second order harmonic generation. (a) appropriate arrangement of polar chromophores; (b) apolar chromophore polarized in the field environment.

TESTING FUNCTIONAL COMPONENTS

In attempting to construct functionalized monolayers, careful characterization is critically important to make sure that the intended architecture is actually present.[33] Defect visualization by optical methods including Nomarski microscopy, and by scanning electron microscopy is important. EXAFS (near edge X-ray absorption fine structure) is useful for examining orientation and bonding of molecules in monolayers.[34] In checking the monolayer architecture, linear dichroism in UV, visible and infrared, and resonance Raman in polarized light are informative.[35] So is electron spin resonance.[36] In addition to surface pressure area isotherms, reflection spectra of monolayers at the air-water interface with polarized light at zero and non-zero incident angles are important for analyzing the monolayer architecture, since the spectra provide information about the arrangement of chromophores.[37]

SIMPLE SUPRAMOLECULAR MACHINERIES IN THE SEARCH FOR DESIGN PRINCIPLES

Photoinduced Energy Transfer Coupled with Electron Transfer to Semiconductor Triggered by Light, Protons, or Reductant

Now let us consider some organized assemblies with special layers used as building blocks. The arrangement discussed in the beginning (Fig. 1) can be made more complex by using proton-sensitive or photochromic donor or acceptor molecules. The molecules then act as switches which can be turned on and off by protons or by light pulses of an appropriate wavelength. The energy acceptor can be coupled to a semiconductor. It acts as an electron donor, after having received the excitation energy from the energy donor. The oxidized electron donor is restored by a reductant in the solution phase. A photocurrent, which is triggered by light, protons, or reductant, can be measured in this electrochemical cell.[38]

In principle, molecules in the present systems act as specific chemical or optical sensors, switches, signal transducers, amplifiers, or information storage elements.

Photoinduced Electron Transfer through Monolayers

As shown above (Fig. 1), an excited dye molecule can transfer its excitation energy to an energy acceptor molecule through a hydrocarbon portion of 50 Å. In contrast an electron acceptor at the same distance has no effect. However electron transfer can be easily observed at a distance of 10 to 20 Å between the donor and the acceptor, and the effect has been investigated by using spacer layers of various chain lengths.[39]

In competition with radiative and radiationless de-excitation, electron transfers do not require collisions but takes place at a distance of 20 Å. This was a surprising observation but the results were later confirmed (radiolysis in rigid glasses,[40] donor acceptor pairs connected by rigid bridges[41]).

In the arrangement in Fig. 10a,[42] a stilbene derivative is built into a fatty acid matrix layer and this layer is covered by a bipyridinium layer. Stilbene is excited by UV light and an electron is transferred to bipyridinium, which acts as an electron acceptor, as revealed from the quenching of the fluorescence of stilbene. The length of the hydrocarbon chain in the stilbene molecule can be varied and, in this way, the distance between the electron donor and the electron acceptor can be varied. The rate of electron transfer decreases roughly exponentially with the increase of this distance, as expected from a quantum mechanical tunneling process.

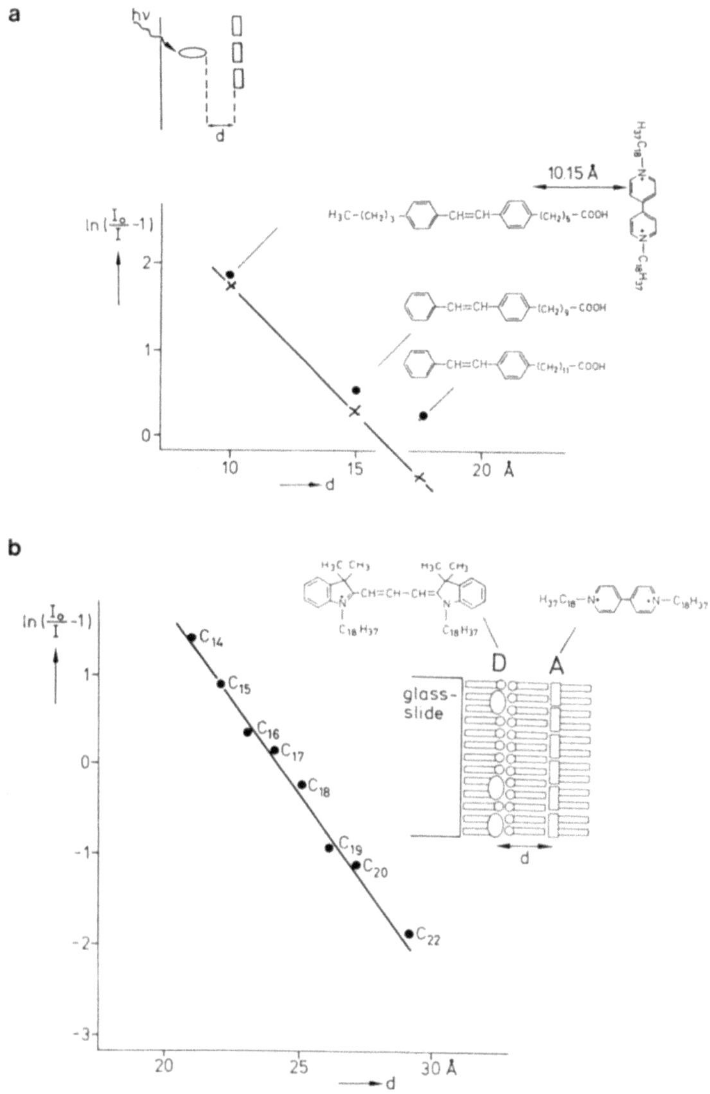

Fig. 10. Electron transfer rate (measured by donor fluorescence quenching) versus distance d. a) donor stilbene, acceptor bipyridinium; thermally activated electron transfer can not be neglected. Its contribution to the rate can be considered. The residual rate (crosses) obeys the linear relationship indicating tunneling. b) donor carbocyanine, acceptor bipyridinium.

In the arrangement in Fig. 10b, the spacer between the excited dye and the electron acceptor is a fatty acid layer, and by using fatty acids with 14 to 22 carbon atoms the exponential dependence of the electron transfer rate on the distance can be tested over a wide range. The experimental points are on a straight line in a semilogarithmic plot.[43]

To transfer electrons from an excited donor to an electron acceptor over 50 Å, a π-electron system acting as a molecular electron conductor connecting the donor and the acceptor must be incorporated in the fatty acid layers. In the arrangement shown in Fig. 11a,[44] quinquethienyl is used as the conductor to channel electrons from the excited donor D to the acceptor A.

The self-assembled system of interlocking molecules D and W considered in Fig. 6 can be used to investigate the action of a molecular wire in detail. The long axis of the chromophore of D is in the layer plane, while W is oriented perpendicularly (Fig. 11b). The monolayer is sandwiched between fatty acid monolayers and semitransparent metal electrodes and a voltage is applied. The excited donor D is excited with light and a photocurrent is observed to be proportional to the light intensity. When the applied voltage is reversed the current becomes smaller by a factor of three. The arrangement then acts as a rectifier. The photocurrent in the absence of W is smaller by an order of magnitude.[45]

The voltage and temperature dependence was investigated in detail. The results are rationalized by assuming that the photoexcited electron is transferred from D to W (barrier of 0.26 eV) by tunneling (at low temperature) or by thermal activation (at higher temperature). The electron in W tunnels to acceptor site A1 in the next interface in a fast process, and from there it tunnels through the next fatty acid layer to interface state A2 and then hops to the positively biased electrode. D^+ is regenerated to D by accepting an electron supplied by the negatively biased electrode. The model can be easily checked by specifically changing the tunneling barrier between states A1 and A2.

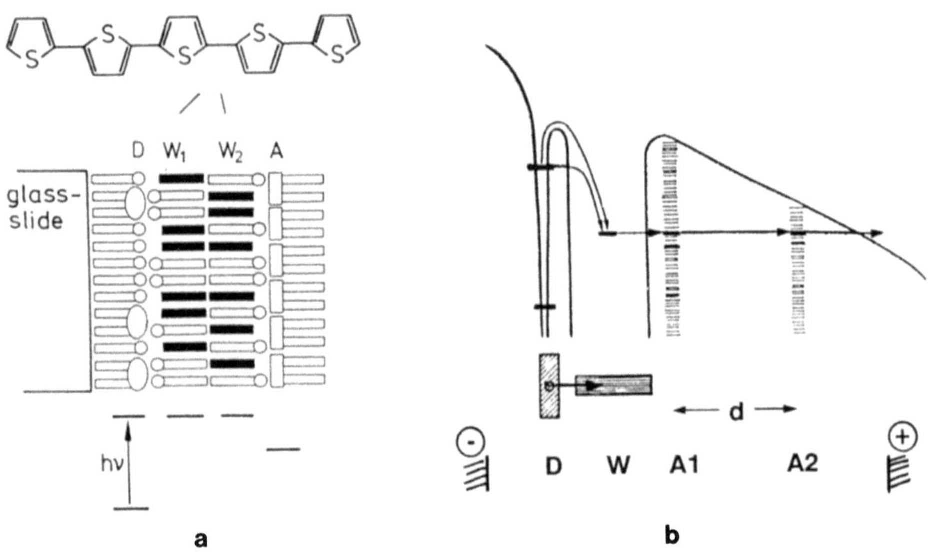

Fig. 11. Electron transfer from the donor D to the acceptor A via the electron conductor W. (a) D = carbocyanine, W1, W2 = quinquethienyl, A = bipyridinium. (b) D = carbocyanine, W = azo dye, A1 = acceptor site, A2 = interface state.

ADDRESSABILITY OF SUPRAMOLECULAR DEVICES AND ATTEMPTS TO PROCESS INFORMATION AT MOLECULAR DIMENSIONS

Addressing Molecular Devices by Light

In the cases considered, light is used as a trigger, and we may ask if a supramolecular switching element is actually addressable by light. The size of an object to be addressed is not limited to the wave length of light. A smaller object is addressable if it is in the near field of a light source.[46] The light source can be a small aperture at a sufficiently small distance from the object to be addressed. Apertures of 500 Å in diameter in a tungsten layer of 200 Å thickness were obtained by shadow imaging of a latex sphere, and simple objects were scanned.[47] Similar results were obtained with differently constructed apertures.[48] So it seems possible to address a single molecular device incorporated in a monolayer in sufficient dilution and to read and write information at the molecular level by optical means (Fig. 12).

Addressing Molecular Devices by Electrons

It is promising to address a molecular device with the scanning tunneling microscope. A good candidate may be an ultra-microelectrode obtained by covering a gold electrode with an insulating monolayer with pinholes of 50-100 Å in diameter at a distance of 500-1,000 Å and by filling the pinholes with a compound which binds to the gold surface and acts as a receptor of copper ions by forming a chelate.[49] This is indicated by the appearance of a strong voltametric signal. The arrangement acts as a tiny electrode which can recognize a selected metal ion in the presence of other ions. It would be interesting to demonstrate, with scanning tunneling microscopy, the response of a single element of this type.

An attempt to demonstrate molecular rectification in a similar arrangement[50] has not been confirmed.

Copy Information Stored at the Molecular Level

Let us consider a fatty acid monolayer with incorporated dye molecules which have their chromophore axes oriented. A second monolayer with incorporated dye molecules is deposited on top of this first monolayer. Dimers are formed between the chromophores in the first and in the second layer; the chromophores in the second layer are oriented with respect to the chromophores in the first layer. The two layers can be separated from each other.

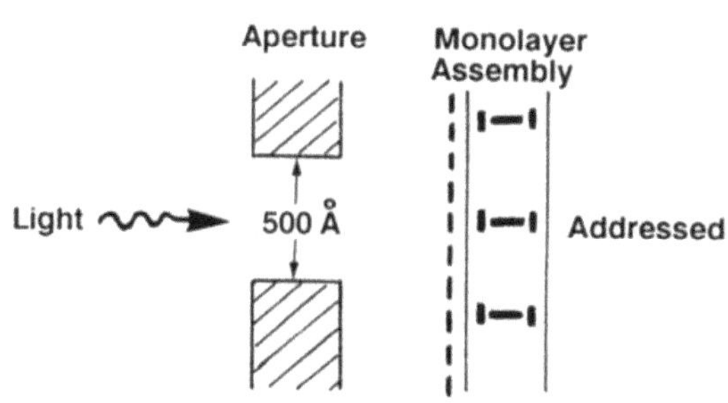

Fig. 12. Possibility to address a molecular device by light in the near field region of an aperture of 500 Å in diameter. An antenna system (shown as a broken line on the monolayer assembly) is required to absorb the incident light quanta and to transfer its energy to the embedded molecular device.

The chromophores will preserve their orientation if conditions are appropriate. This indicates molecular information transfer from one matrix layer to the other layer.[51]

Amplification of Information Stored at the Molecular Level

Monolayers of enantiomeric amino acids with long hydrocarbon substituents are formed at the surface of water with dissolved glycine.[52] As soon as a condensed monolayer was formed, glycine crystallized immediately, but only if the molecular arrangement at the interface was matching exactly with the arrangement in the growing crystal. Bulky groups in the monolayer (e.g. cholesterol) prevented the occurrence of glycine crystals, showing the sensitivity on the matching conditions. Information stored in two dimensions can be developed in three dimensions, demonstrating molecular signal amplification.

VESICLES AS POSSIBLE ELEMENTS OF SUPRAMOLECULAR DEVICES

The monolayer assembly technique allows for the construction of functional units which act as elements for information processing, but planned coupling of many different such elements to form an intricate network is an unsolved problem. Components of sufficient complexity with specific recognition and binding sites for assembly into planned superstructures should be available. Promising results were obtained with vesicles from specifically interacting molecules stabilized by polymerization.[53] They can be equipped with receptor sites and surface recognition regions, and thus may be useful as building blocks of complex systems developing, let us say, at a planned pattern made by electron beam writing on a solid surface. They may be candidates for the future development of supramolecular computers. The self-organization of elements acting as cellular automata can be important in such developments. Attempts to construct amphiphilic molecules that self-organize in well-defined and planned structures have a great potential in future developments of complex assemblies. These molecules consist of several sections with different distinct functions in the self-assembly process (hydrophobic stiff sections for aggregate formation, hydrophilic flexible spacer sections, such as side group or main chain spacers, for adjusting[53]).

PLANNED SELF-ORGANIZATION OF SUPRAMOLECULAR DEVICES

Another attempt to build organized molecular assemblies is by using the biological principle of building an organized assembly of molecules by binding the molecules to a messenger in a sequence determined by that of the monomers in the messenger. A single stranded oligonucleotide can be used as

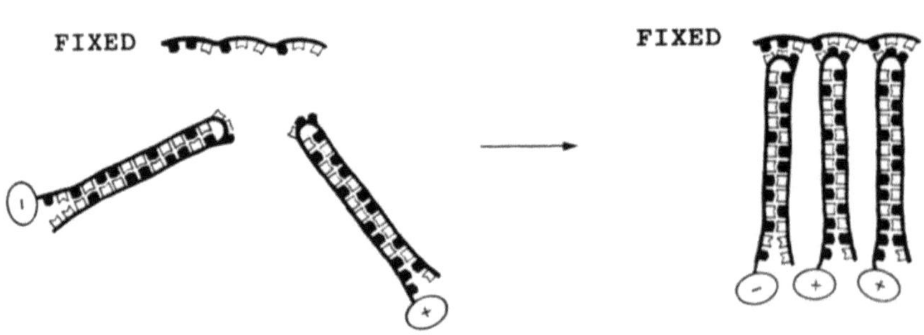

Fig. 13. Building an organized assembly by binding hairpins to a collector strand (messenger). The sequence is determined by that of the monomers in the collector strand.

the messenger (collector strand) and oliginucleotides in hairpin conformation as adapters that bind to the collector strand by complementary base pairing with the hairpin loop (Fig. 13). In aiming at this goal, it could be demonstrated that hairpin loops bind to a collector strand by triplets of complementary base pairs. The collector strand was immobilized (fixed to cellulose), and the hairpins bound at 0°C. The strength of the binding was measured by warming up and measuring the temperature of elution. The result was compared with the elution temperature of corresponding single stranded oligonucleotides.[54] It is hopeful to construct a translation machine by binding amino acids to the hairpins and, in this way, to simulate a critical step of modeling the origin of life. But the principle can also be of more general interest in fabricating supramolecular devices.

TWO ASPECTS OF MOLECULAR ENGINEERING: DEVELOPING TOOLS TO FABRICATE SUPRAMOLECULAR DEVICES AND SEARCHING FOR POSSIBILITIES AND DESIGN PRINCIPLES OF SUPRAMOLECULAR MACHINES

The different approaches, such as LB films, controlled adsorption, planned synthesis of molecules and polymers, polymerized vesicles, etc., should be seen as different complementary parts in the challenging game of constructing supramolecular machinery of increasing sophistication and increasing usefulness.

The technical achievements and possibilities of fabricating molecular assemblies were considered above, and molecular and supramolecular engineering were viewed as a tool to construct organized aggregates. Supramolecular engineering is also seen as a style of thinking in terms of supramolecular machines: to reflect on the design of molecular assemblies in the broadest sense, on realizing possibilities in the near and far future, to search for principal types of supramolecular machines, and to study possibilities of how machinery can emerge and evolve.[55,56] To illustrate this aspect it may be asked: How can energy be stored by photoinduced vectorial charge separation? How can the yield be optimized?

DESIGN OF A LIGHT-DRIVEN ELECTRON PUMP

Design Principles and Comparison with the Photosynthetic Reaction Center

To design a light-driven electron pump, a light-excitable molecule D is required and the electron should move to an acceptor A and remain at a high energetic level until it is used to drive chemical reactions. The acceptor must be at a certain distance (~30 Å) from D to avoid back transfer to D by quantum mechanical tunneling within a reasonable time (a tenth of a second). A π-electron system acting as molecular conductor (W) is required to lead an electron from the excited D to the acceptor A (Fig. 14).

A secondary electron donor D′ must be present to convert D^+ back to D. Its optimum distance from D is given by a compromise; it must be sufficiently close to allow electron tunneling to D^+ in the required time interval but sufficiently far not to hinder the first step - the transfer of electron from D^* to A via W.

To minimize energy loss, thermally activated back-transfer must be avoided, and for that reason the energy level of W^- must be 0. 17 eV below the level of excited D, the level of A^- 0.4 eV below the level of W^- (Fig. 14). Thus the system must be designed so that the electron transfers from excited D to W and also from D′ to D^+ is almost loss free. This means that the environment must be solid and non-polar (to avoid environmental reorganization), and D′, D and W should be an extended π-electron system (to avoid

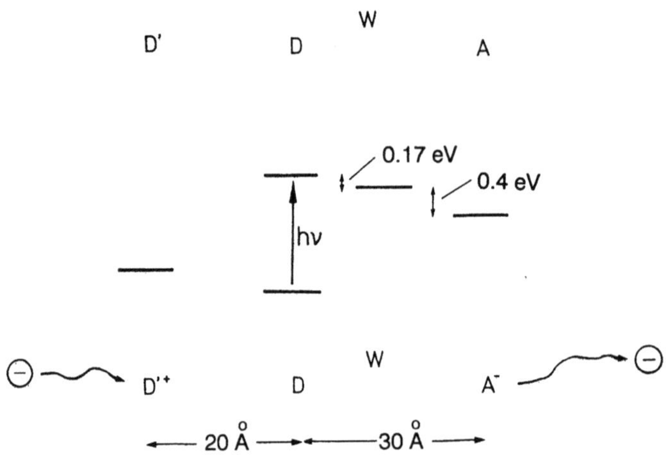

Fig. 14. Light-driven electron pump. Electron transfer proceeds from the photoexcited electron donor D to the acceptor A via the molecular wire W. The secondary electron donor D' converts D^+ back to D. Mobile carriers supply an electron to D'^+ and remove it from A^-. The arrangement of chromophores and the energy levels are optimized to prevent back transfer by tunneling and thermal activation, and to prevent excessive loss by reorganization.

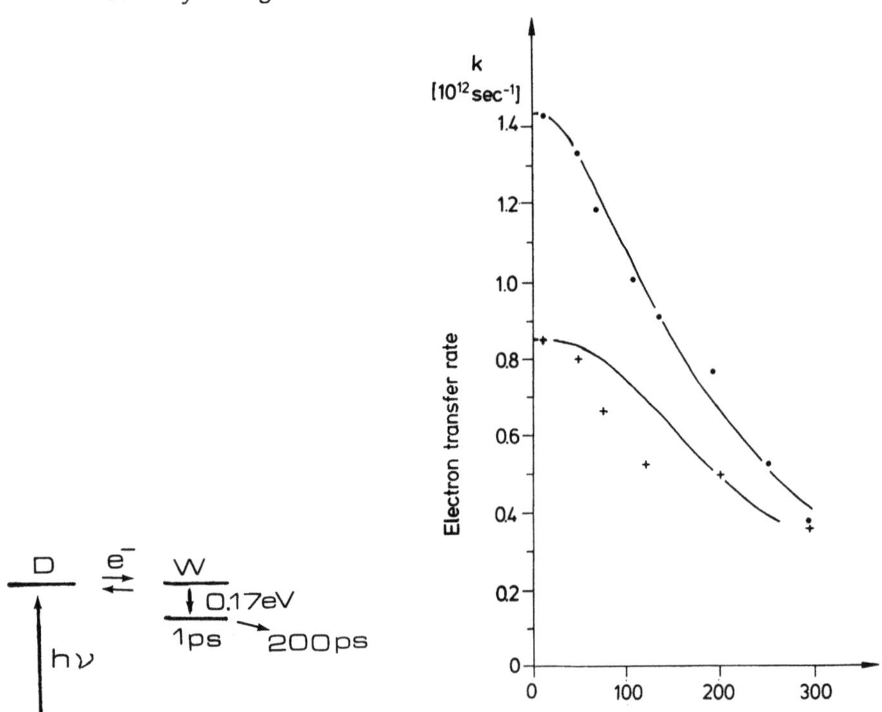

Fig. 15. Rate of electron transfer from the excited D^* to W as a function of temperature. The level of D^* coincides with the vibronically excited level of W^- at $T = 0$ (maximum rate). The match gets poorer with increasing T due to thermal noise (slower rate).

loss of free energy by intramolecular reorganization). For the electron transfer from W^- to A, a considerable energy dissipation is required. The acceptor A should be loosely bound to allow for environmental reorganization, but all four components D', D, W and A are required to be in a distinct steric arrangement. To supply the electron to D'^+ and to remove it from A^- requires mobile carriers. D' may dissociate and reassociate, thus taking over the function of a carrier, but in this case D' needs to occupy its distinct position sufficiently often, so that electron transfer from D' to D^+ is fast in comparison with back-transfer from A^- to D^+.

A detailed study, based on what has been learned from the monolayer experiments on electron tunneling (see Figs. 4, 10, and 11), has shown[57] that the machinery can work in the intended manner only if the properties of the component molecules, their energetic position, and their arrangement agree with given requirements within narrow limits. It is remarkable that the actual arrangement of the chromophores in the bacterial reaction center[58] is in accordance with the derived design principles, whereby W is represented by two strongly coupled molecules W_1 (monomeric bacteriochlorophyll) and W_2 (bacteriopheophytin).

To render possible the required fast electron transfer from the excited D to W these components must be particularly well fixed in their position and the energy level of D^* should match with the vibronically excited level of W^- (~0.17 eV above ground state level) (Fig. 15). Then the electron, after D is excited, moves between D and W for about a picosecond. Then the vibronic energy, by collisions, is transferred to the environment: the electron is trapped in W for about 200 ps when it is further transferred to A. At temperature T = 0 the level of D^* and the vibronically excited level of W^- are assumed to coincide, and differ at temperature T, on the average, by $\pm\frac{1}{2}k_BT$, where k_B is Boltzmann's constant. Then the rate of electron transfer from D^* to W is given by the equation

$$\text{rate} = \frac{1}{2\tau} \cdot \frac{1}{1 + (k_BT/4\varepsilon)^2}, \tag{1}$$

where ε is the perturbation energy for the electron transfer from D^* to W, and τ is the vibronic relaxation time.

The rate was recently measured as a function of T (Fig. 15, experimental points: (•) for *Rps. viridis*, (+) for *Rb. sphaeroides*).[59] The result cannot be reasonably explained by conventional theories.[60] From Eq. 1, using the values $\varepsilon = 4\cdot10^{-3}$ eV, $\tau = 3.5\cdot10^{-13}$ s for *Rps. viridis*, $\varepsilon = 5\cdot10^{-3}$ eV, $\tau = 5.8\cdot10^{-13}$ s for *Rb. sphaeroides*, the theoretical curves in Fig. 15 are shown to be in agreement with the experimental points. The values of ε and τ used in the computation are within the theoretically expected range.[57] With increasing temperature the energy levels, due to thermal noise, match less perfectly, and the rate decreases.

The reaction center consists of two branches (L and M branch), but the electron flow is restricted to the L branch. This is expected from the present model since the level of W^- in the M branch is lower by 0.1 eV than the corresponding level in the L branch (the value was obtained[61] by considering the electrostatic interaction with the protein polar groups and asymmetric medium polarization interactions). Therefore the above mentioned coincidence of levels is not present in the M branch. It follows from Eq. 1 that the rate at room temperature should be 35 times higher for the L branch than for the M branch. In contrast, the unidirectionality cannot be explained by this energy difference of 0.1 eV modifying the Franck-Condon factor.[61]

The proposed energetic coincidence is thought to be of evolutionary significance; the originally symmetric protein dimer has developed into the asymmetric form by gene duplication and mutations. There was a strong evolutionary pressure to preserve electron flow (and thus the energetic coincidence) in the L branch but not in the M branch.[57]

Attempts to Realize Photoinduced Electron Transfer

The real challenge in supramolecular engineering is not to copy the natural system, but to construct other, hopefully useful, systems based on the derived design principles.

Several attempts were made.[62,63] Monolayer assemblies with the cyanine dye D and the molecular wire W of Fig. 6 were sandwiched between aluminum, constituting D', and barium, constituting A (Fig. 16a). The mobile carriers are electrons in the electrodes of this battery.[62]

An interesting photoinduced electron transfer system is obtained by incorporating a molecule with three functional components into a fatty acid monolayer and by depositing the layer on a gold electrode.[16] In Fig. 16b, the photocurrent is measured for this system in a solution with a saturated mercury counter electrode. The donor D (pyrene) is excited by light and the electron tunnels to W (viologen) and from there to A (the gold electrode). D is converted back to D by charge movement from D' (ferrocene) to D^+. The mobile carriers are ions in the electrolyte and the electrons in the gold electrode. The position of D' can be optimized by altering the length of th hydrocarbon chain between D' and W. When proceeding from 16 to 11 carbon atoms in the chain, the photocurrent decreases.

MOLECULAR MACHINERY BY DARWINIAN EVOLUTION - A CHALLENGE IN MOLECULAR ENGINEERING

A particularly relevant question in molecular engineering is: How can a self-organizing molecular replication/translation machine (i.e. a machine that carries the recipe for its own reproduction) be designed? Self-reproducing machinery can not be formed in one step, but may result in a many-ste

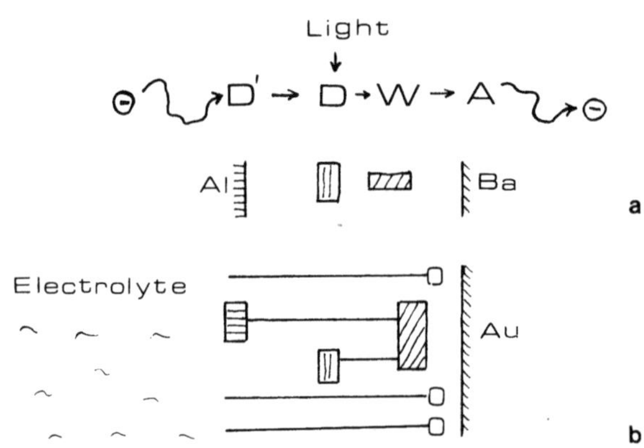

Fig. 16. Light-driven electron pump constructed by assembling monolayers. (a) D and W from Fig. 6 are sandwiched between an Al electrode (representing D') and a Ba electrode (representing A). (b) D', D and W are represented by a single molecule between the electrolyte solution (supplying charge to D'^+) and the gold electrode (representing A). (Arrangement according to Ref. 16)

Darwinian evolution process. The principle of a Darwinian process should be seen as an important future design principle in engineering intricate molecular devices.[56]

This problem is related to the problem of the origin of life and is intimately connected with the question: How does one design a 'knowledge' accumulating (intelligent and creative) molecular machine?. It is most challenging to reflect on practical supramolecular machines based on the same principles.

It was my intention to indicate that molecular and supramolecular engineering is a worthwhile endeavor as an attempt to construct molecular assemblies that function as supramolecular machines, and is also a useful style of thinking about supramolecular machines.

REFERENCES

1. F. L. Carter (ed.), "Molecular Electronic Devices II," Marcel Dekker, New York (1987).
2. D. Möbius (ed.), "Proc. 3rd Int. Symp. Langmuir-Blodgett Films, July 26-31, 1987, Göttingen, FRG," Thin Solid Films 159, Nos. 1-2 (1988) and *ibid.* 160, Nos. 1-2 (1988); M. Sugi, Thin Solid Films 152:305 (1987); M. Sugi, in: "3rd Int. Symp. Molecular Electronic Devices, October 6-8, Arlington, USA" (1986); G. G. Roberts, Adv. Phys. 34:475 (1985); L. M. Blinov, Russ. Chem. Rev. 52:713 (1983); H. Kuhn, D. Möbius and H. Bucher, in: "Physical Methods of Chemistry, Vol. I, Part 3B," A. Weissberger and B. Rossiter, eds., p. 577, Wiley, New York (1972); H. Kuhn and D. Möbius, Angew. Chem. Int. Ed. 10:620 (1971).
3. K. E. Blodgett and I. Langmuir, Phys. Rev. 51:964 (1937).
4. H. Kuhn, Verhandlungen der Schweizerischen Naturforschenden Gesellschaft p. 245 (1965); Pure Appl. Chem. 11:345 (1965); Naturwissenschaften 54:429 (1967).
5. O. Inacker, H. Kuhn, D. Möbius, and G. Debuch, Z. Physik. Chem. (NF) 101:337 (1976).
6. O. Inacker, H. Kuhn, Chem. Phys. Lett. 27:317 (1974); *ibid.* 27:471 (1974); *ibid.* 28:15 (1974).
7. H. Kuhn, J. Chem. Phys. 53:101 (1970); K. H. Tews, O. Inacker, H. Kuhn, Nature (London) 228:276 (1970); K. H. Drexhage, in: "Progress in Optics," Vol. XII, E. Wolf, ed., p. 165, North-Holland, Amsterdam (1974); R. R. Chance, A. Prock, and R. Silbey, Adv. Chem. Phys. 37:1 (1978).
8. H. Bucher, H. Kuhn, Z. Naturforschung. 25b:1323 (1970); J. Hesemann and H. P. Zingsheim, in: "Membrane Spectroscopy", E. Grell, ed., p. 172, Springer, Berlin (1981).
9. G. Wahling, H. Raether, and D. Möbius, Thin Solid Films 58:391 (1979); I. Pockrand, A. Brillante, and D. Möbius, Chem. Phys. Lett. 69:499 (1980); J. Chem. Phys. 77:6289 (1982); I. Pockrand and J. D. Swalen, J. Phys. Chem. 83:1438 (1979).
10. B. Mann and H. Kuhn, J. Appl. Phys. 42:4398 (1971); E. E. Polymeropoulos, J. Appl. Phys. 48:2404 (1977); Solid State Commun. 28:883 (1978).
11. J. Sagiv and E. E. Polymeropoulos, J. Chem. Phys. 69:1836 (1978).
12. M. Sugi, K. Nembach, D Möbius, and H. Kuhn, Solid State Commun. 15:1867 (1974); M. Sugi, T. Fukui, and S. Iizima, Appl. Phys. Lett. 27:559 (1975); S. Iizima and M. Sugi, Appl. Phys. Lett. 28:548 (1976); M. Sugi and S. Iizima, Thin Solid Films 68:199 (1980); M. Sugi, M. Saito, T. Fukui, and S. Iizima, Thin Solid Films 99:17 (1983).
13. M. Sugi, T. Fukui, and S. Iizima, Phys. Rev. B 18:725 (1978); M. Sugi and S. Iizima, Appl. Phys. Lett. 34:290 (1979).
14. R. H. Tregold, Rep. Prog. Phys. 50 1609 (1987).
15. R. Steiger, Photogr. Sci. Eng. 28:35 (1984); R. Steiger, H. Hediger, P.

Junod, H. Kuhn, and D. Möbius, Photogr. Sci. Eng. 24:185 (1980); L. v. Szentpaly, D. Möbius, and H. Kuhn, J. Chem. Phys. 52:4618 (1970).
16. M. Fujihira and H. Yamada, Thin Solid Films 160:125 (1988); M. Fujihira, K. Nishiyama, and H. Yamada, Thin Solid Films 132:77 (1985).
17. H. Bucher and H. Kuhn, Chem. Phys. Lett. 6:183 (1970); E. E. Polymeropoulos, D. Möbius, and H. Kuhn, J. Chem. Phys. 68:3918 (1978); J. Heesemann, J. Am. Chem. Soc. 102:2176 (1980); X. Xu, M. Era, T. Tsutsui and S. Saito, in press.
18. H. Gruniger, D. Möbius, and H. Meyer, J. Chem. Phys. 79:3701 (1983); D. Möbius and H. Gruniger, Bioelectrochem. Bioenerg. 12:375 (1984).
19. U. Lehmann, Thin Solid Films 160:257 (1988).
20. H. Möhwald, Thin Solid Films 159:1 (1988); A. Miller and H. Möhwald, Europhys. Lett. 2:67 (1986); A. Miller, W. Knoll, and H. Möhwald, Phys. Rev. Lett. 56:2633 (1986).
21. J. Sagiv, J. Am. Chem. Soc. 102:92 (1980); R. Maoz and J. Sagiv, J. Colloid Interface Sci. 100:465 (1984); J. Gun and J. Sagiv, J. Colloid Interface Sci. 112:457 (1986).
22. J. Sagiv, Isr. J. Chem. 18:339 (1979); *ibid.* 18:346 (1979).
23. J. H. Kim, T. M. Cotton, and R. A. Uphaus, Thin Solid Films 160:389 (1988).
24. A. Ruaudel-Teixier, M. Vandevyver, and A. Barraud, Mol. Cryst. Liq. Cryst. 120:319 (1985); A. Barraud, P. Lesieur, J. Richard, A. Ruaudel-Teixier, M. Vandevyver, M. Lequan, and R. M. Lequan, Thin Solid Films 160:81 (1988); T. Nakamura, M. Matsumoto, F. Takei, M. Tanaka, T. Sekiguchi, E. Manda, and Y. Kawabata, Chem. Lett. 709 (1986); K. Ikegami, S. Kuroda, M. Saito, K. Saito, M. Sugi, T. Nakamura, M. Matsumoto, and Y. Kawabata, Phys. Rev. B, 35:3667 (1987); T. Nakamura, F. Takei, M. Matsumoto, M. Tanaka, T. Sekuguchi, E. Manda, Y. Kawabata, and G. Saito, Synthetic Metals 19:681 (1987).
25. E. E. Polymeropoulos, D. Möbius, and H. Kuhn, J. Chem. Phys. 68:3918 (1978); D. Möbius in: "Photochem. Conversion and Storage of Solar Energy, Part A," J. Rabani, ed., p. 139, Weizmann Science Press of Israel, Jerusalem (1982).
26. D. Möbius and H. Kuhn, J. Appl. Phys. 64:5138 (1988).
27. I. Ledoux, D. Josse, P. Fremaux, J. P. Piel, G. Post, J. Zyss, T. McLean, R. A. Hann, P. W. Gordon, and S. Allen, Thin Solid Films 160:217 (1988).
28. L. Chi and D. Möbius, in press.
29. C. A. Jones and M. C. Petty, and G. G. Roberts, Thin Solid Films 160:117 (1988); G. H. Davies, Y. Yarwood, M. C. Petty, and C. A. Jones, Thin Solid Films 159:461 (1988).
30. H. Kuhn, in: "Molecular Electronic Devices II," F. L. Carter, ed., p. 411, Dekker, New York (1987).
31. E. E. Polymeropoulos and D. Möbius, Ber. Bunsenges. Phys. Chem. 83:1215 (1979); D. Möbius, H. Bucher, H. Kuhn, and J. Sondermann, Ber. Bunsenges. Phys. Chem. 73:845 (1969); G. G. Roberts, T. M. McGinnity, W. A. Barlow, and P. S. Vincett, Solid State Commun. 32:683 (1979); M. Era, S. Hayashi, T. Tsutsui, and S. Saito, J. Chem. Soc. Comm. 1985:557 (1985).
32. A. Yabe, Y. Kawabata, H. Niino, M. Matsumoto, A. Ouchi, H. Takahashi, S. Tamura, W. Tagaki, H. Nakahara, and K. Fukuda, Thin Solid Films 160:33 (1988).
33. M. Vandevyver, Thin Solid Films 159:243 (1988).
34. D. A. Outka, J. Stohr, J. P. Rabe, and J. D. Swalen, J. Chem. Phys. 88:4076 (1988).
35. J. F. Rabolt, M. Jurich, and J. D. Swalen, Appl. Spectroscopy 39:269 (1985); J. D. Swalen, J. Molec. Electronics, 2:155 (1986).
36. S. I. Kuroda, K. Ikegami, K. Saito, M. Saito, and M. Sugi, Thin Solid Films 159:285 (1988).
37. M. Orrit, O. Möbius, U. Lehmann, and H. Meyer J. Chem. Phys. 85:4966 (1986).

38. W. Arden and P. Fromherz, Ber. Bunsenges. Phys. Chem. 82:868 (1978); J. Elektrochem. Soc. 127:370 (1980); J. Am. Chem. Soc. 102:6211 (1980).
39. H. Kuhn, J. Photochem. 10:111 (1979); D. Möbius, Acc. Chem. Res. 14:63 (1981).
40. J. R. Miller, J. V. Beitz, and R. K. Huddlestone, J. Am. Chem. Soc. 106:5057 (1984); R. K. Huddlestone and J. R. Miller, J. Phys. Chem. 85:2292 (1981); *ibid.* 86:1347 (1982); *ibid.*87:4867 (1983).
41. J. W. Verhoeven, Pure and Appl. Chem. 58:1285 (1986); A. D. Joran, B. A. Leland, G. G. Geller, J,J. Hopfield, and P. B. Dervan, J. Am. Chem. Soc. 106:6090 (1984); A. D. Joran, B. A. Leland, P. M. Felker, A. H. Zewail, J. J. Hopfield, and P. B. Dervan, Nature (London) 327:508 (1987).
42. W. F. Mooney and D. G. Whitten, J. Am. Chem. Soc. 108:5712 (1986).
43. D. Möbius and G. Debuch, unpublished data, cited in H. Kuhn, in: "Proc. Robert A. Welch Foundation Conferences on Chemical Research XXX, Advances in Electrochemistry," p. 359, Houston, TX, November 3-5 (1986).
44. L. Chi and D. Möbius, in press.
45. E. E. Polymeropoulos, D. Möbius, H. Kuhn, J. Chem. Phys. 68:3918 (1978).
46. H. Kuhn, in: "Structural Chemistry and Molecular Biology," A. Rich and N. Davidson, eds., p. 571, Freeman, San Francisco (1968).
47. U. Ch. Fischer, reported at the Second meeting on Molecular Electronic Devices in 1983 (see Ref. 30, p. 417); U. Ch. Fischer, J. Vac. Sci. Techn. B3(1):386 (1985); U. Ch. Fischer, J. Opt. Soc. Am. B, in press.
48. D. W. Pohl, W. Denk, and M. Lanz, Appl. Phys. Lett. 44:651 (1984); U. Durig, D. W. Pohl, and F. Rohner, J. Appl. Phys. 59:3318 (1986); IBM J. Res. Develop. 30:478 (1986); U. Ch. Fischer, U. T. Durig, and D. W. Pohl, Appl. Phys. Lett. 52:249 (1988).
49. I. Rubinstein, S. Steinberg, Y. Tor, A. Shanzer, and J. Sagiv, Nature (London) 332:426 (1988).
50. A. Aviram, C. Joachim, and M. Pomerantz, Chem. Phys. Lett. 146:490 (1988). See, however, J. Chimie Phys. 85, No. 11/12 (1988) (special issue on Langmuir-Blodgett films and related structures, pannel discussions)
51. D. Möbius and G. Debuch, Ber. Bunsenges. Phys. Chem. 80:1185 (1976); H. Kuhn, Pure Appl. Chem. 53:2105 (1981).
52. E. M. Landau, M. Levanon, L. Leiserowizt, M. Lahav, and J. Sagiv, Nature (London) 318:353 (1985); S. G. Wolf, E. M. Landau, M. Lahav, L. Leiserowitz, M. Deutsch, K. Kjaer, and J. Als-Nielsen, Thin Solid Films 159:29 (1988).
53. H. Ringsdorf, B. Schlarb, and J. Venzmer, Angew. Chem. lnt. Ed. Engl. 27:113 (1980); A. Laschewsky, H. Ringsdorf, G. Schmidt, and J. Schneider, J. Am. Chem. Soc. 109:788 (1987); T. Kunitake, in: "Modern Trends of Colloid Science in Chemistry and Biology," H. F. Eicke, ed., p. 34, Birkhauser, Basel (1985); J. H. Fuhrhop, K. Ellermann, H. H. David, and J. Matheu, Angew. Chem. Int. Ed. Engl. 21:440 (1982).
54. U. Baumann, U. Lehmann, K. Schwellnuss, J. H. van Boom, and H. Kuhn, Eur. J. Biochem. 170:267 (1987).
55. M. Conrad, Eur. J. Operational Research 30:280 (1987); J. J. Hopfield, J. N. Onuchic, and D. N. Beratan, Science 241:818 (1988).
56. H. Kuhn, IBM J. Res. Develop. 32:37 (1988).
57. H. Kuhn, Phys. Rev. A 34:3409 (1986).
58. J. Deisenhofer, O. Epp, K. Miki, R. Huber, and H. Michel, J. Mol. Bol. 180:385 (1984).
59. G. R. Fleming,J. L. Martin, and J. Breton, Nature (London) 333:190 (1988).
60. J. Barber, Nature (London) 333:114 (1988).
61. M. E. Michel-Beyerle, M. Plato, J. Deisenhofer, H. Michel, M. Bixon, and J. Jortner, Biochim. Biophys. Acta 932:52 (1988); L. I. Krishtalik, in: "Bioelctrochemistry Today and Tomorrow, 4th Int. Frumkin Symp.," Moscow-Suzdal, Oct. 24-28 (1988).

62. E. E. Polymeropoulos, D. Möbius, and H. Kuhn, Thin Solid Films 68:173 (1980).
63. K. Saito, M. Yoneyama, M. Saito, K. Ikegami, M. Sugi, T. Nakamura, M. Matsumoto, and Y. Kawabata, Thin Solid Films 160:133 (1988); M. Saito, M. Sugi, T. Fukui, and S. Iizima, Thin Solid Films 100:117 (1983); H. Schreiber, H. Kuhn, and H. D. Forsterling, Ber. Bunsenges. Phys. Chem. 91:798 (1987).

SYNTHESES OF ELECTRICALLY CONDUCTIVE LB FILMS AND HOST-GUEST LB FILMS OF CYCLODEXTRINS AND AZOBENZENES

Yasujiro Kawabata

National Chemical Laboratory for Industry
Higashi, Tsukuba, Ibaraki 305
Japan

INTRODUCTION

The planned arrangement of functional molecules is of great interest in developing molecular electronic devices. The molecules are to be in exact locations and to cooperate with each other in the molecular assembly. The Langmuir-Blodgett (LB) technique is one of the best approaches to fabricate molecular assemblies at the supramolecular level. The syntheses of functional molecules as well as the techniques of controlling the architecture of molecular organizates are important in obtaining well defined and organized structures. Not only biomolecules but also man-made molecules can be used to construct these organizates. In this paper, electrical properties of the LB films of man-made molecules and optical properties of the host-guest LB films of biomolecules and dyes will be presented.

ELECTRICAL CONDUCTIVE LB FILMS

Recently, conductive LB films have attracted attention as precursors of organic devices at the molecular level. More than ten conductive LB films have been reported using charge transfer (CT) complexes and polymers. They are characterized in several categories. Anisotropic and isotropic conductors were prepared with and without alkyl chains in the LB films. These two major groups are classified into five subgroups.

(1) Anisotropic Conductors

i. Initially insulating LB films were made conductive by doping with iodine.[1,2]

ii. Intrinsically conductive LB films were prepared from amphiphilic CT complexes.[3,4]

iii. Initially insulating LB films of monomers were made conductive by electropolymerization accompanied by the doping of ClO_4.[5]

(2) Isotropic conductors

iv. Polymers with long alkyl chains were synthesized. Then the alkyl chains were removed by heating, resulting in conductive polymer LB films by doping.[6]

v. Initially conductive LB films were prepared from CT complexes without long alkyl chains.[7]

Two kinds of anisotropically conductive LB films using amphiphilic mixed valence complexes were investigated in this paper.

LB films of TMTTF-alkyl TCNQ

We have reported[4] that a CT complex of TMTTF and octadecyl TCNQ forms highly conductive LB films with conductivity as high as 0.1 S/cm. To further explore the possibility of incorporating CT complexes in LB films, we studied the formation of monolayers and LB films of CT complexes by varying the length of the alkyl chain of the TCNQ moiety in order to elucidate the effect of the hydrophobic interaction between alkyl chains on the physical properties of the materials. Figure 1 shows the chemical structures of the amphiphilic charge transfer complexes.

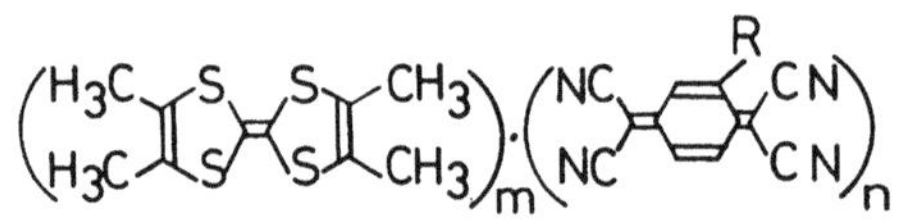

1: R = $C_{14}H_{29}$, m=3, m=2, 2: R=$C_{18}H_{37}$, M=1,n=1, 3: R = $C_{22}H_{45}$, m=1, n=2

Fig. 1. Structures of amphiphilic CT complexes.

Surface pressure-area isotherms of these complexes show that the limiting areas are almost the same (0.45-0.50 nm^2) irrespective of the donor to acceptor ratio and of the length of alkyl chain, indicating that the limiting areas are governed by the TCNQ moiety. Table 1 summarizes the average of bulk conductivity of compaction, monolayers at 25 mNm^{-1} at the air-glycerin interface,[8] and LB films. It is clear that the monolayer assembly technique is useful for the samples of Complex 1 and Complex 2, but not suitable for Complex 3. The conductivity of the compaction samples are almost the same, irrespective of the donor to acceptor ratio and of the length of the alkyl chain. Complex 1 and Complex 2 show an increase in lateral conductivity when they form monolayers at the air-water interface or LB films, but Complex 3 does not. The reason is not clear, but may be related to the balance between the hydrophilicity and the hydrophobicity or even to the dissociation or decomposition of the complex at the air-liquid interface. The bulk conductivity of LB films is as high as 1.0 and 0.1 S/cm for Complex 1 and Complex 2, respectively.

Table 1. Bulk conductivity of amphiphilic CT complexes.

Complex	Conductivity (S/cm)		
	Compacted	Monolayer[a]	LB film[b]
1	1.4×10^{-2}	0.3	1.0
2	1.2×10^{-2}	0.3	0.1
3	2.4×10^{-2}	$<10^{-3}$	$<10^{-3}$

a. At the air-glycerin interface. Lateral conductivity at 25 mNm^{-1}.
b. Lateral conductivity.

These results demonstrate the effect of the length of the alkyl chain on the physical properties of monolayers at the air-liquid interface and LB films on a solid substrate. High conductivity of up to 1 S/cm was obtained, which was the highest among anisotropic conductors.

The microscopic nature of the conductive LB films of Complex 1 was examined by electron spin resonance (ESR) spectroscopy.[9] Anomalous behavior has been detected in the temperature range 100-200 K. Three striking features were observed: a decrease in the spin susceptibility with decreasing temperature, a prominent maximum of the linewidth, and a maximum of the g value. These features are very similar to those observed in the single crystal of TTF-TCNQ around Peierls transition temperature, which indicates the important role of Peierls instability in this system. Moreover, the temperature dependent conductivity was found to show a cross-over around 200 K, indicating that the conduction mechanisms in the higher and lower temperature regions are different from each other. At high temperatures the observed activation energy was fairly small ($E_a = 0.057$ eV) and could be governed by the boundaries between metallic domains.

<u>LB films of dialkyldimethylammonium Ni(dmit)$_2$</u>

The conductive LB films reported so far are characterized as one-dimensional systems, including TMTTF-alkyl TCNQ mentioned above. Therefore, they are sensitive to defects and disorders and have intrinsic instability as one-dimensional conductors. Conductive LB films of higher dimensionality are expected to suppress the instability and to achieve higher conductivity.

Recently, it has been reported that the mixed valence complex of $(Et_2Me_2N)[Ni(dmit)_2]_2$ is a two-dimensional conductor.[10] A superconductor transition at 5 K under 7 kbar has been observed for an analogous complex,[11] $(Me_4N)[Ni(dmit)_2]_2$.

The LB films of dialkyldimethylammonium-Ni(dmit)$_2$ (4) were, therefore, studied (Fig. 2). The alkyl chain lengths were varied from C_{10} to C_{22}.

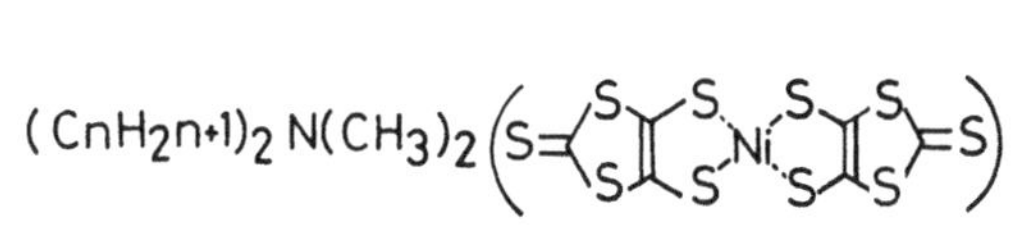

Fig. 2. Structure of dialkyldimethylammonium Ni(dmit)$_2$.

The complexes were prepared according to procedures described previously.[10] These 1 to 1 complexes were insulating and further oxidation turned them into green-black solids, which were insoluble in common solvents and not appropriate for preparing LB films. As an alternative way to obtain conductive LB films, the monolayers of the precursor 1:1 complex were transferred onto a solid substrate, and then the films were made conductive by oxidation.

Surface pressure-area isotherms of the Ni(dmit)$_2$ complex (n=18) and the mixed monolayers with icosanoic acid indicate that the pure Ni complex forms a collapsed film. On the other hand, the 2:1 and 1:1 mixtures with icosanoic acid form stable monolayers at the air-water interface. Figure 3 shows the F-A isotherms of the 1:1 mixture of Ni complexes and icosanoic acid. As the carbon number of the side chains increases, the slope corresponding to the condensed phase becomes steeper in these curves. This indicates that the longer alkyl chains are effective for monolayer stability at the air-water interface.

Oxidation of LB films was performed by bromine gas diluted with nitrogen gas. The bulk conductivity of the oxidized films at room temperature are summarized in Fig. 4. These values were obtained by assuming a monolayer thickness of 3 nm. As the carbon number of the alkyl chain becomes smaller, the conductivity becomes larger. The value of 0.1 S/cm was obtained for n=8 of Complex 4, which is almost the same as that of TMTTF-octadecyl TCNQ and is one of the most conductive LB films. To attain higher conductivity the oxidation state of the complex should be controlled. Along this line, several investigations are in progress.

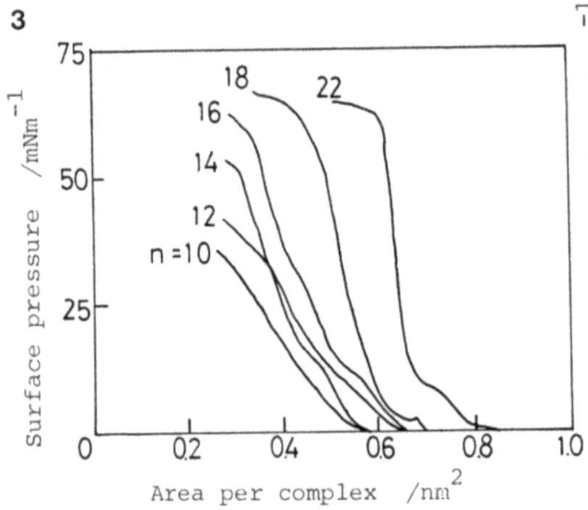

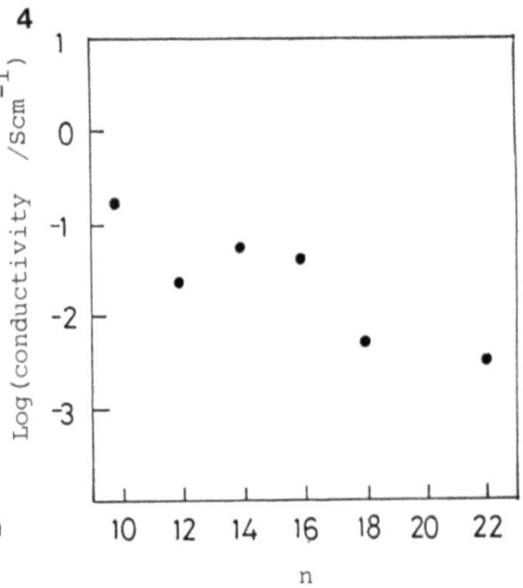

Fig. 3. Surface pressure-area isotherms of 1:1 mixed monolayers of 4 and icosanoic acid.

Fig. 4. Conductivity of oxidized 1:1 mixed LB films of 4 and icosanoic acid.

Host-Guest LB Films of Cyclodextrins and Azobenzenes

One approach towards utilizing the LB technique is to fabricate very thin organic films from a single kind of functional molecule, e.g., insulating or conductive LB films. The other approach is to assemble different molecular species in a complex functional unit, so that different species of molecules cooperate with each other, e.g., interlocked complexes with cyanine dye and azo dye.[13] Although several excellent studies were done based on the latter approach by H. Kuhn and his coworkers in 1970s, the amount of work and the number of systems studied is much less compared with those based on the former approach.[13] It is indispensable to fabricate highly organized organic assemblies made up of different molecular species for future molecular electronics.

Along this line we have fabricated host-guest LB films with amphiphilic cyclodextrins and azo dyes, and have demonstrated that these molecular assemblies show specific photochemical behaviors. Cyclodextrins (CDs) are biomolecules that consist of several glucose units which have a cylindrical shape with a cavity inside of the molecule. A variety of organic molecules can be incorporated into the cavity in an ordered state to form host-guest complexes without any covalent bonds. The LB films of CDs will provide new molecular systems owing to free volume of the cavity and the formation of a host-guest complex with functional molecules.

Chemical modifications of hydrophobic groups to CDs are required for the formation of stable monolayers at the air-water interface because CDs themselves are soluble in water. Long alkyl sulfinyl, sulfide or amino derivatives of CDs were synthesized and fabricated as LB films (Fig. 5). The values of the limiting areas of the monolayers at the air-water interface and polarized infrared spectra of the LB films indicated that the CD rings are parallel to the film plane and long alkyl chains are perpendicular to it.[14,15]

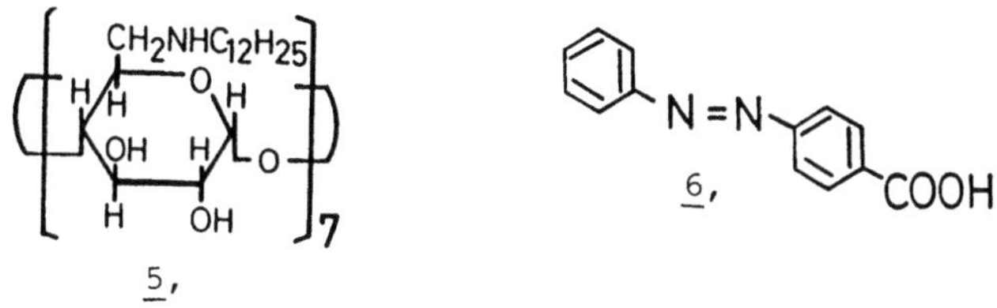

Fig. 5. Structures of amphiphilic CD (5) and an azobenzene (6).

Host-guest LB films were fabricated from chloroform solution of equimolar mixture of amphiphilic CD (5) and azobenzene derivatives through the formation of monolayers at the air-water interface. Azobenzenes, without long alkyl chains, were incorporated in the LB films accompanied by the salt formation of the acidic residue of the azobenzene with the amino residue of the CD. Azobenzene (6) are included in the cavity of CD in such a manner that the long axis is almost perpendicular to the film plane[16] as shown in Fig. 6.

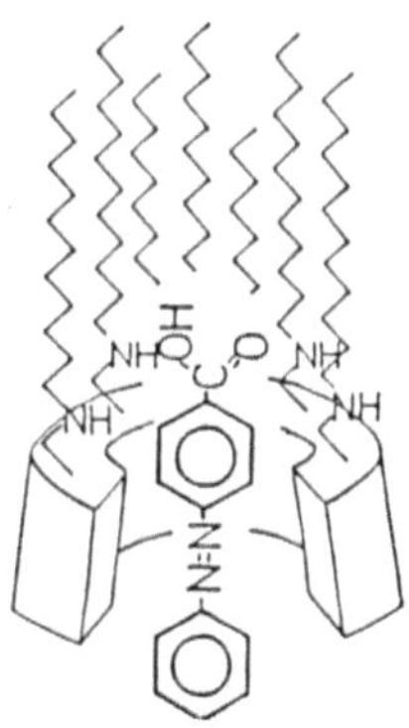

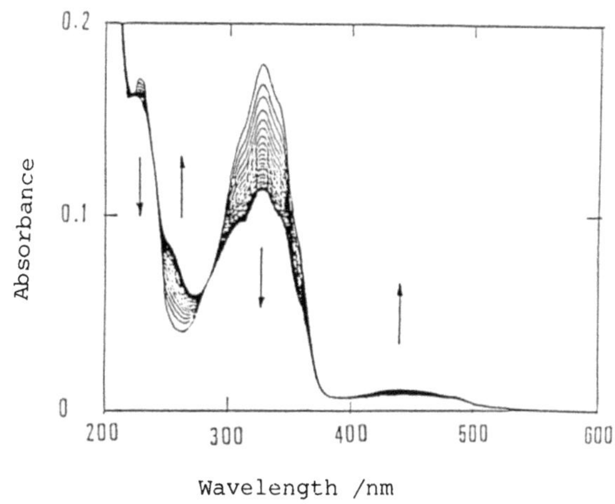

Fig. 6. Host-guest LB film of 5 and 6.

Fig. 7. UV-VIS spectral change of host-guest LB film of 5 and 6 irradiated at 360 nm.

Azobenzene (6) in the host-guest LB films with highly ordered structure exhibits quite different photochemical behaviors compared with the amphiphilic azobenzene derivative with long alkyl chains in the LB films. The former shows reversible *cis-trans* isomerization but the latter does not isomerize reversibly in the LB films. This is probably due to strong interactions between alkyl chains attached to the amphiphilic azobenzene or the bulkiness of the *cis* isomer causing steric hindrance in the compressed assemblies.[17] Figure 7 shows changes in the electronic absorption spectra as a function of irradiation time for the host-guest LB film of Complexes 5 and 6. Upon irradiation with 360-nm light, the absorption bands at 228 and 327 nm decreased while the bands at 255 and 445 nm increased, indicating the *trans*-to-*cis* isomerization of Complex 6. After the photostationary state with 360-nm irradiation was reached, the host-guest LB films was irradiated with 254-nm light. The reversal spectral change was observed, indicating the occurrence of the *cis*-to-*trans* isomerization.[18]

Host-guest LB films offer several advantages for constructing molecular assemblies.[19]

(1) Amphiphilic CD is a novel molecule capable of incorporating functional molecules without long alkyl chains and capable of being assembled in a highly ordered state.

(2) The cavity of CD provides a free volume for functional molecules and enables them to react even in this compressed assembly.

(3) A variety of functional molecules can be introduced in the LB films depending on the size of the cavity.

Therefore, these systems will provide more examples of complex functional units utilizing the cooperation of different molecular species.

CONCLUSION

LB films of TMTTF-C_{14}TCNQ show conductivity of up to 1.0 S/cm without doping or further treatment. The results of the characterization of the films suggest that the LB films are made up of metallic domains. The dialkyldimethylammonium-Ni(dmit)$_2$ have conductivity from 0.01 to 0.1 S/cm, depending on the alkyl chain length, after oxidation by bromine. This system will be one of the most promising candidates for LB films of larger conductivity and higher dimensionality.

Host-guest LB films of amphiphilic CD and azo dyes exhibit excellent *cis-trans* photoreversible isomerization due to the lightly substituted azo dyes and the appropriate cavity in the CD molecule.

ACKNOWLEDGEMENTS

The author wishes to thank Drs. M. Sugi, A. Yabe and M. Matsumoto for informative comments and helpful discussion.

REFERENCES

1. A. Ruaudel-Teixier, M. Vandevyver, and A. Barraud, Mol. Cryst. Liq. Cryst. 120:319 (1985).
2. M. Vandevyer, J. Richard, A. Barraud, A. Ruaudel-Teixier, M. Lequan, and R. M. Lequan, J. Chem. Phys. 87:6754 (1987).
3. T. Nakamura, M. Matsumoto, F. Takei, M. Tanaka, T. Sekiguchi, E. Manda, and Y. Kawabata, Chem. Lett. 709 (1986).
4. Y. Kawabata, T. Nakamura, M. Matsumoto, M. Tanaka, T. Sekiguchi, H. Komizu, E. Manda, and G. Saito, Syn. Met. 19:663 (1987).
5. T. Iyoda, M. Ando, T. Kaneko, A. Ohtani, T. Shimizu, and K. Honda, Tetrahedron Lett. 27:5633 (1986).
6. Y. Nishikata, M. Kakimoto, and Y. Imai, Paper presented at "6th Int.

Conf. on Surface and Colloid Science," Abst. No. 4B06, June (1988) Hakone, JAPAN.
7. M. Fujiki and H. Tabei, Syn. Met. 18:815 (1987).
8. T. Nakamura, F. Takei, M. Matsumoto, M. Tanaka, T. Sekiguchi, E. Manda, Y. Kawabata, and G. Saito, *ibid.* 19:681 (1987).
9. K. Ikegami, S. Kuroda, K. Saito, M. Saito, M. Sugi, T. Nakamura, M. Matsumoto, and G. Saito, Syn. Met. 27:B587 (1988).
10. T. Nakamura, H. Tanaka, M. Matsumoto, H. Tachibana, E. Manda, and Y. Kawabata, *ibid.* 27:B601 (1988).
11. R. Kato, H. Kobayashi, H. Kim, A. Kobayashi, Y. Sasaki, T. Mori, and H. Inokuchi, Chem. Lett. 865 (1988).
12. H. Kim, A. Kobayashi, Y. Sasaki, R. Kato, and H. Kobayashi, *ibid.* 1799 (1988).
13. D. Möbius, Acc. Chem. Res. 14:63 (1981).
14. Y. Kawabata, M. Matsumoto, M. Tanaka, H. Takahashi, Y. Irinatsu, S. Tamura, W. Tagaki, H. Nakahara, and K. Fukuda, Chem. Lett. 1833 (1986).
15. Y. Kawabata, M. Matsumoto, T. Nakamura, M. Tanaka, H. Takahashi, S. Tamura, W. Tagaki, H. Nakahara, and K. Fukuda, Thin Solid Films 159:353 (1988).
16. M. Tanaka, Y. Ishizuka, M. Matsumoto, T. Nakamura, A. Yabe, H. Nakanishi, Y. Kawabata, H. Takahashi, S.Tamura, W. Tagaki, H. Nakahara, and K. Fukuda, Chem. Lett. 1307 (1987).
17. H. Nakahara, K. Fukuda, M. Shimomura, and T. Kunitake, Nippon Kagaku Kaishi, *ibid.* 1001 (1988).
18. A. Yabe, Y. Kawabata, H. Niino, M. Tanaka, A. Ouchi, H. Takahashi, S. Tamura, W. Tagaki, H. Nakahara, and K. Fukuda, Chem. Lett. 1 (1988).
19. A. Yabe, Y. Kawabata, M. Matsumoto, A. Ouchi, H. Takahashi, S. Tamura, W. Tagaki, H. Nakahara, and K. Fukuda, Thin Solid Films 160:33 (1988).

ARTIFICIAL STRUCTURE OF METAL COMPLEXES BY SELECTIVE COORDINATION EPITAXY

Hiroshi Tanino, Kazuhiro Takahashi, and Takafumi Yao

Electrotechnical Laboratory
Tsukuba, Ibaraki 305, Japan

ABSTRACT

This paper describes a novel fabrication technique to construct molecular or bioelectronic devices. Selective coordination epitaxy (SCE), which utilizes the chemical nature of the coordination bond of metal complexes, is introduced. This method enables us to assemble organic molecules including metal complexes into various artificial structures for molecular electronic devices. The characteristics and applications of SCE are discussed in comparison with various epitaxial methods for inorganic semiconductors and the Langmuir-Blodgett method for organic polymers.

INTRODUCTION

The proposal and the realization of the superlattice of inorganic semiconductors is one of the greatest advances in the history of materials science (Esaki and Tsu, 1970). New materials have been created by chemical synthesis for a long time, and new phenomena or new applications have been studied for synthesized materials. However, the superlattice, which never exists in nature, is not created by ordinary chemical syntheses, but only by well-controlled technological processes. We may say that the notion of new materials itself has been changed by this invention. Various new techniques, such as liquid phase epitaxy (LPE) by Nelson (1963), metal organic chemical vapor deposition (MOCVD) by Manasevit (1968), molecular beam epitaxy (MBE) by Arthur and LePore (1969), and atomic layer epitaxy (ALE) by Suntola *et al.* (1977; 1980), have been proposed, developed, and improved. At present, we can control growth on a monoatomic-layer scale, and further investigation to realize quantum wires and quantum dots is in progress (Petroff *et al.*, 1982). Microscopic growth techniques will be improved further by utilizing recent advanced lithography techniques which enable the building of three-dimensional structures (e.g. Foxon and Harris, 1987; Bate, 1988).

On the other hand, a new approach to extremely small devices is important in molecular electronics, organic electronics, and bioelectronics, where a wide variety of functions of organic or biomolecular systems will be utilized. One important and powerful technique for constructing organized molecular systems is the well-known Langmuir-Blodgett (LB) method, first realized by Blodgett (1935). Monolayer superstructures of organic polymers are constructed by this method. Recently, the old-fashioned dream of organic semiconductors is revived as a realization of practical molecular devices (e.g. Carter, 1982; Kuhn, 1985).

Between these two main streams of the study of the superlattices of semiconductors and the LB layers of organic molecules, we propose a third way to construct superstructures where metal complexes are utilized. Metal complexes are classified as inorganic materials in chemistry (e.g. Huheey, 1978), but they include both inorganic and organic components as ligands of central metals. Thus, they have intermediate characteristics between inorganic semiconductors and organic molecules. The structures of a metal complex units are very simple and they are smaller than organic polymers used in the LB method. Crystal growth is easier than for organic materials consisting of large molecules. They are to some extent similar in nature to organic compounds, since they have various organic ligands.

In this paper, we introduce a novel fabrication technique called **selective coordination epitaxy** (SCE) (Tanino *et al.*, 1986), which utilizes the chemical nature of metal complexes. We suggest metal complexes as new materials to construct extremely small molecular devices. It is possible to realize three-dimensionally ordered artificial structures, if metal complexes are appropriately introduced as connecting units of organized molecular assemblies (Takahashi *et al.*, 1987). We present some typical examples of artificial structures of one-dimensional mixed-valence metal complexes. The characteristics and future applications of SCE are reviewed.

HALOGEN-BRIDGED MIXED-VALENCE METAL COMPLEXES (HMMC's)

Recently, a group of metal complexes called Wolffram's red salt analogs, quasi-one-dimensional halogen-bridged mixed-valence metal complexes (HMMC's), have attracted wide interest from various viewpoints of physics and chemistry

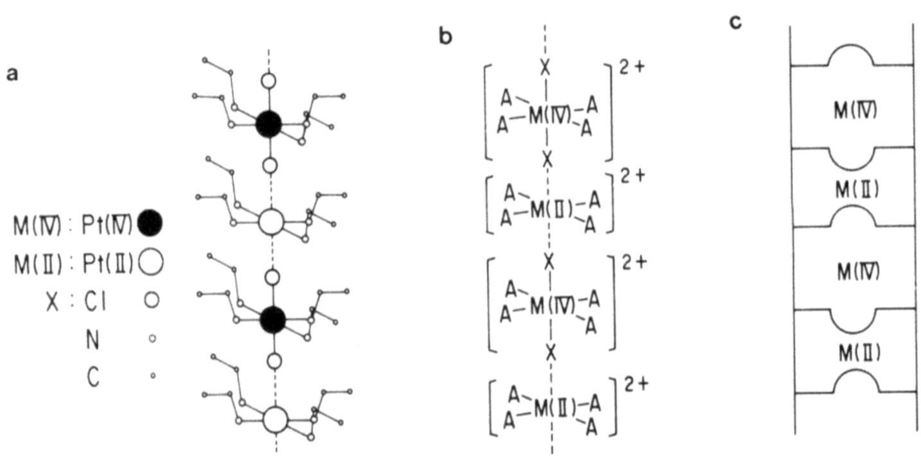

Fig. 1. Structure of halogen-bridged mixed-valence metal complexes (HMMC's). (a) Structure of a chain of Wolffram's red salt, $[Pt(II)(etn)_4][Pt(IV)Cl_2(etn)_4]Cl_4 \cdot 4H_2O$, a typical HMMC. (b) General representation of a chain of HMMC's, $[M(II)A_4][M(IV)X_2A_4]Y_4$, where M, X, A and Y represent central metals, bridging halogens, neutral ligands (such as amines), and counter anions, respectively. (c) Schematic model of HMMC's describing a key-lock relationship between $[M(II)A_4]^{2+}$ and $[M(II)X_2A_4]^{2+}$ ions.

(e.g. Day, 1975; Kobayashi, 1986). Figure 1 shows the structure of one chain of HMMC's $[M(II)A_4][M(IV)X_2A_4]Y_4$, where M are transition metals such as Pt, Pd, and Ni; X are halogens such as Cl, Br, and I; A are neutral ligands of amines such as ethylamine, 1,2-diaminoethane, and 1,2-diaminopropane; and Y are counter anions such as halogens, ClO_4, and BF_4. More than 100 compounds have been synthesized to date. HMMC's consist of linear chains of alternating metals and halogens of ...M(II)...X-M(IV)-X...M(II)...X-M(IV)-X... . Since each chain is surrounded by neutral ligands of amines and is set apart from neighboring chains, the one-dimensional character is observed in electronic states and lattice vibrations. As shown in Fig. 1b, their structure is understood as an alternating stack of two types of complex ions, $[M(II)A_4]^{2+}$ and $[M(IV)X_2A_4]^{2+}$. In practice, they are grown from solutions in which these two stable ions are dissolved in an appropriate solvent such as methanol and water.

If we imagine a hypothetical chain of -M(III)-X-M(III)-X-, where all halogens X are at the mid-point between metals, the chain becomes a one-dimensional metal, since M(III) (M= Pt, Pd, and Ni) has seven d-electrons. However, because of strong electron-phonon coupling in one-dimensional systems, the electronic state relaxes to the ground state of a charge density wave, where the charge of metals alternate along the chain, and the periodicity is doubled by the $2k_F$ Peierls instability. From this viewpoint, these complexes are interesting models of a one-dimensional electron-phonon system, and their optical properties in particular have been investigated by solid state physicists collaborating with inorganic chemists. Further details of HMMC's are reviewed in literature (Kobayashi, 1986). We only comment here that they show special optical properties because of their strong interaction with visible light, and are expected to be applied as new optical switching devices.

Let us discuss the growth process of these compounds in relation to the crystal structure. Their growing and functional units are complex ions such as $[M(II)A_4]^{2+}$ of square planar configuration and $[M(IV)X_2A_4]^{2+}$ of elongated octahedral configuration. Neutral ligands of amines, A, are strongly bound to the central metals, M, at one end, but are chemically inactive at the other end and interact only with weak molecular force such as van der Waals force. Halogens, X, interact with other metal complex ions. If a $[M(IV)X_2A_4]^{2+}$ ion comes close to a $[M(II)A_4]^{2+}$ ion, they are connected to each other to form M(II)...X-M(IV) due to a charge transfer interaction. Since the coupling of M(II)...X-M(IV) is much stronger than the coupling of M(II)-M(II) or M(IV)-M(IV), the same kinds of ions will not be linked to each other. Thus, the alternative structure of...M(II)...X-M(IV)-X...M(II)...X-M(IV)-X... will grow by stages. This process can easily be understood by a **key-lock relationship** similar to enzymatic reactions as shown in Fig. 1c. The coupling of M(II)...X-M(IV) is possible for various metals M, halogens X, neutral ligands A, and counter anions Y. For example, the so-called mixed-metal compounds such as $[Pd(II)(en)_2][Pt(IV)Cl_2(en)_2](ClO_4)_4$, which have the coupling of Pd(II)...X-Pt (IV), are synthesized.

In ordinary inorganic semiconductors, atoms such as Si, Ga, and As are the constituents, but their functional units are much larger than metal complexes because the extended electronic states are used as the information units. The minimum size of the three-dimensionally confined extended electronic states would be of the order of the electron wavelength of 30 nm. The organic polymers used for the LB method are long chain molecules of hydrous or anhydrous nature at each end with a dimension of more than 10 nm. If we can use the localized electronic states of metal complexes, whose functional units are of 1 - 2 nm, the degree of integration can be much greater than the theoretical limit for devices made of inorganic semiconductors or LB layers.

SELECTIVE COORDINATION EPITAXY

A schematic representation of the principle of growth by selective coordination epitaxy is shown in Fig. 2. Let us assume that the $[M(II)A_4]$ unit (abbreviated as M2) is at the surface of the substrate being held in a solution with two types of stable ions, $[M(IV)X_2'A_4]^{2+}$(M4') and $[M(II)A_4]^{2+}$(M2). When a $[M(IV)X_2'A_4]^{2+}$(M4') ion approaches the substrate, it will be attracted and adsorbed on the surface because of the bonding force between M(II) and X'. However, M2 ions will not be adsorbed there because the M(II)-M(II) coupling force is rather weak. In the next step, an M2 ion will be adsorbed on this M4' unit. The chain direction will be strictly conserved because the square planes must stack parallel to each other regulated by the neutral ligands A and counter anions Y. As a result, we will

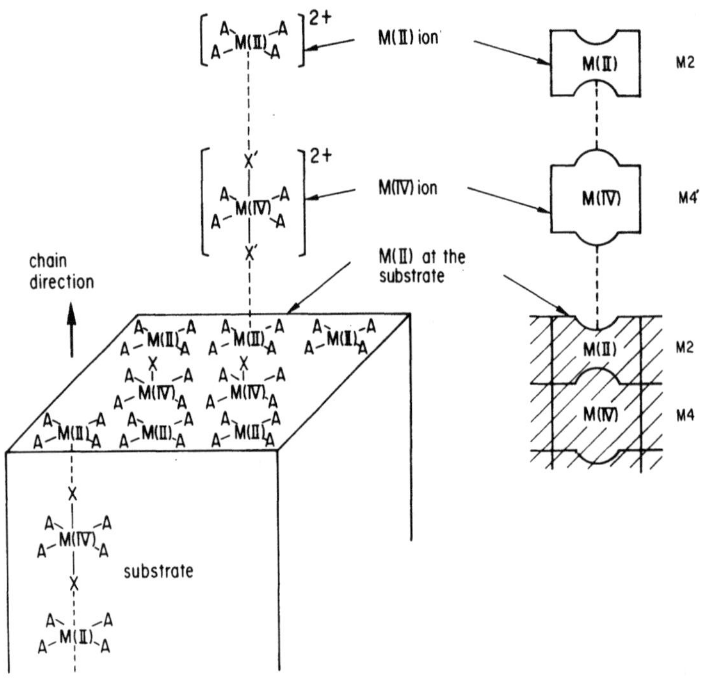

Fig. 2. Principle of selective coordination epitaxy.

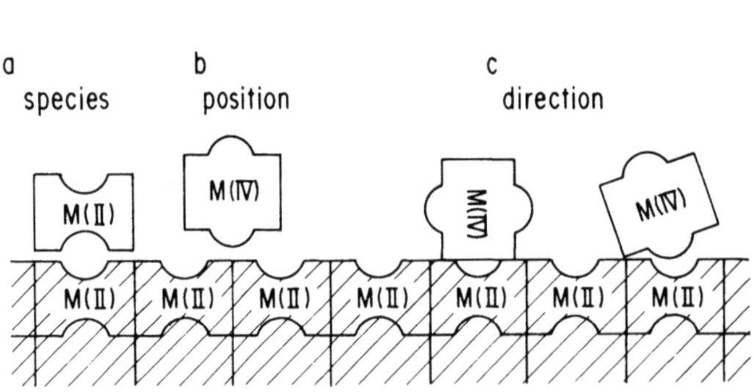

Fig. 3. Examples of forbidden structures by the selectivity of the coordination bonds of metal complexes.

obtain a heterostructure of M(II)...X'-M(IV)-X'...M(II)...X'-M(IV)-X'... . This new growth method utilizes the chemical nature of the coordination bonds of metal complexes. If M2 ions come close to M(II) units at the surface, there will be hardly any connection between M(II) and M(IV). Or if ligands A of M4' ions come close to the surface, the ends of the neutral ligands are chemically inactive. Examples of forbidden structures are shown in Fig. 3. Thus, the coordinating species, sites, and directions are selected exclusively by the chemical functions of the active sites of metal complexes. Hence the technique has been named selective coordination epitaxy.

This growth process is different from other methods. In the case of inorganic semiconductors, growth progresses by nucleation and step propagation at the surface. This is recognized as a difficulty associated with using atomic point epitaxy to construct quantum well dots. The problem with LB layers grown by weak molecular forces is that they cannot be crystallized well. It is very difficult to obtain lattice matching between the neighboring layers because of the construction principle of the LB method itself. Epitaxial deposition of polymers on alkali metal halide crystals (e.g. Uyeda, 1974), and small organic molecules or some metal complexes such as phthalocyanine on metal substrates (e.g. Firment, 1979; Buchholz, 1977) has been reported by many groups. Here also, weak molecular forces are driving the epitaxial growth, and a difficulty analogous to that inherent in the LB method is again present. However, in the case of SCE, growth proceeds by the chemical connection between the metal complex units. We may call it a non-fractional or digital snapping technique.

ARTIFICIAL STRUCTURES OF METAL COMPLEXES

An example of a single heterostructure grown by SCE is shown in Fig. 4. $[Pt(II)(en)_2][Pt(IV)Cl_2(en)_2](BF_4)_4$ was grown epitaxially on the substrate $[Pt(II)(en)_2][Pt(IV)Br_2(en)_2](BF_4)_4$ in a liquid phase. HMMC's show strong absorption of charge transfer excitons, highly polarized in the chain direction. Polarized pictures show that the chain directions of the substrate and the epi-layer are completely parallel to each other. The three-dimensional ordering of the crystals grown by SCE was confirmed by an x-ray diffraction pattern revealed using the Buerger precession method (Tanino *et al.*, 1987a).

We also constructed a double heterostructure of HMMC's by SCE (Takahashi *et al.*, 1986) as shown in Fig. 5. On a bromide substrate, a chloride complex was grown as the first epi-layer. A bromide complex was then grown atop the chloride complex as the second layer. The result indicates that superlattices of HMMC's can be constructed by SCE in principle, if the growth rate is controlled appropriately.

One interesting feature of the superstructure of HMMC's is the low dimensionality. Quantum wells of inorganic semiconductors are confined one-dimensionally, but quantum wells of HMMC's are confined three-dimensionally, since the intrinsic electronic states of HMMC's are confined two-dimensionally. Recently, we studied optical properties of mixed crystals of $[Pt(II)(en)_2][Pt(IV)(Cl_{1-x}Br_x)_2(en)_2](ClO_4)_4$ (Takahashi *et al.*, 1987; Tanino *et al.*, 1987b). They have been shown to be made of one-dimensional chains including both Cl and Br randomly, and the optical absorption band is 'amalgamated'. Introducing these mixed crystals into the superstructure as shown in Fig. 6a, we can construct alloy superlattices. The well height is controlled by changing the composition x of epi-layers in a similar manner as the superlattice of $GaAs\text{-}Ga_{1-x}Al_xAs$ (Takahashi *et al.*, 1987).

Using this method, we can construct various artificial structures of HMMC's by changing not only the bridging halogens but also all other components (Takahashi *et al.*, 1988). As an example, a heterostructure of

$[Pt(II)(en)_2][Pt(IV)Cl_2(en)_2](ClO_4)_4$ on $[Ni(II)(en)_2][Pt(IV)Br_2(en)_2](ClO_4)_4$ achieved by changing central metals is shown in Fig. 6b.

FUTURE APPLICATIONS OF SELECTIVE COORDINATION EPITAXY

SCE is not limited to HMMC's, but is applicable to almost all metal complexes, e.g. porphyrins, macrocyclic ligands, dialkylglyoximes and β-diketones. Some strong complexes such as phthalocyanines can be grown by vapor-phase SCE. It will be a powerful tool because several techniques such as *in-situ* RHEED (reflection high energy electron diffraction) observation developed in molecular beam epitaxy (MBE) may be used. The principal concept of SCE, i.e., the utilization of a key-lock relationship of chemical reaction, can be applied to organic systems.

As an advanced variation, we proposed a new method, named **complex layer epitaxy** (CLE) (Tanino *et al.*, 1986; Tanino *et al.*, 1988). This is a monolayer epitaxial growth technique similar to atomic layer epitaxy used for inorganic semiconductors. If we dip a substrate with $[M(II)A_4]$ units (M2) or the surface into a solution of $[M(II)X_2A_4]^{2+}$ ions (M4), only one layer of M4 will be adsorbed because of M-X coupling. A second layer of M4 will not be adsorbed on the first layer of M4 as M(IV)-M(IV) coupling is very weak. Using this method, we can make a complex monolayered yakitori-like* heterostructure, e.g. $[Pd(II)(pn)_2]/[Pt(IV)I_2(en)_2]/[Ni(II)(tn)_2]/[Pd(IV)Br_2(chxn)_2]$.

If we imagine one atom of Si, Ga or As, it has no internal plan where the atom let itself be adsorbed. Just after it comes close to the surface, the chemical bonds between the atom and the neighboring atoms on the surface will be created. However, in the case of metal complexes, the species, direction, and position of neighboring complexes are indicated by the complex itself. In this sense, the complex unit is the smallest unit which has an internal plan of conformation, and it exhibits fundamental functions of self-assembly independent of other complex units. A famous example is the case of the Fe complex of hemoglobin, which operates as one device unit. Thus, by SCE, we can construct extremely small tailored structures by using the self-reproducing mechanisms of metal complexes without the need for microlithography. Structural information such as the orientation of the complexes can be transferred from unit to unit. Hence, if we assume that one complex unit acts as one device unit, we will be able to construct extremely small, but precisely ordered, devices. We proposed to call them **coordination devices** (Tanino *et al.* 1987a; Tanino *et al.*, 1988), a type of molecular device, but much smaller than ordinary semiconductor devices or conventional molecular devices.

Another important application of metal complexes and SCE is to use them to connect organic molecules by coordination bonds (Tanino *et al.*, 1988) in the following sequence:

(molecule A)-[complex a].....[complex b]-(molecule B).

Here, the metal complexes are called **molecular snaps**.

In Fig. 7, we have summarized several models of artificial structures of metal complexes. We have realized single heterostructures of HMMC's (Fig. 7a) such as the red chloride complex on the green bromide complex and double heterostructures of HMMC's (Fig. 7b) such as green on red on green. The advanced technique CLE is proposed for the construction of monolayered superstructures of HMMC's (Fig. 7c). Coordination devices, e.g. a branching circuit made of metal complexes (Fig. 7d), will be constructed by SCE in the future. A difficulty in constructing three-dimensional devices (Fig. 7e) is

---

* Japanese skewered chicken and vegetables.

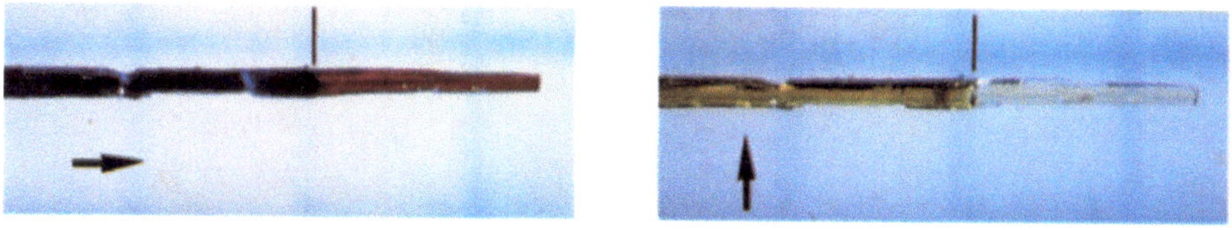

Fig. 4. Single heterostructures of halogen-bridged mixed-valence metal complexes $[Pt(II)(en)_2][Pt(IV)Cl_2(en)_2](BF_4)_4$ on $[Pt(II)(en)_2]$ $[Pt(IV)Br_2(en)_2](BF_4)_4$ constructed by liquid-phase SCE. The polarization of light is indicated by arrows.

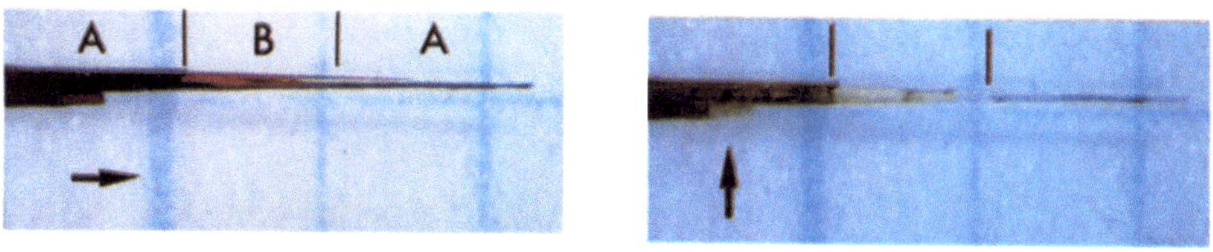

Fig. 5. Double heterostructures of A on B on A, where A= $[Pt(II)(en)_2]$ $[Pt(IV)Br_2(en)_2](BF_4)_4$, and B= $[Pt(II)(en)_2][Pt(IV)Cl_2(en)_2](BF_4)_4$.

Fig. 6. Examples of single heterostructures (a) by introducing mixed crystals, $[Pt(II)(en)_2][Pt(IV)(Cl_{0.95}Br_{0.05})_2(en)_2](ClO_4)_4$ on $[Pt(II)(en)_2][Pt(IV)Br_2(en)_2](ClO_4)_4$; and (b) by changing central metals, $[Pt(II)(en)_2][Pt(IV)Cl_2(en)_2](ClO_4)_4$ on $[Ni(II)(en)_2]$ $[Pt(IV)Br_2(en)_2](ClO_4)_4)$.

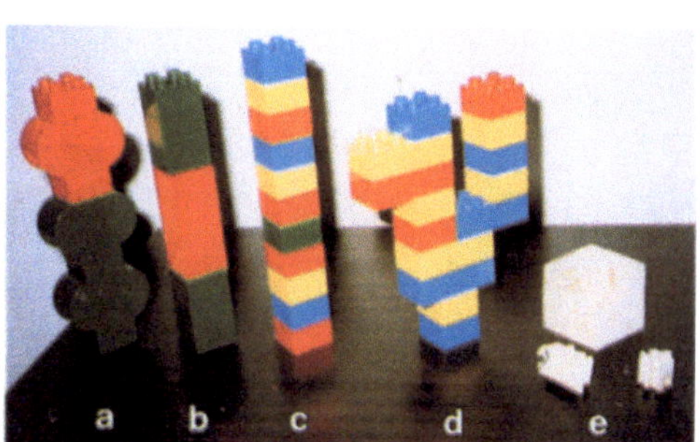

Fig. 7. Models of various artificial structures. (a) Single heterostructure, (b) double heterostructure, (c) mono-layered yakitori-like structure, (d) a branching circuit as an example of a coordination device, and (e) three-dimensional structure.

how to solve the problem of cross-networks with spatial conflicts between molecules. However, this problem is also common to inorganic semiconductor superlattices and organic molecular superstructures.

CONCLUSION

Selective coordination epitaxy, a novel fabrication technique for constructing molecular devices, can manipulate a single molecule, recognizing each molecule by the key-lock relationship of a coordination bond. This is a prototype of the construction of self-assembling 'intelligent' devices by using the internal plans of metal complex units.

ACKNOWLEDGEMENTS

The authors are grateful to Drs. M. Sugi, K. Hoh, T. Ishiguro, and S. Maekawa for their continued interest and encouragement.

REFERENCES

Arthur, J. R., and LePore, J. J., 1969, J. Vac. Sci. Technol. 6:545.
Bate, R. T., 1988, Sci. Amer. 258(3):78.
Buchholz, J. C., and Somorjai, 1977, J. Chem. Phys. 66:573.
Blodgett, K. B., 1935, J. Am. Chem. Soc. 57:1007.
Carter, F. L., ed., 1982, "Molecular Electronic Devices," Dekker, New York.
Day, P., 1975, in: "Low-Dimensional Cooperative Phenomena," H.J. Keller, ed., p. 191, Plenum, New York.
Esaki, L., and Tsu, R., 1970, IBM J. Res. Dev. 14:61.
Firment, L. E., and Somorjai, G. A., 1979, Surface Sci. 84:275.
Foxon, C. T., and Harris, J. J., ed., 1987, "Proc. 4th Int. Conf. on Molecular Beam Epitaxy 1986," in: "J. Cryst. Growth 81".
Huheey, J. E., 1978, "Inorganic Chemistry," 2nd ed., Harper Row, New York.
Kobayashi, K., 1986, in: "Excitonic Processes in Solids, Solid State Physics 60," M. Ueta, H. Kanzaki, K. Kobayashi, Y. Toyozawa, and E. Hanamura, ed., p.475, Springer, Berlin.
Kuhn, H., 1985, "Proc. Int. Symp. on Future Electron Devices - Bioelectronic and Molecular Electronic Devices," p. 1, Res. Dev. Assoc. for Future Electron Devices, Tokyo.
Manasevit, H. M., 1968, Appl. Phys. Lett. 12:156.
Nelson, H., 1963, RCA Rev. 24:603.
Petroff, P. M., Gossard, A. C., Logan, R. A., and Wiegmann, W., 1982, Appl. Phys. Lett. 41:635.
Suntola, T., and Antson, J., 1977, US Patent 4 058 430.
Suntola, T., Antson, J., Pakkala, A., and Lindfors, S., 1980, Society for Information Display Digest 80:108.
Takahashi, K., Tanino, H., Yao, T., 1987, Jpn. J. Appl. Phys. 26:L97.
Takahashi, K., Tanino, H., Yao, T., 1988, in: "Proc. 4th Int. Conf. on Langmuir-Blodgett Films, Tsukuba, Japan, April, 1989," in press.
Tanino, H., Takahashi, K., Yao, T., 1986, Jpn. J. Appl. Phys. 25:L571.
Tanino, H., Takahashi, K., Yao, T., 1987a, Jpn. J. Appl. Phys. 26:L983.
Tanino, H., Takahashi, K., Kato, M., Yao, T., 1987b, Solid State Commun. 65:643.
Tanino, H., Takahashi, K., Yao, T., 1988, J. Mol. Electronics 4:17.
Uyeda, N., Murata, Y., Kobayashi, T., and Suito, E., 1974, J. Cryst. Growth 26:267.

INDUSTRIAL SCALE PRODUCTION OF L-B LAYERS

O. Albrecht, T. Ginnai, A. Harrington, D. Marr-Leisy and V. Rodov

Molecular Electronics Corporation
4030 Spencer St., MS 108
Torrance, CA 90503, U.S.A.

INTRODUCTION

To date, the fast growing field of L-B films has not found any industrial, large scale application. Of the obstacles remaining in the drive towards commercialization, the following may be perceived to be important.

a) No versatile and scalable volume fabrication technology available.
b) The absence of clean and reliable methods; consequential repeatability problems.
c) Incompatibility with standard processing methods. For electronics applications, this generally means semiconductor processing methods.
c) Inadequate chemistry. There is at present no real basis of understanding of structure-function relationships and little attention has been given to materials suitable for practical applications.

Clearly, to a very considerable extent, these problems point out the need to address from an industrial perspective the basic issues involved in the fabrication of L-B films and understand the prime requirement of obtaining a reliable, volume fabrication technology.

In this article, we discuss some of the basic ideas and results we have obtained so far in taking these themes a step further with the development of machinery for the continuous production of these ultra-thin films. The continuous production approach (as opposed to the standard batch process), quite apart from possibly being the only practical, viable, option has a number of merits. Not the least important are the following.

a) Productivity; coating of endless tapes or object attached to continuous tapes.
b) The film itself can be used to clean the machine and the subphase by removing the first square meters of film.
c) The residence time (age of the monolayer, time between first compressing and depositing the film) can be controlled by removing excess film in the right proportion to the rate of deposition. This results in films with the same properties for all layers, independent of the dipping speed or number of layers.

These and other relevant issues, together with some results obtained in the coating of 5-in. chromium photomask blanks with e-beam sensitive L-B films form the subject matter of this article.

PRINCIPLE

The basic idea for the continuous production of L-B layers utilized in the machines described in this article is the following.

Friction between the monolayer and the moving medium above or below the monolayer allows the transportation and continuous compression of a monolayer from a lower pressure area to a higher pressure area. Fig. 1 shows two implementations of that idea. The increase in pressure is proportional to the viscosity of the moving medium and the velocity gradient at the interface. There are a variety of geometries possible for using the basic idea including gradual compression to the desired pressure through several stages. These are described in more detail elsewhere[1].

Work done by the moving medium is dissipated (partially) in changing the phase of the monolayer.

$$W = \int_{A_1}^{A_0} \pi \cdot dA \cdot (m,T)$$

where $\pi = F(A,m,T)$ is the equation of state of the monolayer and where T is temperature and other symbols are defined in Fig. 1.

SYSTEM CAPABILITIES

We have built two prototypes based on the above idea. Data for the larger one are given in Table 1. Some of the issues are discussed further below.

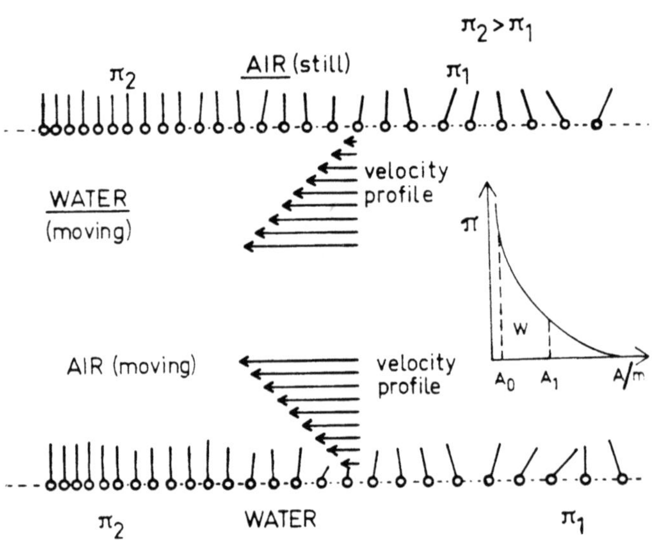

Fig. 1. Schematic of two basic implementations of the idea used here for continuous L-B film fabrication. The inset shows a generalized pressure v. area/molecule curve.

Table 1 Selected data for one of Molecular Electronic's continuous L-B film fabrication prototypes.

Size of trough (water surface)	1.4 m x 0.8 m
Productivity (film)	15 m^2/hour
Pressure range	10 to 50 mN/m
Pressure stability	± 1.5 mN/m
Air Cleanliness	Sub class - 10
Maximum object size	7-in. x 7-in. x 1-in.
Lift speed range	4 cm/sec to ~0.5 mm/min Ramping, different speeds for each layer, step pattern
Productivity (plates)	14-16 5-in. mask blanks/hour
Residence time control	2 min to 30 min

Productivity

The main problem with existing deposition equipment is that it is too small for commercial use and simply building them bigger is not sufficient. For example, to coat a 5-in. photomask blank with 30 layers of L-B film, in one step (without respreading) requires a trough area of around 1.4m^2. The daily productivity with such equipment would be very poor due to the need to repeatedly respread and clean (actual coating would be taking place only ≥ 10% of the total operating time!). In addition, currently there is no lift available which is suitable for handling large and heavy samples like 5-in plates.

We have, to-date, found only two publications by others that deal with these problems[2,3]. The obvious answer is to build a machine that provides the film continuously i.e. spreading, compression, deposition and cleaning-removal all take place simultaneously. With this approach, it is possible to increase the productivity by an order of magnitude depending upon the application.

Surface Pressure Control/Film Uniformity

All existing deposition equipment depends on a mechanical barrier of some sort to confine or manipulate the L-B film. This gives rise to at least three problems.

a) Film breakdown at the barrier and trough edges.
b) Unstable feedback/control of surface pressure. Often there are large pressure differentials and a considerable lag involved in equilibrating the surface pressures near the barrier and further away from it where the deposition actually takes place.
c) Non-uniform flow and differentially aged film.

Although especially severe for certain materials, these problems do raise the possibility of depositing highly non-uniform films generally and underscore the desirability of obtaining barrier-less compression such as achieved in the systems discussed in this article.

System Cleanliness

Conventional systems accumulate particulates, bacteria and impurities (from leaching) with time[4]. In addition, accumulated film, especially those that are viscous, are difficult to remove completely and quickly. It is well known that the spreading, compression and removal of film helps "getter" surface-active contaminants. In the present continuous system, this "gettering" and removal of excess film is on-going and very efficient.

Another important consideration is the composition and purity of the subphase. A good approach is to avoid the use of chemical additives altogether (since they are difficult to obtain in sufficient purity) and to try to use pure water only. This, of course means that close attention has to be given to the choice of materials as well as to their purity. The stagnant subphase typical of conventional systems, which leads to rapid accumulation of contaminants, the absence generally of rigorous process-fluids filtration and little or no control of airborne particulates are additional important considerations which require addressing in a fabrication system intended for high volume, industrial work. All of these have been incorporated in the present system.

Dependence of Film Properties on Residence Time

It is well established that the residence time (age of the monolayer; time elapsed between first compressing and depositing the film) can dramatically affect film properties[5]. We have found that for L-B resists for example the development time (time required to remove unexposed resist in a suitable solvent) is strongly dependent upon the viscosity of the monolayer (see Fig. 2) which can be related back to residence time. The variation in development time can be as much as from 20 secs to 3 mins for residence time from 3 mins to 20 mins. This is probably an effect due to trace impurities in the system which are unfortunately a real life problem. Obviously, because they operate only in batch-mode, conventional systems are not able to deposit layers with constant residence time. With the present continuous deposition method however, the residence time (or monolayer viscosity) can be easily controlled.

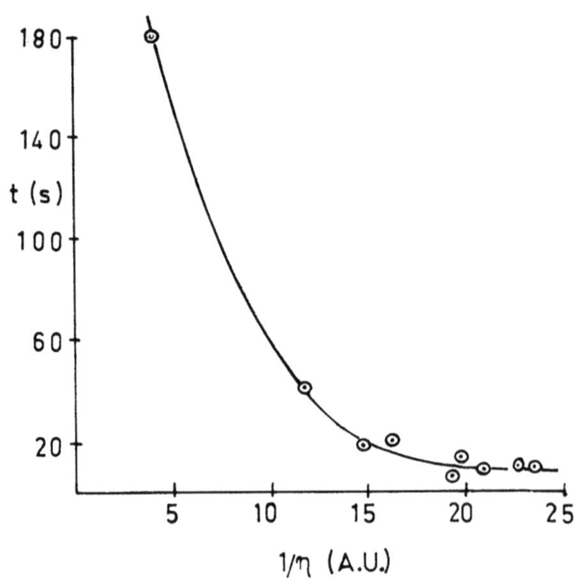

Fig. 2. Development time (t) v. reciprocal surface viscosity ($1/\eta$) for 22-tricosenoic acid films. Initial thickness was 90 nm. Development was carried out in 1-1-1,trichloroethane at room temperature.

Deposition Speeds

We try to do as much as possible of our deposition using clean water (resistivity >17 MOhm-cm and Total Organic Carbon content < 35 ppb) as subphase for the reasons mentioned already.

Under these conditions almost all standard, pure materials can be deposited at speeds of the order of magnitude of cm/sec. We have even (in one case) deposited a prepolymerized material at that speed.

Defect Levels

The present defect level is limited by the cleaning procedures for the objects to be coated.

Counting all visible defects and particles on a mask-blank or Si-wafer with 30 layers under a Normarsky type microscope at a magnification of 400 results in defect levels of 10 per cm^2. It is difficult to compare this with what is typically achievable in conventional systems because of the scarcity of such data. Our own estimates indicate that our defect level is at least three orders of magnitude smaller than generally found using conventional systems. Our sucess has to be attributed to the rigorous attention paid in our system to the issues discussed previously regarding cleanliness and surface pressure control.

Vibrations and Waves

A potential source of problems with the described technology (where water or air flow occurs during deposition) is the generation of waves on the surface. In practice however, we have not found these waves, or vibrations of the substrates to be coated, to be a contributing factor to film quality degradation, within a very comfortable margin.

At a very low deposition speeds, we find that the L-B films tend to show folded-over steps and irregular deposition. At much higher speeds, above what constitutes an upper deposition speed limit, the films begin to appear rough under the microscope. To give some idea of this range, for a standard material such as 22-tricosenoic acid, these limits are about 1mm/s and 1.5cm/s respectively.

We belive that the behavior found experimentally can be explained as follows. At high deposition speeds, actual deposition of the monolayer occurs at the end of a long meniscus which is pulled several millimeters, either above (for a up-stroke) or below the baseline water surface. Typically, the cohesion of the film appears to be sufficient for the relaxation of any strain caused by waves to occur at a place well remote from the actual line of deposition.

For upward motion of the substrate this effect is easily understood. The water layer between the monolayer and substrate is adequately thin and adherent to the substrate, that it follows any movements of the latter. And since liquids do not slip at a solid surface, the waves are effectively dampened.

The low deposition speed limit occurs when with decreased speed, the meniscus tends to relax such that it almost makes an angle of 90 degrees with the substrate. Under these conditions, waves with a sufficient amplitude and speed can effectively stop and reverse the relative movement and cause the deposition of additional double-layer steps.

At very high speeds, the meniscus is pulled so high and thin that the contact angle approches zero (or 180 degrees for a down-stroke). When this occurs, it is possible for a wave to cause contact of the monolayer with the substrate a short distance remote from the ideal line of deposition. This leads to (depending on whether the substrate is being withdrawn or immersed) very thin layers of either trapped water or air. These "bubbles" cause an increasing number of defects which soon hinder additional deposition. At high particulate levels, this effect is expected to be even more prominent and is clearly observed to be so experimentally.

According to the explanation, we expect the lower deposition speed limit to be approximately of the same order of magnitude as the maximum velocity of the waves whereas the highest useful speed on an ideal, smooth substrate should approach the maximum drainage speed minus the maximum velocity of the waves.

STRUCTURE OF THE FILMS

We see the same domain structures as published in the literature. To date, we have not found that the structures interfere with any of our intended applications except for waveguiding. There are monolayer forming materials available which do not show these structures (see Fig. 3).

However very thick films (many hundred layers), at least at our present level of cleanliness,do tend to show some "milkiness". Under the microscope, these very thick films look like "orange-peel" (see Fig. 4). The degree of the effect can be controlled to some extent by optimizing the deposition conditions and is very much dependent upon the cleanliness of the system.

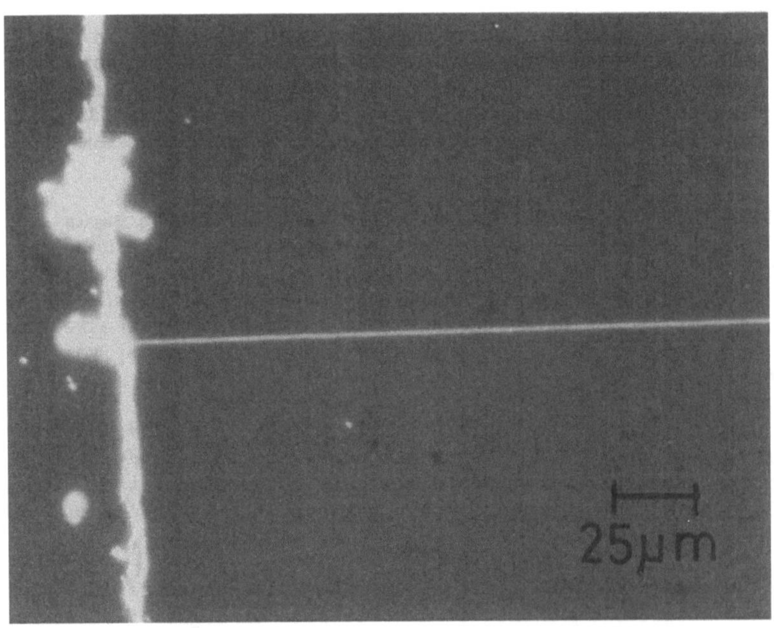

Fig. 3. A step in a thick L-B film (800 to 960 layers) of a proprietary material without the common domain structure usually observed. The substrate (silicon) is visible in the scratched area at left.

Even under otherwise ideal conditions, rapid deposition may produce a pressure drop between bulk film and film close to the sample. Thus, it is necessary for the film-forming material to possess sufficient latitude with respect to pressure variations. This generally requires that the material demonstrate a phase that can be deposited over a pressure range of at least 10 mN/m together with a surface viscosity low enough to allow the film to flow uniformly to the sample.

Fig. 4. Structure observed in 320 layers, 960 nm (top) and 800 layers (bottom) films of a proprietary material on silicon.

In order to get an understanding of the extent of the problem, we carried out an experiment where we deposited 22-tricosenoic acid at different pressures (~27-40mN/m) on silicon wafers, producing an 8-step pattern with a maximum of 176 layers. The samples showed clearly different interference colors for the same number of layers in the thick parts indicating that for this material the thickness of the layers depends upon the pressure chosen for deposition. Similar results for 22-tricosenoic acid have been reported previously[6]. Surface pressure effects may be expected to be more marked for this material on account of the fact that deposition normally has to be undertaken at pressures below the apparent secondary transition.

In a subsequent experiment we tried to find the pressure drop in the monolayer during fast deposition. For this the speed during deposition was "ramped" (slow at the lower edge and high at the upper edge of the sample). The resulting difference in thickness could be found by ellipsometric measurements only. The difference between slow dipping (< 1 mm/sec and fast (~14 mm/sec) was about 4.6 nm or 2 layers out of 200. Combining this with the above results gives an estimate for the upper limit of the pressure drop due to fast deposition. It was in this case less than 3 mN/m.

L-B FILM E-BEAM RESIST APPLICATION

We developed an electron beam resist that shows high sensitivity (~2μC/cm^2) and very good resolution (Fig. 5). The resolution at present seems to be limited by the resolution of the e-beam writing machines (e.g. we see quite often a discrepancy between vertical and horizontal lines).

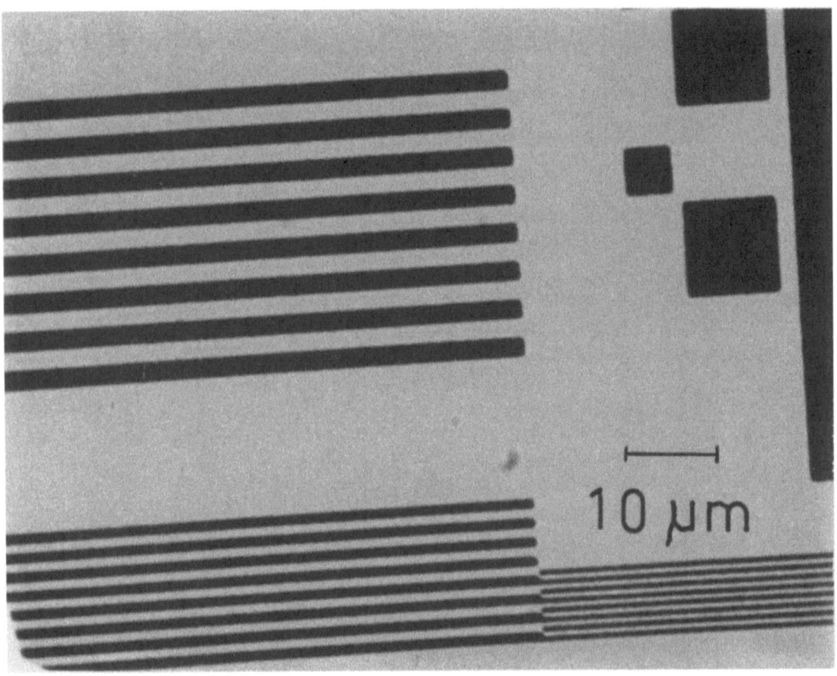

Fig. 5. Polaroid photograph of part of a mask with an e-beam written test pattern. The finest features are 0.5 μm lines/spaces. A commercial, raster scan type machine was used for the writing.

We have processed up to 5-in. mask blanks from cleaning, coating (30 layers, 90nm), writing, developing to etching (wet etch). Recent developments include dry (heat) developability and a resist without the common domain structure.

CONCLUSIONS

It has been shown that it is possible to build a continuous L-B deposition machine with the necessary up-scaling to accomodate the throughput levels and sample size requirements which are of interest for industrial production. The machine has demonstrated that it is realistic for instance, to entertain the possibility of using L-B films as e-beam resists to produce commercial chromium masks with at least 0.5 μm resolution and that repeatable quality can be achieved.

REFERENCES

1. O. Albrecht, M. Bartur and V. Rodov, U.S. Patent 4,722,856 (1988).
2. A. Barraud and M. Vandevyver, Thin Solid Films, 99 (1983) 221-225.
3. A Barraud, R. Gras and M. Vandevyver, U.S. Patent 4,093,757 (1978).
4. N. Carr, Chemtronics, 1 (1986) 167.
5. I.R. Peterson, G.J. Russell and G.G. Roberts, Thin Solid Films, 109 (1983) 371.
6. W.L. Barnes and J.R. Sambles, Surface Science, 187 (1987) 144-152.

DEPTH PROFILING IN ORGANIC THIN FILMS WITH OPTICAL GUIDED WAVES

D.R. Miller and P.W. Bohn

Department of Chemistry and
Materials Research Laboratory
University of Illinois at Urbana-Champaign

Introduction

Many of the devices and schemes of importance in molecular electronics rely intimately on the precise manipulation of molecular entities with control of positioning and orientation on size scales ranging from molecular dimensions to many thousands of molecular diameters. Unfortunately the correct positioning of these assemblies must often be inferred from the macroscopic behavior of the device. It would be most beneficial if there were some way to obtain chemical structural information *in situ*. What we would like is a tool to tell us not only about the structure and orientations of individual layers of molecules, but also how these layers are related one to the other. Ideally the experiment would give us molecular information, it would be capable of both depth and lateral resolution, and it would be capable of being applied *in situ*. One such approach, developed in our laboratory makes use of the sample of interest as the active layer in an asymmetric slab dielectric optical waveguide [1].

The basis of the experiment is illustrated in Figure 1. When radiation is coupled from the laboratory into an eigenmode of an optical waveguide, the radiation propagates along the film as a travelling wave, but the electric field distribution perpendicular to the film plane is a standing wave [2]. As shown on the left side of the figure, the electric field distributions for each eigenmode are distinct. In addition these electric field intensity distributions can sample a molecular distribution in a spatially anisotropic manner. For example in a linear scattering experiment, the signal obtained at normal incidence would be related to the product of the electric field intensity distribution and the molecular distribution (shown in the center of the figure), integrated over the film thickness,

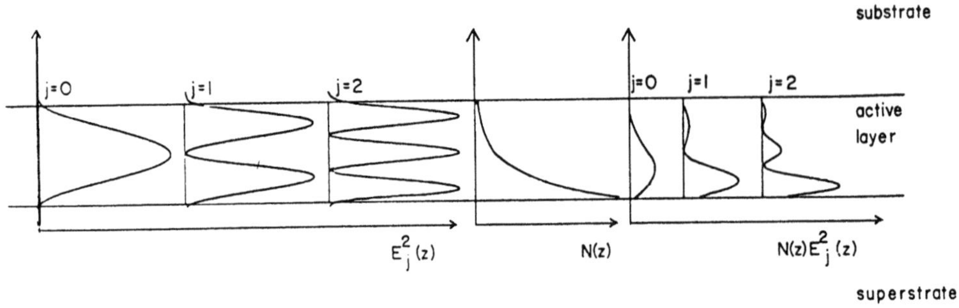

Figure 1. Left: Plot of the electric field intensity distribution vs. depth in the film for the first three eigenmodes. Center: Hypothetical molecular number density distribution. Right: Product of distributions for the first three eigenmodes.

$$I_j = K_j \int_0^t E_j^2(z) N(z) \, dz \qquad (1)$$

where j is a subscript defining which eigenmode is excited, K_j is a mode dependent constant, E(z) is the electric field distribution, N(z) is the molecular distribution, and I is the measured spectroscopic intensity [3-5]. The goal of optical depth profiling experiments, then, is to invert this set of equations and obtain the molecular distribution N(z) as some function of the measured intensities and the electric field amplitude distributions.

Theory

The task of inverting the equations described above is made difficult by three factors. First, the electric amplitude distributions are very nearly symmetric, relying only on the small difference in refractive index discontinuity between film-substrate and film-superstrate interfaces for the asymmetry in the distribution. This means it is difficult to distinguish the upper and lower halves of the film. Second the signal equations are first order Fredholm integral equations, which are classic examples of ill-posed systems [6,7]. Thus small errors in the measured signal intensities, I_j, tend to be amplified into large errors in the recovered profile. Finally the system is severely underdetermined. While we may have, at most, 10 - 15 eigenmodes with which to work, we would like to recover a continuous function characterized by an infinite number of concentration values. Clearly some ingenuity is required in dealing with these difficulties.

Having identified the problems, we can begin to deal with them by first recognizing that all distribution recovery problems fall into one of two classes. Either the functional form of the distribution is known *a priori*, or it is not.

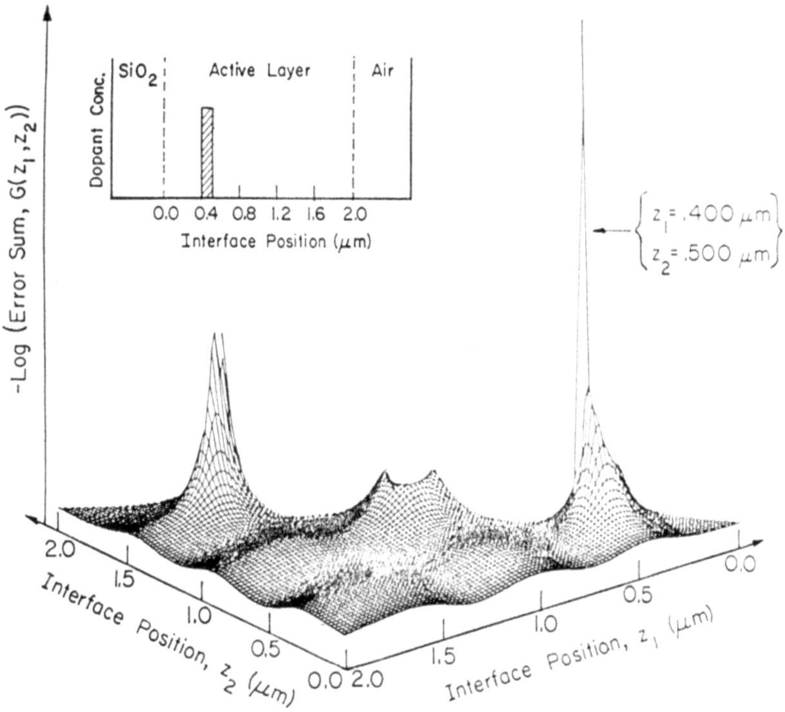

Figure 2. Error surface for the recovery of the π-function distribution shown in the inset. Values on the back half-plane are identical to those on the front, and thus are not shown.

Typically this information might be expected to come from some knowledge about the mechanism of incorporation of spectroscopically active sites. If the form of the distribution is known beforehand, then the problem reduces to one of non-linear model fitting, i.e. the distribution is described by some function $f(z; \alpha_1, \alpha_2, \ldots \alpha_n)$, and the task is to recover the parameters, α_i, which specify the distribution. Standard methods of analysis may be applied to effect this parameter fitting. At worst this means mapping an (n+1)-dimensional error surface. In practice this procedure involves measuring a characteristic spectroscopic band for the species of interest for each of the eigenmodes of the waveguide structure. Then an assumed distribution is used to calculate the same spectroscopic features. Calculated and observed spectra are then compared, and an error function is defined which characterizes the differences summed over all modes. This error function is then plotted for all values of the α_i. It can readily be seen from Figure 2 that plotting the entire error surface is necessary in this scheme, because there is a strong secondary minimum, corresponding to rotation of the correct distribution about the center line of the film. This is due to the symmetry of the electric field distributions about the center, mentioned above. However with careful attention to experimental detail reasonable recoveries can be effected [8].

In the case where the functional form of the distribution is not known *a priori*, the recovery is much more difficult. First it is absolutely necessary to destroy the symmetry of the electric field distribution. This can be accomplished in several different ways, the simplest of which involves using a spectroscopically inert waveguiding layer on which to place the sample of interest. Because the refractive indices and thicknesses of the two layers can be adjusted, the electric field intensities can be made asymmetric in the overlayer (i.e. the film of interest). Then we are left to deal with the ill-posed and underdetermined nature of the recovery.

We follow a modified version of a procedure first developed by Phillips [9]. First eqn. (1) may be rewritten in more illuminating form,

$$I_j + \varepsilon_j = K_j \int_0^t E_j^2(z) N(z)\, dz \tag{2}$$

where ε_j is the error associated with spectroscopic measurement of the jth mode. Since the problem is severely underconstrained, we can obtain an infinite number of possible solutions, $\nu = \{N_1, N_2 \ldots\}$. However if the distribution is reasonably smooth, then the true distribution is the one which is the smoothest according to,

$$\int_{z_1}^{z_2} (N_s'')^2\, dz = \min \int_{z_1}^{z_2} (N_i'')^2\, dz \tag{3}$$

where N_i is a general member of the set ν. To find N_s, first we discretize the integrand,

$$I_j + \varepsilon_j = \sum_{i=0}^{m} w_i K_j E_{ij} N_i \tag{4}$$

where w_i is a weight factor which depends on the quadrature formula chosen. We then look for a function N, which minimizes the discretized second differences. This is obtained by starting with an exact solution of the matrix equation,

$$[N] = [A]^{-1}[I] + [A]^{-1}[\varepsilon] \tag{5}$$

where

$$A_{ij} \equiv w_i K_j E_{ij} \qquad (6)$$

Thus [N] must satisfy,

$$\sum_{i=0}^{m} (N_{i+1} - 2N_i + N_{i-1})(\alpha_{i+1,j} - 2\alpha_{i,j} + \alpha_{i-1,j}) + \gamma^{-1}\varepsilon_j = 0; \quad j = 0,1,2,\ldots,m \qquad (7)$$

where γ is a Lagrangian multiplier, which has the effect of determining the amount of smoothing in the solution.

In our experiments we assume that we know γ, and use it to determine ε. Synthetic data is used to confirm whether or not the choice of γ is reasonable. It is found that, for most distributions likely to be encountered in practice, the recovery is robust in the sense that γ can be off by an order of magnitude or so without significantly affecting the quality of the recovery. In practice, of course, γ is not known. Instead different values of γ are tested until the one which reproduces the experimental error most closely is found. This value is then used to reconstruct the experimental distribution.

Results and Discussion

In order to test the techniques developed for distributions which are not known ahead of time, a poly(styrene)/poly(styrene-d_8) sandwich structure was fabricated. First the deuterated polymer was coated on an SiO_2 substrate. Then a film of the protonated polymer was formed on another substrate and floated off at an air-water interface. The floating film was then picked up on the deuterated film by emersion. By carefully checking the thickness of the film before and after addition of the protonated polymer the film structure was established with high precision. Then the entire structure was heated above the glass transition temperature, after which depth profiling experiments were performed. In these experiments the C-H and C-D stretching vibrations in the Raman spectrum were used as marker bands for the two polymers. Results are shown in Figure 3.

Several features are clear from the recovery. First of all the value of $\gamma = 10^{-2}$ was found to give an error sum closest to the observed experimental error, so this distribution is the correct one, given the criteria we have adopted. Clearly this recovery is intermediate between the overdamped one at $\gamma = 1$ and the underdamped one at $\gamma = 10^{-3}$. In addition the

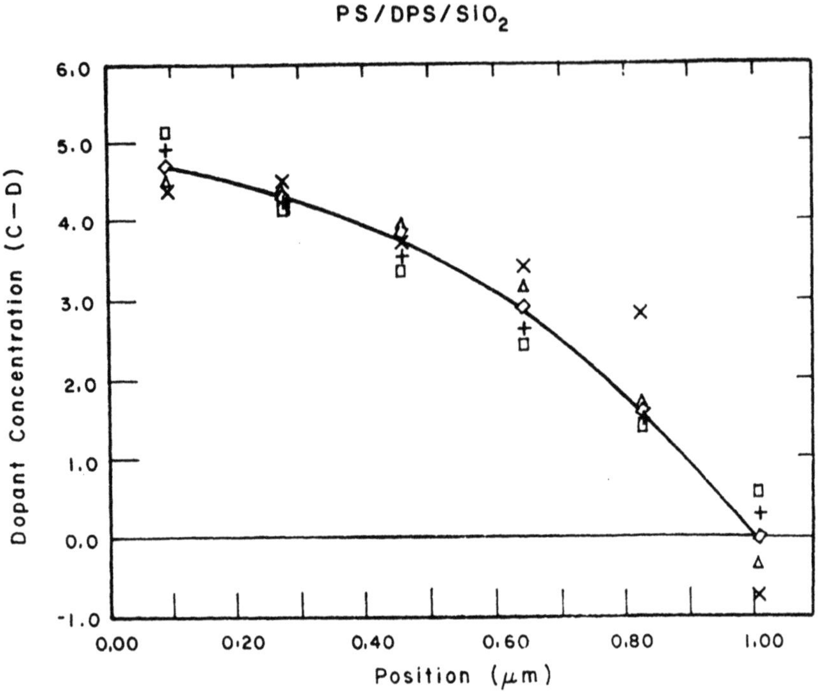

Figure 3. Recovered profile of poly(styrene-d_8) for various values of the smoothing parameter, γ; triangles, $\gamma = 10^{-3}$; diamonds, $\gamma = 10^{-2}$; crosses, $\gamma = 10^{-1}$; squares, $\gamma = 1$.

recovery predicts the general features of the polymer interdiffusion correctly. However for more detailed studies one would like to have better resolution. Fortunately the fundamental resolution limit is imposed in the current experiment, not by optical diffraction, but rather by the precision with which the measurements can be made. Thus with more attention to signal enhancement, and the incorporation of spatial bandwidth narrowing techniques [10], it should be possible to improve the spatial resolution by a factor of 2-4.

Acknowledgement

The work described herein was supported by the National Science Foundation under grant DMR 8612860.

References

(1) P.K. Tien, Rev Mod. Phys. 49, 361 (1977).

(2) J.N. Polky and G.L. Mitchell, J. Opt. Soc. Am. 64, 274 (1974).

(3) P.W. Bohn, Anal. Chem. 57, 1203 (1985).

(4) D.R. Miller, O.H. Han, and P.W. Bohn, Appl. Spec. 41, 245 (1987).

(5) D.R. Miller, O.H. Han, and P.W. Bohn, Appl. Spec. 41, 249 (1987).

(6) R.J. Trumpler and H.F. Weaver, *Statistical Astronomy*, Dover, New York (1962), pp. 95-120.

(7) L. Landweber, Am. J. Math. 73, 615 (1951).

(8) D.R. Miller and P.W. Bohn, Anal. Chem. 60, 407 (1988).

(9) D.L. Phillips, J. Assoc. Comput. Mach. 9, 84 (1962).

(10) P.W. Bohn and N.F. Fell, Proc. SPIE 990, in press.

SYNCHROTRON X-RAY STUDIES ON LIPID MONOLAYERS

C.A. Helm

UCSB, Dept. Chem. Nucl. Eng.
Santa Barbara, CA 93106

ABSTRACT

This report is an overview about newly developed X-ray techniques used to investigate lipid monolayers, either at the air/water interface or on solid substrates. The physical principles of the method are summarized. In order to demonstrate its power and possibilities, experiments on two different molecules (arachidic acid, L-α-dimyristoylphosphatidic acid) at the air/water interface are outlined. Thus several different "crystalline" phases of these self-organized monolayers are characterized.

INTRODUCTION

A study of the properties μ and the molecular arrangement of lipid monolayers is of considerable interest for physical, technical and biological reasons. Water as the supporting phase cannot influence the structure of the monolayer. Rather, the monolayer in such systems is self-organizing and so may be used as a model for two-dimensional "crystallinity", a subject of great physical interest [1,2,3]. Furthermore, arachidic acid is a classical molecule for Langmuir-Blodgett films [4]. The possible technical applications of such films is a major topic of this volume. Since all membranes consist mainly of phospholipid bilayers, experiments on monolayers of these lipids should improve our understanding of biological membranes.

Traditionally, lipid monolayers at the air/water interface are characterized by their isotherms (surface pressure π versus molecular area A). However, only thermodynamic properties are accessible with this method. It would be preferable to obtain information on a molecular level. X-ray scattering techniques are particularly suitable for such studies.

The wavelength of X-rays is of the order of interatomic distances. Therefore, for three-dimensional crystals, X-ray diffraction became a classical analytical technique. However, before this technique could be applied to monolayers, especially those at the air/water interface, it was necessary to solve certain problems: (i) X-rays are not surface sensitive; their penetration depth into the bulk amounts to several µm. Thus, the signal from the surface is swamped by the background signal of the substrate. (ii) The intensities of the surface signals are very weak, therefore necessitating the development of a powerful light source. (iii) A technique suitable for the investigation of the monolayer perpendicular to the surface had to be provided.

Essentially, two publications demonstrated how these three difficulties could be overcome: (i) 1981 Eisenberger and coworkers showed, that the grazing incidence technique drastically reduces the penetration depth [5]. Thus, diffraction peaks of monolayers could be measured. Also, (ii) they demonstrated that the intensity of synchrotron radiation is sufficient. (iii) Als-Nielsen and his group investigated the density distribution perpendicular to a clean water surface by specular reflection [6]. They found deviations from the Fresnel theory, which could be ascribed to the surface roughness caused by capillary waves. Furthermore, Als-Nielsen developed formulae, which made the interpretation of the data simple and reliable [7].

Thus, it became possible to investigate the phases of lipid monolayers on a molecular level.

EXPERIMENTAL

This is only a very rough outline of this new technique. A general description, containing useful formulas and a wide range of examples can be found in ref. [6]. The more special case of lipid monolayers is treated in detail in ref. [8].

A schematic diagram of two different scattering geometries which were employed is presented in Fig.1: 1(a) shows the impulse transfer Q perpendicular to the surface which gives information about the Out-of-Plane structure (electron density profile), while the impulse transfer Q mainly parallel to the surface depicted in 1(b) was used to study the In-Plane structure ("crystalline" lattice).

(a) To measure the impulse transfer perpendicular to the surface, specular reflection at angles above the critical angle α_c was used. The reflected intensity R_f at an infinitely sharp air/water interface can be calculated easily according to the Fresnel theory. However, when a monolayer resides at the interface, the reflected intensity is modulated to some value R. To compute the measured intensity R, one has to know the optical properties of the monolayer. For the applied wavelength ($\lambda \approx 1.4$ Å), the index of refraction n is, in principle, a very simple function of known quantities, i.e. electron mass m, unit charge of the electron e, speed of light c, wavelength λ and the electron density ρ of the material.

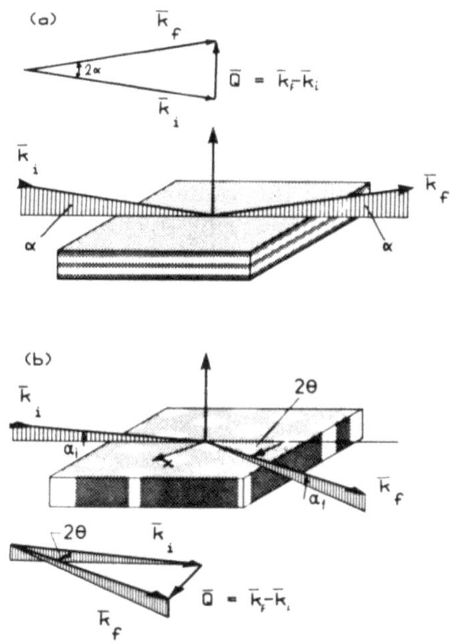

Fig.1. Principle of the employed scattering geometries for (a) the impulse transfer perpendicular to the surface (Out-of-Plane structure) and (b) the impulse mainly parallel to the surface (In-Plane structure). The shading represents typical Fourier components of the electron density, measured in either mode.

$$n = 1 - \frac{e^2}{2\pi\, mc^2}\lambda^2 \rho \tag{1}$$

Since the absorption of the monolayer is negligible, it is possible to obtain an approximate formula for the reflected intensity R. It is a simple Fourier transform of the electron density distribution ρ(z) perpendicular to the surface [7]:

$$\frac{R}{R_f} = \left| \frac{1}{\rho_s} \int dz \frac{d\rho(z)}{dz} e^{iQz} \right|^2 \tag{2}$$

where ρ_s is the electron density of the substrate. It is well established that the hydrophilic headgroup of the lipid is next to the water, while the hydrophobic chains are oriented towards the air. It follows that the most simple model for ρ(z) consists of two slabs each of constant electron density; one slab represents the lipid head-group, while the other represents the chain. Finally, the roughness of the water surface due to capillary waves should be taken into account. This is done by convoluting the entire density distribution with a Gaussian.

In order to find an electron density distribution capable of explaining a particular measured reflectivity curve, least squares fits were executed using $\rho(z)$ as the fitting parameter. Since the chemical structure of the lipid is known, so too is the number of electrons in the chains. The area per chain is determined by the isotherm. Hence, by varying the value for the length of the chains, $\rho(z)$ is effectively varied taking care to maintain $\rho(z)$ within the above specified limits. This is done until the value of R thus calculated corresponds to that experimentally measured. Generally, by compression of the monolayer, its thickness and its electron density is increased and so the same analysis may be performed for a range of molecular areas. Thus, a complete description of a lipid phase on a molecular level becomes possible.

(b) To study the impulse transfer mainly parallel to the surface, we use an incident angle α smaller than the critical angle α_c. Thus, total external refraction occurs, and an evanescent wave arises which is diffracted by the "crystalline" monolayer.

Generally speaking, a peak contains three different pieces of information: (i) the peak position gives the lattice distance according to Bragg's law,

$$d = \frac{2\pi}{Q} = \frac{\lambda}{2 \sin\theta} \tag{3}$$

where θ is defined in Fig.1. (ii) The width correlates with the coherence length and (iii) the integrated intensity is proportional to the covered area of the "crystalline" monolayer and to the structure factor.

The evanescent wave has a small penetration depth into the substrate (i.e $\approx$100 Å for water). Its intensity is enhanced by a factor of 1.6 for water relative to the incident beam. Thus, the technique exhibits the surface sensitive properties necessary to characterize the monolayer.

RESULTS and DISCUSSION

Phospholipids

To illuminate the possibilities of this new method, two successful experiments and their interpretation will be outlined here.

The first example, an In-Plane study, concerns the condensed phases of L-α-dimyristoylphosphatidic acid (DMPA) and L-α-dimyristoylphosphatidyl ethanolamin (DMPE) [9,10]. The Bragg peaks of these lipids are the first ever measured [11] and are depicted in fig.2. The insert of this fig. shows the isotherm. When decreasing the area per molecule, the lipid undergoes various phase transitions, marked by distinct changes of slope of the isotherm. Most remarkable is the break in the isotherm at π_c. This corresponds to the onset of a first order phase transition between a homogeneous fluid and a "gel" phase. The nucleation and the growth of the "gel" phase domains (diameter $\approx$1-10µm) can be observed with fluorescence microscopy [12,13]. The investigation of these domains by means of electron diffraction revealed a hexagonal lattice structure. The bond orientational order was constant

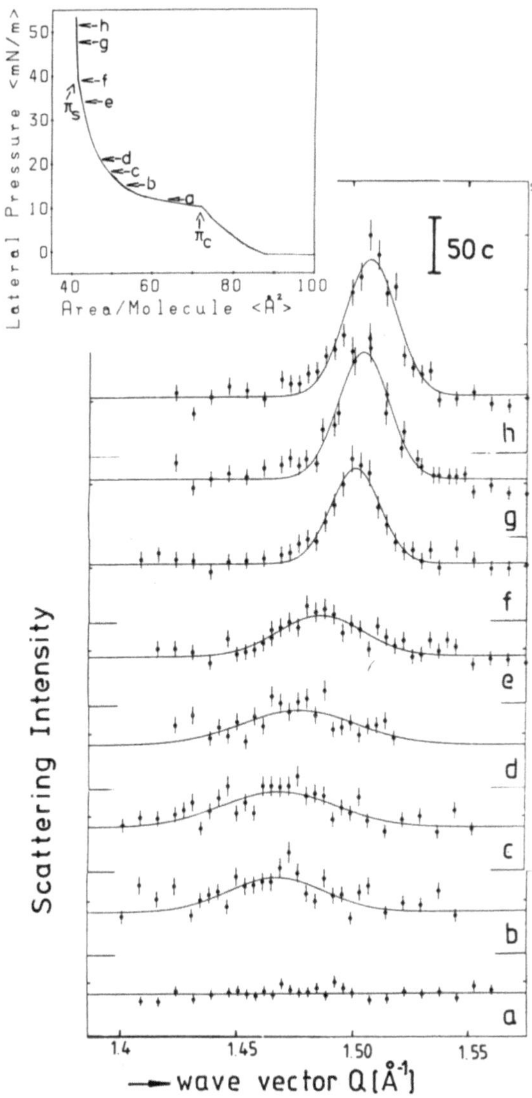

Fig.2 In-Plane diffraction peaks (intensity versus scattering angle 2θ) at various points a-h on the isotherm as given in the insert. T=23°C, DMPA film on water subphase with 10^{-2} M NaCl and $5 \cdot 10^{-5}$ M EDTA.

throughout one domain [14]. Unfortunately, with the X-rays, only a powder diffraction pattern can be observed because of the large illuminated area as compared to the domain size.

Fig. 2 presents the Bragg peaks obtained for DMPA. The peak position moves as the molecular area decreases. Hence, changes in the peak position are indicative of lattice compression. At closer inspection two groups of peaks can be distinguished. All peaks measured above the slight kink in the isotherm, denoted by π_S (see insert), are more intense and more narrow than those measured below π_S. The sharp and distinct increase of coherence length indicates a phase transition. By variation of the headgroup or the ionic conditions of the subphase, the quantitative value of π_S can be changed, but the correlation between thermodynamic and X-ray data is not influenced [10]. Analysis of the lattice distance d as a function of lateral pressure reveals the phase transition to be of second order [15].

In our best measurements and above π_S, the positional correlation length amounts to some hundreds of Å, while electron diffraction [14] and optical polarization experiments [16] demonstrate bond orientational correlation lengths of several μm. Additional information is supplied by line shape analysis of the peaks, which revealed Lorentzian lineshape, indicating an exponential decay of the positional order. Thus the "gel" and the solid phase exhibit properties of a hexatic phase, which was postulated by Nelson [1,2,3]. Also, all previous diffraction experiments report short positional correlation lengths. These measurements concerned phosphatidyl cholines [17], lipids with amino acid head groups [18], fatty acids and long chain alcohols [19,20]. Thus, the order of these self-organized two-dimensional "crystals" is more similar to liquid crystals than to a three-dimensional crystal.

Arachidic Acid

The second experiment is an investigation of arachidic acid, employing both the In-Plane and Out-of-Plane methods [21]. The isotherm is depicted in the insert of fig. 3. Clearly, three different regions can be distinguished: a zero-pressure gaseous phase, a phase with high constant compressibility (henceforth called tilted) and a condensed phase with very low compressibility.

The In-Plane studies again can be correlated very easily to the isotherm. In the zero pressure region, peaks of constant position and increasing intensity were observed. These can be definitely ascribed to solid islands in a gaseous phase, which grow during compression. Completion of island growth marks the onset of the tilted phase, where the lattice constant decreases, while the peak intensity remains quite constant. The positional correlation length is ≈150Å. A big increase (factor 2.5) in peak intensity is observed again in the condensed phase.

Reflectivity measurements, the results of which are shown in fig.3(a), were executed in order to further characterize the "crystalline" structure of the tilted phase. The compression of the monolayer leads to a shift in the position of the minimum of reflectivity (i.e. to lower Q values) which can be understood as an increase of the monolayer height [8]. The least squares fits reveal a constant and high electron density of the chains throughout the whole tilted phase. Only the thickness of the hydrophobic region increases (fig.3.b). Since peaks could be observed in the tilted phase, the lipid chains have to be in an all-

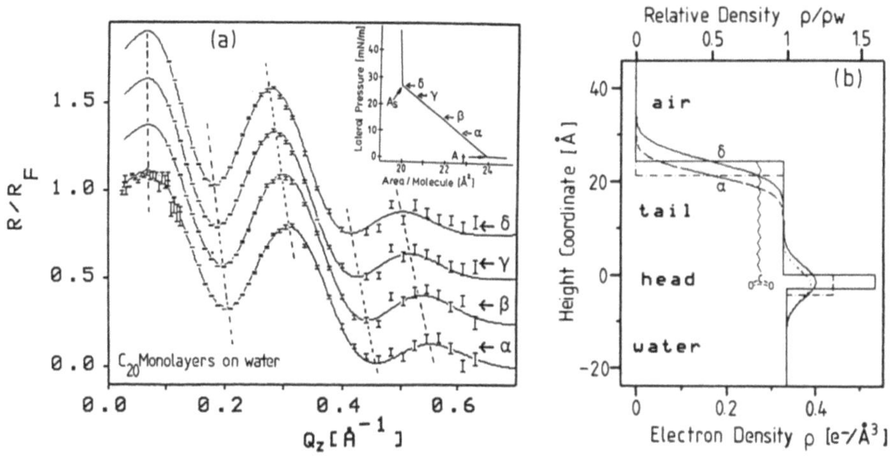

Fig.3 (a) Normalized X-ray reflectivity R/R_f versus vertical scattering vector Q for arachidic acid monolayers on pure water (pH 5.5, T=20°C). The measurements are displaced vertically by 0.25 units and correspond to the surface pressures indicated (α,β,γ,δ) on the isotherm of the insert. The full lines correspond to the fitted densities ρ(z).

(b) Best fit density profiles ρ(z) corresponding to the data of fig.3(a). The straight lines describe the slab model prior to the Gaussian smearing. The full curve corresponds to (δ) (π ≈25 mN/m) and the broken curve to (α).

trans-configuration and are uniformly tilted. During compression, an area change of ≈20% occurs, thus the tilt angle decreases continuously from 33° to 0°.

In order to establish a structural model in the tilted phase, two different tilt directions within the symmetry of a hexagonal lattice have to be considered: (i) tilt in the direction of the nearest neighbor (Model I), or (ii) the next nearest neighbor (Model II), as depicted in fig.4. Let us consider the case of high tilt angles (>15°). It can be shown [21] that, in order to measure a high peak intensity, the impulse transfer has to be mainly parallel to the surface. This condition is not fulfilled for any of the lattice planes of Model II since this model does not provide any diffraction intensity for high tilt angles. Consequently, Model II must be ruled out as a possible model of the system. The condition is, however, satisfied by the d_{1-1} lattice plane of Model I. When the tilt angle is decreased to less than 5°, the d_{10} and d_{01} lattice planes of Model I now make a contribution to the peak intensity which, consequently, increases by a factor of 3. This corresponds to what was observed experimentally for arachidic acid as mentioned above.

However, according to Model I, the value of the d_{1-1} - lattice distance is not influenced by the tilt angle, in contradiction to our measurements.

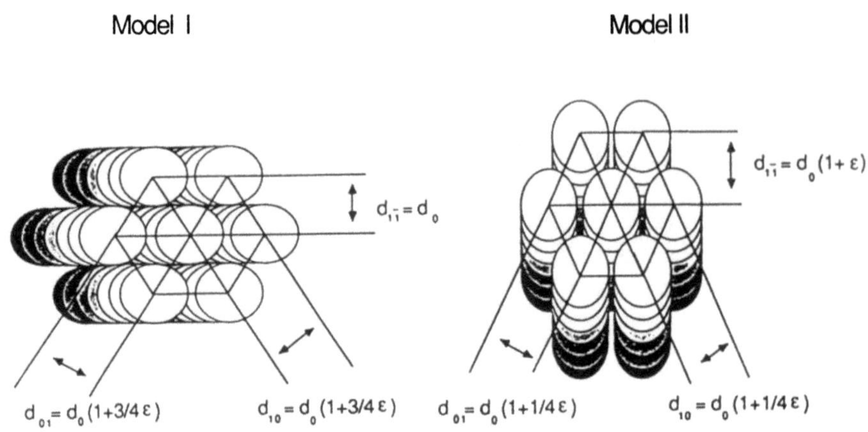

Fig.4. Model of possible tilt directions in a hexagonal lattice: Mode I (tilt towards a nearest neighbor) and Model II (tilt towards the next nearest neighbor).

Therefore, the chains must be forming a centered orthorhombic (pseudo-hexagonal) lattice, tilted towards their nearest neighbors.

A phase somewhat similar to that seen here for arachidic acid has very recently been observed in an amino acid molecule with an attached fluorocarbon chain [22].

Future Developments

Divalent ions in the subphase can be chemically bound to the arachidic acid, thus drastically changing the electron density of the headgroup, while the hydrophobic region stays unaffected [21,23,24]. Even more difficult is the interpretation of reflectivity measurements of charged lipid monolayers which give rise to a diffuse double layer of counterions in the water. Measurements of the electron density profile of the double layer are possible using the presented technique. Such verifications of the DLVO-theory [25] have important implications to Colloid and Interface Science. Furthermore, fluid lipid phases can be characterized by reflectivity measurements. Very little is known about these biologically pertinent phases. Studies on this topic are currently in progress.

ACKNOWLEDGEMENT

I acknowledge collaboration with H. Möhwald and P. Tippmann-Krayer, Mainz, and with J. Als-Nielsen and K. Kjær, Risø, as well as support by the BMFT.

LITERATURE

[1] D.R. Nelson, M. Rubinstein, and F. Spaepen. 1982. Order in two-dimensional binary random arrays, Phil. Mag. A 46:105.

[2] D.R. Nelson. 1983. Reetrant melting in solid films with quenched random impurities, Phys. Rev. B27:2902.

[3] P.R. Nelson and B.J. Halperin,.1979. Dislocation-mediated melting in two dimensions, Phys. Rev. B19:2457.

[4] G. Gaines. 1966. Insoluble Monolayers at Liquid-Gas Interfaces. Ch. 3. John Wiley & Sons, Inc., New York, London, Sidney.

[5] P. Eisenberger and W.C. Marra. 1981. X-ray diffraction study of the Ge(001) reconstructed surface, Phys. Rev. Lett. 46:1081

[6] A.M. Braslau, M. Deutsch, P.S. Pershan, A. Weiss, J. Als-Nielsen and J. Bohr. 1985. Surface roughness of water measured by X-ray reflectivity, Phys. Rev. Lett. 54:114

[7] J. Als-Nielsen. 1986. Solid and liquid surfaces studied by synchrotron x-ray diffraction. In Curr. Top. Phys.Ch. 4. W. Schommers, P. Blanckenhagen, editors. Springer Verlag, Berlin.

[8] K. Kjær, J. Als-Nielsen, C.A. Helm, P. Tippmann-Krayer and H. Möhwald. 1988. An X-ray scattering study of lipid monolayers at the air-water interface and on solid supports, Thin Solid Films, 159:17

[9] K. Kjær, J. Als-Nielsen, C.A. Helm, L.A. Laxhuber and H. Möhwald. 1987. Ordering in Lipid monolayers studied by synchrotron x-ray diffraction and fluorescence microscopy, Phys. Rev. Lett, 58:2224

[10] C.A. Helm, H. Möhwald, K. Kjær, J. Als-Nielsen. 1987. Phospholipid monolayers between fluid and solid states, Biophys. J. 52:381

[11] H. Möhwald. 1986. Microstructure of organic mono- and multilayers. In The Physics and Fabrication of Microstructures and Microdevices. M.J.Kelly and C. Weisbuch, editors. Springer Verlag, Berlin, Heidelberg.

[12] M. Lösche, E. Sackmann and H. Möhwald. 1983. Fluorescence microscopic study concerning the phase diagram of phospholipids, Ber. Bunsenges. Phys. Chem. 87:848

[13] R.M. Weiss, H.M. McConnell. 1984,.Two-dimensional chiral crystals of phospholipids, Nature (Lond.). 310:5972

[14] A. Fischer and E. Sackmann. 1984, Electron microscopy and diffraction study of phospholipid monolayers transferred from water to solid substrates, J. Phys. (Paris). 45:269

[15] C.A. Helm, Thesis, TU Munich 1988

[16] V.T Moy, D. Keller, H.E. Gaub and H. M. McConnell. 1986. Long range molecular orientational order in monolayer solid domains of phospolipids, J. Phys. Chem. 90:3198

[17] C.A. Helm, H. Möhwald, K. Kjær and J. Als-Nielsen. 1987. Phospholipid monolayer density distribution perpendicular to the water surface: a synchrotron X-ray reflectivity study, Europhys. Lett, 4:697

[18] S. Grayer Wolf, L. Leiserowitz, M. Lahav, M. Deutsch, K. Kjær and J. Als-Nielsen. 1987. Elucidation of the two-dimensional structure of an α-amino acid surfactant monolayer on water using synchrotron x-ray diffraction, Nature (Lond.), 328:63

[19] P. Dutta, J.B. Peng, B. Lin, J.B. Ketterson, M. Prakash, P. Georgopoulos, S. Ehrlich. 1987. X-ray diffraction study of an organic monolayer on the surface of water, Phys. Rev. Lett. 58:2228
[20] S.W. Barton, B.N. Thomas, E.B. Flom, S.A. Rice, B. Lin, J.B. Peng, J.B. Ketterson, P. Dutta. X-ray diffraction study of a Langmuir monolayer of $C_{21}H_{43}OH$, to be submitted
[21] K. Kjær, J. Als-Nielsen, C.A. Helm, P. Tippmann-Krayer and H Möhwald. 1988. Synchrotron x-ray diffraction and reflection studies of arachidic and monolayers at the air-water interface, accepted for publication
[22] S.G. Wolf, M. Deutsch, E.M. Landau, M. Lahav, L. Leiserowitz, K. Kjær and J. Als-Nielsen. 1988. Synchrotron X-ray study of a solid-solid phase transition in a two-dimensional crystal, submitted to Science
[23] R.M. Richardson and S.J. Roser. 1987. X-ray reflectivity from insoluble monolayers spread on aqueous subphases. Liq. Crystals. 2:797
[24] R.M. Richardson and S.J. Roser. 1988. X-ray and neutron reflectivity from spread monolayers, Thin Solid Films 159:43
[25] S. McLaughlin. 1977. Electrostatic potentials at mebrane-solution interfaces, In Curr. Top. Membr. Transp. 9:71

FLUORESCENCE IMAGE ANALYSIS OF THE SPATIAL ORGANIZATION OF LIPID LAYERS

F. Caratozzolo (+), A. Chiabrera (+), M. Esposito(+), A. Gliozzi (*), M. Grattarola (+), G. Mazzoni (*), R. Paradiso (+), and M. Parodi (#)

(+) Biophysical and Electronic Engng. Dept., University of Genova, Italy
(*) Department of Physics, University of Genova, Italy
(#) Institute of Electrical Engng, University of Cagliari, Italy

INTRODUCTION

The study of self-organized two-dimensional structures can furnish new ideas for future molecular devices, especially if the self-organization of the structures can be modulated by external agents, like an exogenous electric field.

The generation of two-dimensional structures in organic monolayers can easily be analyzed by means of optical techniques. The possibility of transferring them onto solid substrates suggests their possible use in the field of integrated biosensors (1). From a more speculative point of view, the possibility of controlling the formation of bidimensional patterns in a matrix of organic material leads to regarding these structures as elements of memory which could be utilized to process information (2).

Monolayers made of lipids (in particular, phospholipids) are among the most extensively studied structures, where the presence of organized patterns has been identified by means of fluorescence techniques both at the air/water interface and after their transfer onto alkylated supports (3,4).

The formation of organized structures is related to the coexistence of two phases, a gel-like phase and a fluid one.

The characteristic features of each pattern are a function of various physico-chemical parameters, such as ionic strength, surface pressure, lipid chemical composition and electrical properties.

The exact mechanism by which these structures are formed is not well characterized and a quantitative description of their geometrical features (e.g., shape, dimensions, and mean relative interdistance) is required.

In this paper, the use of digital fluorescence microscopy for a quantitative characterization of such patterns is described. Moreover, the importance of the lipid dipole component for pattern formation is briefly discussed.

MATERIALS AND METHODS

Preparation of lipid monolayers

Phospholipids

L-dipalmitoyl-phosphatidylcholine (DPPC) was obtained from SIGMA chemie Gmbh, and used without further purification.

1-Acyl-2 (N-4-nitrobenzo-2-Oxa-1, 3-diazole) aminocaproyl Phosphatidylcholine (NBD-PC) was supplied by Avanti Polar Lipids, Birmingham,UK.

Monolayer techniques

Monolayers were formed in a multicompartment circular trough designed by P. Fromhertz and realized by Mayer Feintechnik. The trough consists of two components: a circular trough, combined with a double-barrier driving assembly, and an electronic control unit. Surface pressure was measured with a Wilhelmy balance.

The phospholipids were dissolved in 1 mg/ml chloroform. The solutions were mixed before spreading to establish a well defined (2%) NBD -PC ratio.

The water used was distilled and filtered. After deposition of the solution onto the water surface, the monolayers were left for a sufficient period of time in the expanded state, to allow solvent evaporation and stabilization of the surface pressure. During the formation of monolayers, one of the barriers of the assembly was fixed, while the other compressed the monolayer at a speed of 0.8 cm^2/s. The apparatus has an electronic feedback system that keeps the surface pressure constant. In this condition, the transfer of monolayers onto alkylated glass slides was performed. Glass coverslips (Zeus Super, 20 x 20 mm^2) were alkylated according to the procedure described in (5).

If a monolayer is under sufficient surface pressure so that its molecules are closed together, and if the solid is clean and wet when the monolayer is transferred onto the solid, the polar groups stick to the solid support. Otherwise, if a hydrophobic solid is used, then, when the monolayer is transferred onto the slide, the hydrocarbon chains remain attached to the solid surface. The monolayers deposit onto the solid when the slide is dipped into or out of the liquid-gas interface.

Fluoresence digital microscopy

An epifluorescence microscope (Leitz, W. Germany) equipped with a Xenon lamp was utilized. The microscope was interfaced with a high-sensitivity ISIT TV camera (RCA, USA). The images generated by the TV camera were digitized (DT2851 Data Translation board) and saved on the floppy disks of a personal computer (AT, IBM). The experimental apparatus is shown in Figure 1.

The glass coverslips onto which the lipid layers were deposited were viewed through a 50 X air-objective.

Ten images were acquired for each viewed pattern. A shutter along the light pathway was closed after each acquisition, in order to reduce photodamage to the lipids. All processings were performed on the image obtained by averaging the ten original ones.

EXPERIMENTAL RESULTS

Lipid monolayers at different surface pressures were viewed, according to the pressure-area isotherm in Figure 2.

Two typical images of patterns observed at two different pressures are shown in Figures 3 and 4. The dark regions, which exclude the

Fig. 1. Experimental apparatus.

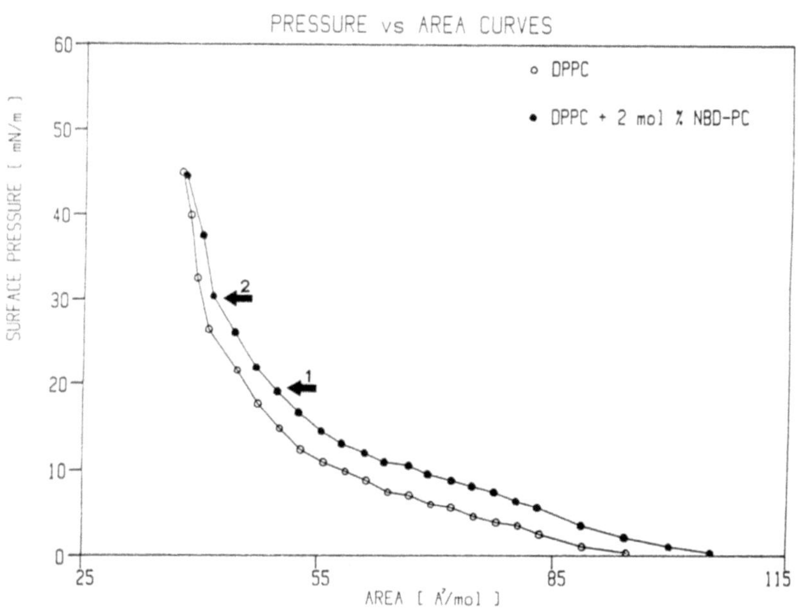

Fig. 2. Experimental pressure-area isotherm. Numbered arrows refer to the approximate values chosen for image acquisition.

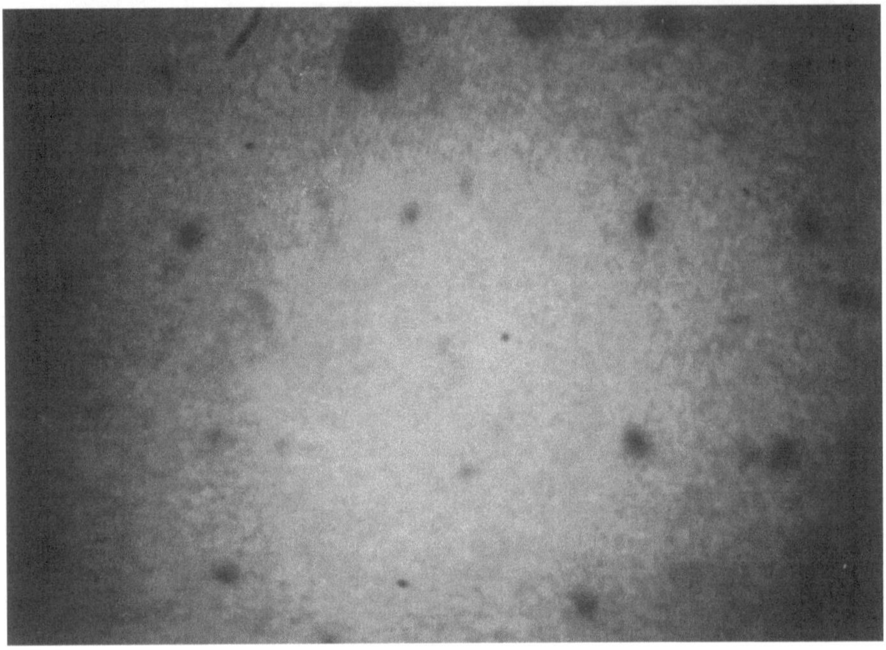

Fig. 3. Digital fluorescence image of a lipid monolayer transferred onto a solid support under the pressure conditions indicated by arrow # 1 in Figure 2.

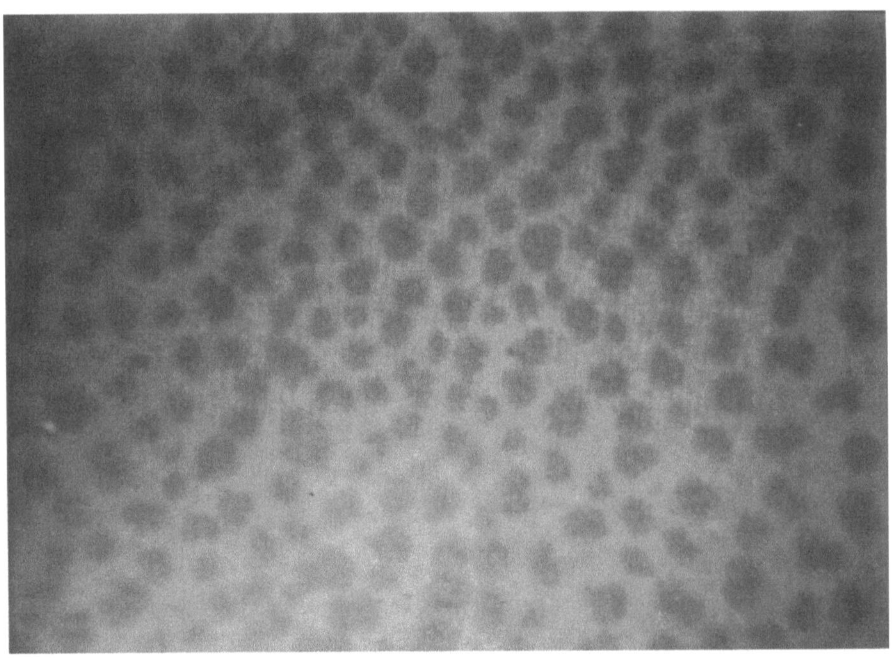

Fig. 4. Digital fluorescence image of a lipid monolayer transferred to a solid support under the pressure conditions indicated by arrow # 2 in Figure 2.

fluorescent lipids, are interpreted as gel-like islands, while the bright-fluorescent regions are interpreted as fluid regions (4). The digital images were processed according to the following sequence:

i) Correction for uneven distribution of light intensity. The acquired images typically display a brighter central region than the periphery. In order to eliminate this inhomogeneity, each image was corrected pixel by pixel using a radially distributed gaussian weight function.
ii) A low-pass filter was applied to each image in order to eliminate noise.
iii) Finally each image was segmented, (i.e, reduced to a two-level image), in order to make geometrical measurements.

The segmented images were utilized to measure the area of each dark island and the corresponding center of mass. Figures 5 and 6 are examples of the result of such processings. They refer to Figures 3 and 4, respectively.

In each figure, the original image is shown in the left top panel. The right top panel shows the result of the correction for uneven distribution of light. The left bottom panel is the result of the two-level segmentation. The right bottom image shows the distribution of the centers of mass of the islands.

The histograms of the area distributions for the patterns in Figures 3 and 4 are shown in Figure 7.

Future work will also be devoted to the analysis of the mean distance among neighbouring islands, as a function of the imposed physico-chemical conditions. The possibility of describing the patterns according to the theory of cellular automata will also be considered (6).

MODELING

The electrical properties of a lipid layer can be studied through the electrical dipole moment of the lipid heads.

The surface where the heads are constrained separates two regions where the electrical potentials φ_1, φ_2 fulfil the Laplace equation

$$\nabla^2 \varphi_i = 0 \qquad (i=1,2)$$

The polar heads' orientation over the surface depends on the electrostatic interactions between the dipoles. A mathematical treatment of these interactions can be developed through a continuous representation of the dipole distribution. This is done by introducing a dipole momentum per unit surface $\underline{\tau}$. According to this representation, the dipoles are regarded as being confined to an infinitesimal region close to the surface. Then the dipole distribution equations are treated as "boundary conditions" linking the general field solutions of the Laplace equation inside the two media.

In rectangular coordinates, let x=0 be the equation of the surface. The first boundary condition is

$$(\varphi_2 - \varphi_1)\Big|_{x=0} = \frac{\tau_x}{\varepsilon_0}$$

where τ_x is the component of $\underline{\tau}$ normal to the surface and ε_0 is the dielectric permittivity of the vacuum (7).

The second condition (8) is

$$(D_{2x} - D_{1x})\Big|_{x=0} + \nabla_t \cdot \underline{\tau}_t = 0$$

This condition links the normal components D_{2x} and D_{1x} of the electrical displacement vectors at x=0 to the projection $\underline{\tau}_t$ of $\underline{\tau}$ on the surface. The term $\nabla_t \cdot \underline{\tau}_t$ is the divergence of $\underline{\tau}_t$ on the surface.

These equations are based on a continuous approximation taken for the dipole distribution. In the original physical problem, however, discrete dipoles are confined to a thin slab near x=0. Close to each dipole vector inside the slab, the electric field vector is expected to assume the same orientation. Then, according to the continuous approximation, the following equation holds

$$\underline{\tau} \wedge \underline{E} = 0$$

where $\underline{E}$ is the limit value of the local electric field at the surface as

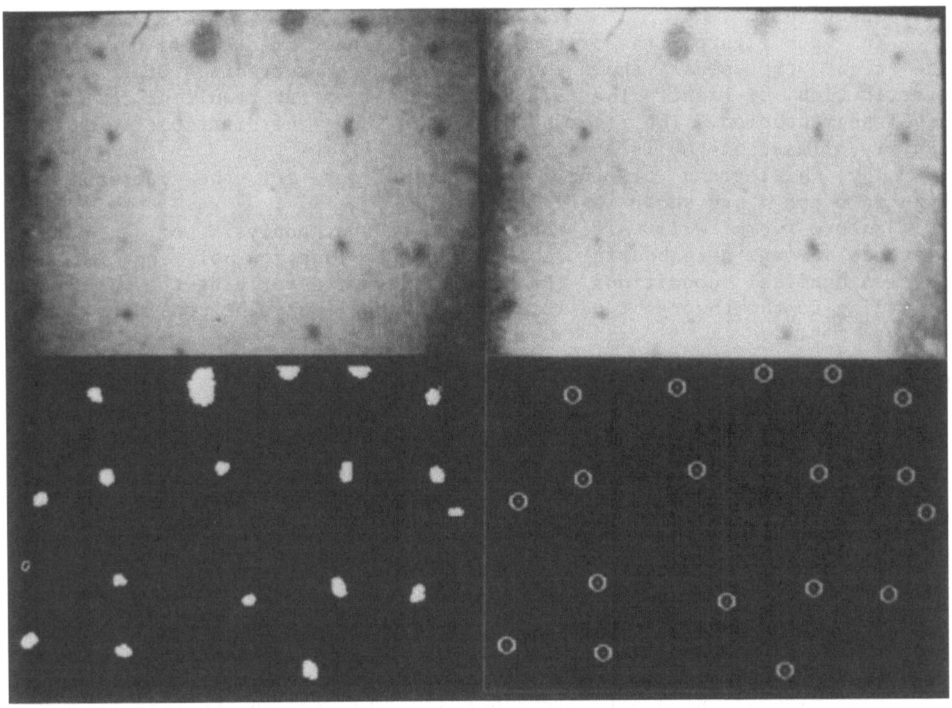

Fig. 5. Processing of Figure 3. For more details, see the text.

the slab thickness becomes zero.

The definition of the mathematical model requires a further equation, which should take into account the mechano-elastic properties of the dipoles. So far, this equation has not been given a well-defined structure. However, calculations based on reasonable tentative assumptions show that periodical dipole distributions are compatible with the set of equations discussed previously.

Work supported by CNR, PF MADESS

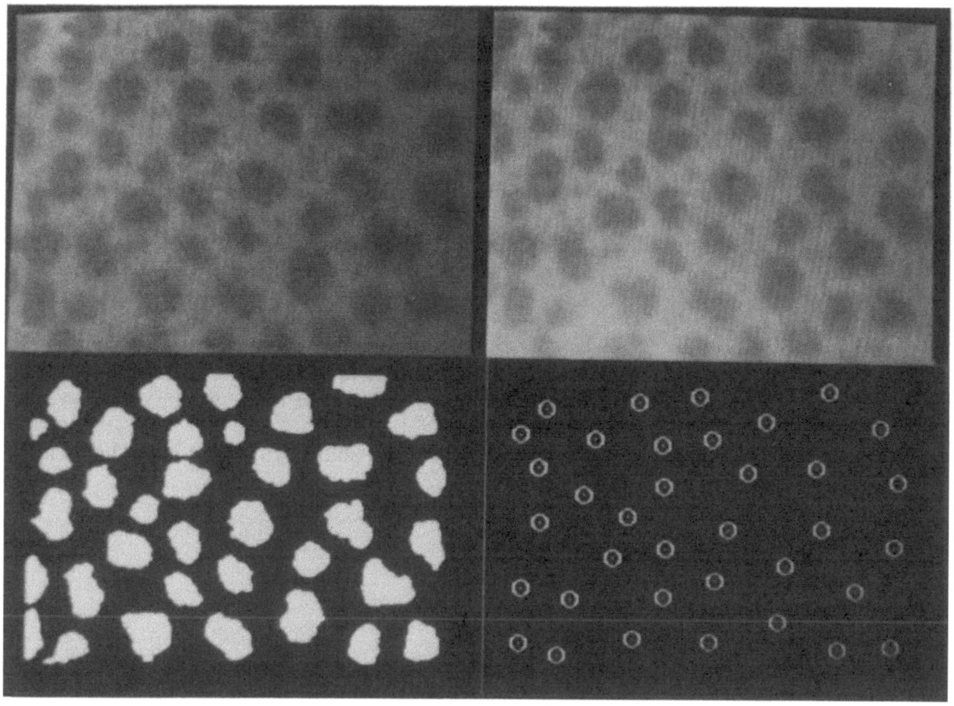

Fig. 6. Processing of Figure 4. For more details, see the text.

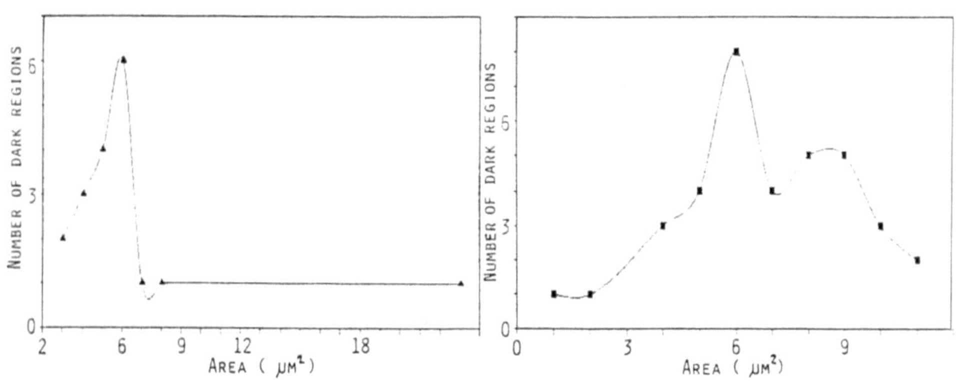

Fig. 7. Area distributions of the dark islands in Figures 5 (left) and 6 (right).

REFERENCES

1. G. G. Roberts, "An applied science perspective of Langmuir-Blodgett films", Adv.in Phys., 34:475 (1985)
2. H. Mohwald, "Lateral molecular organization and order in monomolecular layers", J. Mol. Electr., 4:47 (1988)
3. H. M. McConnell, L. K. Tamm and R. M. Weis, "Periodic structures in lipid monolayer phase transitions", Proc. Natl. Acad. Sci. USA, 81:3249 (1984)
4. M. Losche and A. Mohwald, "Fluorescence microscope to observe dynamical processes in monomolecular layers at the air/water interface", Rev. Sci. Instrum., 55:1968 (1984)
5. V. Von Tschaner and H. McConnell, "Physical properties of lipid monolayers on alkylated planar glass surfaces", Biophys. J., 36:421 (1981)
6. J. R. Barker, "Complex networks in molecular electronics and semiconductor systems", in: "Molecular Electronic Devices II", F.L. Carter Ed., M. Dekker, New York (1987)
7. J. A. Stratton, "Electromagnetic Theory", McGraw Hill, New York (1941)
8. A. Chiabrera and M. Parodi, "Electrical modelling of lipid layers", submitted to J. of Electrost.

ELECTROCHEMICAL PROCESSES IN MEMBRANE CHANNELS

AND BIOSYNTHETIC STRUCTURES

Martin Blank

Department of Physiology & Cellular Biophysics
Columbia University, College of Physicians and Surgeons
630 West 168th Street, New York, NY 10032

INTRODUCTION

In the development of microelectronic devices for use in sensory and information storage/retrieval applications, it is important to understand the mechanisms of these processes in biological systems. Over the course of evolution, living systems have developed highly specialized biopolymer structures with advantages (e.g., small size, speed, sensitivity, selectivity and stability) that make biomaterials desirable for direct use in microelectronic applications. Even when biopolymers cannot be used, the knowledge of how biological structures achieve their unusual properties can help in the development of non-biologically based materials.

In this paper we shall present a brief review of an electrochemical model that we have developed and tested for understanding the function of biological structures involved in information processing and storage/retrieval. The model has been helpful in explaining the operation of voltage gated ion channels in excitable membranes, as well as the mechanism of excitation itself. The same model has recently proved useful in analyzing the stimulating effects of electromagnetic signals on biosynthetic structures involved in transcription and translation.

MEMBRANE CHANNEL ENERGETICS

The specialized proteins in membranes that control the flow of materials into and out of cells have different biochemical structures and compositions, but a common overall design. They all appear to be cylindrical aggregates of oligomeric proteins or the associated helices of a single protein, formed from many strings of about 20 hydrophobic amino acids that traverse the lipid matrix of the membrane. The various cylindrical assemblies, which range in size from about 25-250 kD, carry out the fundamental membrane processes of living cells (e.g. generation of ion gradients, sensory transduction, conduction of impulses, energy transduction) by opening and closing in response to external factors, such as the electrical potential across the membrane or the binding of signaling (neurotransmitter, hormone) molecules to specific receptor or enzyme sites. The structural similarities between the different protein assemblies suggest that the opening and closing reactions can be described in terms of oligomeric equilibria.

Aggregation/disaggregation reactions of oligomeric proteins generally are not accompanied by changes in internal structure, so the total free energy change of a reaction can be approximated by the change in surface free energy. We have used surface free energy calculations to demonstrate the disaggregation of hemoglobin tetramers into dimers at pH's removed from the isoelectric point in either direction,[1] as well as the variations of the oxygenation equilibrium constant with pH and with ionic strength.[2] The surface energy model has explained the linkage between the various reactions (i.e. ligand binding and oligomer aggregation) and has shown that the Hill coefficient, a commonly used measure of cooperativity, is a classic surface excess quantity.[3] This model of oligomeric equilibria has been extended to hemoglobin aggregation in concentrated solutions, where it can account for the very high viscosities observed.[4]

We have recently tested the predictions of the surface free energy model for the disaggregation of hemoglobin tetramers into dimers as a function of the surface charge. In the alkaline range, the disaggregation constant varies with pH as predicted, and the quantitative agreement is excellent.[5]

Oligomer aggregation reactions are generally described as "entropy driven", because they proceed spontaneously with the evolution of heat, i.e. a positive ΔH. The reverse reaction, disaggregation, is accompanied by cooling. Both forward and reverse reaction heats are consequences of changes in the protein interfacial area in contact with the aqueous solution, and when similar reactions occur in the protein assemblies of channels, opening is accompanied by cooling and closing by heating. This model of channel opening/closing helps to account for the observed cooling/heating in active membrane systems.

VOLTAGE GATED CHANNELS IN EXCITABLE MEMBRANES

The ionic flows during excitation appear not to follow the equations of electrodiffusion. Excitable cells have two types of channels with different kinetics that show an apparent specificity for sodium and potassium ions, respectively. The faster responding channel is selective for sodium ions and the slower one for potassium. The selectivity for a single ionic species is difficult to explain because lithium can pass easily through the sodium channel and thallium in the potassium channel. Also, there is no relation between the selectivity and the conductance of a channel as would be expected with a bottleneck type of structure.

Our understanding of channel function and ion flows during excitation is considerably enhanced if we abandon the over-simplified conventional simple barrier model and consider the ionic processes that occur at the charged surfaces of membrane channels.[6] When this is done by using the Surface Compartment Model (SCM) approximation for ionic processes in the electrical double layer regions, it is possible to show that the unusual ion flows characteristic of electrical excitation phenomena arise from the same electrodiffusion equations that apply to ions in solution.[7] A typical result showing the "voltage clamp" response of SCM voltage gated channels is given in figure 1. The ionic currents in both fast ("sodium") and slow ("potassium") channels can be described by electrodiffusion equations. Both cations pass through each channel, but the ion gradients change with time, and the observed fluxes reflect the electrochemical gradients at different times. At short times the greater gradient is due to sodium, while at longer times the greater gradient is due to potassium. A sodium channel is characterized by rapid gating currents that open the channel at shorter times, and a potassium channel by slower gating currents, as observed in excitable membranes.

A kinetic basis for selectivity helps to answer the specific questions raised earlier, as well as the theoretical problem of having a Maxwell demon-like device with an ability to discriminate absolutely between sodium and potassium ions. The channel model also can account for the effects of the poison batrachotoxin on sodium channels, where the binding of one molecule per sodium channel causes it to respond more like a potassium channel. The selectivity for sodium decreases by a factor of about five and the characteristic inactivation is lost. This all appears to be correlated with the decrease in the channel gating current.[8] The switch from a

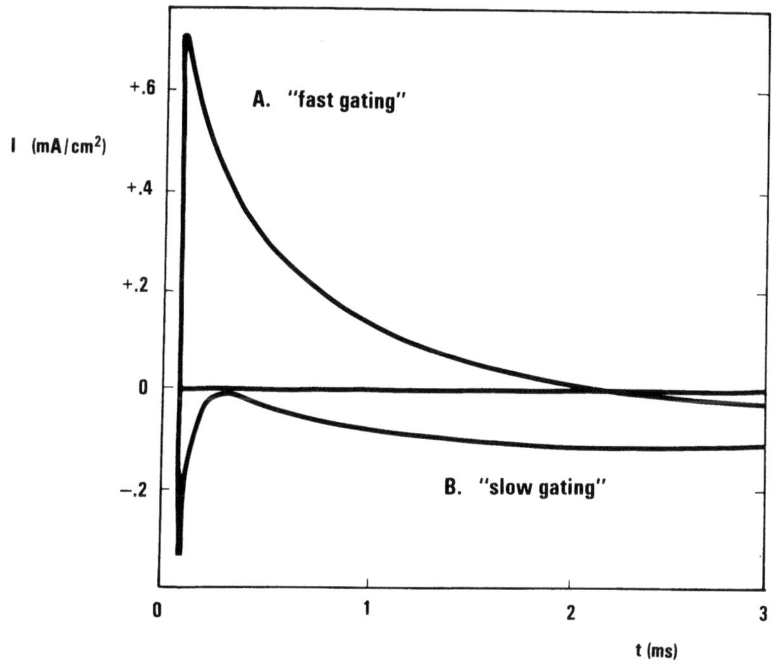

Figure 1. The ionic current I (in mA/cm^2) vs. the time t (in ms.) for different gating rates. Curve A for fast gating is 15 times faster than curve B for slow gating. The parameters of curves A and B are identical except for the one that controls the gating current. (Reproduced with permission from Bioelectrochem. Bioenerg.)

sodium channel to a potassium-like channel that is correlated with the decrease in the gating current is a practical demonstration of the SCM theory calculation shown in figure 1.

Following our studies of how DC electric fields cause membrane charges to shift, channels to open and close, and ions to flow, we have demonstrated similar effects by applying AC fields to the same theoretical model. Oscillating electric fields cause all of the ionic concentrations in the electrical double layers to change as functions of the frequency, but the relative change is greatest in those concentrations with the lowest steady state values. Since these ions normally control the activity of the ion pump enzyme, the Na-K ATPase of cell membranes, the calculated increases at particular frequencies suggest that the ions stimulate the enzyme and enhance ion transport in the range where biological effects of AC have been demonstrated.[9] The mechanism that leads to low frequency optima resulting from ion transport between electrical double layers, could account for the additive effects of very low amplitude signals over time.

The SCM-oligomeric channel model, derived from first principles and utilizing physical-chemical concepts that apply to proteins and ions in solution, has been successful in explaining a number of important physiological observations. With this theoretical framework we have shown that:

- gating currents lead to changes in charge and the partial disaggregation of the proteins (i.e., opening) of voltage gated channels.
- the cooling/heating observed in excitable membranes can be explained by the opening/closing reactions of channels that vary only a small fraction (about 3%) of the protein/water interface.
- the ionic currents calculated from Nernst-Planck equations for fast and slow gating channels approximate those observed in sodium and potassium channels, respectively (as shown in figure 1).
- the ion selectivity of the channel depends upon the kinetics of channel opening and varies with the gating current.
- the changes in ionic concentrations at the channel surfaces with AC stimulation are frequency dependent and show optima in the range where physiological effects have been reported.

BIOSYNTHETIC STRUCTURES

The electrochemical model for the opening/closing reactions of voltage gated channels and the ion flows during nerve excitation predicts that the charged surfaces involved in biosynthesis should experience frequency dependent changes as a result of the currents induced in alternating electromagnetic (EM) fields. The distributions of molecular weights in EM stimulated protein synthesis and RNA transcription show the predicted increase in lower molecular weight components as well as the dependence on the frequency of the EM field[10].

Additional support for the electrochemical model comes from an analysis of the properties of the new proteins synthesized in response to EM stimuli and not present in the controls. The new proteins appear to be the result of an interruption or interference with normal biosynthetic processes due to the induced electric field. Proportionally more new low molecular weight proteins are synthesized and the distribution of their mass is skewed to lower molecular weights. Whereas the proteins in control experiments are largely negative for all molecular weights, the new proteins are distributed symmetrically with regard to zero charge and the smaller molecules are more highly charged, both positively and negatively. The number of new proteins divided by the amount of new protein, a measure of diversity, increases with the frequency or repeat rate of the EM stimulation. These data agree with the predictions of the electrochemical model that interference should increase with both the EM frequency and the charge on the molecules, and result in shorter polypeptides.

The frequency dependence and the higher charge on the new proteins synthesized during EM stimulation suggest a rationale for the learning that results from endogenous electrical stimulation in the nervous system. The ionic fluxes that are the basis of the electrical signals in nerve axons and cell bodies are mediated by channels, that open and close according to the charge on the protein molecule[6]. Also, the greater the number of channels, the larger the change in polarization at dendrites, and the faster the conduction of an impulse along an axon.

The new proteins synthesized under EM stimulation increase with an increase in EM frequency, have lower molecular weights than the control proteins, and usually have a higher molecular charge. If some of the new proteins formed as a result of endogenous stimulation in the nervous system were incorporated into the membrane matrix as channels, this would be a molecular basis for habituation or learning that accompanies use. As the frequency of impulses increases, more channels would be created to speed up the conduction process. Furthermore, the higher molecular charge that would result from the faster process would increase the speed of opening of the channel and accelerate the response. These processes would facilitate or accelerate the responses in reaction to the increased stimulus. In the central nervous system, long term potentiation apparently requires high frequency stimulation of the nerves that form synapses.

SUMMARY

Channels in biological membranes are cylindrical assemblies of proteins that open and close in response to electrical and chemical stimuli. The opening/closing process is related to aggregation/disaggregation reactions of oligomeric proteins that respond to changes in surface charge. We have shown this for the disaggregation of hemoglobin tetramers, and have used this model to develope a physical/chemical account of the molecular events during excitation. The model describes the ionic fluxes and shows that the kinetics of opening in voltage gated channels determine ion selectivity. The same processes that lead to the opening/closing of channels in membranes apparently occur in biosynthetic structures under the influence of imposed electromagnetic fields and lead to changes in the patterns of transcription and translation.

ACKNOWLEDGEMENT

We thank the Office of Naval Research for its support.

REFERENCES

1. M. Blank, A Surface Free Energy Model for Protein Structure in Solution: Hemoglobin Equilibria, Colloids and Surfaces 1:139 (1980).
2. M. Blank, A Model for Calculating the Bohr Effect in Hemoglobin Equilibria, J. theoret. Biol. 51:127 (1975).
3. M. Blank, Hemoglobin Reactions as Interfacial Phenomena, J. Electrochem. Soc. 123:1653 (1976).
4. M. Blank, Molecular Association and the Viscosity of Hemoglobin Solutions, J. theoret. Biol. 108:55 (1984).
5. M. Blank and L. Soo, Surface Free Energy as the Potential in Oligomeric Equilibria: Prediction of Hemoglobin Disaggregation Constant, Bioelectrochem. Bioenerg. 17:349 (1987).
6. M. Blank, The Surface Compartment Model (SCM) - Role of Surface Charge in Membrane Permeability Changes, Bioelectrochem. Bioenerg. 9:615 (1982).
7. M. Blank, The Surface Compartment Model: A theory of ion transport focused on ionic processes in the electrical double layers at membrane protein surfaces, Biochem. Biophys. Acta 906:277 (1987).
8. B.I. Khodorov and S.V. Revenko, Further Analysis of the Mechanisms of Action of Batrachotoxin on the Membrane of Myelinated Nerve, Neuroscience 4:1315 (1979).
9. M. Blank, Theory of frequency-dependent ion concentration changes in oscillating electric fields, J. Electrochem. Soc. 134:1112 (1987).
10. M. Blank and R. Goodman, An Electrochemical Model for the Stimulation of Biosynthesis by External Electric Fields, Bioelectrochem. Bioenerg., in press.

ELECTROCONFORMATIONAL COUPLING: A FUNDAMENTAL PROCESS OF BIOMOLECULAR ELECTRONICS FOR SIGNAL TRANSDUCTIONS

Tian Yow Tsong

Department of Biochemistry
University of Minnesota College of Biological Sciences
St. Paul, MN 55108, U.S.A.

Summary: Biomolecules are like electronic elements and are built to process information. Proteins of a cell membrane are adept to receive and to decipher even the very weak electromagnetic signals. Experiments on ion channel activity and on electric activation of [Na,K]-ATPase and mitochondrial ATPase reveal a fundamental process of the biomolecular electronics, the "Resonance Electroconformational Coupling". An alternating electric field of a defined amplitude and frequency is shown to induce oscillation of protein conformers. The process can turn the activity of the protein on and off much like a molecular switch. When coupled to ligand binding it also allows the molecule to use the binding energy of a ligand to drive an endergonic reaction.

I. Introduction

A cell is a fine device of molecular electronics. It has not only the capacity to receive, to process, to storage, and to transmit energy and information, but also the capacity to self-program for adjusting to a sudden and unexpected change in its environment and to self-replicate for preserving its genetics information. Even in the simplest life form, a virus particle, the complexity, the flexibility and the efficiency of its molecular circuitry far surpass those of the human designs. Biochemists have been studying mechanisms of action of these molecules for more than half a century now, and there is a wealth of information concerning how these molecules work individually and collectively. Some classes of these biomolecules are directly responsible for transforming energy from one form to another or for performing transcellular signalings (Bullock & Heiligenberg, 1986; Hong & Montal, 1979; Hopkins, 1988; Nicholls, 1982).

These molecules are analogous to many simple electronic elements or sensors and we will use two enzymes of this type, namely [Na,K]-ATPase of human erythrocytes and F_0F_1ATPase of mitochondria, to illustrate certain fundamental concepts in bioelectronics designs. [Na,K]-ATPase utilizes the hydrolytic energy of ATP for pumping Na^+ and K^+ against their respective ion concentration gradient and mitochondrial ATPase converts metabolic energy into the γ-phosphorous bond energy of ATP. In recent experiments we have found that these two enzymes are capable of absorbing free energy from applied electric fields to perform the work function they are designated to do (Serpersu & Tsong, 1983, 1984; Teissie et al., 1981; Tsong et al., 1988b; Chauvin et al., 1987; Liu et al., 1988; Tsong, 1983; Tsong & Astumian, 1986). Since the applied fields would generate effective field strength comparable to that they experience, in vivo, it is logical to assume that these enzymes should also respond to the endogenous electric fields. As such, study of mechanisms of the electric activation is a valid approach for deciphering the mechanisms of action of these enzymes. In this article, I will also discuss how a weak electromagnetic signal is amplified and registered by membrane enzymes and receptors and how this amplification greatly improves the capability of these biomolecular sensors in transcellular, organ-to-organ, and organism-to-organism communications.

II. Some Characteristics of Membrane Proteins

Macromolecules distinguish themselves from simple inorganic or organic molecules by the fact that in their structures there are many more degrees of freedom. The conformational energy is a term used to describe internal energy preserved in these modes of motion within the structure of the macromolecule. In addition, membrane proteins have certain characteristics which make them especially suitable for interacting with an electric field. These characteristics are:

1). A membrane proteins is relatively immobile and its incorporation into lipid bilayer has a specific orientation. An otherwise scalar reaction when taking place in a solution becomes vectorial via its catalytic or modulating action. A freely rotating catalyst can not achieve such a task. As a consequence of the orientation of molecules in the bilayer, most membrane channels possess rectifier property.

2). A cell membrane can amplify an electric signal because of its non-conductive character relative to the cytoplasmic and the extracellular fluids. For example, if microelectrodes were impeded on both sides of the membrane and an electric field of 1 V/cm were imposed, the potential drop across the membrane (of thickness d) would be nearly 1 V. The effective field strength across the membrane would thus be, 1/d V/cm. The thickness of the cell membrane is about 5 nm, and 1/d is therefore 2×10^6. That is to say the field is amplified by 2 million times. For a cell of spherical shape, the amplification factor according to the Maxwell relation ($\Delta\psi = 1.5\ R_o\ E \cos\theta$) is $1.5\ R_o/d$, where R_o is the outer radius of the cell and $\Delta\psi$ is the transmembrane potential (Tsong, 1983; Kinosita & Tsong, 1977). For a cell the size of a human red cell, R_o/d is approximately 1000. If an organ is enclosed by a membrane, the value for R_o/d can be very large, indeed.

3). The interior of a protein and the bilayer of lipid have low dielectric constants. Dielectric constants of lipid bilayer and protein structures are in the range 2 to 4 compared to 80 for water (Pethig, 1979). For a protein in the lipid environment, effect of an electric field would be 20-40 times greater than the effect it would experience in a solubilized form in an aqueous solution.

4). A membrane protein is rich in α-helices. It also carries net charges at neutral pH. An α-helix is an electric dipole of 3.6 n Debye, n being the number of amino acid that constitutes the helix. The N-terminal corresponds to the positive pole and the C-terminal the negative pole of the helix dipole. Other prosthetic groups of a protein usually have net charges. Phosphorylation of a protein also introduces charges into the molecule.

5). Roughly 10% of membrane lipids bear negative charges. Water inside a lipid bilayer or a protein structure can also form oriented dipoles. These charged lipids and oriented dipoles of water influence the environment and add to the electric environment of a membrane protein.

III. Electroconformational Change of Proteins And the Molecular Switch

Because of these properties of a membrane integral protein it is especially susceptible to influence by an electric field (Tsong & Astumian, 1986, 1987, 1988; Tsong et al., 1988a). The interaction is Coulombic and is governed by the generalized van't Hoff Equation.

$$\delta\ (\ln K) = \frac{\Delta M}{R\ T}\ \delta E \qquad (1)$$

where K is the equilibrium constant, ΔM, the change in the molar electric moment, R, the gas constant and T, the Kelvin temperature. From here on δE will represent an effective electric field change, in most cases, a transmembrane electric field, i.e. $\Delta\psi/d$. In other words, if a molecule

has different conformational states, an electric field will shift the equilibrium towards the conformer of the highest molar electric moment. Let us consider a protein molecule which is embedded in a membrane and which has two conformational states at equilibrium under a low field condition.

$$P_1\ (M_1) \rightleftharpoons P_2\ (M_2) \qquad (2)$$

The equilibrium constant (The use of this term is only for convenience. The phenomenon discussed below involves large amplitude perturbation of an equilibrium and is better treated using non-equilibrium thermodynamics.) at a low field, $K(1) = [P_2]_1/[P_1]_1 = k_1(1)/k_{-1}(1)$ and at a high field, $K(h) = [P_2]_h/[P_1]_h = k_1(h)/k_{-1}(h)$.

$$\Delta M = M_2 - M_1 = [(\sum_i z_i\ \Delta d_i) + \Delta\mu + \Delta\alpha \cdot E] \qquad (3)$$

where Δd_i is the vector displacement of a charge i between the two states, $\Delta\mu$, change in permanent dipole moments, and $\Delta\alpha$, change in the polarizeability. When a high field condition is imposed, the equilibrium will shift towards P_2 if ΔM has a positive value and towards P_1 if ΔM has a negative value. The extent of change is given in Eq. (1). Eqs. (1) and (2) mean that it is possible to design a molecule with two conformational states, one active and one latent, which can then be switched on and off by an electric field.

The above analysis is quite familiar to electrochemists. Wien Effects and Onsager's Dissociation Field Effects are two examples in which an electric field causes a shift in a chemical equilibrium. However, what is unusual when a membrane protein is considered is that it usually displays a large ΔM of several hundred Debye and one can impose a large change in electric field, δE, so that the transition is complete. This is not possible for a reaction taking place in an aqueous solution because of the difficulty in attaining a high field and also because of the rapid rotation and tumbling of the molecule, as we mentioned earlier.

There are many experimental observations of the electro-conformational change of membrane proteins. The opening and closing of most membrane channel proteins are dependent on the gating voltage. Fig. 1A gives an example of the opening and closing of Na^+-channel-batrachotoxin complex as a function of the gating voltage. The data is adapted from the work of Hartshorne et al. (1985). Other work have shown that electroconformational change can alter ligand affinity, enzyme activity...etc. The electroconformational change is used, in vivo, to

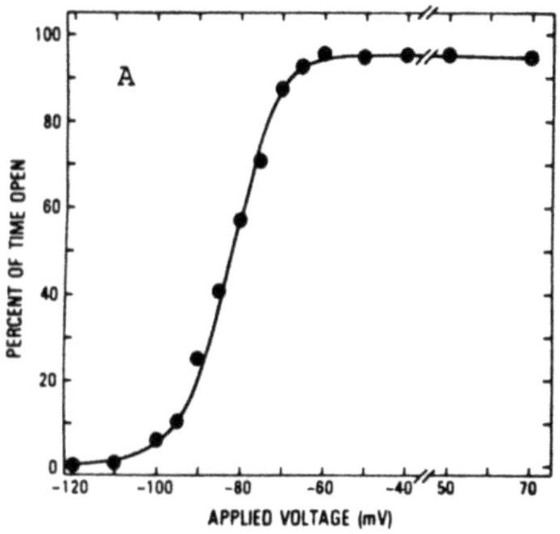

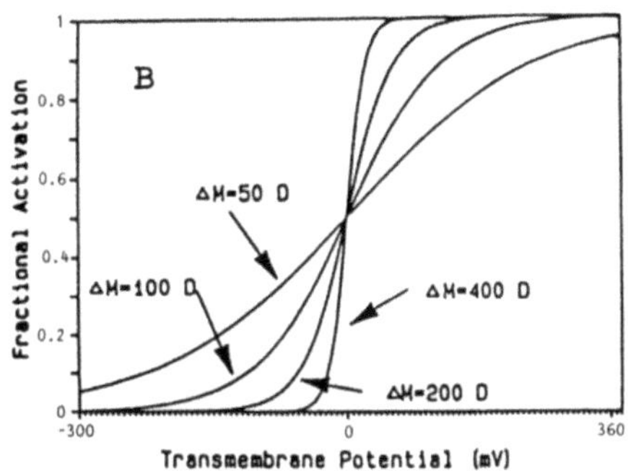

Fig. 1 (A) Voltage dependence of Na^+-channel opening (Data from Hartshorne et al., 1985). (B) Calculation based on Eq. (1) of the dependence of the equilibrium of Eq. (2) on the transmembrane potential in mV. Membrane thickness of 5 nm was assumed. Fraction of molecules in P_2 state is plotted against $\Delta\psi$ for different values of ΔM.

turn on and off of certain functions of a cell, and this property of a protein can be used to design a molecular switch. The effectiveness of the switch is quite variable and is dependent on the value of ΔM between the two conformers. Fig. 1B illustrates how the transition of the two conformers depends on the transmembrane potential, $\Delta\psi$, with different ΔM values, based on Eq. (1). It is clear that the transition from one form to another can be achieved within a narrow range of $\Delta\psi$ if ΔM is large. The range of ΔM for biomolecules is quite large, from 10 to 50000 D (Debye), and the biomolecular switch can be made to suit a wide range of conditions.

IV. Resonance Electroconformational Coupling for Energy Transduction

But, the most unique feature of a membrane protein is its ability to absorb energy from an external source and convert it to another form. This property is due to the unique orientation of membrane proteins. An enzyme in an homogeneous aqueous solution can only catalyze a chemical reaction towards its equilibrium. A membrane embedded enzyme can also catalyze a reaction away from its equilibrium position if it is coupled to a free energy source. In term of molecular electronics, this means, a membrane protein can be a receptor and a transmitter of energy or signals. A mechanism derived from the concept of electroconformational change has been proposed and experimental results presented in many occasions (Tsong & Astumian, 1986, 1987, 1988; Tsong et al., 1988a; Westerhoff et al., 1986; Astumian et al., 1987; Chen, 1987; Kemp et al., 1988). Here we will summarize the salient features. Other models may

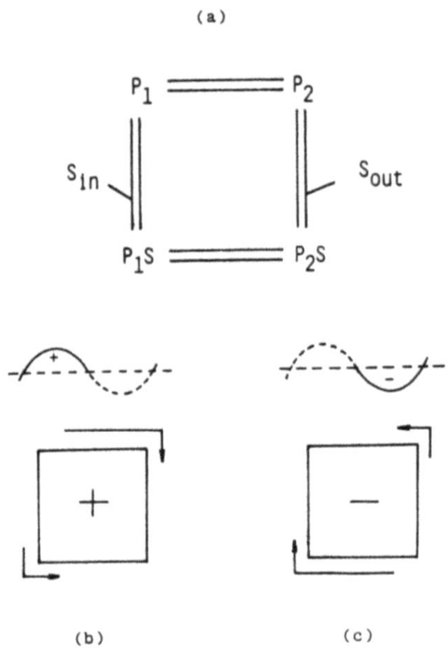

Fig. 2 A molecular pump based on the concept of the resonance electroconformational coupling.

also explain some features of our data (Blank, 1987; Edmonts, 1986).

Let us consider a cyclic kinetic model shown in Fig. 2A. This simple system is designed to use the binding energy of the enzyme for a ligand, S, to perform work function, in this case the pumping of the ligand from a low concentration compartment (out) to a high concentration compartment (in). Let us design a molecule, P, so that when it is in a low field (P_1) it has a low affinity for S, and when it is in a high field (P_2) it has a high affinity for S. The transition from P_2S to P_1S involves the translocation of the bound ligand and its detailed mechanism will not be considered here. There are many publications which describe possible structural changes of a protein for such a process. We also assume that the ligand binding and dissociation reactions are much faster than the conformational transitions. At a low field the system shifts to the left and the dominating species is P_1. At a high field the system shifts to the right and the dominating species is P_2S. This asymmetric property confers the system the ability to use the binding energy [$\Delta G = - R\ T\ \ln\ ([P_2S]/[P_2][S])$] to pump S against its concentration gradient.

Computer simulation shows that when the system is exposed to an positive electric pulse, the field induces fluxes indicated by the arrows in Fig. 2B, and when the system is exposed to a negative electric pulse, the field induces fluxes indicated by the arrows in Fig. 2C. Here, the length of the arrow is proportional to the amplitude of the flux. What is interesting here is that regardless of the field direction, it induces a clockwise fluxes of the reactions. If an alternating electric field is imposed, the system will turn clockwise until $R\,T\,\ln\,[S]_{out}/[S]_{in} = \Delta M \cdot E$. A molecular pump is shown in action. It is conceivable that a net work of membrane receptors or charge relay systems, such as redox protein complexes, can be fabricated and within this ensemble each class of molecules will respond to an electric signal of a defined frequency and waveform. This is likely the case in the natural design of a cell membrane.

V. Electroactivation of [Na,K]-ATPase and Mitochondrial ATPase

These calculations are based not only on thought experiments but that they have been done to interpret the results of electric field activation of [Na,K]-ATPase of human erythrocytes and mitochondrial ATPase of beef heart and rat liver. In the [Na,K]-ATPase experiments, human red cells in an isotonic suspension were exposed to applied alternating electric field (a.c.) of strength up to 50 V/cm and of frequency between 1 Hz and 20 MHz. Movements of Na^+ and Rb^+ (a substitute for K^+) into and out of the red cells were then monitored by radioactive tracers and compared to control samples in which no a.c. stimulation was imposed. Under the normal physiological and other broad experimental conditions, the enzyme hydrolyzes each ATP to pump 2 K^+ (or Rb^+) into the red cell and 3 Na^+ out of the red cell. Since, the cytoplasmic Na^+ concentration (5-10 mM) is generally only 1/20 of the external Na^+ concentration the ATP linked Na^+ transport is an active transport. K^+ transport is also an active transport, although, many authors consider distribution of K^+ to be governed by transmembrane potential.

The main findings of these experiments are as follows (Serpersu & Tsong, 1883, 1984; Tsong et al., 1988a; Liu et al., 1988). 1). At 4°C, the Rb^+-pump and the Na^+-pump of the enzyme have an identical optimum field strength of 20 V/cm. This electric field can generate a 20 kV/cm effective transmembrane electric field. 2). The optimum frequency using a 20 V/cm a.c. field is 1.0 kHz for the Rb^+-pump and is 1.0 MHz for the Na^+-pump (Fig. 3). Both are transport activities against their respective concentration gradient. 3). At the optimum conditions, the rate of the Rb^+-pump is between 15 and 20 /s /pump and of the Na^+-pump is between 20 and 30 /s /pump, at 4°C. The ratio is roughly 2/3 for Rb^+/Na^+. 4). These a.c. stimulated activities do not depend on ATP hydrolytic activity of the enzyme, and they are completely inhibited by ouabain and other potent inhibitors of [Na,K]-ATPase. 5). The Rb^+

efflux and Na^+ uptake are not stimulated by the applied a.c. field for up to 60 V/cm, in the frequency range 1 to 20 MHz. 6). A person whose erythrocytes exhibits twice basal [Na,K]-ATPase activity also showed

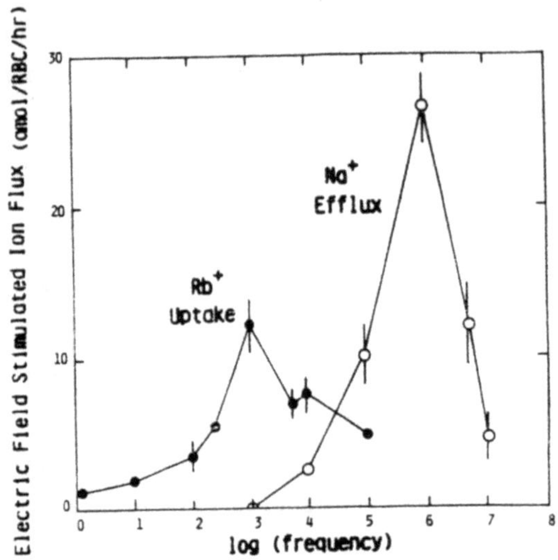

Fig. 3 Activation of the Na^+- and the Rb^+ (K^+)-pumps of [Na,K]-ATPase of human erythrocytes by an oscillating electric field of 20 V/cm, at 4°C. The ouabain sensitive Na^+ efflux (O) and Rb^+ uptake (●) are plotted against the frequency of the applied field. These a.c. stimulated ion movements were active transports against their respective concentration gradients. See Liu et al. (1988) for experimental details.

twice the a.c. stimulated activity. 7). The a.c. stimulated activity diminishes as the temperature is raised to 37°C. At this temperature the ATP-linked pumping activity of the enzyme is at its maximum. This observation implies that a maximally functioning enzyme can not be further stimulated by the a.c. field. 8). The activity observed here is purely electrical not thermal.

Experiment with mitochondrial ATPase is technically more demanding owing to a much higher transmembrane potential required for ATP synthesis than that required for the pumping of Rb^+ and Na^+ by [Na,K]-ATPase. It is believed that the in vivo transmembrane electric field of mitochondrial inner membrane is about 200 mV. To generate such a potential in submitochondrial particles of about 100 nm requires an electric field of 30 kV/cm. With this field strength, Joule heating becomes very severe. To circumvent this difficulty, we use pulsed electric field method (PEF). Short exponentially decaying electric pulses of initial strength up to 30 kV/cm and of decaying time up to 100 μs are used. The result is rather astonishing. Indeed ATP is synthesized from ADP and P_i even when the electron transport chain is completely inhibited by cyanide and/or rotenone. The unexpected finding is that this activity also requires the presence of DTT, a -SH group reducing reagent. Without DTT, the yield is less than 1 ATP /enzyme /pulse (Teissie et al., 1981). With it the yield reaches 5-10 ATPs /enzyme /pulse (Chauvin et al., 1987). Apparently, DTT can facilitate the enzyme turnover. The reason of this requirement is not clear. We propose that at each turnover certain -SH group of the enzyme goes through an oxidation-reduction cycle.

The above results on [Na,K]-ATPase are in a complete accord with the prediction of the RECC model. The model predicts that for each enzyme that can absorb and transmit free energy from an electric field, there would be an optimum field strength and optimum frequency for it to attain electroconformational resonance and hence to achieve the maximum efficiency. The results on mitochondrial ATPase are, however, less congruent to the concept of conformational resonance. For one thing the applied pulsed electric field was d.c. in nature, and it is demonstrated that a d.c. field is unsuited for inducing conformational resonance. Notwithstanding, these experiments allow us to prove further yet another important property of the cellular mechanisms of signal and energy transductions.

VI. Localized And Delocalized Transmembrane Potential

A constant transmembrane electric field normally reported in the literature represents a time average and spatial average transmembrane potential of a cell, and its value reflects a delocalized, overall transmembrane potential. In considering the electrostatic interaction between a membrane protein and an electric field, such number is irrelevant. It is the characteristics of the electric field at the immediate vicinity of a protein that is important (Tsong & Astumian, 1986, 1987, 1988; Tsong et al., 1988). A constant stationary transmembrane electric field may be modulated to become locally oscillatory. Here we will deal with how such modulation can be accomplished by a membrane channel or by an electron transport reaction. In Fig. 4 it is shown how a stationary transmembrane electric field may be modulated to become locally oscillatory by the opening/closing of an ion channel nearby an energy transducing protein. Channel opens due to the transmembrane electric potential, but closes on the passing of ions which locally neutralize the transmembrane potential. Recurrent opening/closing may modulate the field to become oscillatory. However, its effect is limited to its vicinity.

The question is whether in a cell modulation of membrane potential is a general feature. We propose that this indeed is the case. The situation is portrayed in Fig. 5. Here it is shown that how one can position two microelectrodes at different locations relative to the plasma membrane to measure a transmembrane potential. When they approach closer and closer to the membrane surface, one expects the recorded electric field to become noisier and noisier. This effect is to be distinguished from that of the electronic noise of the instrument. Local

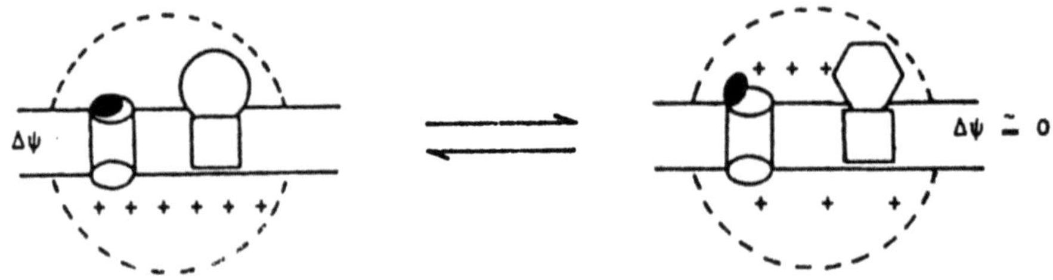

Fig. 4 Modulation of local electric potential by an ion channel. A constant $\Delta\psi$ maintained by the metabolic reactions is locally modulated by the opening/closing of an ion channel. This modulation has certain frequency window according to the chemical rate of the opening/closing of the channel.

electric field fluctuation should depend on channel activities, electron transport, and/or redox reactions of the cell membrane, and should change on induction or inhibition of one or more of these activities. Thus, these noises are not featureless and they should exhibit certain kinetic characteristics.

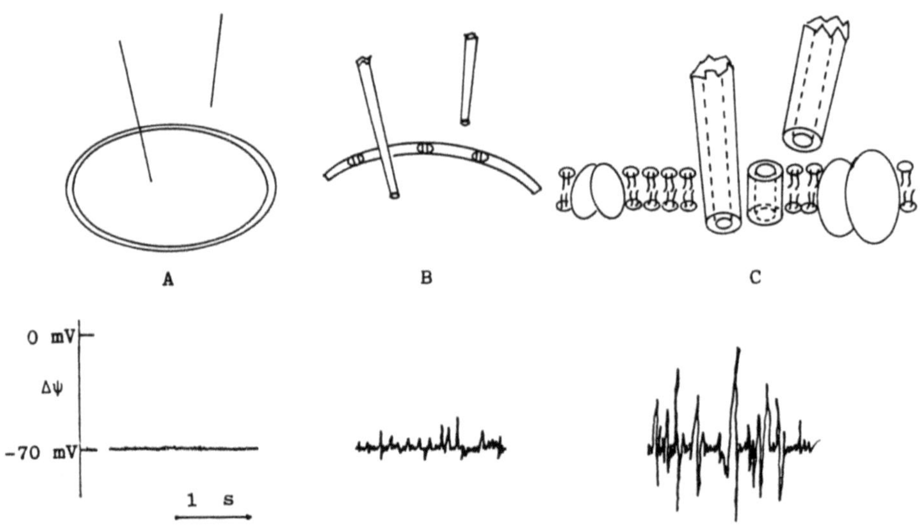

Fig. 5 Schematic illustration of how the measured transmembrane potential as function of time will depend on the positioning of the electrodes. As the two electrodes move closer and closer to the surface of the membrane the measured potential becomes more and more fluctuating. These fluctuating patterns reflect local ion and electron transport activities and contain signals of cells.

VII. Information Content in Electric Noises of A Cell Membrane

The above analyses imply that the noises of a cell membrane are not the result of equilibrium fluctuations and should not be treated as such (Tsong & Astumian, 1987, 1988; Astumian et al., 1987; Tsong et al., 1988). Membrane noises are likely sustained by energy dissipating reactions and in them are free energy and signals for fueling or regulating cellular reactions via resonance electroconformational coupling of membrane receptors and enzymes. In other words, the transmembrane electric field which appears noisy may in fact carry signals or words of a cell (Tsong, 1988). Intracellular, organ-to-organ, and organism-to-organism communications are accomplished through these electrical "noises".

Let us now consider another important means of cellular signalings, namely, the transmission of an action potential in a neuron. Most textbooks of neurobiology attribute the propagation of an action potential to the controlled ion fluxes, e.g. Ca^{++}, Na^+ and K^+ fluxes, and fluxes of neurotransmitters. These interpretations may be fashionable but they are not in accord with many basic chemistry of ions in an aqueous solution. We know that the propagation of an action potential in major neurons can be as fast as 25-100 m/s. No ions can transfer that fast in an aqueous medium under a reasonable lateral gradient of electric potential. Transport of a proton in an ice is perhaps the fastest, but it can only reach 4 cm/s under a 100 V/m electric potential gradient. This value is only slightly greater than 1/100 the speed of an action potential. With such a wide discrepancy in speed, it would be difficult to explain the rapid propagation of an action potential by movement of ions no matter what kind of physical model is constructed. This deficiency in speed can be ignored if we accept that many of the reactions we are dealing with here are electrical and resonance electroconformational coupling is a fundamental process. A cell as a molecular electronic device will then be governed by such a process.

VIII. High Sensitivity of Resonance Electroconformational Signaling

Another important concept in understanding transcellular signalings is the distinction between a signal and an informationless noise. A cell or an organism is capable of sensing very weak signals, such as electromagnetic, acoustic or thermal signals (Bullock & Heiligenberg, 1986; Hopkins, 1988; Altschuler et al., 1986; Tsong, 1988; Tsong et al., 1988a). In many cases, these signals are so weak, one wonders how they can be differentiated from the environmental noises. At least four mechanisms may be conceived. The first one is the amplification of signal through the cell membrane by the Maxwell relationship. This mechanism was mentioned earlier. The second one is by the use of Eq. (1). Shift in the equilibrium of Eq. (2) can be made infinitively sensitive to δE if ΔM has an infinitively large value. Many membrane proteins has a large ΔM. For example, Na^+-channel is reported to have a ΔM of about 900 D when changing from the closed state to the open state (Hartshorne et al., 1985). Bateriorhodopsin is reported to have a ΔM of 2000 D when exposed to an electric field. ΔM values as high as 12000 to 20000 D are observed for biomacromolecules. Shift in the equilibrium constant, K, changes the relative concentration of P_1 and P_2 and the extent of changes can be made much higher than the equilibrium concentration fluctuation even with modest ΔM and δE.

The third one is the distinction between an autonomous noise and an equilibrium noise. The former characterizes a noise which waveform is not altered by its interaction with a receptor, and the latter characterizes a noise which waveform is altered owing to the reciprocal interaction with the receptor (Astumian et al., 1987; Tsong & Astumian, 1987). Only a signal sustained by an energy source can be treated as autonomous. A cell is endowed with mechanisms to differentiate these two types of potential fluctuations.

The fourth one is that a weak signal may be time averaged to screen from the background noises. A cell or an organism shall be able to use this mechanism to sort out signals from incidental noises. In all these cases, the reception of signals are by the resonance conformational coupling of membrane proteins.

Acknowledgements -- This work was supported by a NSF grant and an ONR contract. I thank my past and present colleagues and my collaborators for their contributions to this project.

References

Altschuler, R.A., Hoffman, D.W. & Bobbin, R.P., eds. (1986). "Neurobiology of Hearing: The Cochlea", Raven Press, New York.

Astumian, R.D., Chock, P.B., Tsong, T.Y. & Westerhoff, H.V. (1987). Proc. Natl. Acad. Sci., 84, 434-438.

Blank, M. (1987). J. Electrochem. Soc., 134, 1112-1117.

Bullock, T.H. & Heiligenberg, W. (1986). "Electroreception", Wiley & Son, New York.

Chauvin, F., Astumian, R.D. & Tsong, T.Y. (1987). Biophys. J., 51, 243a.

Chen, Y.-D. (1987). Proc. Natl. Acad. Sci., 84, 729-731.

Edmonds, D.T. (1986). Proc. R. Soc. London, Ser. B, 228, 71-84.

Hartshorne, R., Keeler, B.U., Talvenheimo, J.A., Catterall, W.A. & Montal, M. (1985). Proc. Natl. Acad. Sci., 82, 240-244.

Hong, F.T. & Montal, M. (1979). Biophys. J., 25, 465-472.

Hopkins, C.D. (1988). Ann. Rev. Neurosci., 11, 497-535.

Kemp, F., Astumian, R.D. & Westerhoff, H.V. (1988). Proc. Natl. Acad. Sci., 85, 3792-3796.

Kinosita, K. Jr. & Tsong, T.Y. (1977). Proc. Natl. Acad. Sci. 74, 1923-1927.

Liu, D.-S., Astumian, R.D. & Tsong, T.Y. (1988). Submitted.

Nicholls, D.G. (1982). "Bioenergetics", Academic Press, New York.

Pethig, R. (1979). "Dielectric and Electronic Properties of Biological Materials", Wiley & Sons, New York.

Serpersu, E.H. & Tsong, T.Y. (1983). J. Membrane Biol., 74, 191-201.

Serpersu, E.H. & Tsong, T.Y. (1984). J. Biol. Chem., 259, 7155-7162.

Teissie, J., Knox, B.E., Tsong, T.Y., Wehrle, J. (1981). Proc. Natl. Acad. Sci., 78, 7473-7477.

Tsong, T.Y. (1983). Bioscience Reports, 3, 487-505.

Tsong, T.Y. (1988). Submitted.

Tsong, T.Y. & Astumian, R.D. (1986). Bioelectrochem. Bioenerg., 15, 457-476.

Tsong, T.Y. & Astumian, R.D. (1987). Prog. Biophys. Mole. Biol., 50, 1-45.

Tsong, T.Y. & Astumian, R.D. (1988). Ann. Rev. Physiol., 50, 273-290.

Tsong, T.Y., Liu, D.-S., Chauvin, F. & Astumian, R.D. (1988a). Bioscience Reports, In press.

Tsong, T.Y., Astumian, R.D. & Liu, D.-S. (1988b). Biophys. J., 53, 623a.

Westerhoff, H.V., Tsong, T.Y., Boon, P.B., Chen, Y.-D. & Astumian, R.D. (1986). Proc. Natl. Acad. Sci., 84, 434-438.

THE ROLE OF SOLVENT PERMITTIVITY DISPERSION IN ELECTROSTATIC INTERACTIONS IN PROTEINS

Vladimir N. Buravtsev, Pavel I. Lazarev, Victor S. Sivozhelezov

Research Computing Center
U.S.S.R. Academy of Sciences
Pushchino, Moscow Region 142292 U.S.S.R.

and Alexander I. Rubinstein

Institute of Experimental Endocrinology and Hormone Chemistry
U.S.S.R. Academy of Medical Sciences
11 Dmitri Ulyanova St., Moscow 117162 U.S.S.R.

INTRODUCTION

The electrostatic field of a protein molecule is an important physicochemical characteristic, and in many cases, determines the rates of protein complex formation and complex stabilities.[1] In view of the progress currently made in protein engineering, mathematical modeling of protein interaction with various ligands including other proteins is getting more important. Since the modeling gives estimates of thermodynamic and kinetic characteristics of the interaction under study it is capable of reducing the number of parameters that must be measured experimentally. This makes modeling a pragmatic task and naturally brings about the need to compare the cost of the experiments to that of the modeling before deciding which is to be done first.

Molecular electronics seems to be a very promising field where protein engineering can be applied. Specifically, electron transfer proteins could be used as materials for electronics.[4] The one-electron circuit design that is currently being developed requires non-physiological protein complexes with predetermined structures and electronic properties to be built.[4] Since the interactions between electron-transfer protein are in most cases controlled by electrostatics,[2,3] the electrostatic field/potential calculations are necessary for the design of the required complexes.

Here we introduce an approach to protein electrostatics calculations based on the theory of Dogonadze, Ulstrup and Kornyshev[11] that takes into account the non-locality of the dielectric response in various media caused by correlations in the orientation of dipoles constituting the medium. In water, for instance, the correlation is due to the hydrogen bonds between water molecules that form a quasi-crystalline structure in water and prevent the molecular dipoles from complete reorientation in the field. The resulting model was tested on Cu,Zn-superoxide dismutase (SOD). The latter has been chosen because a variety of electrostatic models have been applied by other groups[17-19] to this same protein.

BACKGROUND

The electrostatic potential of a protein molecule is caused by charges of ionizable side chains of the following amino acid residues: aspartic acid, glutamic acid (Asp, Glu - negative), lysine and arginine (Lys, Arg - positive). There is a large body of experimental data[2,3,6] showing that the electrostatic interactions between electron-transfer proteins play a key role in the formation of protein-protein complexes. It was shown that the approximation of a protein electrostatic potential by the potential of a single charge equal to the total charge of a protein or by a dipole potential is insufficient to explain the observed ionic strength dependence of complex stabilities and formation rates.[5] The same applies to SOD (see below).

This brings about the need to calculate the potential of the actual charge distribution in a protein. Since in modeling the protein interactions the potential must be repeatedly recalculated, the method has to be fast. Below we briefly review the methods currently used for protein electrostatic potential calculations.

Those methods can be divided into (a) microscopic methods which consider the actual dipoles constituting the system (i.e. those of each protein atom and water molecule), and simulate their behavior in the field of protein charges, and (b) macroscopic methods in which the field is averaged over a small volume element and the water-protein system is considered as a continuous medium (or two continuous media), applying therefore the whole formalism of classical electrostatics to proteins.

The microscopic methods are very expensive and cannot be applied to the protein-protein interaction simulation.

In the macroscopic methods, the main equation of electrostatics

$$\text{div}\,\vec{D} = 4\pi\rho \tag{1}$$

is traditionally solved under the assumption that the electric induction $\vec{D}$ is related to field $\vec{E}$, by

$$\vec{D}(\vec{r}) = \epsilon(\vec{r})\vec{E}(\vec{r}), \tag{2}$$

where ϵ is the dielectric permittivity. It is constant for the monophasic homogeneous system. But in the case of the water-protein system, ϵ can be between 2-6 for points within the protein molecule interior and about 80 for points outside of it. Under these assumptions, Eq. 1 reduces to

$$\text{div}(\epsilon\,\text{grad}\,\phi) = -4\pi\rho. \tag{3}$$

Furthermore, when ϵ is constant, Eq. 3 becomes $\nabla^2\phi = -4\pi\rho$ which is solved in the Debye-Hückel theory. In the two-dielectric-constant model, Eq. 3 can be solved directly by a finite-difference iterative procedure[7] in which space is divided into cubical volume elements by a coordinate grid and new potential values at a certain grid point are computed from old potential values at neighboring points (6 in a rectangular grid) by Eq. 3 rewritten in the finite-difference formulation.

An alternative way, called the finite-element method,[6] is to rewrite Eq. 3 in terms of surface charges resulting from surface polarization and divide the water-protein interface into planar area elements. This yields a system of linear algebraic equations by solving which the element can be determined and the total field can be computed as a superposition of the field resulting from the initial, the so-called real charges, of the system and the induced charges.

In the case of a planar boundary, the induced charge density distribution can be represented by one additional charge per real charge. This gives rise to the method of images.[9] The spherical boundary also allows an analytical solution of Eq. 3 which is used in the Tanford-Kirkwood method and its various modifications (Ref. 10 and refs. therein). In the last three methods the boundary, whose shape is very complex (most commonly, the boundary is defined as the solvent-accessible surface of the protein), has to be approximated by planar or spherical fragments. This reduces the accuracy of the calculations.[8-10]

The common feature of all the models mentioned above is the dielectric permittivity defined as a coordinate function, as in Eq. 2. Its meaning is the dielectric response, i.e. the reorientation of the medium dipoles in the field, of a certain volume element, being determined only by dipoles contained within the element.

One factor these models do not take into account is the correlations in the dipole orientations due to non-electrostatic interactions that arrange them into structures characterized by a spatial range viz. secondary and tertiary structures in protein and the local structures in water (see Introduction section). Note that the range of these interactions is exactly that of the electrostatic interactions in proteins, i.e. 5-10 Å. The response of the volume element then is determined, generally speaking, by all the dipoles of the system, not only those within the element.

To account for these effects, the formula Eq. 2 must be replaced by a more general integral equation

$$\vec{D}(\vec{R}) = \int \epsilon(\vec{R},\vec{R}')\vec{E}(\vec{R}')d\vec{R}', \qquad (4)$$

where $\epsilon(\vec{R},\vec{R}')$ is the dielectric response function determined by the medium structure. Since Eq. 3 is known as the local approximation, the models using Eq. 4 are termed "non-local". This approach has led the authors of Refs. 12-14 to revise the basic concepts of electrolyte solutions properties. It was found that the solvation energy of small ions in solution is lower than predicted by the Born theory, the revised theory being in quantitative agreement with the experimental data.[12] The pairwise interaction energy at distances of about 1 nm is an order of magnitude greater than estimated by the Coulomb's law.[13] The apparent reduction of dielectric permittivity in water near surfaces that follows from the theory can account for the low effective permittivity of double layers at electrodes.[14] The reduction originates from the correlations in the medium: correlation makes the dipoles reorient in concert with their neighbors making it impossible for all the dipoles to orient themselves along the applied field.

Since the range of these "non-local" effects is several angstroms (i.e. exactly the range of protein electrostatic field spatial change), we investigated those effects in a protein.

THE MODEL

The solution of Eqs. 1-4 for a point charge in a homogeneous medium with spatial dispersion is

$$\phi(\vec{R}) = (Q/R) \cdot \{(1/\epsilon_1 - 1/\epsilon_s)\exp(-R/L) + 1/\epsilon_s)\}, \qquad (5)$$

where R is the distance from the charge, ϵ_1 is the short-range dielectric permittivity, ϵ_s is the long-range one, and L is the correlation length characterizing the spatial range of the correlation in the medium. It is shown on Fig. 1.

We have applied it to SOD without taking into account the difference in dielectric permittivities in water and protein, that is, using Eq. 5 with ϵ_s, ϵ_1 and L of water.

The electrostatic potential at a certain point was calculated as a superposition of potentials of the charges of the system. The latter was calculated from Eq. 5. This means that we use a macroscopic Coulomb's law with a particular kind of distance-dependent dielectric function. From the results of Ref. 18 where an empirical dielectric function was introduced that fits the experimental data on the ionization of side chains in proteins, it is evident that the function we use in our calculations is better than any other analytical dielectric function. The use of this dielectric function is also justified by the theoretical and experimental studies of thermodynamic properties of electrolyte solutions (Refs. 14-16 and refs. therein). The proposed model was tested on a common object of methodical studies in protein electrostatics, Cu,Zn-superoxide dismutase.

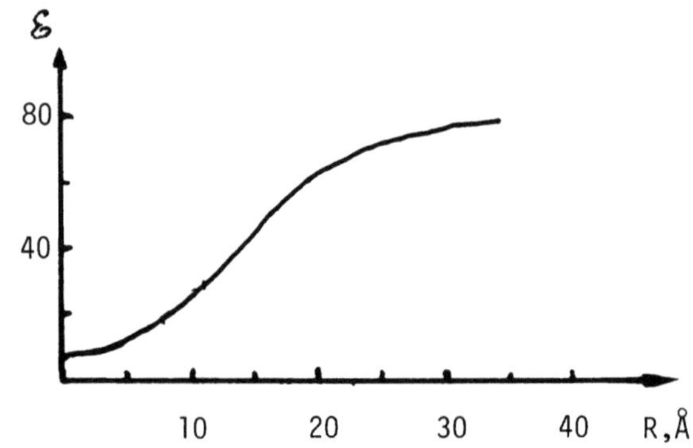

Fig. 1. The distance-dependent dielectric function (Eq. 5). The parameters are: ϵ_1 = 5, ϵ_s = 60, L = 5 Å.

THE PROTEIN

Cu,Zn-superoxide dismutase is a dimer consisting of two practically identical subunits spatially related by a two-fold symmetry axis. The area of the inter-subunit contact as about 15% of the whole surface of the subunits and consists mainly of hydrophobic amino acid residues, the enzyme being a dimer to preserve its structure in water.[15] The molecule has two active sites with a copper ion and a zinc atom in each.

The function of SOD is to convert the superoxide radical, O_2^-, into oxygen, O_2, and peroxide, O_2^{--}, by a cyclic process in which a superoxide molecule alternately oxidizes and reduces the copper ion of the enzyme:

$$O_2^- + Cu^+ \rightleftharpoons O_2^{--} + Cu^{++}$$

$$O_2^- + Cu^{++} \rightleftharpoons O_2 + Cu^+.$$

The subunits function independently of each other.[18] The key functioning feature of SOD is the electrostatic recognition between SOD and superoxide.[16] These react with a bimolecular rate constant of 10^9 $M^{-1}s^{-1}$, which is typical of diffusion-controlled reactions, i.e. those in which every collision of the two molecules leads to reaction. However, for a reaction to occur a superoxide radical must presumably collide with a copper atom, the exposed surface area of which is about 0.01% of that of the whole protein. The electrostatic forces are thought to guide the superoxide to its destination, the copper.[16] The forces are derived from electrostatic interaction between the superoxide bearing a negative charge and the charged residues of the protein. Those situated near the active site are mainly positively charged, as are the copper and zinc ions. The total charge of the protein, however, is -4, because of the large number of negative residues scattered all over the protein. Once the superoxide gets close enough to the active site it is sure to land on the copper ion. But how it gets to the active site is not yet clear.

The probability of finding a molecule at a certain point should naturally depend on (a) energy of the molecule at the point and (b) height of the barrier a molecule has to overcome to get to that point. Both are strongly dependent on electrostatics.

To calculate the reaction rate, the Brownian dynamics method was used[17,19] in which the motion of superoxide relative to SOD in the force field of the latter was simulated by numerically solving the Newton equation, complicated by a random term added to a displacement vector (whose length is proportional to the square root of the diffusion coefficient of superoxide accounting for Brownian motion of the superoxide molecule). The force field was taken to be purely electrostatic. It was found[17] that, if electrostatic effects are approximated by Coulomb's law, they cannot compensate for the reduction of reaction rate caused by small size of the active site. Moreover, the positive charge of the active site is not enough to compensate for the overall negative charge within this model. This is why in Ref. 7 the electrostatic potential was calculated by solving the Poisson equation so as to take into account the difference in dielectric constants of water and protein. In that work, various electrostatic models have been applied to the calculation of the electrostatic potential (force field can be obtained by differentiation) including:

1) Coulomb's law (Model 1):

$$\phi(\vec{r}) = \sum_i Q_i/DR_i, \tag{6}$$

where Q_i is the value of the ith charge (protein residue or metal), R_i is the distance from the ith charge to the point at which potential is calculated, and D is the dielectric constant of the medium, taken to be that of water (78.3 at 293 K).

2) Coulomb's law modified by the empirical distance-dependent dielectric function $D(R_i)$ (Model 2):

$$\phi(\vec{r}) = \sum_i Q_i/D(R_i)R_i, \tag{7}$$

where $D(R_i)$ was taken to be equal to R_i in angstroms.

3) Numerical solution of the Poisson-Boltzmann equation (Model 3):

$$\mathrm{div}(D(\vec{r})\mathrm{grad}\phi(\vec{r})) = -4\pi\rho(\vec{r}), \tag{8}$$

where D is now a function of the coordinate rather than that of the distance from a charge and ρ for the discrete charge distribution is

$$\rho(\vec{r}) = Q_i \delta(\vec{r} - \vec{R}_i). \quad (9)$$

Since dielectric constants of water (78) and protein (2-6 by estimates[7,16]) differ, the last method seems the most exact. Its application

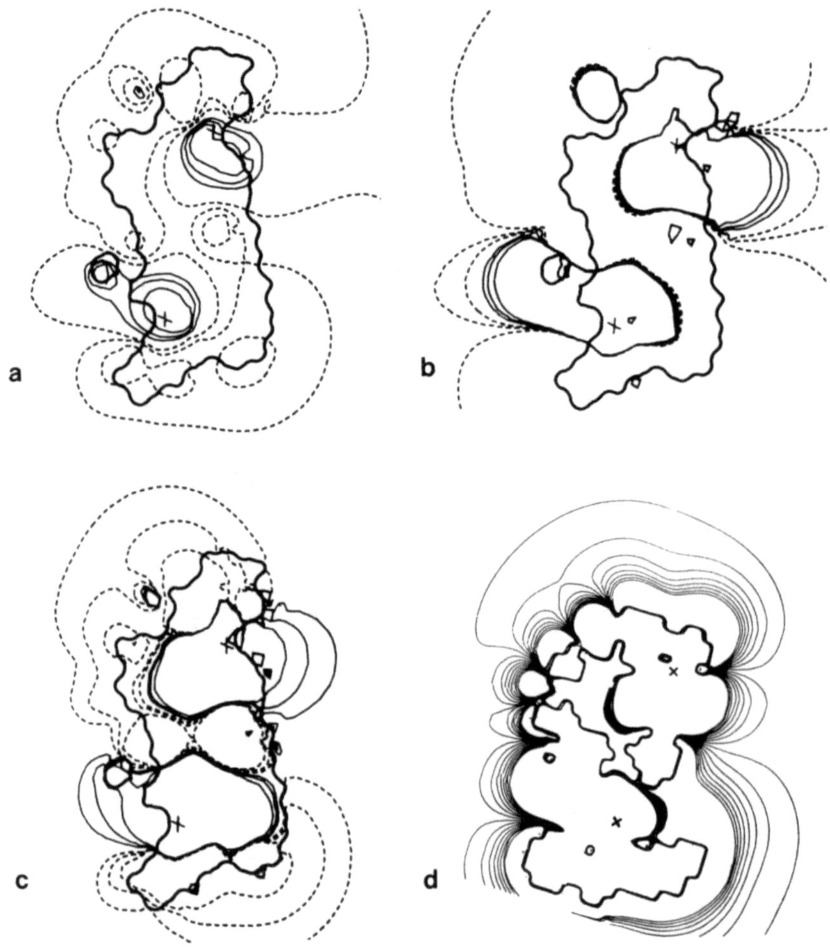

Fig. 2. (a) The electrostatic potential of SOD calculated by Eq. 6 (Coulomb's law; Model 1). Potential lines are shown for -0.5, 1, 2kT. Solid lines denote positive potential, and dashed lines denote negative potential. Bold line is the water-protein boundary. Copper ions are marked with X (similarly in (b), (c), and (d)).
(b) SOD potential for a linear distance-dependent dielectric function (Solution of Eq. 7; Model 2).
(c) SOD potential calculated by a finite-difference procedure (solution of Eq. 8; Model 3).
(d) SOD potential obtained with a "non-local" distance-dependent dielectric function (Model 4). Potential lines for 0, -0.5, 1, 1.5, 2 kT are shown.

to BD simulations[19] gave better agreement with the observed reaction rate value than the use of Coulomb's law.

RESULTS AND DISCUSSION

The electrostatic potential of SOD calculated as described under The Model section (hence termed Model 4) is presented in Fig. 2d as a chart of isopotential lines in the same planar slice through the active site as used in the Models 1-3 of the paper Ref. 7 (Figs. 2a-2c). It is to be noted that the potential distribution on all the figures has some features of symmetry due to the protein being a dimer. The authors of Ref. 7 say that the incomplete symmetry is because the protein itself is not symmetrical enough. We have found that the normal to the slice in which the potential is presented is not parallel to the protein's symmetry axis.

Large regions of negative potential are visible on all the figures. X-ray data analysis revealed that the negatively charged residues Asp-40,122,123, Glu-130,131 are situated in those regions. The shape of their external boundaries is practically the same in all the figures but their sizes differ, namely in the succession of models the sizes are 2>1=4>3. This means that the "non-local" Model 4 gives potential distribution at large distances from the protein similar to that of Model 1. The similarity is quite natural because at $R \gg L$, Eq. 5 reduces to Coulomb's law. In contrast, Model 2 does not show this similarity as it does not coincide with Coulomb's law in the limit $R \gg L$ which demonstrates its artificialness and physical unreliability.

Near the active sites, positive potential regions can be seen. That potential is caused by the charges of metal ions as well as the residues Lys-120,134,151, Arg-141 (upper right and lower left corners of the figures). The sizes of those regions are almost the same on Figs. 2b-2d but about three times lower on Fig. 2a. This shows that the "target" size for the superoxide increases in those models compared to that based on Coulomb's law. The same applies to potential values at the active sites marked "X" on the figures. This means that the superoxide binding to SOD is tighter in Models 2-4 than in Model 1. However, these factors are not sufficient for a proper enzyme-substrate interaction. Since the enzyme SOD has a negative total charge the negatively charged superoxide has to overcome a barrier on its way to the active site. The barrier height[7] can be estimated as the greatest potential value on the "easiest" way, i.e. at a saddle point of the potential in the planar slice. The values of potential at active sites, barrier heights, target areas and distances from saddle points to active sites for all models are listed in Table 1. They show that, in view of SOD-superoxide interaction rate, Model 4 gives no worse barrier height than Coulomb's law whereas the rest of parameters are improved and approach those of the Poisson equation solution.

It is now evident that inclusion into the model of "non-local" effects alone cannot fully explain the electrostatic enhancement of SOD-superoxide interaction rate. To do that, one has to consider the effects of dielectric boundary in the non-local formalism. The corresponding improvement of the approximate "non-local" model is now in progress.

At the center of every figure, the inter-subunit contact region is located. It is known[1] that, at the contact of two macromolecules that are physiological partners, there are complementary electrostatic interactions. This leads to charge neutralization which causes the appearance of the zero potential line in the region of inter-protein contact. Models 1 and 2 give no such line in that region. Instead, they have negative potential regions due to the charges of the residues Glu-21, Asp-50, and the C terminals of

Table 1

Model	1	2	3	4
Site potential, kT/e	1.8	4.5	2.7	2.6
Barrier height, kT/e	0.8	1.7	0.15	0.8
Barrier length, Å	5	22	12	16
Target area, $Å^2$	1,050	2,300	2,600	2,200

both subunits. However, if the low dielectric constant of the protein (Model 3) or dispersion of dielectric permeability (Model 4) is taken into account, the positive charges of the active sites (see above) seem to contribute more to the potential. This results in the positive potential regions being larger, relative to the negative, for Models 3 and 4 than for Models 1 and 2.

Furthermore, in Model 4, there are two patches of positive potential in the inter-subunit contact region, resulting probably from the residues Lys-3 and 9. In the other models, they presumably do not influence the potential because of the large number of neighboring negatively charged residues. The "non-local" effects cause deeper penetration of zero potential lines into the contact area. Thus, Models 3 and 4 give potential distributions in the contact region consistent with the general physical concept of protein-protein interaction.

REFERENCES

1. A. Ferscht, "Enzyme Structure and Mechanism," p. 197, Academic Press, Reading/San Francisco (1977).
2. H. R. Bechtold and R. Bosshard, J. Biol. Chem. 260:5192 (1985).
3. R. E. Dickerson and R. Timkovich, The Enzymes 11:397 (1975).
4. R. I. Gilmanshin and P. I. Lazarev, J. Molecular Electronics (in press).
5. E. Margoliash and H. R. Bosshard, Trends in Biochem. Sci. 9:216 (1983).
6. M. R. Mauk and A. G. Mauk, Biochemistry 21:4730 (1982).
7. I. Klapper, R. Hagstrom, R. Fine, K. Sharp, and B. Honig, Proteins 1:47 (1986).
8. R. J. Zauhar and R. S. Morgan, J. Biol. Chem. 186:815 (1985).
9. H. L. Friedman, Mol. Phys. 29:1533 (1975).
10. M. Gilson, A. Rashin, R. Fine, and B. Honig, J. Mol. Biol. 183:503 (1985).
11. A. A. Kornyshev, R. Dogonadze, and J. Ulstrup, In: "The Chemical Physics of Solvation," Pt. A, pp. 70-112, Elsevier, Amsterdam (1985).
12. A. I. Rubinstein, Elektrokhimia (Russ.) 22:210 (1985).
13. L. I. Krishtalik, "Electrode Reactions: Mechanism of the Elementary Act," pp. 80-86, Nauka, Moscow (1979).
14. A. A. Kornyshev, M. A. Vorotyntsev, H. Nielsen, and J. Ulstrup, J. Chem. Soc. Faraday Trans. Pt. 2, 78:217 (1982).
15. A. Cudd and I. Fridovich, J. Biol. Chem. 257:11443 (1982).
16. E. D. Getzoff, J. A. Tainer, P. K. Weiner, P. A. Kollman, J. S. Richardson and D. S. Richardson, Nature (London) 206:287 (1983).
17. S. A. Allison, G. Ganti, and J. A. McCammon, Biopolymers 24:1323 (1985).
18. A. Warshel and M. Levitt, J. Mol. Biol. 103:227 (1976).
19. S. A. Allison, R. J. Bacquet, and J. A. McCammon, Biopolymers 27:251 (1986).

AN ELECTROCHEMICAL APPROACH TO THE DESIGN OF MEMBRANE-BASED MOLECULAR OPTOELECTRONIC DEVICES

Felix T. Hong

Department of Physiology
Wayne State University
Detroit, Michigan 48201 U.S.A.

INTRODUCTION

Many prototype molecular optoelectronic devices utilize a membrane or a thin film as the substrate in which photoactive elements are embedded.[1,2] This type of design strategy is implemented in nature as photosynthetic and visual membranes (for reviews, see Refs. 3-6). Thus, insights into the operating mechanisms in these structures via "reverse engineering" may be useful in the development of molecular devices. Not only can the associated biopigments be exploited as bioelectronic materials, but a better understanding of the fundamental design principle of these structures can also inspire new strategies to build novel devices from synthetic organic materials.

The photosynthetic membrane is a light energy transducer, which transforms absorbed photon energy into electrochemical energy whereas the visual membrane is a light signal transducer, which detects the level of photon energy for the purpose of information processing. Despite the apparent structural and functional differences, photosynthetic membranes in higher plants and visual membranes in animals do have something in common. Their active centers comprise membrane-bound pigment proteins asymmetrically oriented with respect to the membrane and their common primary event is light-induced charge separation. This latter action leads to a vectorial charge movement in the direction perpendicular to the plane of the membrane and is observable as a photovoltage or photocurrent.

In the case of the photosynthetic membrane, illumination with continuous light leads to the generation of a sustained photovoltage by virtue of a net proton transport across the membrane, i.e., it is a photocell (for example, see Ref. 7). In the case of the visual membrane, only a transient photovoltage is detectable. There is neither sustained generation of photovoltage nor a net proton transport across the membrane.[8,9] Instead, the absorbed photon energy is used to trigger a biochemical process known as the cyclic GMP cascade[5,6] and to unleash previously stored energy. In so doing, the original photon energy is amplified by a factor of 100,000. Thus, the visual membrane is a light-activated switch, similar to a phototransistor.

The conceptual link between the photosynthetic membrane of higher plants and the visual membrane of animals is provided by the purple membrane system of *Halobacterium halobium*.[10] The purple membrane contains only a single protein component, bacteriorhodopsin, which resembles the visual pigment rhodopsin; both pigments contain retinal as the chromophore. However, the purple membrane functions as a photosynthetic membrane; absorbed photon energy is utilized to translocate protons from the intracellular space to the extracellular space so as to form an electrochemical gradient of protons. Unlike the more elaborate photosynthetic membranes in purple phototrophic bacteria and in higher plants, proton translocation is achieved without the participation of electron transfer reactions. In this sense, the purple membrane can be regarded as the simplest photosynthetic membrane which fulfills the minimum requirements of a photon energy transducer.

Of particular relevance to our inquiry is a class of bioelectric signals induced by flash light in both photosynthetic and visual membranes. These signals are known collectively as fast photovoltages, fast photoelectric effects, or displacement photocurrents. The best-known fast photoelectric signal is the early receptor potential (ERP), which was discovered in monkey retina about a quarter century ago.[11] A similar fast photosignal was subsequently found in reconstituted bacteriorhodopsin membranes.[12,13] The similarity of the two signals is so striking that similar terminology was used to describe the individual components.[8,14,15] Both signals have a fast component with unresolved rise time and with little temperature sensitivity (the R1 component in the visual membrane and the B1 component in the purple membrane) and a slower component which has an opposite polarity and a significant temperature sensitivity (the R2 component and the B2 component).

In this article, a reductionist's strategy will be adhered to in our analysis using a combined electrochemical and electrophysiological approach. The advantage of this approach is its generality; we believe that our analysis is also applicable to the development of prototype molecular optoelectronic devices using either biomaterials or synthetic organic materials.

A SIMPLE MECHANISTIC MODEL OF LIGHT-INDUCED PROTON TRANSPORT

Our analysis of the fast photoelectric effect will be based on a simple model which seems to best summarize the essential features of a photon energy transducer, as exemplified by the purple membrane and the thylakoid membrane of higher plants. We shall treat electron movement and proton movement equivalently, despite important differences between an electron and a proton.

For light-induced proton translocation across the purple membrane, we shall consider three distinct steps: (a) the binding of protons to the intracellular domain of bacteriorhodopsin, (b) the light-assisted transmembrane proton movement from the intracellular domain to the extracellular domain of bacteriorhodopsin, and (c) the release (dissociation) of protons from the extracellular domain of bacteriorhodopsin. In brief, we are considering two interfacial proton transfer reactions coupled by a light-assisted transmembrane proton movement.

On the basis of our analysis, it may be concluded that the rate-limiting step of light-driven transmembrane proton translocation is the transmembrane step itself rather than the two interfacial steps.[16] Therefore, on the time scale of the fast photoelectric relaxation, the two interfacial proton transfer processes can be considered chemically decoupled and are therefore independent of each other. This leads to the concept of <u>local reaction conditions</u>:[16] each interfacial proton transfer process can be regarded as a heterogeneous bimolecular reaction, which is regulated by the aqueous composition of the adjacent aqueous phase but not by that of the aqueous phase on the opposite side of the membrane. Such reactions are expected to be governed also by the law of mass action. Thus, the interfacial proton uptake at the cytosol domain of bacteriorhodopsin is expected to be sensitive to intracellular pH but not to extracellular pH, and vice versa for the interfacial proton release at the extracellular domain. If one can link the displacement photocurrent component to one of these processes, the interpretation of data can be vastly simplified.

MOLECULAR MECHANISMS OF DISPLACEMENT PHOTOCURRENT GENERATION

The primary event in both the purple membrane and the visual membrane is a *cis-trans* isomerization of the chromophore, retinal. In association with this step, considerable photon energy is stored for subsequent use in driving the ensuing thermal reactions: ~16 kcal/mol for bacteriorhodopsin[17] and ~35 kcal/mol for rhodopsin.[18] As is inevitable in many chemical reactions, a considerable fraction of these separated charges subsequently recombine. Because of the asymmetric spatial orientation of bacteriorhodopsin and rhodopsin in the membrane, the processes of charge separation and recombination present a transient array of electric dipoles. This charge separation and

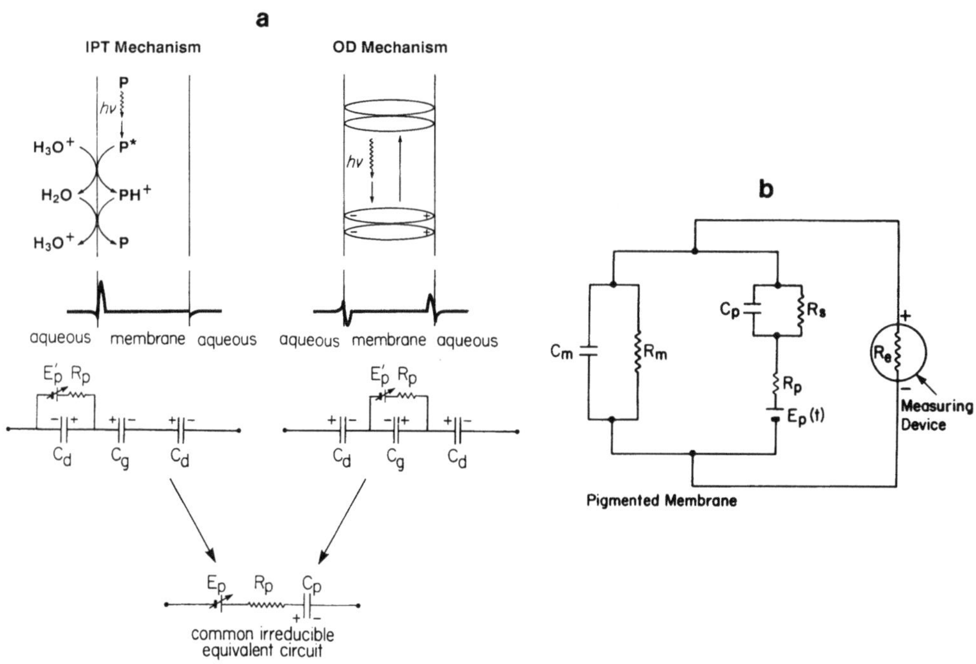

Fig. 1. Two types of light-induced rapid charge separation (a) and the associated equivalent circuits (a,b). In the oriented dipole (OD) mechanism, the light-induced internal charge separation inside the oriented pigment molecule generates a transient array of electric dipoles, which vanish upon charge recombination. In the interfacial charge (proton) transfer (IPT) mechanism, the charge separation occurs at the membrane-solution interface upon binding of a proton. The charge recombination is achieved by the reverse proton transfer at the same interface. E_p and E_p' are the photoemfs. The chemical capacitance (C_p) is a composite of the dielectric capacitance (C_g) and two diffuse double layer capacitances (C_d). R_p is the internal resistance of the photogenerator. R_s is the transmembrane resistance on the proton path. R_m and C_m are the ordinary membrane resistance and capacitance, respectively. R_e is the access impedance. (Modified from Ref. 21)

recombination is thus equivalent to charging and discharging of a capacitor and is a mechanism of displacement photocurrent generation (oriented dipole (OD) mechanism; Fig. 1a). This mechanism, commonly referred to as the dipole mechanism, was accepted as the molecular mechanism of the ERP in the literature (for reviews, see Refs. 19-21). But is it the only possible mechanism for a displacement photocurrent?

Our previous model system study with a membrane-bound magnesium porphyrin indicates that a light-induced interfacial electron transfer can also generate a displacement photocurrent that possesses all the major characteristics of the ERP and yet there is no molecular basis for the formation of an array of oriented electric dipoles inside the membrane.[22] Rather, our analysis shows that it is possible to generate a displacement photocurrent by interfacial electron transfer. In the case of the purple membrane, the obligatory process of intracellular proton uptake and extracellular proton release would be natural candidates for the mechanism of displacement photocurrent generation. In fact, for each proton bound to bacteriorhodopsin at its intracellular domain, a counter ion must be left behind in the diffuse electrical double layer. This constitutes another kind of charge separation, which, together with the subsequent reverse (back) reaction, generates a capacitative current (interfacial proton transfer (IPT) mechanism; Fig. 1a).

The link between the two above-mentioned molecular mechanisms and the measured photoelectric signals is established by an equivalent circuit which can be derived from the analysis of electrochemical models of these two mechanisms,[21] using the diffuse double layer theory of Gouy-Chapman. We found that the two molecular mechanisms correspond to two slightly different equivalent circuits shown in Fig. 1a. Basically, the photoreactions generate current to charge the three fundamental capacitors usually associated with a membrane: one dielectric capacitance and two double layer capacitances. There exists one distinction between the two models. The photogenerator in an oriented dipole process resides inside the membrane, whereas that in an interfacial proton transfer process resides at one membrane-solution interface. However, the two distinct equivalent circuits are equivalent to the same irreducible equivalent circuit (Fig. 1a), in which the photogenerator is connected in series with a capacitor, C_p. That is, the photocurrent is ac-coupled via C_p. It has been shown that this capacitor, C_p, is physically distinct from the ordinary membrane capacitor, C_m. Therefore, it was named chemical capacitance.[23] We further emphasize that the ordinary membrane capacitance is connected in parallel with the photogenerator (Fig. 1b). This is because the current source utilized to charge a membrane capacitor must always reside outside of the membrane. Thus, an equivalent circuit that describes the time course of a displacement photocurrent originating from a single first order or pseudo-first order process must include both the ordinary membrane capacitance and the chemical capacitance.

MODULATION OF PHOTOSIGNALS BY MANIPULATING ELECTRICAL PARAMETERS

The presence of the chemical capacitance has a profound effect on the interpretation of the displacement photocurrent. This effect is most likely at the root of discrepancies in the relaxation data of fast photosignals of reconstituted purple membranes reported by various laboratories (discussed in Refs. 24,25). We have previously demonstrated that the apparent waveforms of the B1 and B2 components vary according to how they are measured.[15] Under open-circuit conditions, the amplitude of B2 appears larger than that of B1. As the measurement condition approaches short-circuit, this amplitude ratio eventually reverses and the B1 amplitude becomes larger than the B2 amplitude and the decay times also decrease. That is, the apparent relaxation time course is strongly dependent upon the access impedance, which is effectively zero for short-circuit conditions or infinite for open-circuit conditions.

This waveform dependence on the access impedance can be shown to be a direct consequence of the interaction between the chemical capacitance and the ordinary membrane capacitance.[26]

Our analysis of the equivalent circuit indicates that undistorted photoelectric relaxation reflecting the molecular process can be directly observed only if a true short-circuit condition is fulfilled. Otherwise, the photocurrent splits into two fractions (Fig. 1b): one to reach the measuring device and the other to charge the ordinary membrane capacitance. Thus, the relaxation of a displacement photocurrent is a result of the interaction between the discharging process of two capacitances; each first order or pseudo-first order process is reflected in two apparent exponential decays. The true relaxation time is therefore not directly observed but can be recovered by deconvolution using the equivalent circuit analysis, provided that the access impedance of the measurement is also included in the analysis. In the extreme case of open-circuit measurements, all of the photocurrent is diverted to charge C_m and the resultant relaxation largely reflects the discharge of the membrane RC (with the relaxation time constant equal to R_mC_m).

Figure 2a shows the time courses of the B1 component of the ERP-like bacteriorhodopsin signal measured with two different access impedances. The variation of the apparent decays are completely predictable by the equivalent circuit.[24,25] In fact, as the access impedance varies, the deconvoluted relaxation time constant remains the same, despite the differences in apparent relaxation. Our analysis thus shows that, given a characteristic photochemical relaxation, the time course of the photosignal can be modulated by the access impedance, i.e., the load resistance (impedance) relative to the source impedance.

In the methods of membrane reconstitution used in our study (known as the Trissl-Montal method; Ref. 12), oriented purple membrane fragments are attached to a thin Teflon film instead of being incorporated into a lipid bilayer. Therefore, consistent interpretation of relaxation data in the context of chemical capacitance can be obtained only if the Teflon film is

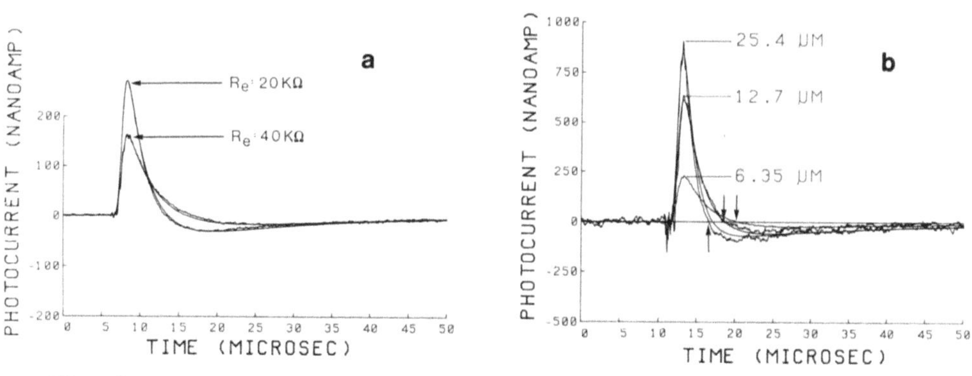

Fig. 2. Variations of the photosignal waveforms as a result of changing the access impedance (a) or changing the thickness of the Teflon films (b). The reconstitution method which produces an oriented purple membrane film with a pure B1 component is described in Ref. 24. The light source was a 590-nm dye laser flash (flash duration about 1 μs). In a, two different access impedances (20 KΩ and 40 KΩ) were used to record signals from the same film (thickness 6.35 μm). In b, three different Teflon film thicknesses (6.35 μm, 12.6 μm and 25.4 μm) were used and the access impedance was 40 KΩ. (Reproduced from Refs. 24 and 27)

considered as part of the signal generating device. In fact, the apparent relaxation of photosignals elicited in these films can be modulated by the thickness of the Teflon films, as seen in Fig. 2b. Although the apparent relaxation is retarded by the thicker Teflon support, the true relaxation time constant obtained by deconvolution remains unchanged.[27] The effect is again quantitatively predictable from the equivalent circuit analysis.

Thus, the chemically inert supporting structure can be modified to custom design the relaxation of the displacement photocurrent. This demonstrates that a molecular electronic device is governed by the same rules encountered by a conventional microelectronics designer.

MODULATION OF PHOTOSIGNALS BY MANIPULATING CHEMICAL REACTIONS

Our analysis of the two prototype generation mechanisms of a displacement photocurrent indicates that a displacement photocurrent is associated with each light-induced interfacial charge transfer reaction in a membrane. For a transport protein such as bacteriorhodopsin, the binding of charges (protons) at one interface and the release of charges at the opposite surface are necessary steps. Therefore, a third component reflecting proton release from the extracellular surface of the purple membrane is expected. Such a hypothetical component (B2′) would have been abolished in a Trissl-Montal membrane because the Teflon film prevents bacteriorhodopsin its access to the extracellular aqueous phase. This component may be rendered observable in a genuine lipid bilayer in which bacteriorhodopsin is fully incorporated and is fully exposed to both aqueous phases.[28] However, the B2 and B2′ components would not be immediately distinguishable because both would have the same electric polarity. The two components may be distinguished on the basis of the concept of local reaction conditions: the B2 component is sensitive to intracellular pH whereas the B2′ component may be sensitive to extracellular

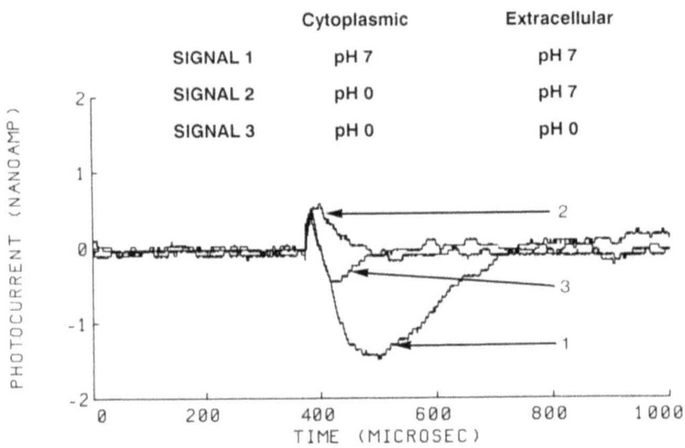

Fig. 3. Effect of acidifying either of the two aqueous phases on the fast photoelectric signal in a purple-membrane-impregnated collodion film, prepared according to the method of Drachev *et al.*[29] The light source was the same as in Fig. 2. The access impedance was 40 KΩ. The signal 1 (B1 + B2) was first obtained with both aqueous phases maintained at pH 7. Acidification of the cytoplasmic side suppressed the B2 component but left the B1 component intact (signal 2). Subsequent acidification of the extracellular side restored a negative component which is identified as the hypothetical B2′ component (signal 3). (Reproduced from Ref. 28).

pH. Such a strategy was implemented in an experiment shown in Fig. 3. If neutral pH is maintained in both aqueous phases, a photosignal can be observed to be composed of both positive and negative peaks, resembling the B1 and the B2 components as observed in a Trissl-Montal film. The negative peak can be reversibly suppressed by acidifying the intracellular aqueous phase. However, subsequent acidification of the extracellular aqueous phase reveals a new negative peak. A simple explanation can be obtained if we regard the new negative peak as the B2′ component, which has a pH dependence opposite to that of the B2 component. If so, the membrane presents an interesting configuration in designing optoelectronic devices. The two surfaces of the device are sensing local pH in different fashions. The resultant photosignal is the algebraic sum of the two components.

The above analysis indicates that the diversity of chemical reactions can be exploited to modulate the signal waveform. The possibility of manipulating chemical reactions *per se* as a means to modulate electric signals is a distinct advantage in molecular electronics.

SOME CONSIDERATION ABOUT PHOTOCELL DESIGNS

Comparison of the purple membrane and the thylakoid membrane reveals the essence of Nature's design of a photocell. The relatively impermeable membrane affords a supporting structure to implement light-inducing vectorial charge transfer and to store the converted energy as a transmembrane electrochemical gradient of protons. This can be accomplished by an array of components spatially arranged to move electrons across the membrane via coupled redox reactions. But it can also be accomplished by a single protein component via the not fully understood translocation through a transmembrane proton passage (a "proton wire" in bacteriorhodopsin). In the recently elucidated structure of the bacterial reaction center,[4] the electron transport chain components possess optimally matched redox levels and optimal spatial localization. The end result is an enhanced forward reaction and a diminished reverse reaction.[30] An ideal photocell is one with a very efficient forward charge transfer and with virtually no reverse transfer, i.e., a rectified proton photoconduction "channel". Does this ideal construction exist in the purple membrane? Is the light-pumped step of bacteriorhodopsin photocycle devoid of a reverse reaction? Our analysis of the displacement photocurrent indicates that the bacteriorhodopsin photocycle is no exception to the plague of reverse reactions; all three components of the bacteriorhodopsin fast photosignal exhibit reverse reactions (charge recombination). Furthermore, the rate-determining step of proton pumping is one of the intramembrane steps rather than one of the two interfacial steps. Determination of the photoconductance also indicates that the interfacial resistance is small compared with the total transmembrane resistance (unpublished observation, B. Fuller, T. L. Okajima and F. T. Hong). Therefore, the translocation of protons from the intracellular binding site to the extracellular release site is most crucial in determining the efficiency of the photocell.

VISUAL PHOTORECEPTOR MEMBRANE AS A PHOTOTRANSISTOR

A pertinent question to ask is whether the ERP plays any physiological role. Because of its small amplitude and its ubiquitous presence in many pigment-containing membranes, the ERP has been regarded as an epiphenomenon - an evolutionary vestige playing no role in the visual phototransduction. We believe this conclusion may be premature. The ubiquity of ERP-like signals is a consequence of the fact that many pigments respond to light by a rapid charge separation. We have previously suggested that the R2 component may be associated with a surface potential.[21] Because the ERP is generated by a current source which is either inside the membrane (R1) or at the membrane-solution interface (R2), the electric field so generated tends to be highly localized and therefore the externally measured photosignal tends to appear small. This large surface potential associated with the R2 component was experimentally observed by Cafiso and Hubbell.[31] In the light of this additional information, the role of the ERP must be reevaluated.

A mechanistic model of the visual phototransduction may be proposed on the basis of the following observations. (a) The cyclic GMP cascade is initiated upon binding of a peripheral protein known as transducin (or G-protein) to the cytoplasmic domain of photoactivated rhodopsin (metarhodopsin II), resulting in the activation of transducin.[5] (b) The rise phase of the R2 component of the ERP coincides with the formation of metarhodopsin II.[19] (c) The formation of metarhodopsin II from its precursor metarhodopsin I involves the net uptake of one proton from the cytoplasmic surface of the photoreceptor disc membrane.[9] This reaction is expected to generate a positive surface potential. (d) The R2 component may be generated by an interfacial proton uptake, following our analysis. (e) The polarity of the R2 component as determined by use of a penetrating anion is consistent with the proton uptake from the cytoplasmic surface.[32] (f) Photoactivated rhodopsin is extensively photophosphorylated on its cytoplasmic domain following transducin activation.[33] This latter reaction is expected to generate a large negative surface potential at the cytoplasmic side. Unless one is willing to dismiss all these as mere coincidences, a visual transduction mechanism can be postulated as follows.[34] The positive surface potential associated with proton binding serves as the "turn on" switch and the negative surface potential associated with the photophosphorylation of rhodopsin serves as the "turn off" switch. An intriguing observation was reported by Ostrovsky's group:[9] there is no proton release at the intradiscal surface and therefore no transmembrane proton transport. This latter observation is consistent with the functional differences between bacteriorhodopsin as a photon energy converter and rhodopsin as a photon sensor.

CONCLUSION

The main problem addressed in this article is the mechanistic aspect of light-induced rapid charge separation - a common primary process in both photosynthetic and visual membranes. Our combined electrochemical and electrophysiological analysis exploits the striking similarity of the fast photoelectric signals exhibited by the photoreceptor membrane and the purple membrane of *Halobacterium halobium*. This formulation is based on a simple mechanistic model, and the associated electrochemical concepts provide simple explanations for a number of otherwise inexplicable phenomena. Our investigation of the interaction between photoactive elements and chemically inert supporting structures in reconstituted bacteriorhodopsin membranes defines the factors that control the waveform of the photosignals: (a) the external load resistance influences the apparent relaxation of the photosignal; (b) the chemically inert supporting structure modulates the photosignal by presenting a parallel path for the photocurrent; (c) the ability to regulate chemical reactions at both surfaces of the membrane permits additional control of the photosignal. We believe that the electrochemical analysis presented in this article is sufficiently general to be applicable to more complex systems such as the thylakoid membrane in higher plants.[26] With appropriate modifications, the analysis may even be applied to molecular devices constructed as Langmuir-Blodgett films.

Our parallel treatment of the photoreceptor membrane and the purple membrane in terms of the corresponding photosignal components offers an interesting comparison. For the purple membrane which transport protons, the proton uptake at the cytoplasmic surface and the proton release at the extracellular surface are obligatory molecular processes. For the visual membrane, there is no need to efficiently convert photon energy into a proton gradient. The absence of the corresponding proton release at the intradiscal surface (and therefore the absence of the corresponding R2′ component) suggests that the proton uptake process in the visual membrane and the associated R2 component of the ERP may exist for an entirely different reason, namely, to serve as a photoactivated switch. Although the possibility of the ERP as an epiphenomenon cannot presently be ruled out, the proposed electrostatic regulatory mechanism can be experimentally tested and may be used in the design of molecular electronics regardless of what actually is operative in the photoreceptor membrane.

This analysis demonstrates that the electrochemical analysis is a useful approach in "reverse engineering" of naturally occurring molecular optoelectronic devices. Our parallel treatment of the photosynthetic membrane and the visual membrane provides a unified perspective for the design of a photocell and a phototransistor. We further suspect that various photobiological membranes may be the variations of a common design scheme perfected by Nature over billions of years of evolution.

ACKNOWLEDGMENT

This work was supported by an ONR contract (N-00014-87-K-0047).

REFERENCES

1. M. Sugi, Langmuir-Blodgett films for molecular electronics - recent trends in Japan, in: "Molecular Electronic Devices," F. L. Carter, R. E. Siatkowski, and H. Wohltjen, eds., pp. 441-463, North-Holland, Amsterdam (1988).
2. H. T. Tien, Ultrathin bilayer film: an experimental approach to biomolecular electronic devices, in: "Molecular Electronic Devices," F. L. Carter, R. E. Siatkowski, and H. Wohltjen, eds., pp. 209-226, North-Holland, Amsterdam (1988).
3. J. Barber, Photosynthetic electron transport in relation to thylakoid membrane composition and organization, Plant Cell Environ. 6:311-322 (1983).
4. J. Deisenhofer, H. Michel, and R. Huber, The structural basis of photosynthetic light reactions in bacteria, Trends Biochem. Sci. 10:243-248 (1985).
5. L. Stryer, The molecules of visual excitation, Sci. Am. 257(1) 42-50 (1987).
6. H. Shichi, Visual phototransduction: biochemical aspects, this volume.
7. L. A. Drachev, A. Yu. Semenov, V. P. Skulachev, I. A. Smirnova, S. K. Chamorovsky, A. A. Kononenko, A. B. Rubin, and N. Ya. Uspenskaya, Fast stages of photoelectric processes in biological membranes, III. bacterial photosynthetic redox system, Eur. J. Biochem. 117, 483-489 (1981).
8. L. A. Drachev, G. R. Kalamkarov, A. D. Kaulen, M. A. Ostrovsky, and V. P. Skulachev, Fast stages of photoelectric processes in biological membranes, II. visual rhodopsin, Eur. J. Biochem. 117:471-481 (1981).
9. T. F. Shevchenko, G. R. Kalamkarov, and M. A. Ostrovsky, The lack of H^+ transfer across the photoreceptor membrane during rhodopsin photolysis, Sensory Systems (USSR Acad. Sci.) 1:117-126 (1987).
10. W. Stoeckenius, Light energy transducing and signal transducing rhodopsins of *Halobacteria*, this volume.
11. K. T. Brown and M. Murakami, A new receptor potential of the monkey retina with no detectable latency, Nature (London) 201:626-628 (1964).
12. H.-W. Trissl and M. Montal, Electrical demonstration of rapid light-induced conformational changes in bacteriorhodopsin, Nature (London) 266:655-657 (1977).

13. S.-B. Hwang, J. I. Korenbrot, and W. Stoeckenius, Transient photovoltages in purple membrane multilayers: charge displacement in bacteriorhodopsin and its photointermediates, Biochim. Biophys. Acta 509:300-317 (1978).
14. L. A. Drachev, A. D. Kaulen, L. V. Khitrina, and V. P. Skulachev, Fast stages of photoelectric processes in biological membranes, I. bacteriorhodopsin, Eur. J. Biochem. 117:461-470 (1981).
15. F. T. Hong and M. Montal, Bacteriorhodopsin in model membranes: a new component of the displacement photocurrent in the microsecond time scale, Biophys. J. 25:465-472 (1979).
16. F. T. Hong, Effect of local conditions on heterogeneous reactions in the bacteriorhodopsin membrane: an electrochemical view, J. Electrochem. Soc. 134:3044-3052 (1987).
17. R. R. Birge and T. M. Cooper, Energy storage in the primary step of the photocycle of bacteriorhodopsin, Biophys. J. 42:61-69 (1983).
18. A. Cooper, Energy uptake in the first step of visual excitation, Nature (London) 282:531-533 (1979).
19. R. A. Cone and W. L. Pak, The early receptor potential, in: "Handbook of Sensory Physiology, Vol. I. Principles of Receptor Physiology," W. R. Loewenstein, ed., pp. 345-365, Springer-Verlag, Berlin (1971).
20. S. E. Ostroy, Rhodopsin and the visual process, Biochim. Biophys. Acta 463:91-125 (1977).
21. F. T. Hong, Mechanisms of generation of the early receptor potential revisited, Bioelectrochem. Bioenerg. 5:425-455 (1978).
22. F. T. Hong, Charge transfer across pigmented bilayer lipid membrane and its interfaces, Photochem. Photobiol. 24:155-189 (1976).
23. F. T. Hong and D. Mauzerall, Interfacial photoreactions and chemical capacitance in lipid bilayers, Proc. Natl. Acad. Sci. USA 71:1564-1568 (1974).
24. T. L. Okajima and F. T. Hong, Kinetic analysis of displacement photocurrents elicited in two types of bacteriorhodopsin model membranes, Biophys. J. 50:901-921 (1986).
25. F. T. Hong and T. L. Okajima, Electrical double layers in pigment-containing biomembranes, in: "Electrical Double Layers in Biology," M. Blank, ed., pp. 129-147, Plenum Press, New York (1986).
26. F. T. Hong, Displacement photocurrents in pigment-containing biomembranes: artificial and natural systems, ACS Adv. Chem. Ser. 188:211-237 (1980).
27. F. T. Hong, The bacteriorhodopsin model membrane system as a prototype molecular computing element, BioSystems 19:223-236 (1986).
28. F. T. Hong and T. L. Okajima, Rapid light-induced charge displacements in bacteriorhodopsin membranes: an electrochemical and electrophysiological study, in: "Biophysical Studies of Retinal Proteins," T. G. Ebrey, H. Frauenfelder, B. Honig, and K. Nakanishi, eds., pp. 188-198, University of Illinois Press, Urbana, IL (1987).
29. L. A. Drachev, A. D. Kaulen, and V. P. Skulachev, Time resolution of the intermediate steps in the bacteriorhodopsin-linked electrogenesis, FEBS Lett. 87:161-167 (1978).
30. H. Kuhn, Forrest L. Carter Lecture: organized monolayers - building blocks in constructing supramolecular devices, this volume.
31. D. S. Cafiso and W. L. Hubbell, Light-induced interfacial potentials in photoreceptor membranes, Biophys. J. 30:243-263 (1980).
32. V. I. Bolshakov, G. R. Kalamkarov, and M. A. Ostrovsky, Photoinduced generation of potential on disc membrane of photoreceptor cell, Dokl. Akad. Nauk USSR 240(5):1241-1244 (1978).
33. U. Wilden and H. Kühn, Light-dependent phosphorylation of rhodopsin: number of phosphorylation sites, Biochemistry 21:3014-3022 (1982).
34. F. T. Hong, Relevance of light-induced charge displacements in molecular electronics: design principles at the supramolecular level, J. Molec. Electron., in press.

BIOMOLECULAR ELECTRONICS: STRUCTURE ↔ FUNCTION RELATIONSHIP

L. Powers

Center for Bio-catalysis Science and Technology
Utah State University
Logan, UT 84322-4630

INTRODUCTION

Biological microstructures perform a variety of electronic functions, some of which are unparalleled by current solid state technology. In addition, the size and energy requirements for these bioelectronic devices are substantially less and beyond the capability of current solid state technology. Two divergent and complementary approaches have emerged to accomplish the design of molecular electronic devices: the first is to essentially duplicate the powerful information processes of biological systems, while the second is to duplicate existing electronic functions using molecular components. Although the problems integral to both approaches are significant, they both are predicated on biological mechanisms and rely on the important relationship of structure ↔ function ↔ energy for molecular design.

Metallo-proteins and -enzymes perform many of the primary electronic functions associated with life processes: electron transfer, voltage sensitive gating, proton translocation, and charge separation. These molecules alter or direct the flow of electrons or ions by electron tunneling, voltage sensitive conformation changes or ligand exchanges which store energy in chemical bonds, and formation of transient intermediates.[1] The active sites contain Mn, Fe, Co, Ni, Cu or Zn ions surrounded by an environment capable of delocalizing charge and are sometimes a multi-nuclear combination when a stoichiometric multiple of electrons are required. These molecules or complexes are embedded in a highly resistive, low capacity lipid bilayer matrix which separates aqueous regions containing ions and a host of other small molecules.

Although many of these proteins and enzymes are difficult to study and are not well understood, some details of the structure ↔ function relationship have been explored by a host of physical techniques, especially x-ray absorption spectroscopy. These are described below and expose a few of the basic design principles.

DISCUSSION

Myoglobin, the oxygen transport protein in muscle, has often been called "the biophysicist's hydrogen atom." It is a relatively small protein which has the active site Fe atom incorporated in a porphyrin heme group, a large planar structure which delocalizes the charge and provides four pyrrole nitrogen atoms (N_p) as ligands. Like most Fe-heme proteins and enzymes, the fifth axial ligand is N from a histidine amino acid (N_ε) residue which provides the only linkage to the protein ultrastructure. Under physiological conditions, the sixth axial ligand or proximal position is either vacant or occupied by an O of O_2. The crystal structures of the deoxy-,[2] oxy-,[3] and other liganded[4,5] states have been determined and the biochemical properties well characterized. In addition, model compounds have been synthesized which mimic many of these properties.[6,7,8] Thus, myoglobin and these model compounds provide an excellent opportunity to examine some basic structure ↔ function aspects of the large class of biological molecules containing Fe-heme active sites.

The chemical nature of the sixth axial ligand significantly affects the biochemical and spectroscopic properties of the active site. Weak field or electron donating ligands such as F and formate produce high spin derivatives in met-myoglobin, while strong field or electron withdrawing ligands such as azide and CN produce low spin derivatives. The resulting structure changes in the Fe-heme active site[9] are shown in Fig. 1. Ligands which direct electrons toward the heme π-electron cloud cause an increased Fe-N_p average distance and Fe-N_ε distance when compared to those ligands which withdraw electrons. In addition, the Fe atom is out of the porphyrin plane for the expanded structures. Thus the entire site expands or contracts to move charge toward or away from the heme π-electron cloud.

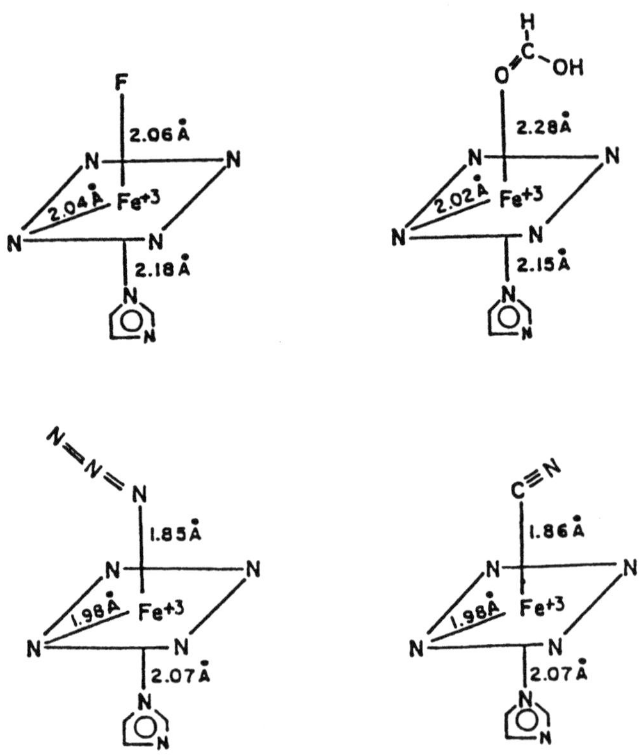

Fig. 1. Effects of ligand binding on the Fe-heme active site structure of met-myoglobin.[9] Weak field or electron donating axial ligands, fluoride and formate [top], and strong field or electron withdrawing axial ligands, azide and cyanide [bottom]. Errors are ±0.01 Å for Fe-N_p and ±0.02 Å for axial ligands.

Although the fifth or proximal ligand is histidine in most of the proteins and enzymes having Fe-heme active sites, this ligand is the only linkage to the protein ultrastructure. Steric constraints on this ligand result in a lower O_2 affinity and the accompanying structure changes[10] are shown in Fig. 2 for the oxy-"picket fence"[6] model compounds. Two Fe-O distances are available to the adducts with axial hindered bases (top right and bottom, Fig. 2): one with the Fe-O distance similar to that for the unhindered base (top left, Fig. 2) but a longer $Fe-N_\varepsilon$ distance by ~0.1 Å and vice versa. Thus, the steric constraint can change the distance of either axial ligand. It is worthwhile to note that these structures offer a plausible explanation for the cooperativity in O_2 binding observed in hemoglobin.

Steric constraints can also be exerted on the sixth axial ligand by the architecture of the pocket. For example, in oxy-myoglobin, a hydrogen bond exists between a histidine residue that comprises a portion of the pocket and the O_2 ligand.[3] However, there is no hydrogen bond when CO is a ligand.[11] In model compounds having no pocket or cavity around the sixth ligand binding site, the Fe-O-O bond angle is ~120° (anti π-bonding) while the Fe-C-O is nearly linear (σ-bonding).[8,12] In myoglobin and the "pocket" porphyrin model compounds, the Fe-O-O bond angle is relatively unchanged, but the Fe-C-O angle is ~135°; suggesting that the pocket architecture plays a role in the position of the CO ligand, i.e. the CO ligand is forced to bind in a position similar to that of O_2.[13] Figure 3 compares the axial ligand distances for these ligands and shows that both axial ligand distances are increased by ~0.1 Å.[13,14]

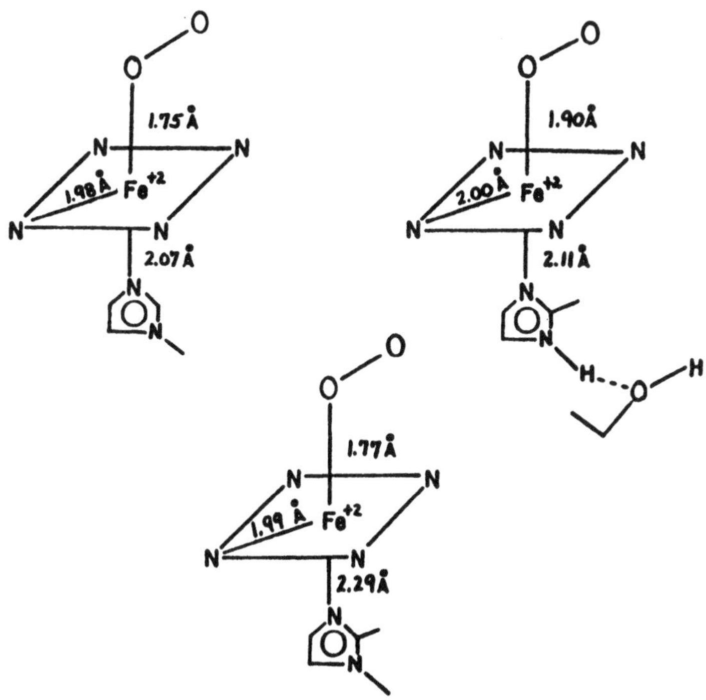

Fig. 2. Effect of steric constraint of the proximal ligand on the Fe-heme active site structure of the oxygen adducts of "picket fence" porphyrin compounds.[10] $Fe(O_2)$ picket fence (1-MeIm) [top left], $Fe(O_2)$ picket fence (2-MeIm). C_2H_5OH [top right], and $Fe(O_2)$ picket (1,2-Me_2Im) [bottom]. 1-MeIm = 1-methylimidazole, 2-MeIm = 2-methylimidazole, and 1,2-Me_2Im = 1,2-dimethylimidazole. Errors are ±0.02 Å.

Another important aspect of ligand binding to proteins and enzymes is the existence of intermediate states. Kinetic Studies[15,16] have identified such states for the recombination of CO with myoglobin at low temperatures. The fact that the recombination rate becomes temperature independent below ~20°K[15,16] suggests a possible tunneling mechanism. The existence or the role of such intermediate states under physiological conditions is a matter of current investigation and debate. Regardless, the structure of these intermediate states are important for establishing the parameters and energy requirements for tunneling and ligand recombination.

The structure of the photoproducts observed at 4° and 40°K are shown in Fig. 4 together with deoxy-, oxy- and carboxy-myoglobin.[13,17] It is clear that neither photoproduct is similar to deoxy-myoglobin. Although the Fe-C bond is broken on photolysis, the Fe-N_p average distance does not expand to the full deoxy- value and the Fe-N_ε distance remains unchanged. Apparently, the heme and proximal histidine cannot relax at these temperatures, and this conformation remains until ~100°K or greater. At 4°K, the CO has moved only 0.05±0.02 Å away from the Fe, but the Fe-C-O angle is decreased by 15-20°. This significantly decreases the overlap of the molecular orbitals of the Fe with those of CO since CO requires σ-bonding. The photoproduct at 40°K differs from that at 4°K only in the distance the CO has moved away from the Fe. Recent high resolution crystallographic and molecular dynamics analysis show a crevice in the pocket of myoglobin at this distance which is just large enough for the CO molecule end-on.[5] Taken together these results indicate that discrete way stations or "docking sites" exist on the

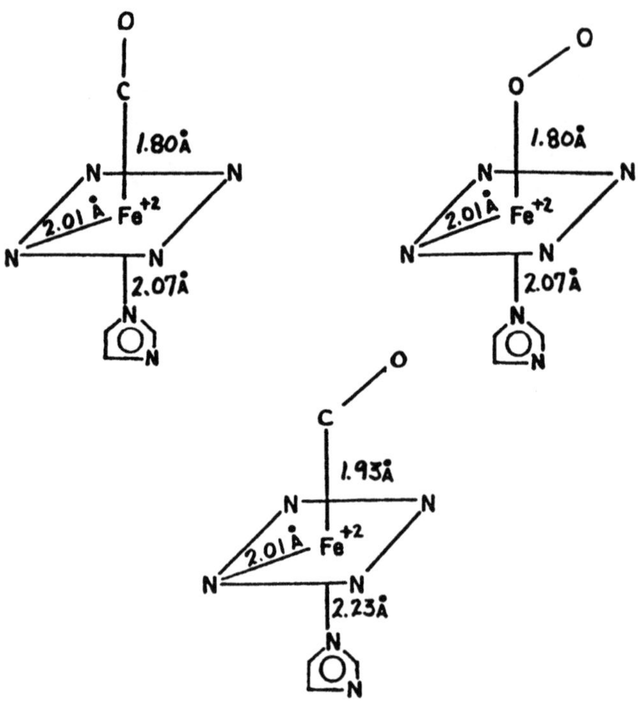

Fig. 3. Effect of steric constraints of ligand biding on the Fe-heme active site structure of proteins and model compounds.[13,14] Fe(CO) picket fence (1-MeIm) [top left], Fe(O_2) pocket (1-MeIm) and oxy-myoglobin [top right], and Fe(CO) pocket (1-MeIm) and carboxy-myoglobin [bottom]. Errors are ±0.02 Å.

ligand route to the Fe-heme active site which represent low-affinity binding sites on the reaction pathway. As such, these docking sites likely alter the ligand pathway and modify recombination barriers.

Enzymes differ in function from other proteins in that enzymes perform as catalysts, capable of greatly enhancing the rate of specific chemical reactions. As both proteins and enzymes contain Fe-heme with proximal histidine active sites, how is it that the same structure can be employed for such diverse functions? For example, myoglobin reacts with peroxide to form a compound that has served as a model for the intermediate states formed in the reaction of peroxide with the peroxidase enzymes, especially horse radish peroxidase, since the first studies by George and Irvine.[18] While, under physiological conditions, the reaction with peroxidases proceeds through two ferryl ion intermediate states termed Compound I and II, resulting in peroxidase and water, the reaction with myoglobin forms ferryl ion peroxide com-

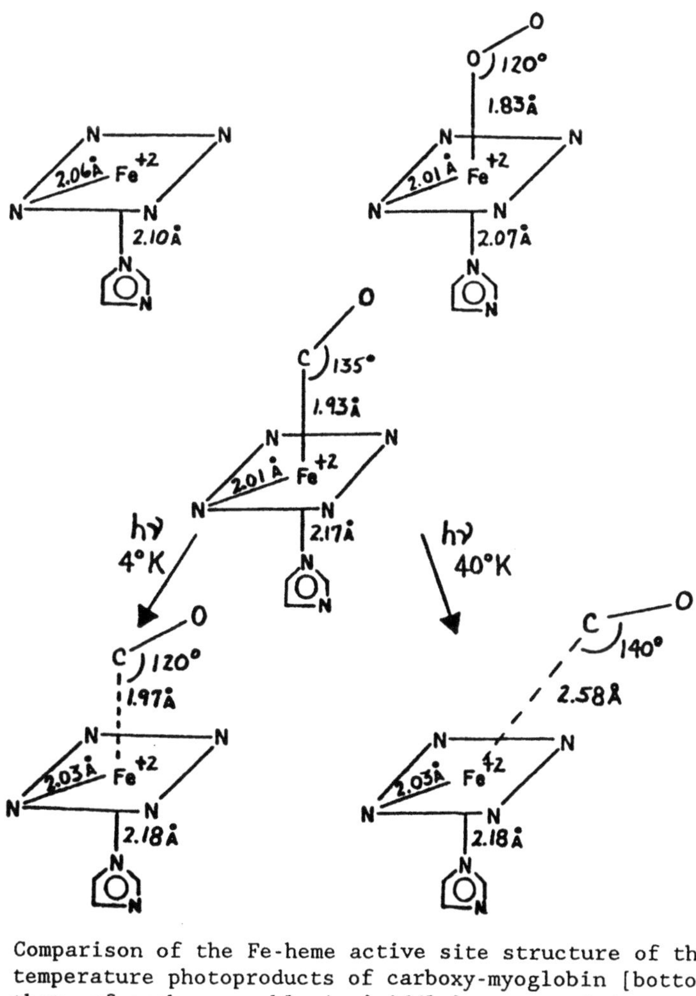

Fig. 4. Comparison of the Fe-heme active site structure of the low temperature photoproducts of carboxy-myoglobin [bottom] with that of carboxy-myoblogin [middle], oxy- and deoxy-myoglobin [top].[13,17] Errors are ±0.02 Å.

pound which results in degraded myoglobin. Figure 5 compares the native, intermediate, and oxy forms of myoglobin[13,19] and peroxidase.[20,21] The distinguishing structure feature of the Fe-heme active site is the Fe-N_ε bond length: it is ~0.15 Å shorter for the peroxidase enzymes and intermediate states. Crystallographic studies show that the linkage of the proximal histidine to the protein ultrastructure is stiffer in peroxidase than in myoglobin. This shorter Fe-N_ε distance increases the charge density in the heme π-electron cloud and, hence, the average Fe-N_P is expanded. Although the reduced peroxidase can bind O_2 and CO, the rate for CO binding is several orders of magnitude slower than myoglobin. Thus for the active site, the constrained Fe-N_ε distance and resulting long average Fe-N_p distance distinguish the mechanism and separate the function of peroxidase enzymes from that of oxygen transport proteins.

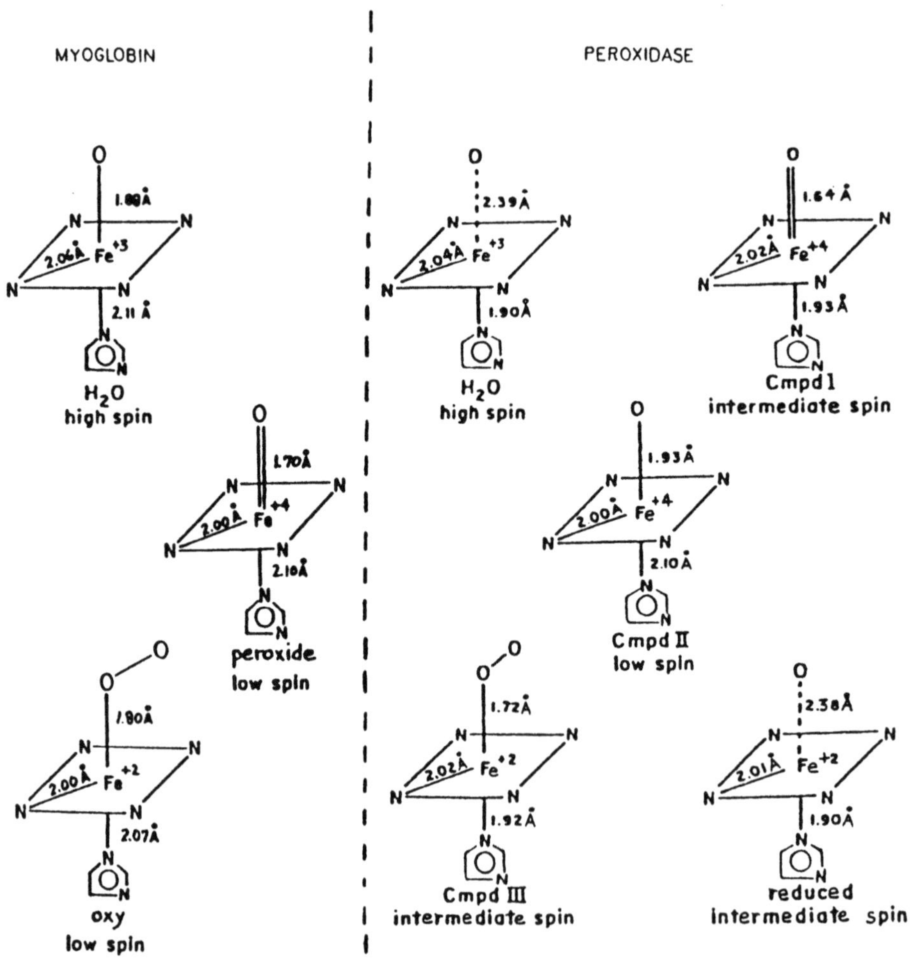

Fig. 5. Comparison of the Fe-heme active site structures of peroxidase and myoglobin.[13,19,20,21] Intermediate states formed in the reaction with peroxide are compounds I and II and myoglobin peroxide. The oxy-compounds are Compound III for peroxidase and oxy-myoglobin. Errors are ±0.01 Å for Fe-N_p and ±0.02 Å for axial ligands.

A binuclear active site consists of two metal atoms in close enough proximity to be magnetically coupled and functions as a single site. For example, the binuclear active site of the mollusk oxygen transport protein, hemocyanin, contains two Cu atoms in an environment of N and/or O ligands. O_2 is bound by forming a bridge between the Cu atoms such that each has one O of O_2 as a ligand. Under these conditions, the Cu atoms are separated[22] by 3.7 Å and charge is delocalized from the Cu atoms, largely to the dioxygen comprising the bridge.[23] Although the Cu atoms in metallothionein, a copper storage protein are bridged by a sulfur atom, they are similarly separated[24] by 3.7 Å. The binuclear active site of cytochrome *c* oxidase, the terminal enzyme in the respiratory chain, contains both Cu and Fe atoms, the latter as an Fe-heme with a proximal histidine ligand. In the fully oxidized forms, the metal atoms are also separated[25,26,27] by ~3.7 Å and the resting form has a sulfur bridge[25] as shown in Fig. 6. The sulfur bridge delocalizes charge from the Cu atom.

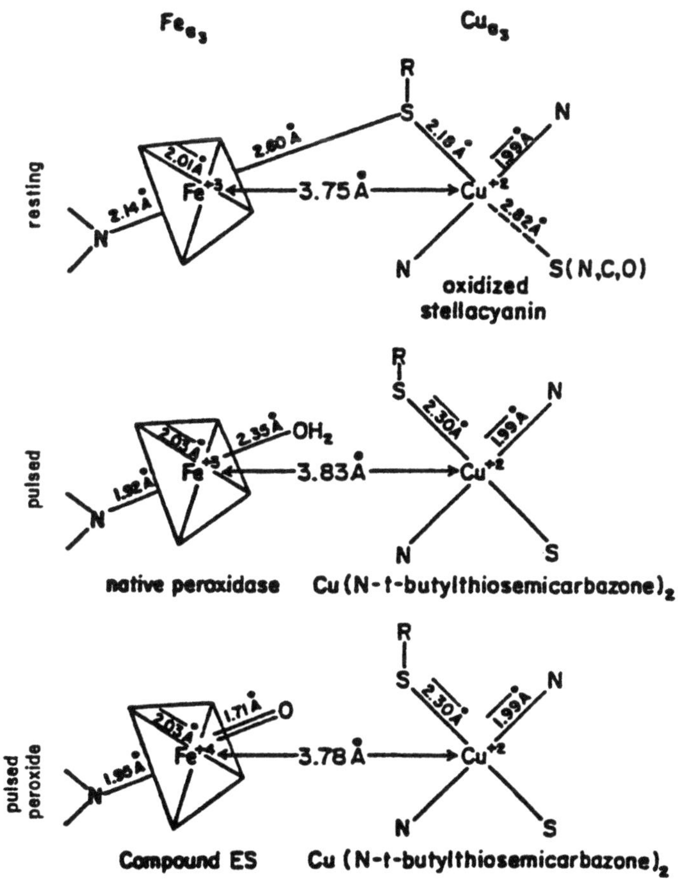

Fig. 6. Comparison of the binuclear active site structures of oxidized cytochrome oxidase.[25,26,27] Model compounds for each metal site are given below the site. Bars over the distances indicate the average distance for all those metal-ligand types in that metal site. Errors are ±0.02 Å except Fe-Cu which is ±0.05 Å.

It is worthwhile to point out the similarities of this binuclear active site to the active sites of other oxygen transport and electron transfer proteins. The Cu environment in the resting form is similar to that of the blue or Type I copper proteins, e.g. stellacyanin, which have electron transfer function and the heme is similar to myoglobin ligated by a weak field or electron donating group. Reduction breaks the sulfur bridge with the sulfur remaining a ligand of the Cu. Similarly, the reduced Cu environment is like that of the reduced blue copper proteins. The reduced Fe environment is like that of deoxy-myoglobin and on binding O_2 is like that of oxy-myoglobin. In the oxidized pulsed form where the sulfur bridge is also broken, this heme is not similar to met-myoglobin as might be expected. Instead, it is similar to that of peroxidases. Indeed, peroxide reacts at this metal site to form the

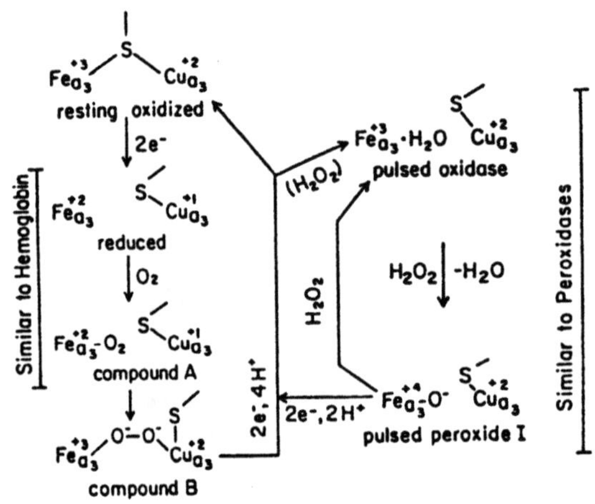

Fig. 7. Structure changes in the binuclear active site of cytochrome oxidase in the cyclic reduction of O_2 [left] and peroxide [right].[28]

pulsed peroxide form resulting in a structure similar to that of Compound I of the peroxidase intermediates. The Cu environment does not participate in this reaction but adopts a more planar geometry than that of the resting form. In the process of reduction of O_2 to H_2O, an intermediate state is formed, Compound B, in which dioxygen bridges the metal atoms. Preliminary studies suggest the metal atoms of compound B are likewise separated by ~3.7 Å, exhibiting a remarkable similarity to hemocyanin.[28] Figure 7 summarizes these results for the binuclear active site of cytochrome oxidase. It appears that this enzyme is a veritable chameleon of enzymes exhibiting several functionally reactive forms, each with an appropriate structure different from each other, yet having analogy with the electron transport proteins, oxygen transport proteins, and peroxidase enzymes and their intermediate states.

SUMMARY

If these few examples are representative of Nature's design, more complex life processes can be achieved from combination of the molecular architecture engineered for simpler functions. Thus a detailed understanding of the structure ↔ function relationship can not only supply us with new materials but with a prescription of how to modify specific properties.

REFERENCES

1. B. Chance, P. Mueller, D. DeVault, and L. Powers, Physics Today, 32, October (1980).
2. T. Takano, J. Mol. Biol. 110:569 (1977); S. E. V. Phillips, Brookhaven Protein Data Bank (1981); S. E. V. Phillips and B. Schoenborn, Nature (London) 292:81 (1981).
3. S. E. V. Phillips, J. Mol. Biol. 147:531 (1980).
4. T. Takano, J. Mol. Biol. 110:537 (1977).
5. J. Kurujan, S. Wilz, M. Karplus, and G. Petsco, J. Mol. Biol. 192:133 (1986).
6. J. Collman, R. Gagne, C. Reed, T. Halbert, G. Lang, and W. Robinson, J. Am. Chem. Soc. 97:1427 (1975).
7. J. Collman, J. Brauman, T. Collins, B. Iverson, G. Lang, R. Pettman, J. Sessler, and M. Walters, J. Am. Chem. Soc. 105:3038 (1983).
8. J. L. Hoard, in: "Porphyrins and Metalloporphyrins," K. Smith, ed., Chap. 8, p. 317, Elsevier, Amsterdam, New York and Oxford (1975).
9. L. Garcia-Iniguez, and L. Powers, unpublished data.
10. G. Wooley, M. Walters, K. Suslick, L. Powers, and T. Spiro, J. Am. Chem. Soc. 107:2370 (1985).
11. J. Norvell, A. Numes, and B. Schoenborn, Science (Wash.) 190:568 (1975).
12. G. Jameson, G. Rodley, W. Robenson, R. Gagne, C. Reed, and J. Collman, Inorg. Chem. 17:850 (1978).
13. L. Powers, J. Sessler, G. Woolery, and B. Chance, Biochem. 23:5519 (1984).
14. P. Eisenberger, R. Shulman, B. Kincaid, G. Brown, and S. Ogawa, Nature (London) 274:30 (1978).
15. R. Austin, K. Beeson, L. Eisenstein, H. Frauenfelder, I. Gunsalus, and V. Marshall, Science (Wash.) 181:541 (1973).
16. T. Iizuka, H. Yamamoto, M. Kotani, and T. Yonetani, Biochem. Biophys. Acta 351:182 (1974).
17. L. Powers, B. Chance, M. Chance, B. Campbell, J. Friedman, K. Khalid, A. Naqui, K. Reddy, and Y. Zhou, Biochem. 26:4785 (1987).
18. P. George and D. H. Irvine, Biochem. J. 52:511 (1952).
19. M. Chance, L. Powers, C. Kumar, and B. Chance, Biochem. 25:1259 (1986).
20. B. Chance, L. Powers, Y. Ching, K. G. Paul, T. Poulos, and I. Yamazaki, Arch. Biochem. Biophys. 235:596 (1984).
21. M. Chance, L. Powers, T. Poulos, and B. Chance, Biochem. 25:1266 (1986).
22. J. Brown, L. Powers, B. Kincaid, J. Larrabee, and T. Spiro, J. Am. Chem. Soc. 102:4210 (1980); G. Woolery, L. Powers, M. Winkler, E. Solomon, and T. Spiro, J. Am. Chem. Soc. 106:86 (1984).
23. J. Larrabee and T. Spiro, J. Am. Chem. Soc. 106:4217 (1984).
24. J. Freedman, L. Powers, and J. Peisach, Biochem. 23:2342 (1986).
25. L. Powers, B. Chance, Y. Ching, and P. Angiolillo, Biophys. J. 34:465 (1981).
26. B. Chance, L. Powers, Y. Ching, and C. Kumar, Biophys. J. 44:353 (1983).
27. L. Power, A. Naqui, C. Kumar, Y. Ching, and B. Chance, J. Biol. Chem. 263:7159 (1988).
28. L. Powers and B. Chance, J. Inorg. Biochem. 23:207 (1985).

SYNTHETIC MODELS FOR PHOTOINDUCED ELECTRON TRANSFER REACTIONS

A. D. Joran

Department of Chemistry
Case Western Reserve University
Cleveland, Ohio 44106

Electron transfer processes occur widely in Nature. The defining characteristic of these reactions is the oxidation of one species coupled to the reduction of another. The highly efficient vectorial electron transfer chains present in photosynthetic and mitochondrial membranes[1,2] offer a natural source of inspiration to the chemist in quest of molecular electronic devices. A better understanding of these reactions would be a boon to the design and construction of electronic devices on the molecular scale. In particular, the ability to predict rate constants for electron transfer is essential for controlling stable charge separation within a given molecular structure. Information about relative and ideally absolute electron transfer kinetics would equip the synthetic chemist with important input parameters useful in the engineering of molecular electronic circuits.[3,4]

Over thirty years of theoretical studies have been rewarded by many insights into the role of such parameters as temperature, distance, driving force, and vibrational coupling in regulating the kinetics of electron transfer reactions.[1,2,5] Recent experiments have lent some credence to Marcus's important prediction of reaction slowdown at sufficiently large exothermicity,[6] which would account for the unidirectionality observed in bacterial photosynthetic reaction centers.[7] This theory might provide a fruitful starting point for developing design criteria for molecular electronic devices.[3,4]

One important experimental approach to the verification of electron transfer theory has focused on the chemical synthesis of organic compounds, with the corollary intention of modeling the charge separation which occurs in biological membranes.[8] A variety of molecules containing electron donors and acceptors coupled by saturated hydrocarbon bridges have been prepared.[9,17] However, because of the flexible bridges commonly used in them, for example, polymethylenes, the geometric orientation of donor and acceptor is not fixed. Since the possibility of translational and rotational diffusion of donor and acceptor generates an infinite number of reacting conformational structures, electron transfer reactions occurring in each of these structures exhibit different characteristic rate constants;[9,10] kinetic measurements give rates averaged over the geometry sampled.The available techniques of observation cannot resolve the microscopic correspondence between conformation and absolute rate. Consequently, the use of nonrigid linkages between electron donor and acceptor greatly complicates, and perhaps obviates simple and direct analysis of the observed data in terms of basic theory.[9,17] This chapter reviews recent efforts at understanding electron transfer reactions through the study of synthetic chemical models possessing rigid, well defined structure.[10]

To be capable of addressing specific theoretical issues conveniently, a chemical model must exhibit several properties. In addition to rigidity so as to avoid translational freedom, the bridge would ideally possess cylindrical symmetry, affording the opportunity to study the effect of distance on the electron transfer rate. (The assumption that the repetitive bridge structure establishes a "lattice-like" periodic potential allows application of Bloch's theorem, giving a straightforward way of estimating tunneling matrix element decay.[11]) A photoexcitable donor provides a simple way of initiating the reaction and observing its evolution. Nature utilizes porphyrin derivatives to serve as electron carriers, such as chlorins in chlorophylls. The large size of the macrocycle lowers the vibrational reorganization energy λ_v, making observation of inverted exothermicity effects more probable. The model should also be amenable to chemical substitution to allow changes in reaction exothermicity. Solubility in a wide range of solvents would permit examination of the role played by solvation and reorganization of solvent around donor and acceptor. Porphyrin-bicyclooctane-quinones such as those shown in Fig. 1 appear to fulfil these criteria.

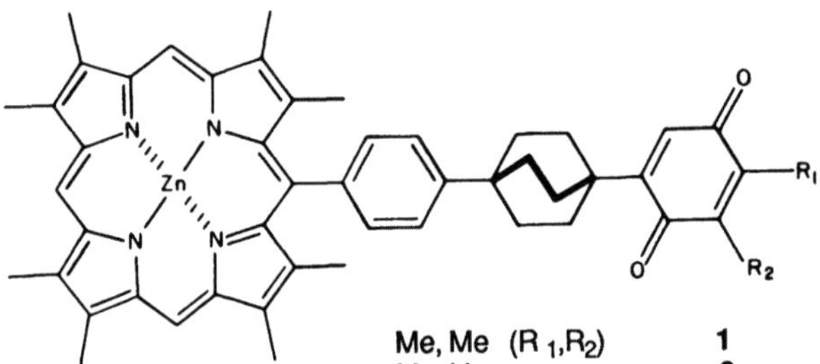

Me, Me (R_1,R_2)	**1**
Me, H	**2**
H,H	**3**
Cl,H	**4**
Br,H	**5**
Cl,Cl	**6**
CN,H	**7**

Fig. 1. Porphyrin-bicyclooctane-quinones.

Distance Effects on Electron Transfer

The rigidity of the bicyclo[2.2.2]octane spacer allows a clear correlation of electron transfer rate and distance between donor and acceptor.[12] The synthesis of homologous porphyrin-quinones with the incremental addition of bicyclooctyl units provides models of transfer at longer distances. The porphyrin moiety is highly absorptive and emissive in the visible range. When a new pathway for deactivation of the porphyrin excited state exists, as for porphyrin-quinones in which electron transfer is feasible, a reduction in emission lifetime is observed. Measurement of the fluorescence lifetime of a reference porphyrin, substituted by a *meso-t*-butylphenyl group instead of the bridge-quinone moiety allows calculation of the rate constant for electron transfer according to the equation $k_{ET} = 1/\tau - 1/\tau_o$.[13] At 10Å (one spacer, edge-to-edge) the rate is about 10^{10} sec^{-1}, whereas at 14Å (two spacers) the rate falls to $\leq 10^7$ sec^{-1}.[13] This rate difference is consistent with the prediction of *exponential* decay of electronic wavefunction overlap, and bolsters the mechanistic expectation that the transferring electron must tunnel through the hydrocarbon spacer. The ultrafast reaction rates observed may be due to the fact that donor and acceptor wavefunction can mix with those of the spacer to give an overall enhancement to the matrix element.[14,23] In comparison with biological electron transfers, electroactive moieties positioned with similar separations must interact through space over part of the pathway in addition to partial mediation through the bonding framework.[14,21]

Solvent Effects on Electron Transfer

The electronic spectrum of the porphyrin-bicyclooctane-quinone **3** was found to be indistinguishable from the superposition of the absorption spectra of the isolated chromophores comprising the molecule.[10] The lack of significant perturbation of the individual electronic transitions suggests that the overlap of electronic wavefunctions for donor and acceptor is small, and therefore supports the assumption that the electron transfer can be described as a nonadiabatic process. Demonstration of nonadiabaticity is important for determining the applicability of long-distance electron transfer theories. The fluorescence lifetimes of the porphyrin-quinone were determined in a series of solvents of widely varying polarity.[10] As is evident from Table I, the electron transfer rate constants depend quite insensitively on solvent polarity, within a factor of three. Nevertheless, a small and consistent trend is clearly present. The reaction rate decreases with increasing solvent polarity. Less noticeable trends are found for viscosity and optical dielectric constant.

Table I. Electron Transfer Rate Constants for **3** Compared with Solvent Parameters[a,b] ϵ, η, and n^2.

Solvent	k_{ET} (sec^{-1})	ϵ	η (mP)	n^2
CH_3CN	5.00×10^9	37.5_{25}	3.449	1.800
DMF	3.69×10^9	36.71	8.0	2.046
n-PrCN	7.84×10^9	20.3_{21}	5.2_{30}	1.910
2MTHF	1.03×10^{10}	6.24_{25}	4.57	1.977
o-Xyl	1.15×10^{10}	2.568_{25}	8.10	2.267
C_6H_6 + 2% Pyr	1.90×10^{10}	NA	NA	NA
C_6H_6 + 0.1M THAP	1.03×10^{10}	NA	NA	NA
C_6H_6	1.25×10^{10}	2.273_{25}	5.961	2.244
p-Dioxane	7.67×10^9	2.21	14.4	2.023

[a] G.J. Janz and R.P.T. Tomkins, *Nonaqueous Electrolytes Handbook,* I, Academic Press, New York (1972).

[b] *Handbook of Chemistry and Physics,* 65th ed., Chemical Rubber Co.

The weak solvent dependence provides further confirmation that the reaction is nonadiabatic. These data contrast with results found for electron transfers more clearly identifiable as adiabatic,[15] *e.g.,* where there are optical indications of direct electronic interactions. In compounds with directly linked phenylamine and naphthalenesulfonamide moieties, where the amine is the donor and an electronic excited state of the arylsulfonamide is the acceptor, the transfer rate correlates with the dielectric relaxation rate of the solvent.[15,16] The solvents examined included a series of unbranched aliphatic alcohols, and the relaxation times associated with the motion of the hydroxyl dipoles gave the best correlation.[15] This remarkably strong dependence of transfer rate on solvent dynamics suggests that electron transfer is coupled to a diffusive (solvent) mode. Transit times through the transition state must be fast relative to motion along the reaction coordinate.[17] In the non-adiabatic case, where the transition is less probable, solvent coordinate motion should play a lesser role in rate-control. The 200-fold range in rate constants observed for a *flexibly*-linked porphyrin-quinone[18] may be due to large solvent-dependent conformational differences.

Temperature Effects on Electron Transfer

The classical Arrhenius relationship predicts an exponential decrease of reaction rate with temperature ($exp(-\Delta G^\dagger/kT)$).[2] Nearly all chemical reactions abide by this prediction. Electron transfer reactions involve the coupled motion of nuclei and electrons, more critically so if the transfers are nonadiabatic, and the observation of the usual Arrhenius relationship seems to depend on whether the classical or quantum limit applies ($\hbar\nu \gg$ *or* $\ll kT$), whether the presence of tunneling phenomena dominate the reaction, or whether thermally activated steps, such as major conformational changes, are expected to take place.[2] The porphyrin-quinone exhibits multiexponential but distinctly quenched fluorescence decay when cooled to 77K in 2-methyltetrahydrofuran (2MTHF), suggesting that electron transfer does not cease at low temperatures.[13] The precise form of the decay curve could be fit to a model assuming free rotation of the

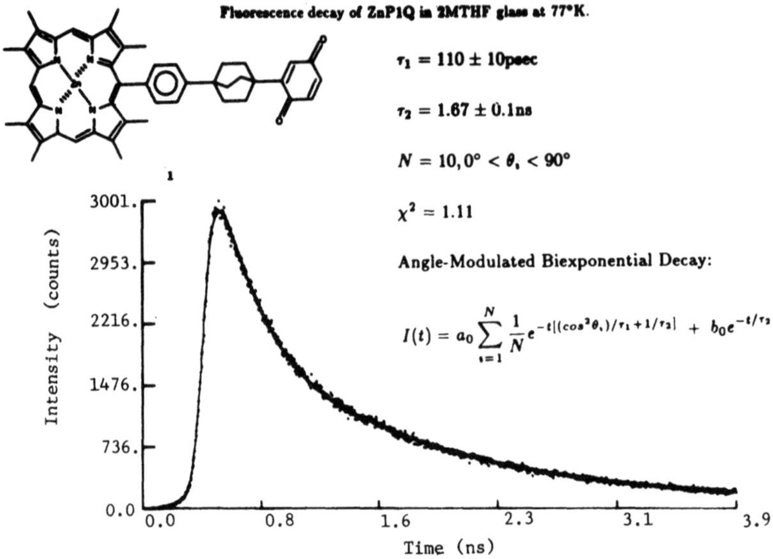

Fig. 2. Fluorescence decay of porphyrin-quinone **3** in 2MTHF (77K).[13]

porphyrin with respect to the quinone at high temperature and a statistical distribution of conformations frozen by the glassy matrix at low temperatures.

Using eq. 1 to fit the data shown in Figure 2, an average rate constant was determined for reaction at 77K to be 4×10^9 sec^{-1}.

$$I(t) = a_o \sum_{i=1}^{N} \frac{1}{N} e^{-t[(cos^2\theta_i)/\tau_1 + 1/\tau_o]} + b_o e^{-t/\tau_o} \tag{1}$$

$I(t)$ is the fluorescence intensity at time t, a_o and b_o are initial amplitudes, N is the number of distinct angles, θ_i are the angles, τ_1 is the lifetime of the porphyrin-quinone, and τ_o is that of the reference porphyrin zinc *meso-t*-butylphenyloctamethylporphyrin. This value is about a factor of two lower than the rate at room temperature, indicating that the process is nearly temperature independent between 77K and 298K.

Several ligand-modified proteins (e.g., heme-modified hemoglobin or myoglobin and $Fe(NH_2)_5$-modified cytochrome *c*) have been shown to exhibit temperature-independent electron transfer at low temperatures, in some cases with classical Arrhenius activation at higher temperatures.[18] In the case of the porphyrin-bicyclooctane-quinone, nuclear tunneling through the potential barrier would be consistent with coupling of electron transfer to the quinone carbonyl stretching vibrational mode ($\sim$ 1600 cm^{-1}) $\hbar\nu \gg kT$). That is, the transfer of an electron is accompanied by the motion of a vibrational mode with higher energy than the thermal energy available to excite the mode fully ($\sim$ 200 cm^{-1} at 298K). The persistence of reactivity at low temperature is thus explained by residual zero-point vibrational motion, which alters the total potential energy sufficiently to produce quasi-degenerate electronic energies of initial and final states.[1]

Driving Force Dependence

A series of porphyrin-quinones possessing different substituents on the quinone moiety (Fig. 1) was prepared from tetrapyrrolic and appropriately substituted aldehydic intermediates to examine the role of driving force in controlling the rate of electron transfer.[20] The structural constancy of the porphyrin, the bicyclooctyl bridge, and the quinone carbon skeleton offers some assurance that the series of compounds really constitute a *homologous* reaction series. Otherwise, significant changes in the carbon skeletal structure are expected to alter the overall vibrational reorganization energy, and therefore make comparison within the series uncertain. Particular care must be taken for this reason in the interpretation of trends between rate constant and driving force. Electrochemical measurements[20] using microvoltammetric techniques provided redox values for the oxidation of the porphyrin and the reduction of the quinone, and together with electronic excitation energies for the porphyrin, gives a way of calculating the reaction driving force.

Fluorescence lifetimes were determined for the porphyrin-quinone series **1-7**.[20] Rate constants calculated from the lifetimes increase rapidly from $\Delta G = -0.8$ to -1.0 eV (in acetonitrile, where coulombic corrections should be small), and are very similar from -1.0 to -1.4 eV (8.09×10^9 to 1.75×10^{10} sec^{-1}). Fitting the data to the classical equation (eq. 2)

$$k = \frac{2\pi}{\hbar} \mid T_{ab} \mid^2 \left(\frac{1}{\sqrt{4\pi\lambda k_B T}}\right) exp\left[\frac{-(\Delta G + \lambda)^2}{4\pi k_B T}\right] \tag{2}$$

where

$$\lambda = \lambda_i + \lambda_o,$$

T_{ab} is the tunneling matrix element, k_B is Boltzmann's constant, λ is the total reorganization energy, λ_i is the inner-sphere reorganization, λ_o is the outer-sphere reorganization energy, and ΔG is the reaction driving force, leads to estimates of the vibrational reorganization parameter that are apparently too high for transfers involving a porphyrin and a quinone.[20] A semiclassical fit which accounts for coupling of the electronic motion to a quantized manifold of acceptor vibrational states[5] (eq. 3) describes somewhat better the rapid rise in rates at low exothermicities and the rate insensitivity noted at higher driving forces.

$$k = \sqrt{\frac{\pi}{\hbar^2 \lambda_s k_B T}} \mid T_{ab} \mid^2 \sum_{m=0}^{\infty} \frac{e^{-S} S^m}{m!} exp\left[\frac{-(\lambda_s + \Delta G^o + mh\nu)^2}{4\lambda_s k_B T}\right] \tag{3}$$

where

$$S = \frac{\lambda_v}{h\nu},$$

λ_s is the solvent reorganization energy, S is the vibronic coupling strength expressed in units of $h\nu$. An example of similar "flattening" in the Marcus plot ($ln\ k_{ET}$ *vs.* ΔG) at high driving forces was observed in quinone-substituted reaction centers.[21] Electron transfer into vibrationally excited states would serve to mask the classical inverted effect predicted by eq. 2.[22]

Conclusions

Through the study of chemical models, much has been learned about the fundamental nature of electron transfer processes. Some of the issues remaining to be resolved include the role of the position of the donor and acceptor in the band gap of the bridge, the precise effects of geometry in vibrational and electronic overlap, and the relative contributions of through-bond and through-space interactions.[23,24] Conformationally defined chemical models are uniquely capable of revealing basic structure-function relationships, with less confusion by adventitious factors, and offer a systematic approach to the design of molecular electronic devices.[3,4]

Acknowledgments

The author is grateful to John Hopfield and Peter Dervan for initiating the approach described, and to Burton Leland, Peter Felker, and A.H. Zewail for contributing to the experiments reviewed.

References

1. M.D. Newton and N. Sutin, *Annu. Rev. Phys. Chem., 35*, 437 (1984).
2. D. Devault, *Quantum-Mechanical Tunnelling in Biological Systems,* 2nd ed., Cambridge University Press, Cambridge, UK (1984).
3. J.J. Hopfield, J.N. Onuchic, and D.N. Beratan, *Science (Washington, D.C.), 241,* 817 (1988).
4. J.J. Hopfield, J.N. Onuchic, and D.N. Beratan, Fine Particle Society Molecular Electronics Symposium (1988).

5. N.S. Hush, *Trans. Far. Soc., 57*, 577 (1961); R.A. Marcus, *Annu. Rev. Phys. Chem., 15*, 155 (1964); J.J. Hopfield, *Proc. Nat. Acad. Sci. U.S.A., 71*, 3640 (1974); J. Jortner, *J. Chem. Phys., 64*, 4860 (1976); V.O. Levich, *Adv. Electrochem. Electrochem. Eng., 4*, 249 (1966).
6. J.R. Miller, L.T. Calcaterra, and G.L. Closs, *J. Amer. Chem. Soc., 106*, 3047 (1984); M.R. Wasielewski, M.P. Niemczyk, W.A. Svec, E.B. Pewitt, *J. Amer. Chem. Soc., 107*, 5562 (1985); R.J. Harrison, B. Pearce, G.S. Beddard, J.A. Cowan, J.K.M. Sanders, *Chem. Phys., 116*, 429 (1987).
7. H. Gerischer and J.J. Katz, eds., *Light-Induced Charge Separation in Biology and Chemistry,* Dahlem Konferenzen, Springer-Verlag, Heidelberg (1979).
8. G.L. Closs and J.R. Miller, *Science (Washington, D.C.), 240*, 440 (1988).
9. A.R. McIntosh, A. Siemiarczuk, J.R. Bolton, M.J. Stillman, T.F. Ho, and A.C. Weedon, *J. Amer. Chem. Soc., 105*, 7215 (1983); N.S. Hush, *et al., Chem. Phys. Lett., 117*, 8 (1985); T. Moore, *et al., Nature (London), 307*, 630 (1984); J.S. Lindsey, D.C. Mauzerall, and H.S. Linschitz, *J. Amer. Chem. Soc., 105*, 6528 (1983); I. Fujita, *et al., J. Phys. Chem., 79*, 5337 (1982); P.A. Loach, J.A. Runquist, J.L.Y. Kong, and T.J. Dannhauser, *Adv. Chem. Ser., 201*, 515 (1982); J.L.Y. Kong and P.A. Loach, *J. Heterocyclic Chem., 17*, 737 (1980).
10. A.D. Joran, *Ph.D. Dissertation,* California Institute of Technology, Pasadena, California (1986).
11. D.N. Beratan and J.J. Hopfield, *J. Amer. Chem. Soc., 106*, 1584 (1984).
12. A.D. Joran, B.A. Leland, G.G. Geller, J.J. Hopfield, and P.B. Dervan, *J. Amer. Chem. Soc., 106*, 6090 (1984).
13. B.A. Leland, A.D. Joran, P.M. Felker, A.H. Zewail, J.J. Hopfield, *J. Phys. Chem., 89*, 5571 (1985); B.A. Leland, *Ph.D. Dissertation,* California Institute of Technology, Pasadena, California (1987).
14. S.L. Mayo, W.R. Ellis, R.J. Crutchley, and H.B. Gray, *Science (Washington, D.C.), 233*, 948 (1985).
15. E.M. Kosower, *J. Amer. Chem. Soc., 107*, 1114 (1985); E.M. Kosower and D. Huppert, *Chem. Phys. Lett., 96*, 433 (1983).
16. H.L. Friedman, *J. Chem. Soc., Faraday Trans., 79*, 1465 (1983).
17. J.A. Schmidt, A. Siemiarczuk, A.C. Weedon, and J.R. Bolton, *J. Amer. Chem. Soc., 107*, 6112 (1985).
18. S.E. Peterson-Kennedy, J.L. McGourty, and B.M. Hoffman, *J. Amer. Chem. Soc., 106*, 5010 (1984); N.M. Kostic, R. Margalit, C.-M. Che, and H.B. Gray, *J. Amer. Chem. Soc., 105*, 7765 (1983).
19. J.N. Onuchic, D.N. Beratan, and J.J. Hopfield, *J. Phys. Chem., 90*, 3707 (1987); W. Bialek, *Ph.D. Dissertation,* University of California, Berkeley (1983).
20. A.D. Joran, B.A. Leland, P.M. Felker, A.H. Zewail, J.J. Hopfield, and P.B. Dervan, *Nature (London), 327*, 508 (1987).
21. M.R. Gunner, D.E. Robertson, and P.L. Dutton, *J. Phys. Chem., 90*, 3787 (1986).
22. D. Rehm and A. Weller, *Isr. J. Chem., 8*, 259 (1970).
23. D.N. Beratan, J.N. Onuchic, and J.J. Hopfield, *J. Chem. Phys., 86*, 4488 (1987).
24. D.N. Beratan, *J. Amer. Chem. Soc., 108*, 4321 (1986).

ELECTRON TRANSFER RATES IN MITOCHONDRIAL PROTEINS:

REGULATION AND SPECIFICITY

G. McLendon[a*], Q. Zhang[a], K. Pardue[a], F. Sherman[b], A. Corin[c], R. Ciacarelli[c], J. Falvo, and D. Holzschu[c]

[a]Dept. of Chemistry, University of Rochester
[b]Dept. of Biochemistry, University of Rochester, Medical Center Rochester, NY 14627
[c]Life Sciences Lab, Eastman Kodak Company, Rochester, NY 14650

INTRODUCTION

Electron transport is the ubiquitous biological mechanism for the translocation of energy. In the sense of this symposium, electron transfer may also be considered as a transducer since the accumulation and distribution of redox equivalents among various pathways both responds to external signals (metabolic "second messengers") and may also serve as a signal to activate metabolic processes.

In the present paper, we consider two key elements of this process: the rate at which electrons flow from one protein to another (analogous to the source of drain current in a solid state device) and the regulation of electron flow which insures the fidelity of information (electron) transfer and which is accomplished in large part by specific biological recognition between complimentary redox partners.

Determinants of Electron Transfer Rates

The theory of electron transfer processes, and its applications both to small molecules and to biomolecules are well established, and have been reviewed in detail elsewhere.[1]

The key insight, first derived by Marcus, is contained in the (deceptively) simple eqn 1:

$$k_{et} = A \exp\left[-(\Delta G-\lambda)^2/4\lambda kT\right] \qquad [1]$$

where, A is an electronic term which depends on orbital overlap. $A \propto 2\pi/h \exp(-\beta R)$ with R = closest distance between the donor and acceptor (Å). The damping parameter, β, has been empirically determined as ca. 1.2 ± 0.3 $Å^{-1}$ between the redox active groups[1]. For most protein reactants, R ranges from ca. 3 Å to ca. 20 Å.

ΔG is the reaction free energy for electron transfer from the donor to acceptor and λ, the "reorganization energy", is a measure of the (frequency weighted) nuclear motions in the molecules and in the surrounding medium, which accompany electron transfer (Fig. 1).

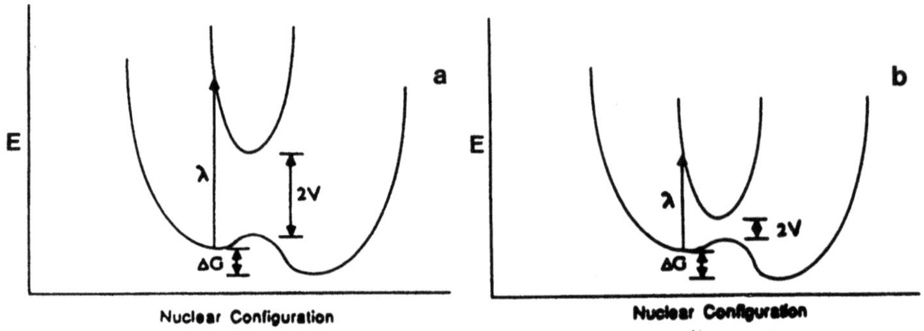

Fig. 1. Potential energy surface for (a) adiabatic electron transfer and for (b) non-adiabatic electron transfer.

For delocalized organic reactants (e.g., anthracene) in polar organic solvents, $\lambda \cong 1.5 \pm 0.5$ V. These ideas have been tested for many molecular systems, with the above result that if λ and R are specified, the rate of an electron transfer process for simple organic molecules can be predicted within about an order of magnitude for a range of rates of ca. 10^{20} observed experimentally. How does the situation compare for the apparently more complex (protein) reactants found in biology?

Although the range of protein to protein reactions so far examined remains small,[1] those examined to date suggest that the same basic physics govern biological electron transfer, and even that values for β and λ are found to be quite similar between proteins and small molecule reactants. For example, for the protein:protein complex formed between cytochrome c (cyt c) and cytochrome b5 (b5),[2] $\Delta G = 0.2$ V, $k = 10^3$ s^{-1}, $\lambda \approx 0.8$ V, and $\beta = 1$. Similarly, for the complex between cyt c and cytochrome c peroxidase (ccp), $\Delta G = 0.4$ V, $k \approx 1$ s^{-1}, $\lambda = 1.4$ V, and $\beta \approx 1$.[3] Thus, biological electron transfer is unusual not in terms of reactivity but in terms of the <u>specificity</u> of pathway. If a variety of small molecule redox reagents are mixed, they will rapidly equilibrate to obtain a minimum free energy. However, in a living cell, a variety of biological redox reagents span a range of redox potentials from ca. -0.8 to 1.0 V. If these biological reagents were to equilibrate, the result would be a disastrous intermediary biological short circuit! Directed energy flow would halt, and the cell would die a "death by entropy". This biological disaster is prevented by insuring that only physiological redox partners can readily access one another. This "limited access" strategy may involve several levels of organization. At the level of individual proteins, the redox active sites are sequestered well within the protein matrix. For example, in ccp, the heme group is buried some 12 Å from the protein surface, thereby reducing its reactivity (relative to free heme) by ca. 10^5. Thus, even when the reaction free energy is optimal ($\Delta G = 1$), a maximum rate $\approx 10^4$ s^{-1} is expected. Therefore a random collision with a reactive molecule would be unlikely to "discharge" the reactive heme, since the contact time between reactants << required reaction time (ca. 10^{-4} sec). The likelihood of reaction is obviously enhanced if the donor and acceptor bind in a favorable configuration.

Thus, ccp has been designed to recognize and tightly bind ($K_b \geq 10^7$ M^{-1}) its physiological electron donor, cytochrome c. A best guess <u>model</u> for this interaction based on molecular modeling studies by Poulos and Kraut,[6] is shown in Fig. 2.

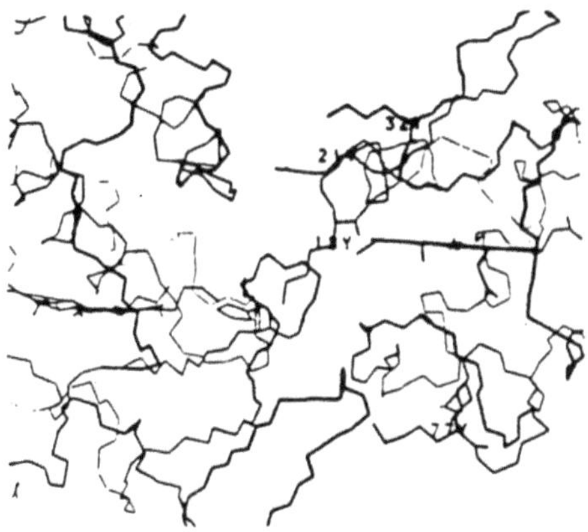

Fig. 2. Hypothetical model of ccp (molecule on left) and cyt c.

How is this recognition accomplished? A leading working hypothesis has been that recognition is governed by stereospecific interactions between cyt c and ccp, involving (evolutionarily invariant) positively charged lysines on cyt c, and negatively charged residues (Glu and Asp) on ccp.[4] However, recent experiments in our lab suggest such "lock and key" ionic interactions do not fully describe the binding and recognition domain. For example, studies of electron transfer within the bound cyt c:ccp complex show that the absolute rate of electron transfer increases significantly when the ionic strength is raised from 10 mM, where ionic interactions are maximal, to 200 mM, where ionic interactions are weakened: that is, the electrostatically most stable complex is "dead end" with respect to electron transfer.[7] Instead, a more weakly bound complex, guided by electrostatics, allows two dimensional diffusion along the protein surface to obtain the optimal orientation for electron transfer.[5]

Site Directed Mutagenesis: A Synthetic Tool for Mechanistic Studies

To gain more specific insight into how specific residues contribute to binding and recognition, we have begun detailed mutagenesis studies in which the otherwise invariant charged residues on cyt c and ccp are converted to neutral, or even oppositely charged, residues. The approaches used in these studies include both conventional mutagenesis (recovering functional revertants of missense mutations in cytochrome c)[8] to oligonucleotide directed site directed mutagenesis. For these studies, we have cloned the genes coding for cyt c (*Saccharomyces cerivisiae* iso-1-cyt c expressed in yeast) and ccp (expressed in *E. coli*). One basic protocol is outlined in Fig. 3. A similar scheme was used for ccp.

The results of these studies are somewhat surprising. Previous studies, including evolutionary analysis,[4] theoretical modeling,[6] and chemical modification[4] had suggested that several key electrostatic contacts mediated the interaction between cyt c and cyt c peroxidase: Lys 27 (c):Asp 79 (ccp), Lys 72 (c):Asp 217 (ccp), Lys 13 (c):Asp 37 (ccp), and Lys 79 (c):Asp 183 (ccp).

We therefore have used protein engineering methods to construct specific mutants at each of these sites. The effects of mutation on binding equilibria (measured via fluorescence energy transfer), binding dynamics (via time resolved energy transfer measurements),[9] and steady state and single turnover kinetic measurements are each being evaluated. Although the complete set of

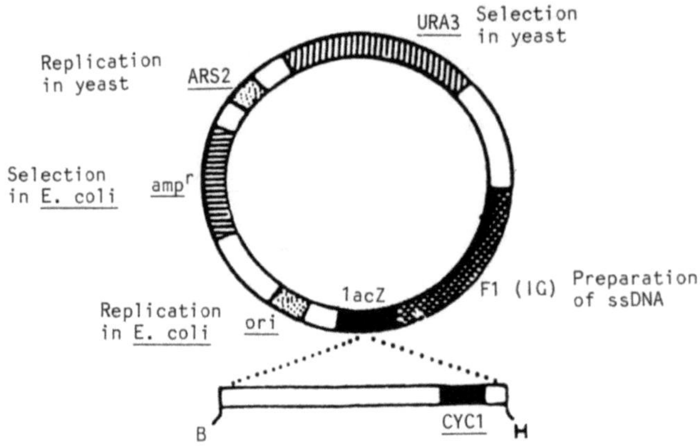

Fig. 3. The regions and corresponding functions of the plasmid used for site-directed mutagenesis. The <u>ARS2</u> and <u>ori</u> regions allow replication in, respectively, yeast and *E. coli*; the <u>URA3</u> and <u>amp</u>^{r} regions allow selections of the plasmids in, respectively, ura3 yeast strains and <u>amp</u>^{s} bacterial strains; the Fl(IG) region (the intragenic region of bacteriophage Fl) allows encapsidation of a single stranded DNA form of plasmid upon superinfection with bacteriophage Fl; lacZ denotes a short segment coding for the α-peptide of β-galactosidase and containing a multiple cloning site (poly-linker); the 2.5 kb yeast fragment with restriction endonucleases sites <u>Bam</u>HI (B) and <u>Hind</u>III (H) encompasses the <u>CYC1</u> locus. The total size of the plasmid is 8.8 kb.

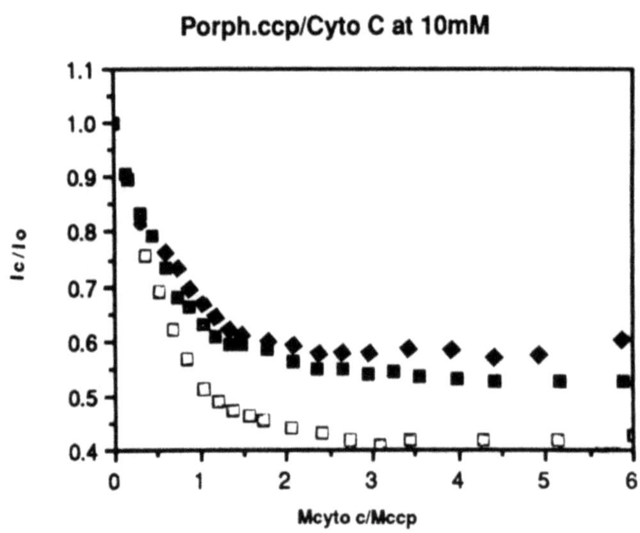

Fig. 4. Fluorescence titration of Mgccp with cyt c. I_f is the fluorescent intensity normalized as (I_f/I_o). Note that a clear break is seen at 1:1 ccp:cyt c. □ = wild type; ◆ = Ile 18; ■ = Gln 32. Conditions: pH 7.2, 25°, μ = 50 μM.

results are not yet available, sufficient data are available to provide some surprising conclusions.

Binding measurements of the fluorescent derivative Mgccp to yeast cytochrome c, and the cytochrome c mutants at positions 18 (Arg→Ile) and 32 (Lys→Gln) are shown in Fig. 4.

The values derived from equilibrium binding analysis, assuming a simple 1:1 isotherm, are as follows:

K_B(wild type) = 9.0×10^7 M^{-1}

K_B(Lys 18→Ile) = 5.0×10^8 M^{-1}

K_B(Lys 32→Gln) = 1.2×10^8 M^{-1}

These binding measurements differ somewhat from measurements made using other techniques, including difference spectral measurements, and NMR. The NMR measurements in particular point out complications which can arise in using a specific technique to monitor binding. At the high protein concentrations normally used for NMR measurements (ca. 1 mM), cytochrome c:ccp binding is characterized by fast exchange kinetics with a surprisingly low binding constant ($K_B = 10^5$ M^{-1}) (Fig. 5a). As we suspected that protein aggregation might occur at such high concentrations, the experiments were rerun at lower (10^{-4} M) concentration, with a resulting binding constant of $K_B > 10^6$ M^{-1} (Fig. 5b). Even more surprising, the NMR exchange kinetics changed from the fast exchange limit to the slow exchange limit.

Comparing all the results from the different techniques strongly suggests that different conformations of the bound complex can exist each of which is sensed differentially by the individual techniques used to study binding (fluorescence, difference spectroscopy, and NMR). Thus the apparent binding constants obtained in any study reflect the percentage of cytochrome which is complexed at any given set of experimental conditions (pH, temperature, ionic strength and concentration) but does not require a unique binding site.

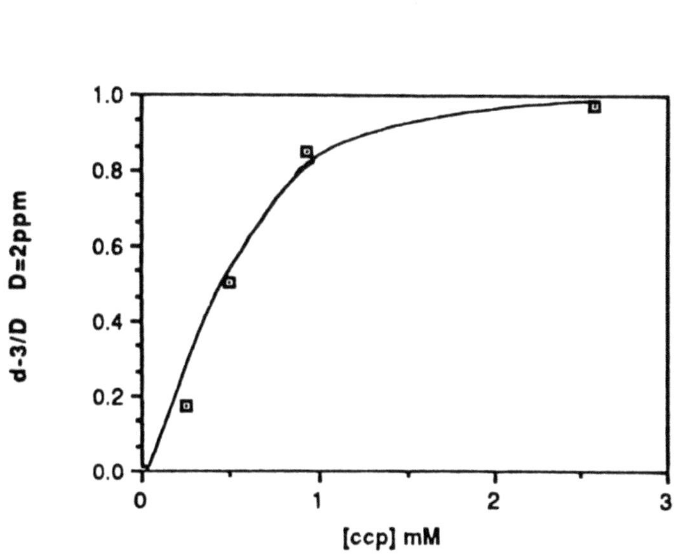

Fig. 5a. High concentration titration of cyt c with ccp(III): μ=50 mM K_{phos}, uncorrected pH_{obs}=6.1-6.2, 24°C, [cyt c]=1.1×10^{-3} M.

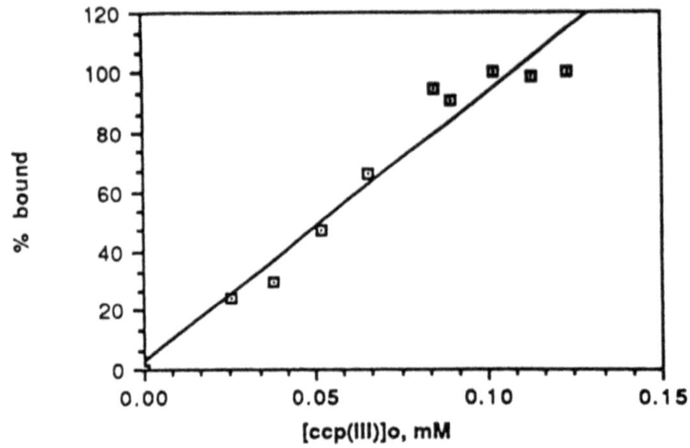

Fig. 5b. Low concentration titration of cyt c with ccp(III): μ=50 mM K_{phos}, uncorrected pH_{obs}=6.00-6.03, 25°C, [cyt c]=10^{-4} M, all samples run on Varian 500MHz NMR.

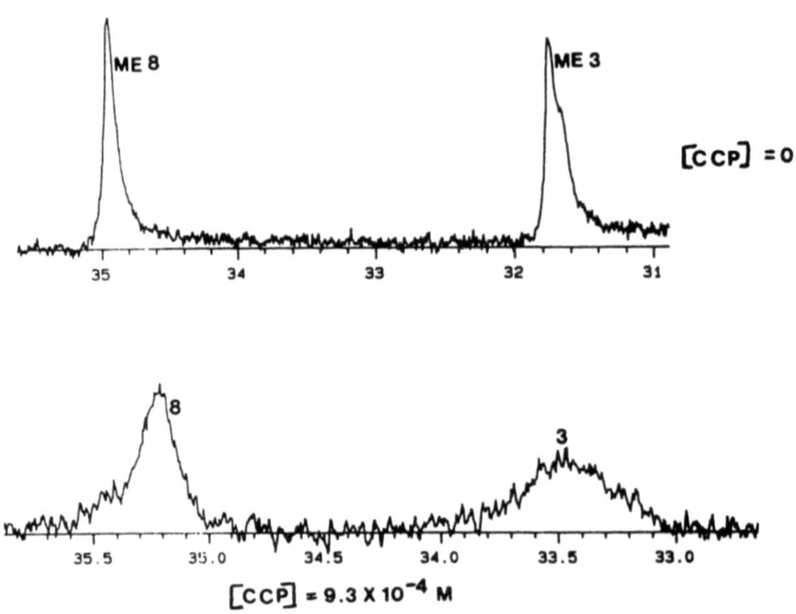

Fig. 5c. NMR titration at 10^{-3} M cyt c; a case of fast exchange.

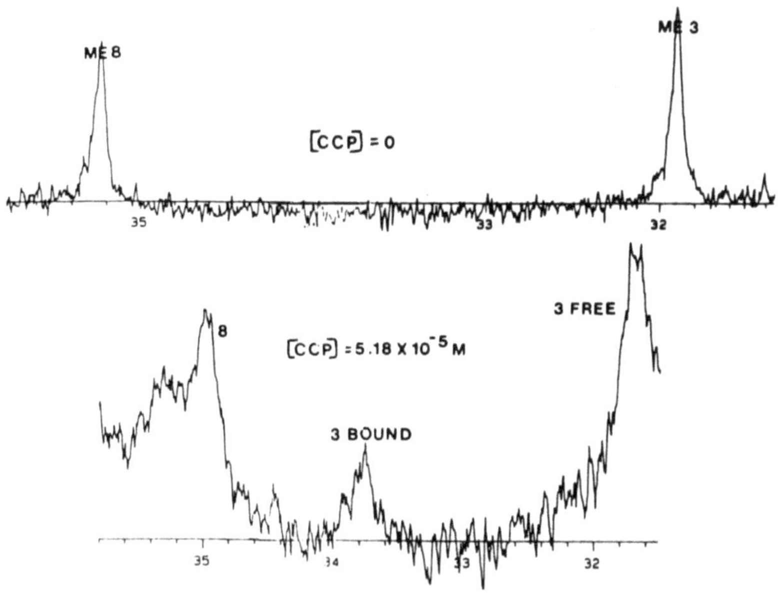

Fig. 5d. NMR titration at 10^{-4} M cyt c; a case of slow exchange.

In comparing binding of ccp with the mutant cytochrome c to that with wild type cytochrome c, mutations at the putative binding site give measurable but quantitatively small decreases in binding affinity between cyt c and ccp. We therefore suggest that while the binding domain between cyt c and ccp is correctly identified by previous chemical modification studies, and that charge interactions clearly contribute to this binding (as noted by the dependence of binding constant on ionic strength), any individual ionic interaction makes only a small quantitative contribution to binding. This is consistent with suggestions that the binding free energy is dominated by hydrophobic interactions.

Finally, it is noteworthy that the modification of individual residues changes the efficiency of energy transfer, which implies that reorientation occurs in the overall bound complex (Fig. 6).

This again suggests a relatively high plasticity of the binding domain. In response to a charge change at the binding site, cytochrome c may move along the ccp surface, and thereby affect the rate constant for energy transfer (which is quite sensitive to donor-acceptor distance and orientation). However, such movement only slightly affects the binding free energy.

Mutations of cytochrome c peroxidase

In order to better characterize the binding interaction of cytochrome c and ccp we have also examined mutant forms of ccp prepared by site directed mutagenesis techniques. Based on the binding site model of Poulos and Kraut shown in Fig. 2, we modified the three negatively charged aspartic acid residues which are key binding sites in this model, to positively charged lysines, thereby producing a double charge change. The binding of native and modified ccp derivatives to cyt c was monitored by fluorescence titrations using a fluorescent derivative of cyt c in which the heme is removed to produce the free base: H_2 porphyrin cytochrome c. As previously established by Kornblatt, the binding of H_2 porphyrin cyt c and ccp is somewhat complex; two cyt c sites exist on ccp as shown by the clear break in the titration curve at 0.66 (Fig. 7).

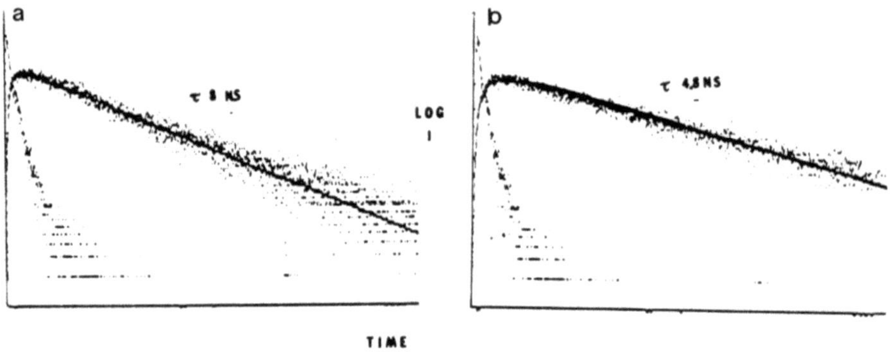

Fig. 6. Comparative luminescence decay kinetics of Mg cytochrome c peroxidase (a) and Mgccp bound to yeast iso-1 Fe(III) cytochrome c (b). Conditions: 10 μM Pi, pH 7.0, Mgccp ≤ cyt c = 10^{-5} M, insuring full binding. The difference in lifetimes between (a) and (b) is attributed to dynamic (Forster) energy transfer over a distance of ca. 20Å.

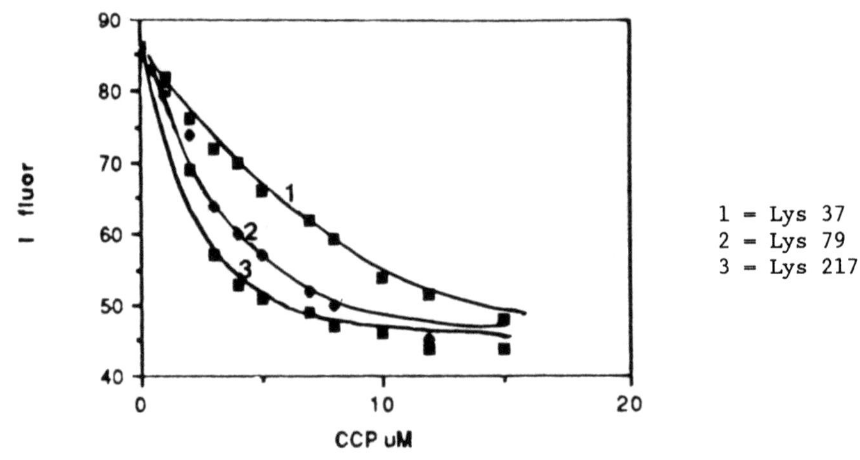

Fig. 7. Fluorescence titration of H_2 porphyrin cytc with ccp.

For these ccp derivatives, mutation can have a more marked effect, consistent with the double changes.

For Lys 217, little change in binding affinity is observed relative to wild type ccp. A modest decrease in binding is observed for the Lys 79 mutant. For the Lys 37 mutant, binding is significantly reduced suggesting that Asp 37 plays a more important role in binding than do the other two aspartates.

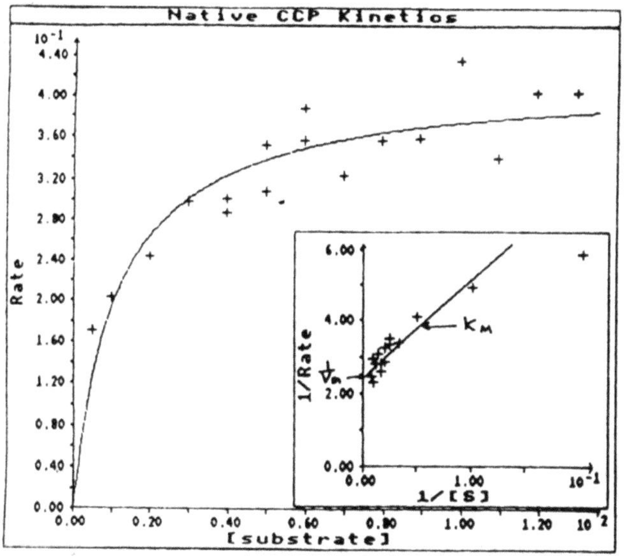

Fig. 8. Initial steady state catalytic rate studies of the reaction $ccp + H_2O_2 + 2cyt\ c(II) \rightarrow H_2O + ccp + 2cyt\ c(III)$

Kinetic Studies

With these binding studies in hand kinetic studies have been undertaken. The results of steady state kinetic analysis (Fig. 8) of the wild type proteins as well as the cyt c mutants and ccp mutants are summarized in Tables I and II.

Table I. Steady State Kinetics

System	K_M (μM)[a]	Turnover Rate (s^{-1} $\times 10^3$)
cyt c:ccp	3[b]	1.0
cyt c:ccp(Lys 79)	51	0.3
cyt c:ccp(Lys 217)	10	0.6
cyt c:ccp(Lys 37)	4.3	0.02
cyt c(Gln 32):ccp	3	1.0
cyt c(Ile 18):ccp	4.5	1.2
cyt c(Arg 77):ccp	6	2.0

a. K_M and V_M are the conventional Michaelis Menten parameters as defined in Fig. 8.
b. Under different conditions, $K_M = 43$ μM, $V_M = 4500\ s^{-1}$ G. Das *et al.* J. Biol. Chem. (1988) in press.

These kinetic characterizations are generally consistent with the equilibrium binding studies. For mutations of ccp the order of the turnover efficiencies is Lys 37 << Lys 79 < Lys 217 ≈ wild type. For the complementary cytochrome c mutants, the steady state turnover rates follow the trend Ile 18 ≥ Gln 32 ≈ wild type. A difficulty in the detailed interpretation of

Table II.[7] Single Turnover Kinetics Of Ccp With Cyt c Mutants

cyt c derivative	k_{obs} (s^{-1}) μ = 8mM[a]	μ = 260mM[b]
iso-1 (wild type)	260[c] 230[d]	1460
iso-2	150[e]	1480
Ser 21	200	1350
Lys 21	100	600
Ile 18	1000	2240
Gln 32	260	780
Asp 77	440	

a. Determined at a single protein concentration with lumiflavin semiquinone as the reductant. It has been shown that the intramolecular electron-transfer rate constant is not dependent on the flavin species utilized at low ionic strength.

b. Determined from nonlinear least-squares fit to the data based on the mechanism given by 3-5 represented as solid lines in Fig. 2 (Hazzard *et al.* Biochem. Biophys. Res. Comm. 1988, 151, 429.)

c. Measured at 30 μM in each component.

d. Measured at 50 μM in each component.

e. Value from Hazzard *et al.* Biochem. 1987, 26, 2836

steady state kinetics is that the observable parameters may be combinations of several fundamental rate constants. In order to better characterize the fundamental kinetics, we have begun detailed studies of conventional ("single turnover") kinetics. The key findings from the studies to date conducted in collaboration with Hazzard, Tollin, and Cusanovich are summarized in Table II. An interesting sidelight of these studies was our observation that contrary to previous expectations, the specific rate constant for intracomplex electron transfer increased with an increase in ionic strength. This observation means that the conformation of the complex which is most stabilized by charge-charge interactions does not correspond to the kinetically most competent conformation, and is consistent with motion at the interface of the protein complex being involved in the electron transfer reaction.

REFERENCES

1. G. McLendon, Accts. Chem. Res. 21:160 (1988).
2. G. McLendon and J. Miller, J. Am. Chem. Soc. 107:437 (1985).
3. a. K. Conklin and G. McLendon, Inor. Chem. 25:4804 (1986).
   b. N. Liang, C. H. Kang, P. S. Ho, E. Margoliash, and B. M. Hoffman, J. Am. Chem. Soc. 108:4665 (1986).
4. E. Margoliash and H. Bosshard, Trends Biochem. Science 8:316 (1983).
5. S. Northrup, J. Boles, and J. Reynolds, Science 241:67 (1988).
6. T. Poulos and J. Kraut, J. Biol. Chem. 255:10322 (1980).
7. J. Hazzard, G. McLendon, M. Cusanovich, G. Das, F. Sherman, and G. Tollin Biochem. 27:4445 (1988).
8. D. Hampsey, G. Das, and F. Sherman, FEBS Lett. 231:275 (1988).
9. G. McLendon, B. Hoffman, *et. al.* submitted to J. Am. Chem. Soc.
10. D. Holzschu, E. Stewart, P. McCoon, and R. Ciacarelli, Rev Indust. Micro. 29:161 (1988).

DYE BINDING TO POLYTRYPTOPHAN AND LYSINE-TRYPTOPHAN COPOLYMER:

ELECTRON TRANSFER QUENCHING BY BIOPOLYMER PENDANT GROUPS

Guilford Jones, II and Cathy Weiss
Department of Chemistry
Boston University
Boston MA 02215 USA

INTRODUCTION

Long range photoinduced electron transfer between potential electron donors and acceptors is of considerable current interest in terms of strategies for artificial photosynthesis and studies regarding the redox properties of proteins.[1] Electron transfer over organized arrays of significant dimension (e.g., 10 nm) may also be important for molecule-based bistable switches or rectifiers. As part of a program of study of long range electron transfer involving biopolymers, we report the results of a study of interaction of the xanthene dye, eosin, (EY), with synthetic peptides constructed of the amino acid residues, lysine (Lys), tryptophan (Trp), and tyrosine (Tyr). The principal interest in these systems has to do with the well ordered secondary structures (e.g., α-helices) adopted by peptide polymers and the capabilities for synthetic modification of peptide side chains and end groups with chromophores or electroactive substituents. In addition to providing mechanisms of charge transport, helical biopolymer backbones may find use in the regular assembly of dipoles or charges necessary for non-linear optical phenomena.

In the present study, eosin has been attached to the N-terminus of the homopolymer, polytryptophan (poly-Trp), or in a random fashion to lysine side chains ($-NH_2$ groups) of the random copolymer of Lys and Trp. Model compounds have also been prepared (Fig. 1) which display the thiourea linkage of dye and Trp residues which is common to the derivatized peptides. Our purposes in this initial investigation have been twofold: (1) to demonstrate a quenching interaction of excited states of eosin (principally the fluorescent singlet state) with neighboring (co-bound) functional groups on the peptide, and (2) to show that independent photooxidation of biopolymer-attached dye with an external electron transfer agent culminates in transfer of the photogenerated "hole" at the peptide terminus to a neighboring peptide group. The selection of amino acids for the synthetic peptides is based on the known electron donor properties of Trp and Tyr side chains (indole or phenol groups).[2]

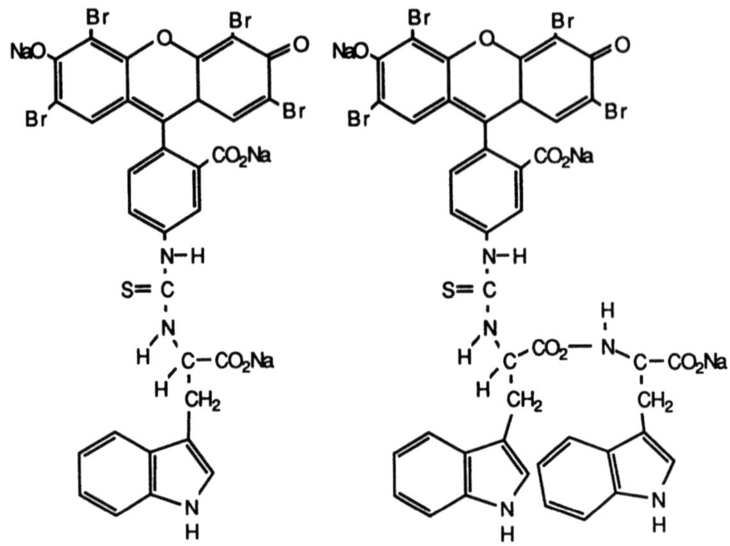

EY-Trp **EY-Trp-Trp**

Fig. 1
Model Compounds Displaying Dye - Trp Thiourea Linkage

The ultimate goal of the project is to demonstrate a long range transport of charge along the length of a peptide, preferably under circumstances in which the biopolymer adopts the α-helix structure. The model of long range electron (hole) transfer involving a terminal eosin dye label, side chains bearing donor groups (D) for hole transport, and a terminal donor (hole trap) is shown in Fig. 2.

EXPERIMENTAL SECTION

Materials

Eosin Y isothiocyanate (Molecular Probes) was used for attachment of eosin Y (EY) via a thiourea linkage to the N-terminus of the following amino acids and peptides (Fig. 2): L-tryptophan (Trp), L-Trp-Trp, and poly-Trp [avg. molecular. mass 4000, (20 residues)] were obtained from Sigma and used as received. EY was also covalently attached to lysine side chains of the random copolymers poly-Lys-Trp(4:1) and poly-Lys-Tyr(4:1) to a dye loading degree of ~ 1 dye molecule/polymer.[3] Tetranitromethane (TNM) (Aldrich) was used as received; acetonitrile was freshly dried over P_2O_5 and distilled, and 2-methoxyethanol was HPLC grade.

Laser Flash Photolysis System

The laser flash photolysis apparatus has been previously described.[4] It consists of a Quantel YG-581-10 Nd:YAG laser with frequency doubling and tripling capability, LeCroy Tr 8818 100-megasamples/s digitizer, an Oriel 150-W xenon monitoring lamp, an RCA 4840 PMT, an H-20 monochromator from Instruments SA, a Kinetic Systems stepping-motor controller for wavelength

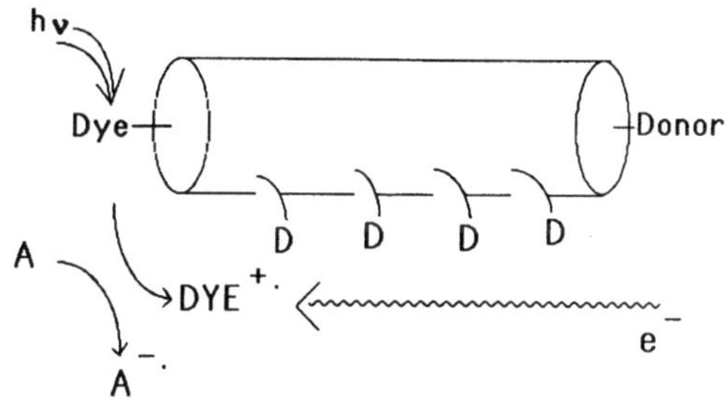

Fig. 2
Side Chain Mediated Electron Transport for a Helical Peptide

selection, a Kinetic Systems CAMAC interface crate, and a PDP-11/23 microcomputer for data acquisition and manipulation.

Transient Detection

All samples employed acetonitrile or 2-methoxyethanol as solvent and were sparged with argon for at least 30 minutes prior to photolysis. The pathlength of the monitoring beam was 2 cm, while the pathlength of the laser was 1 cm, with the monitoring beam perpendicular to the laser beam. Deaerated samples were subjected to the smallest number of laser pulses consistent with the necessary signal averaging (usually 10 pulses for decay curves).

Fluorescence Quenching

Fluorescence spectra for aerated solutions at room temperature were recorded on a Perkin-Elmer MPF-44A fluorescence spectrophotometer.

Quantum Yield Determinations

Quantum yields of photobleaching were determined by measuring the Δabsorption of sample at (λ_{ex} = 546 nm, λ_{mon} = 546 nm) The number of photons incident on the sample was determined by using a quantum counter (Rhodamine B fluorescence in ethanol)[5] to absorb a small constant percentage of the incident light from a 500-W mercury lamp which had passed through a monochromator. The system was calibrated by Aberchrome actinometry[6] at 546 nm (ϕ assumed = .060). Quantum yields were obtained from the slope of a plot of Δ concentration vs Einsteins absorbed and extrapolated to zero time to correct for the change in light absorbed as the chromophore bleached.

Circular Dichroism

Circular dichroic spectra were recorded in 0.1 cm quartz cuvettes at ambient temperature in a Cary Model 61 CD spectropolarimeter calibrated with d-10-camphorsulfonic acid.

RESULTS AND DISCUSSION

Fluorescence Quenching of Eosin Y in H_2O

Aqueous solutions of EY as the free (not covalently bound)

dye and the copolymers of Lys and Trp or Tyr in 4:1 ratio of residues (randomly arranged) were investigated for the quenching of EY fluorescence (λ_{em} = 545 nm, λ_{ex} = 525 nm). The pH of the solution was varied with carbonate and phosphate buffers (<0.01 M). Moderate concentrations of the copolymers (1.0 - 5.0 μM residues) are ineffective in quenching EY fluorescence when the pH of the solution is greater than 10.5, however lowering of the pH between pH 6.6 and 10 gives rise to an increased quenching interaction with dye. The results are consistent with the binding of EY in biopolymer domains due to electrostatic attraction of dye and the positively charged ($-NH_3^+$) pendant groups of Lys residues which remain substantially protonated at pH > 9.

In a second series of experiments, peptides covalently bound with eosin, EY-poly-Trp, EY-poly-Lys-Trp, and EY-poly-Lys-Tyr were examined. The fluorescence of EY is substantially quenched when bound to these polymers at pH 7-13. When fluorescence of EY-poly-Lys-Trp and EY-poly-Lys-Tyr is compared to that of EY-poly-Lys, it is found to be substantially more quenched when Tyr or Trp residues are present. This finding is consistent with the model of electron transfer of dye excited states by neighboring (or bound) aromatic groups associated with Tyr or Trp residues. Electron transfer quenching of EY triplet by free Tyr and Trp has been reported.[7] For the homopolymer, poly-tryptophan, with N-terminal EY, rate constants for an assumed electron transfer from the nearest Trp residue (ca. 1.2 nm distance according to a computerized molecular model)[8] have been estimated from relative fluorescence yield data for several solvents (Table 1.). Rate constants are obtained with the simplifying assumption that electron transfer (with a single rate constant, k_r) competes with radiative (k_f) and nonradiative (k_d) deactivation of excited dye, (i.e. in the simplified model only one dye polymer conformation is assumed).

Table 1

Fluorescence Yields of EY-poly-tryptophan vs Unbound Eosin Dye, EY

Solvent	ε	$\phi_{EY\text{-}pTrp}/\phi_{EY}$	k_r, s^{-1}
acetonitrile	37.5	0.52	2.4×10^8
DMF	36.7	0.73	7.4×10^7
2-MeEtOH	16.9	0.61	1.1×10^8
pyridine	12.4	0.47	2.8×10^8

$\lambda_{ex} = \lambda_{max}$; 515 - 535 nm, $\varepsilon = 9 \times 10^4\ M^{-1}\ cm^{-1}$

$\phi_{EY} = k_f/k_f+k_d$ $\qquad$ $\phi_{EY\text{-}pTrp} = k_f/k_f+k_d+k_r$

Circular Dichroism of Peptides

The conformational properties of poly-Trp investigated by circular dichroism have been reported[9] showing the α-helical nature of this homopolymer in 2-methoxyethanol. It is expected that the EY chromophore bound to the N-terminal of this polymer will participate in electron transfer with the helical peptide conformation intact for 2-methoxyethanol and similar solvents.

Ellipticity measurements have been made of the copolymers poly-Lys-Trp and poly-Lys-Tyr in buffered aqueous solutions from pH 6.6 - 11.5. A comparison with the standard circular dichroism measurements of poly-L-lysine[10] reveal the increased percentage of α-helix in both copolymers at higher pH (e.g. enhanced molar ellipticity at 208 and 222 nm). Like the homopolymer of Lys, the Lys-Trp and Lys-Tyr copolymers are sensitive to the charging of Lys residues in the lower pH range (< 10) (i.e. the highly charged peptides favor non-helical, random coil conformations).

Photooxidation of Bound EY by TNM

On laser flash photolysis (λ 532 nm) of eosin solutions containing the electron acceptor TNM, the dye triplet which absorbs at 580 nm[11] is not observed. The semioxidized eosin, $EY^{+\cdot}$ was formed in an irreversible photooxidation involving the triplet EY and TNM in acetonitrile and 2-methoxyethanol solutions. The semioxidized $EY^{+\cdot}$ has a characteristic transient absorption at 450 nm.[12] A comparison of 450 nm decay lifetime for the $EY^{+\cdot}$ and the EY tagged poly-Trp reveals an increase by a factor of nearly five for $EY^{+\cdot}$-polyTrp compared to $EY^{+\cdot}$ in 2-methoxyethanol and greater than five in acetonitrile (Table 2). The primary route of $EY^{+\cdot}$ radical decay is attributed to hole transfer from $EY^{+\cdot}$ to the neighboring Trp residue (Fig. 3). The kinetic data for the decay of eosin excited triplet species, the decay of the semioxidized intermediate, $EY^{+\cdot}$, and the first order rate constant computed for this hole transfer step are summarized in Table 2.

$$^{3}EY \quad + \quad C(NO_2)_4 \quad \longrightarrow$$

triplet 580 nm   TNM

$$EY^{+\cdot} \quad + \quad C(NO_2)_3^{-} \quad + \quad NO_2\cdot$$

semioxidized dye, 450 nm   trinitromethide anion, 350 nm

Fig. 3. Irreversible Photooxidation of Eosin Y

The quantum efficiencies of EY bleaching and formation of trinitromethide were investigated by steady photolysis of EY, free and conjugated with Trp and poly-Trp. A decrease in ϕ_{EY}(bleach) occurred upon attachment of Trp residues while the ϕ_{TNM}(reduction) remained high throughout the series (Table 3.). This result shows us that although TNM is photoreduced by electron transfer from eosin in all cases, the net photooxidation of dye is significantly diminished due to hole transfer with one or more neighboring biopolymer pendant groups.

Table 2

Flash Photolysis of EY-poly-Trp, EY-poly-Lys-Trp and EY in the Presence of Electron Acceptor, TNM

Solvent	[TNM], μM	triplet decay τ_{580}, μs	$EY^{+\cdot}$ decay τ_{450}, μs	hole trap 10^{-4} k, s^{-1}
2-MeEtOH				
EY	-	220		
EY	380		2900	0.034
EY-polyTrp	-	160		
EY-polyTrp	380		640	0.16
acetonitrile				
EY	-	67		
EY	7.4	13	100	1.0
EY-polyTrp	7.3		19	5.3
EY-polyLysTrp	190		5.7	18

λ_{ex} = 532 nm (OD@532 nm = 0.7); [EY] = 5 - 8 μM

CONCLUSIONS

The fluorescence of eosin in H_2O is quenched on electrostatic attachment of the anionic dye to random peptide copolymers of lysine and tryptophan and lysine and tyrosine (4:1 ratio of amino acid residues). The copolymers in α-helical conformations, with uncharged Lys residues (pH > 10.5), are ineffective in eosin fluorescence quenching. A mechanism of static quenching (eosin counterion binding in the ground state) and reductive quenching by co-bound Trp or Tyr (indole or phenol) groups is proposed.

Eosin has been covalently attached to the homopolymer, poly-tryptophan, and to Lys-Trp and Lys-Tyr copolymers via the thiourea linkage. The fluorescence of covalently bound eosin is substantially (but variably) quenched due to electron transfer interaction with neighboring Trp or Tyr groups. For the

Table 3

Quantum Efficiencies of EY bleaching and $C(NO_2)^{3-}$ Formation.

System	ϕ_{EY}	ϕ_{TNM}
EY/TNM	0.12	0.16
EY-Trp/TNM	0.04	0.31
EY-polyTrp/TNM	0.02	0.22

[EY] = 5 x 10^{-5} M,; [TNM] = 1.1 x 10^{-3} M; steady irradiation at 546 nm of acetonitrile/H_2O (1:1 v/v) solutions

homopolymer, poly-Trp, with N-terminal eosin, rate constants for an assumed electron transfer from a nearest neighbor Trp residue (~ 1.2 nm distance) have been estimated from relative fluorescence yield data for several solvents.

Peptide bound eosin has been photooxidized via dye triplet quenching by tetranitromethane (TNM). Rate constants for hole trapping due to neighboring Trp groups have been measured by laser flash photolysis. Steady state measurements have been made of the quantum yield of (oxidative) bleaching of bound eosin by TNM. The indications are that hole transfer to peptide is effective in suppressing net eosin oxidation due to the activity of electron donor (Trp) side chains.

ACKNOWLEDGEMENTS

The authors gratefully acknowledge the support of the office of Basic Energy Sciences of the Department of Energy and the Donors of the Petroleum Research Fund, administered by the American Chemical Society.

REFERENCES

1 McLendon, G., Acc. Chem. Res., (1988), 21, 160.

2 Merenyi, G., Lind, J., and Shen, Xinhua, J. Phys., Chem., (1988), 92, 134.

3 Cherry, R.J., Cogoli, A., Oppliger, M., Schneider, G., and Semenza, G., Biochemistry, (1976), 15, 3653.

4 Malba, V., Jones, G.II, Poliakoff, E.D., Photochem. and Photobiol., 1985, 42, 451.

5 Jones, G.II,and Becker, W.G., J. Am. Chem. Soc., (1983), 105, 1276.

6 Heller, H.G., Langan, J.R., J. Chem. Soc. Perkin. Trans. II, (1981), 2, 341.

7 Grossweiner, L.I. and Kepka, Photochem. and Photobiol., (1972), 16, 305.

8 Jones, G.II and Oh, C., unpublished results.

9 Muto, K., Mochizuki, H., and Yoshida, R., J. Am. Chem. Soc., 1986, 108, 6416.

10 Greenfield, N. and Fasman, G.D., Biochemistry (1969), 8, 4108.

11 Zwicker, E.F. and Grossweiner, L.I., J. Phys. Chem., (1963), 67, 549.

12 Kasche, V. and Lindqvist, L., Photochem. and Photobio., (1965), 4, 923

PHOTOACTIVE INTERFACIALLY POLYMERIZED POLYPORPHYRIN FILMS

G.W. Rayfield, D.T. Friesen and D. Lorenz
Bend Research, Inc.
Bend, OR 97701-8599

C. Wamser
Chemistry Department
Portland State University
Portland, OR 97207-0751

ABSTRACT

Thin asymmetric polyporphyrin films have been made that show photovoltages upon illumination. The thin films are produced by interfacial polymerization using an acid chloride derivative of tetraphenylpophyrin (TCCPP) in the organic (chloroform) phase and diethylenetriamine (DET) in the aqueous phase. The resulting polymer films are from 70 to 800 nm in thickness. When such a polyporphyrin film is sandwiched between semitransparent electrodes and illuminated, a photovoltage is observed indicating an asymmetry in which the acid chloride side becomes negative. Scanning electron microscopy studies indicate different surface texture of the amine and acid chloride sides.

INTRODUCTION

The objective of this work is to form an asymmetric membrane that under illumination produces efficient charge separation and a net increase in chemical potential. We are attempting to mimic the electron-transfer process of natural photosynthesis in a synthetic-membrane system. It is envisioned that the energy gained by the electrons as they cross the synthetic membrane can ultimately be used to split water or to provide an electrical current through an external circuit.

Our approach to developing a synthetic membrane that mimics photosynthesis involves the novel use of interfacial polymerization to produce an asymmetric polymerized membrane of substituted porphyrins. Interfacial polymerization is a condensation reaction between organic-soluble monomers and aqueous-soluble co-monomers, or crosslinkers, at the interface between the organic solvent and aqueous solvent. In our work, an organic-soluble tetrafunctional porphyrin is reacted with an aqueous-soluble amine at the organic/aqueous interface to form a polyporphyrin film. The concentration gradients of the reactants at the interface cause the film to form in such a way that different porphyrin derivatives of different redox potential are incorporated asymmetrically in the film. Since each porphyrin derivative has a different redox potential, a net gradient in redox potential exists across the film. This gradient provides a driving force for charge separation of the electron-hole pair that is created when light is absorbed by a porphyrin molecule in the film. A photovoltage is produced when the polyporphyrin is sandwiched between identical semitransparent electrodes and illuminated with visible light. This photovoltage is evidence for long-lived charge separation in the illuminated membrane.

MATERIALS AND METHODS

Abbreviations

Tetraphenylporphyrin (TPP) with four X substituents on the phenyl ring is shown in Figure 1. The abbreviations used for symmetrical porphyrins are:

(1) TCPP X = COOH Tetracarboxylphenylporphyrin
(2) TCCPP X = COCl Tetracarboxylchlorophenylporphyrin
(3) TCAPP X = CONHR R = $(CH_2)_2NH(CH_2)_2NH_2$

The abbreviation "C_n" is used for an unsymmetrical porphyrin with n substituents of type CONHR and 4-n substituents of type COOH. Thus C_4 is used to designate CONHR. C_0 or TCPP is not found in the polyporphrin membrane since it has no amide linkage.

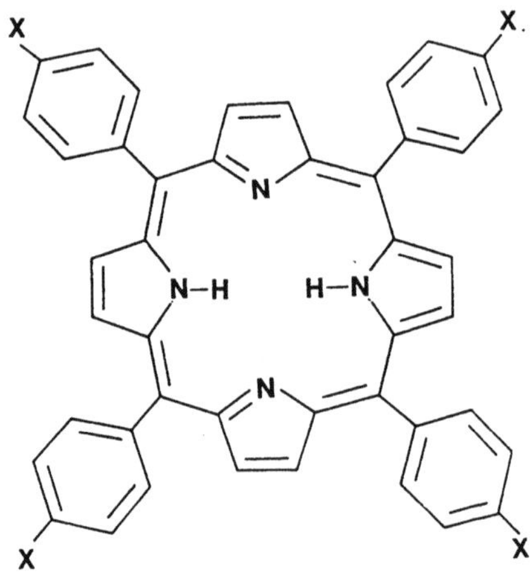

Figure 1. Tetraphenylporphyrin (TPP).

The side of the interfacial film in contact with organic solvent containing polyfunctional acid chloride (TCCPP) is designated as "C". The side of the film in contact with aqueous solution containing polyfunctional amine (DETA) is designated as "N" (see Figure 2). Diethyltriamine is abbreviated as "DETA".

Formation of the Polyporphyrin Film

Figure 2 shows a schematic of the technique used to form the interfacial polymer film. The organic solvent containing 0.1% TCCPP is placed in a Petri dish. An aqueous solution of DETA is placed on top of the organic solvent.

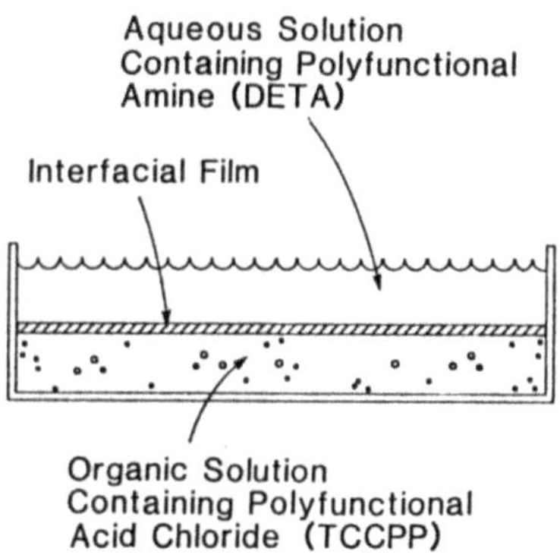

Figure 2. Processes for interfacial polymerization.

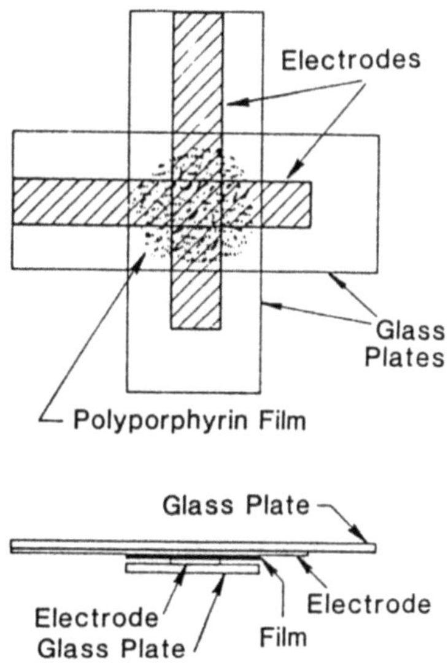

Figure 3. Device for testing the photoresponse of the polyporphyrin film.

At the liquid/liquid interface, an amino group of the DETA reacts with a carboxylchoride group of the TCCPP via a condensation reaction, forming an amide linkage between the porphyrin molecules. Continued reaction of amine and carboxylchoride groups results in the formation of an amide-linked polyporphyrin film. Rapid formation of this polymer film at the liquid/liquid interface limits the thickness of the film by preventing diffusion of the reactants through the film. The film is allowed to form at the interface of the two liquids for 1 to 10 minutes. The film is then transferred to a solution of tetrahydrofuran, followed by a wash in distilled H_2O adjusted to the desired pH (usually pH 5). The wash solutions convert the unreacted carboxylchlorides to COOH and remove unreacted monomers from the polymer.

Photovoltage Measurements

To study the photoresponse of the membranes, a membrane is clamped between two identical semitransparent electrodes, as shown in Figure 3. (Two different materials have been used for the electrodes: tin oxide and paladium-gold [about 60 angstroms thick].) One of the electrodes is supported by a glass substrate (microscope slide). The wet polyporphyrin membrane is lifted onto the electrode support and covered by the other semitransparent electrode. The active area of the membrane, which is determined by the size of the electrodes, is about 0.16 cm^2. The device is vacuum-dried before photoresponse measurements are taken. The only asymmetry in this device is due to the asymmetry built into the polyporphyrin membrane.

Two types of light sources were used for measuring the photoresponse of the polyporphyrin membranes: 1) an incandescent light source of our own design (using a 150-W quartz halogen lamp) with an electric shutter, and 2) a flash-tube pumped dye laser (Phasar) operating at 590 nm. The light intensity at the membrane using the incandescent light source was 20 mW/cm^2. Optical filters were used to filter out wavelengths above 700 nm and below 400 nm. The opening and closing of the electric shutter in the incandescent light source was controlled by a function generator (Krohn-Hite 5200A) and was variable from 1 ms to several seconds. The output pulse from the dye laser had a duration of 0.3 microseconds and an intensity of about 100 mJ per cm^2.

Photovoltages were measured using a voltage-follower circuit with an input impedance of 10^{10} ohms and a rise time of less than 1 microsecond. A digital storage oscilloscope (Tektronix 2220 or Nicolet 4094) connected to the output of the voltage follower was used to record the photovoltage. Data was transferred from the storage oscilloscope to a microcomputer (IBM-PC) and stored on floppy disc.

EXPERIMENTAL RESULTS

General Membrane Characteristics

The thickness of the membrane was determined both by weighing (using the film density) and from capacitance measurements using an estimated value for the dielectric constant. We found the thickness to be approximately 1 micron for a 10-minute reaction time. Shorter reaction times yielded thinner membranes.

The electrical resistance of the membranes in the dark varied from 10^{11} ohms to 10^{12} ohms, depending on membrane thickness.

The absorption spectrum of the film is shown in figure 4. It is approximately the same as the absorption spectrum for monomers.

The capacitance of the membrane was about 30 pF.

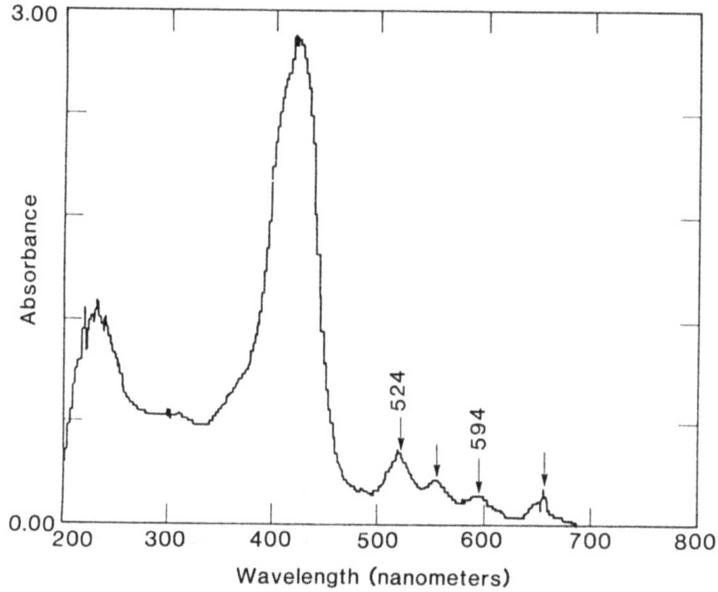

Figure 4. Absorption spectrum of the polyporphyrin film.

The amine surface, N, of the dry membrane was found to be hydrophilic; the acid chloride side, C, is hydrophobic. Electron micrographs of the membrane indicate that the acid chloride side is very "spongy" compared to the amine side.

A typical photovoltage response of a polyporphyrin membrane illuminated with incandescent light is shown in Figure 5. The "up" arrow in Figure 5 indicates "light on," and the "down" arrow "light off." The acid chloride side of the membrane became electrically negative relative to the amine side, N, of the membrane. This result indicates that the asymmetry introduced into the membrane by interfacial polymerization promotes charge separation when the membrane is illuminated. The sign of the photovoltage indicates that electrons move toward side C of the membrane and away from side N.

Photovoltage data taken with the flash-tube pumped dye laser is shown in Figure 6. Consistent with the incandescent-light measurements (Figure 5) the displacement of negative charge is toward the C surface of the membrane. Substituted porphyrins on this side of the membrane function as acceptors relative to other substituted porphyrins.

The continued rise of the photovoltage after the short (1-microsecond) laser-light pulse indicates that charge separation continues for tens of microseconds after the one-microsecond laser-light pulse.

DISCUSSION AND CONCLUSIONS

The photovoltage measurements show that long-lived light-induced charge separation occurs in the polyporphyrin

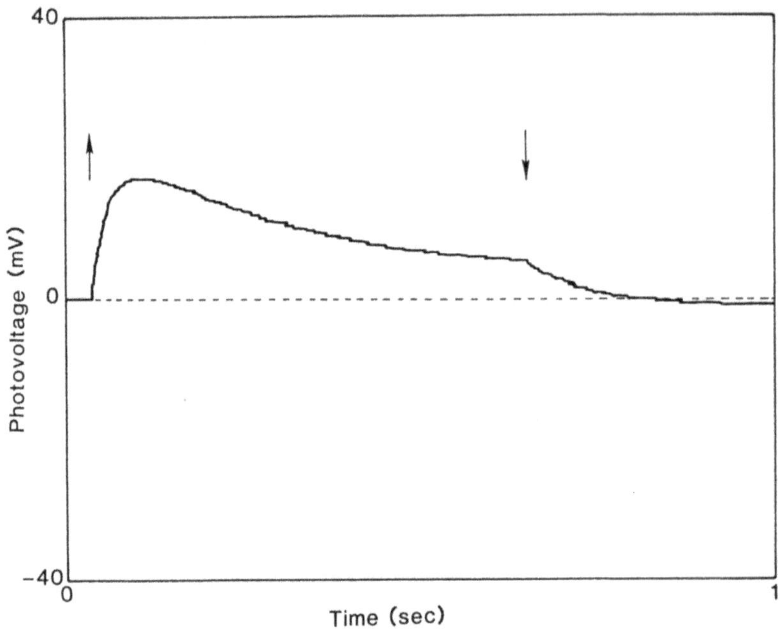

Figure 5. The measured photovoltage response of the polyporphyrin membrane is shown.

membrane. Moreover, the microsecond laser-flash studies indicate that photoactivated electrons continue to move for a relatively long time after the laser flash. Taken together these results provide evidence for a driving force

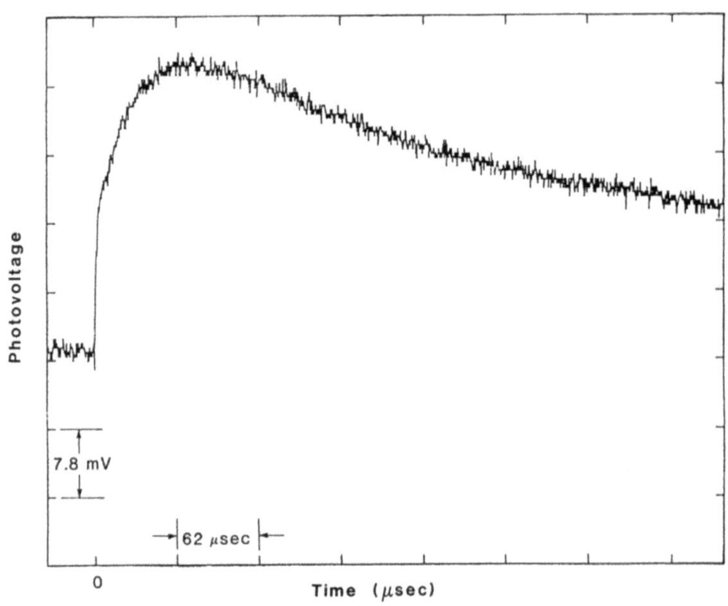

Figure 6. The photovoltage response of the polyporphyrin membrane illuminated by a 1 μsec laser flash is shown.

leading to charge separation in the asymmetric polyporphyrin membrane. Charge separation in the polyporphrin membrane could occur via electron transfer between localized electronic states or through delocalized electronic states similar to those found in amorphous silicon. Current experimental data is insufficient to resolve these two possible electron-transfer mechanisms. At this time, we consider only electron transfer between localized electronic states.

Photoinduced Electron Transfer Between Localized Electronic States

Consider the photoexcitation to the excited state D^* of a particular porphyrin in the polyporphyrin film. Electron transfer from this photoexcited porphyrin to an adjacent (acceptor) molecule (see Figure 7a) is possible. The rate constant for the electron transfer depends on the energy level (or redox potential) of the acceptor state relative to D^*. This two component system may not lead to long-term charge separation due to rapid back reactions. The acceptor state can be made more stable against back reactions by decreasing the driving potential for the back reaction, but then much of the photon's energy goes to useless heat. Increasing the separation between the donor and acceptor also slows the back reaction, but direct recombination dominates over charge separation, and the quantum efficiency is low.

One solution to the problem of attaining long-lived charge separation with good quantum efficiency and energy

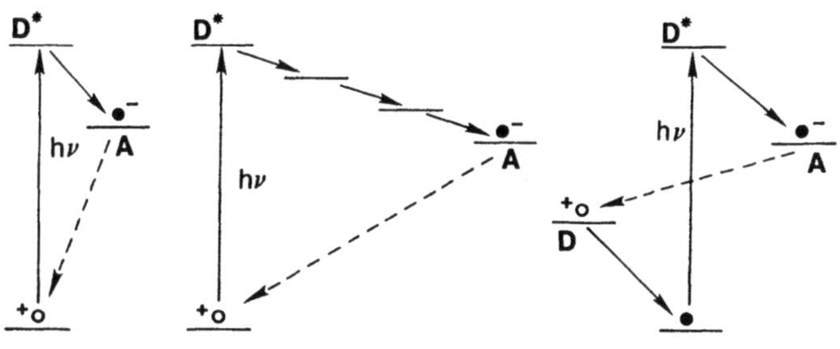

Figure 7. (a) Shows a two-component model system; (b) shows a multistep electron-transfer process; (c) shows electron transfer from a donor to fill a hole.

storage is to introduce a series of electron relay molecules between the photoexcited donor and the ultimate acceptor molecule. Thus, the introduction of intermediate relay molecules, as shown in Figure 7b, can generate long-lived stable charge separation with good quantum efficiency. Photosynthetic reaction centers [2] use a multistep electron-transfer process in a highly ordered supramolecular structure to efficiently convert light energy into chemical energy.

Figure 7c shows another means by which the light-induced charge separation can be efficiently stabilized. The hole left behind by the photoinduced electron transfer is filled by a nearby donor molecule, effectively moving the hole further away from the photoexcited electron and thereby decreasing the back reaction rate.

P-Q Electron Transfer

Recently there has been an effort to construct photosynthetic molecular models consisting of porphyrin-quinone assemblies separated by rigid spacers in order to mimic photosynthetic electron- and energy-transfer processes [3,4]. These P-Q systems are characterized by the model shown in Figure 7a. In this figure, the donor is the photoexcited porphyrin and the acceptor is the quinone. Charge transfer in the P-Q systems has been studied spectroscopically by measuring fluorescene lifetimes.

Solvent, distance, temperature, and exothemicity effects on light-induced electron transfer between the porphyrin and the quinone have been studied. Our assymetric polyporphyrin membrane is much more complex than the two-component, one-dimensional P-Q system and may have efficient charge separation with smaller driving potentials than those required for the P-Q system.

Electron Transfer in the Polyporphyrin Membrane

The interfacial-polymerization technique used to form the polyporphyrin membrane imposes an average long-range order across the membrane. The gradient in substituted porphyrins of Type C4 on the "N" side of the membrane to C1 on the "C" side of the membrane provides a vectorial driving force (mean field) for the multistep electron process that leads to charge separation in the membrane. On a microscopic scale the electron-transfer process between substituted porphyrins is not unidirectional, but rather like a three-dimensional random walk with a higher probability for steps toward the "C" membrane surface than toward the "N" surface.

The P-Q results indicate that back reactions would dominate in the polyporphyrin film, since many of the electron-transfer processes must take place between substituted porphyrins of the same redox potential. Moreover, the driving potential across the membrane due to the asymmetric incorporation of substituted porphyrins is not large (probably only a few tens of millivolts). However, certain features of the polyporphyrin membrane tend to improve electron transfer relative to the P-Q model system. The three dimensionality of our system provides more channels for electron transfer than does the one dimension of the covalently linked P-Q system, and the greater number of channels tends to increase the rate constant for electron transfer. Multistep processes are possible in the polyporphyrin film (Figure 7b). Finally, free amines may donate electrons to D^* which will stabilize the light-induced charge separation (Figure 7c).

Some asymmetric polyporphyrin membranes were soaked in

a methylviologen solution. Methyviologen is known to function well as an electron relay system, and preliminary results indicate that significantly larger photoresponses are obtained from these membranes. Photocurrents have also been observed, and these results will be reported in detail in a future publication.

REFERENCES

1. Morgan, P.W. Condensation Polymers by Interfacial and Solution Methods in Polymer Reviews (1965) Mark, H.F.& Immergut, E.H. eds., vol. 10., Wiley-Interscience, New York.

2. Melandri, B.A. & Venturoli, G. Photosynthetic Electron Transfer in Bioenergetics (1984) Ernster, L. ed. Elsevier Science Publishers.

3. Joran, A.D., Leland, B.A., Felker, P.M., Zewail, A.H., Hopfield, J.J. & Dervan, P.B. (1987) Effect of exothermicity on electron transfer rates in photosynthetic molecular models. Nature 237 508-511.

4. Gust, D. & Moore, T.A., Electron Transfer in Model Systems for Photosynthesis. in Supramolecular Photochemistry (1987) 267-282. Balzani, V. ed. D. Reidel Pub. Co.

LIGHT ENERGY TRANSDUCING AND SIGNAL TRANSDUCING RHODOPSINS OF *HALOBACTERIA*

Walther Stoeckenius

Department of Biochemistry and Biophysics and
Cardiovascular Research Institute, University of California
San Francisco, CA 94143-0130, U.S.A.

Bacteriorhodopsin (bR) is a purple-colored 26 kDa protein with a retinal chromophore and a broad absorption band centered at 570 nm. The molar extinction coefficient is 64,000 M^{-1} cm^{-1}. It forms two-dimensional crystals in the ~5.0 nm-thick cell membrane of *Halobacterium halobium*, which are easily isolated and known as the purple membrane (pm). Halobacteria are found in natural salt lakes and salterns; they require molar concentrations of NaCl for growth, and through bR, can use sunlight as an energy source, because, in the light, bR ejects protons from the cell and generates an electrochemical gradient across the cell membrane. For a brief recent review, see Stoeckenius (1985).*

The amino acid sequence of bR, and its structure to ~7 Å, are known. The polypeptide chain forms seven α-helical segments which span the membrane and have short alternating connections on both sides of the membrane. The α-helices are arranged in two parallel rows and the planar retinal chromophore is inserted between the rows at an angle of ~20° with the membrane plane. It is bound, via a protonated Schiff base, to a lysine residue in the α-helix, next to the carboxyl terminus. Between the protein molecules, lipid molecules are inserted in a bilayer arrangement, but they account for only 25% of the purple membrane mass and, differing from most other biological membranes, do not form a continuous lipid phase. Bacteriorhodopsin in the purple membrane is an unusually stable protein. It can withstand, at least for a limited time, temperatures above 80° C, pH values of 1 or 11, and drying in air or a vacuum. In contrast to most other halobacterial pro-

*Mainly references to recent reviews are given. They may be consulted for access to the primary publications.

teins, it does not require high salt concentrations for preservation of structure or function.

In bright daylight, bR contains only the all-*trans* isomer of retinal. In the dark, the absorption maximum shifts slowly to 558 nm and the absorption decreases by 12%. This dark-adaptation at room temperature and neutral pH takes approximately two hours and is accompanied by isomerization of two thirds of the all-*trans* to 13-*cis* retinal with simultaneous isomerization of the C=N bond to the *syn*-conformation. Only the light-adapted bR translocates protons efficiently.

Since bR functions as a light-energy transducer, it must also undergo a fast cyclic photoreaction. This photocycle of the light-adapted, all-*trans* retinal-containing bR has a halftime of a few milliseconds under physiological conditions and has been extensively analyzed by time-resolved absorption, resonance Raman, and infrared spectroscopy (Smith, et al., 1985; Rothschild, 1988). Recent results support a quantum efficiency for the photocycle of ~0.6 and translocation of one proton/cycle, whereas earlier data indicated values of ~0.3 and two protons/cycle (Kouyama, et al., 1988). Seven intermediates in the photocycle, with life-times in the picosecond to millisecond range, have been characterized, and the early intermediates, K, L, and M, have been trapped at low temperature (Fig. 1). During the photocycle, the retinal chromophore isomerizes around the $C_{13}=C_{14}$ double bond and the Schiff base transiently deprotonates. (The 13-*cis* intermediates of light-adapted bR differ from the 13-*cis* component of dark-adapted bR because the C=N bond remains *anti* in the photocycle intermediates. It should be noted that the 13-*cis* component of dark-adapted bR forms a K-intermediate in most respects very similar to the K of all-*trans* bR (not shown)). In addition, transient protonation changes of aspartate and tyrosine residues located in the membrane interior have been observed. Protons are released from the external membrane surface and replaced by protons taken up at the cytoplasmic surface. In uniformly oriented preparations of pm, transient photocurrents arising from charge shifts within the membrane, as well as photosteady state currents due to the net proton transport, have been measured. They can be resolved into kinetic components which, at least in part, parallel the absorption changes; however, the fastest capacitive current remained unresolved at 20 ps time resolution (Keszthelyi, 1988).

The analysis of the photocycle is facilitated by the fact that the chromophore can relatively easily be replaced by isotopically labelled or chemically modified retinal or retinal analogues. Also, because the bR gene has been cloned and expressed in *E. coli*, it is possible to modify the protein by substituting or deleting single amino acids or whole segments of the polypeptide (Khorana, 1986). Growth on synthetic media allows incorporation

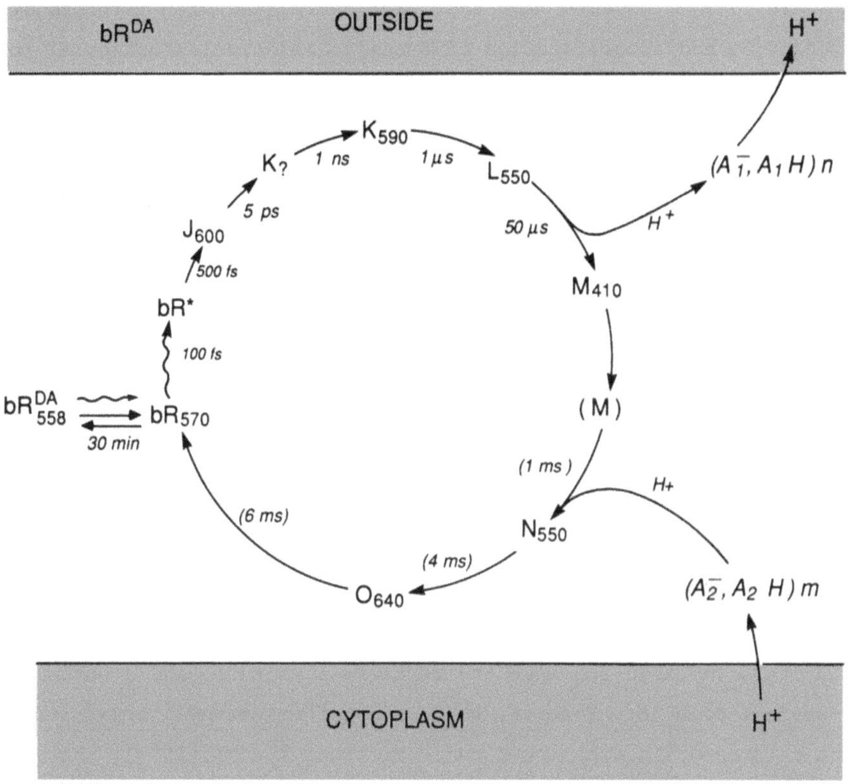

Fig. 1. The photocycle of bR_{570} (simplified). The intermediates are labeled J to O, the subscripts give their calculated visible absorption maxima. The time constants shown are for room temperature and near neutral pH. The time constants in brackets indicate that these values are generally very sensitive, but vary in their sensitivity, to environmental conditions, especially pH. Schiff base deprotonation occurs in the L → M transition and reprotonation during N formation. The 13-*trans*-to-*cis* isomerization of retinal takes place in the excited state and the reisomerization to 13-*trans* in the N → O transition. For most intermediates, it has been shown that absorption of a photon reconverts them to bR faster than the thermal decay.

It is generally assumed that the Schiff base proton is translocated to the external medium via one or more groups in the protein (A_1^-, A_1H) and similarly replaced from the cytoplasmic side (A_2^-, A_2H). This scheme requires two M states, one connected to the exterior surface and the second (M) connected to the cytoplasmic surface. The latter has, so far, not been identified spectroscopically.

Thermal backreactions of late intermediates as well as photoreactions have not been indicated, but may be significant. A K-intermediate is formed and photoreversible, even at or below liquid nitrogen temperature. The L-intermediate may be trapped at 120° K and the M-intermediate at 200° K. Drying pm in air at ~90% humidity has no significant effects on the photoreaction cycle. Reducing the water content further mainly affects the late cycle intermediates and finally stops proton translocation.

of isotopically labeled amino acids. Protonation changes of internal aspartic acid and tyrosine residues have thus been correlated with intermediates in the photocycle and identification of specific amino acids essential for the transport function is rapidly accumulating.

Molecular models for the transport mechanism consistent with these observations have been developed. In general, they assume that a pK change of the Schiff base provides the driving force for the proton transport and that the isomerizations around the 13,14 double bond provide alternating connections to the outer and inner membrane surface. The protonation changes of aspartate and tyrosine residues are interpreted as steps in the transport of the Schiff base proton to the exterior surface and its replacement from the cytoplasmic surface. Alternative models, in which the Schiff base is not in the path of the transport but drives it indirectly, have not been ruled out. They have, however, not been developed in sufficient detail to allow critical tests.

A second, very similar light energy-transducing protein, halorhodopsin (hR), also found in the membranes of the same bacteria, has a 30% homology with bR in its amino acid sequence and tertiary structure, and a chromophore nearly identical to bR's. However, hR does not form extended crystalline arrays and is present in much lower concentrations (Lanyi, 1986). Halorhodopsin translocates chloride ions, which have to be transported into the cell against the membrane potential. The translocating photocycle of hR does not include Schiff base deprotonation, but the other intermediates are almost identical in their chromophore structure and kinetics to those of bR. The retinal pigment is less stable than bR, its photoreactions are more sensitive to pH and salt variations, especially anion species and concentration changes. Halorhodopsin is difficult to isolate in pure form and models explaining the chloride transport mechanism are still highly speculative.

In addition to these two light energy transducers, two signal transducers, sensory rhodopsins I and II (sR-I and sR-II), with absorption maxima at 588 and 548 nm, respectively, have been identified in the same organism (Spudich and Bogomolni, 1988). They mediate the phototactic responses which allow the cells to seek out environments with optimal light conditions. Their most striking differences from bR and hR are the slower photocycles with halftimes of 0.3 and 0.7 s, respectively. Sensory rhodopsins are not electrogenic ion transporters. The nature of the intracellular signal which they transmit, and which controls the flagellar motor, is still unknown. Preliminary evidence indicates that it is generated by a long-lived intermediate analogous to M_{410} in the bR cycle.

In the context of this symposium, it may be interesting to point out that all halobacterial rhodopsins, as well as the visual pigments, show very similar photoreactions. Especially the early intermediates have a very fast risetime, are red-shifted, stable but photoreversible at low temperature, and largly independent of environmental conditions. Introduction of modified chromophores can shift the wavelengths of these pigments and their intermediates and alter the photocycle kinetics. Only the halobacterial pigments undergo cyclic photoreactions, which do not require additional energy input for completion. Among them, bR, in its planar crystal state, is by far the most stable and most easily obtainable in that form, i.e. as purple membrane.

References

Keszthelyi, L., 1988, Primary charge motions and light-energy transduction in bacteriorhodopsin. Biophys. Chem., 29:127.

Khorana, H. G., 1986, Rhodopsin, the visual pigment, and bacteriorhodopsin. Ann. N. Y. Acad. Sci., 471:272.

Kouyama, T., Nasuda-Kouyama, A., Ikegami, A., Mathew, M. K., and Stoeckenius, W., 1988, Bacteriorhodopsin photoreaction: identification of a long-lived intermediate N (P, R_{350}) at high pH and its M-like photoproduct. Biochemistry, 27:5855.

Lanyi, J. K., 1986, Halorhodopsin: a light-driven chloride ion pump. Ann. Rev. Biophys. Biophys. Chem., 15:11.

Rothschild, K. J., 1988, Infrared studies of bacteriorhodopsin. Photochem. Photobiol., 47:883.

Smith, S. O., Lugtenburg, J., and Mathies, R. A., 1985, Determination of retinal chromophore structure in bacteriorhodopsin with resonance Raman spectroscopy. J. Membrane Biol., 85:95.

Spudich, J. L., and Bogomolni, R. A., 1988, Sensory rhodopsins of halobacteria. Ann. Rev. Biophys. Biophys. Chem., 17:193.

Stoeckenius, W., 1985, The rhodopsin-like pigments of halobacteria: light energy and signal transducers in an archaebacterium. Trends Biochem. Sci., 10:483.

VISUAL PHOTOTRANSDUCTION: BIOCHEMICAL ASPECTS

Hitoshi Shichi

Eye Research Institute
of Oakland University
Rochester, Michigan 48309-4401
U.S.A.

This paper is a brief overview of the current status of research on the molecular mechanism of visual phototransduction and does not discuss the diversified aspects of the visual process in detail, for which the reader is referred to other reviews[1-4] and monographs.[5-6]

What is the architectural uniqueness of photoreceptor cells?

In the vertebrate eye a light signal is received and programmed by the retina, a tissue composed of several layers of neural cells. Light is absorbed by the visual (photoreceptor) cells located behind the neural cells which encode and transmit the photosignal to the visual cortex of the brain.

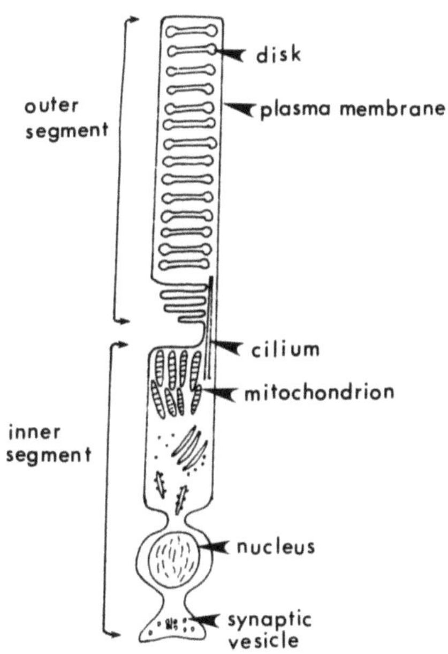

Fig. 1. Rod visual cell

Each visual cell is composed of two parts: the outer segment with visual (photoreceptor) pigment and a set of enzymes involved in the phototransduction process; and the inner segment which contains metabolic machinery and supports the functions of the outer segment (Fig 1). The two segments are connected by the ciliary process (cilium). Thus, the outer segment as the photoreceptor and the inner segment as the metabolic powerhouse are functionally differentiated.

The vertebrate visual cells are classified into two types; the rods responsible for dim light (black and white) vision and the cones essential for daylight (color) vision. The rod outer segment consists of a stack of several hundred disks encased in a sack of plasma membrane. The disks are presumably formed by evagination of the plasma membrane under the control of the ciliary process. Two adjacent evaginates fuse and detach from the plasma membrane to form a disk. Membrane evaginates in the cone outer segment do not seem to fuse and remain continuous with the plasma membrane. Outer segment membranes (>95% disks and <5% plasma membrane) contain high levels of unsaturated fatty acids such as docosahexaenoic acid (C22 and six unsaturated bonds) linked to phospholipids and show high membrane fluidity at physiological temperatures. The fluid environment of membrane may be important for photoactivated rhodopsin to collide quickly with peripheral membrane proteins which are involved in the phototransduction process.

<u>How are the visual pigments disposed within the photoreceptor cell membrane and how are they activated by light?</u>

The rod visual pigment rhodopsin is an integral membrane protein of a molecular weight of about 40,000. Half of the mass is embedded in the hydrophobic region of the membrane bilayer, with the remaining half distributed equally on both surfaces of the disk membrane. The primary structure of bovine,[7,8] sheep,[9] human,[10] fruitfly,[11,12] and octopus[13] rhodopsins has been determined. The three mammalian rhodopsins show a high degree of sequence homology and all possess seven hydrophobic peptide segments separated by hydrophilic segments. The seven hydrophobic sequences are presumed to be in alpha-helical conformation and form a bundle that spans across the membrane (Fig. 2).

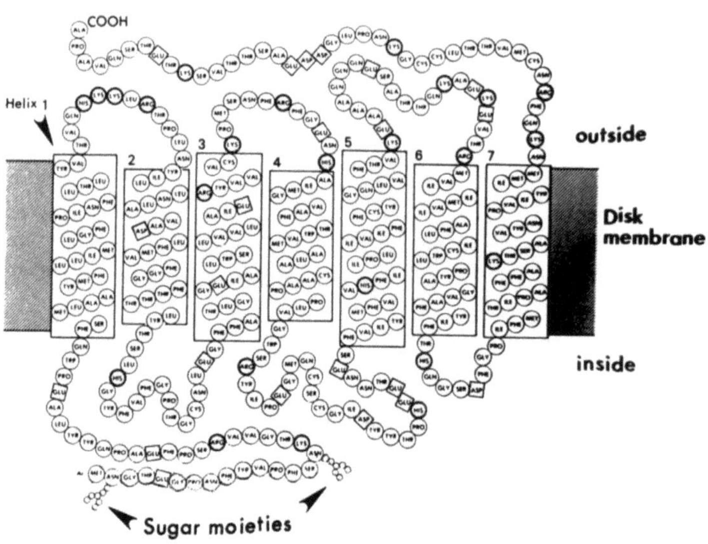

Fig. 2. The amino acid sequence of bovine rhodopsin

The sugar-containing N-terminal sequence is located on the intraluminal surface and the C-terminal sequence with a cluster of hydroxy amino acids that serve as phosphorylation sites is exposed on the cytoplasmic surface of the disk membrane. The primary structures of human cone pigments (blue, red and green), recently deduced from cDNA sequences,[14] also contain seven hydrophobic peptide segments and presumably have a transmembrane configuration similar to that of rhodopsin.

There is general agreement that the chromophore of visual pigment is 11-cis-12-S-trans retinal linked to the ε-amino group of a specific lysine residue (e.g., lysine-296 in bovine rhodopsin) by a protonated Schiff base. Light isomerizes the chromophone to a constrained all trans form. Thermal relaxation of the constrained chromophore results in the release of free retinal from opsin. The process occurs through several spectrally distinguishable intermediates. Rhodopsin, for example, is converted by light to bathorhodopsin which thermally decays to opsin via lumirhodopsin, metarhodopsin I, metarhodopsin II and metarhodopsin III. Decay of metarhodopsin I (lifetime, 3×10^{-4} sec) is correlated with the photocurrent generation.

What is the basis of electrical response evoked by light in photoreceptor cells?

Here, as an example, we choose a frog rod outer segment which is about 6 μm in diameter and about 50 μm in length. Almost one half of the rod membrane permeability is accounted for by permeability toward Na^+ ion (about 2.8×10^{-6}cm/sec). The plasma membrane of the rod outer segment possesses about 4,000 Na^+ channels through which Na^+ ions enter the cytoplasm at a rate of about 2.5×10^4 Na^+ ions per channel per second.[2] Na^+ ions also enter the cell by $3Na^+/Ca^{2+}$ exchange. On the other hand, the plasma membrane of the inner segment has ATP-dependent Na^+/K^+ pumps ($3Na^+/2K^+$) which pump Na^+ ions out of the cell.

Thus Na^+ ions enter the outer segment and exit from the inner segment. Because of the Na^+ current, the dark-adapted rod cell membrane shows a membrane potential of 35-45 mV (inside negative). Absorption of a photon by rhodopsin results in the closure of a few percent of the channels, leading to membrane hyperpolarization within 100 msec. At a saturation light intensity the channels are closed maximally and the membrane is hyperpolarized by as much as 30 mV (i.e. hyperpolarization from -40 mV to -70 mV). Below the saturation light intensity, the following relation holds between the magnitude of light response (I) and energy of irradiating light (A): $I=I_{max}.A(A+K)$ where I_{max} is the magnitude of response at the saturation light intensity and K is light energy when the magnitude of response reaches 50% of the maximum.[2] This equation has the same form as the Michaelis-Menten equation of enzyme kinetics. The analogy leads to the assumption that enzyme(rhodopsin) reacts with substrate(light) to form an enzyme.substrate complex (photoactivated rhodopsin) which then dissociates to enzyme(rhodopsin) and product(intracellular photosignal messenger). The identity of the messenger has been intensively studied during the past 30 years. It is reasonable to assume the generation of a messenger because light absorption by rhodopsin occurs in the disk membrane and its consequence, channel closure, ensues in the plasma membrane.

What is the identity of the photosignal messenger and how does it work?

The first candidate proposed for photosignal messenger was Ca^{2+}.[15] The Ca^{2+} hypothesis assumed that Ca^{++} ions sequestered within the disk in the dark are released into the cytoplasm when a rhodopsin molecule in the disk membrane absorbs a photon, and close the Na^+ channels in the plasma membrane. The hypothesis was supported by the finding that injection of Ca^{2+} into the outer segment in the dark suppressed Na^+ permeability and induced membrane hyperpolarization. However, several findings contradicted the hypothesis. Upon light illumination, cytoplasmic Ca^{2+} concentration in the outer segment decreases rather than increases. There is no evidence that the disks can actively accumulate Ca^{2+} in the dark. Exposure of the cytoplasmic surface of outer segment membrane to Ca^{2+} in patch-clamp experiments does not inhibit Na^+ conductance of the membrane fragment.

Another candidate proposed for messenger was guanine cyclic 3',5'-phosphate (cGMP). The cGMP hypothesis assumes that Na^+ channel is kept open by cGMP and closes when cGMP is hydrolyzed in light. Supporting evidence includes: reversible increase in the Na^+ conductance of outer segment membrane upon exposure to cGMP,[16] rapid decrease in cGMP level in the outer segments irradiated at physiological temperature, light activation of an enzymatic cascade process responsible for the hydrolysis of cGMP,[3] and isolation of the cGMP-sensitive channel protein.[17,18] The channel protein incorporated in lipid membranes increases cation permeability of membrane upon exposure to cGMP. According to the cGMP hypothesis, one photoactivated rhodopsin molecule (R*) activates several hundred GTP binding protein molecules (alias G protein, transducin) each of which, in turn, activates a cGMP phosphodiesterase molecule (Fig. 3).

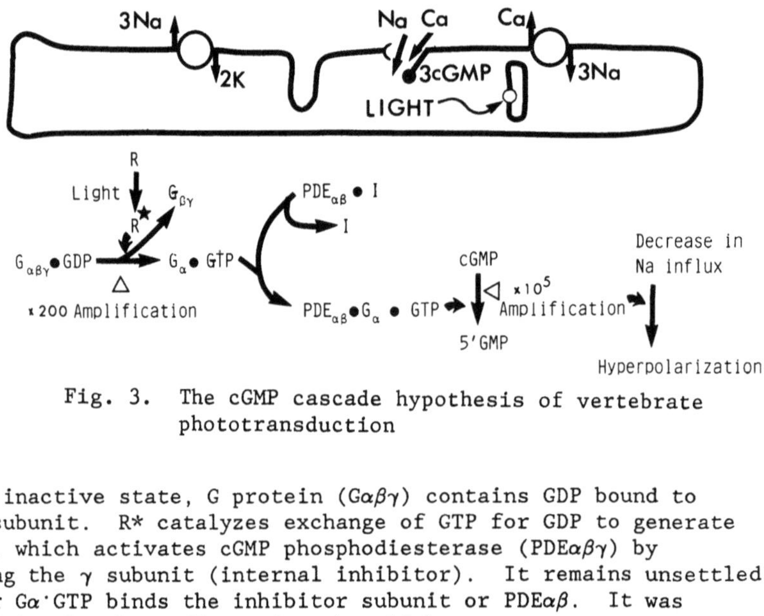

Fig. 3. The cGMP cascade hypothesis of vertebrate phototransduction

In the inactive state, G protein ($G\alpha\beta\gamma$) contains GDP bound to the α subunit. R* catalyzes exchange of GTP for GDP to generate $G\alpha$.GTP, which activates cGMP phosphodiesterase ($PDE\alpha\beta\gamma$) by removing the γ subunit (internal inhibitor). It remains unsettled whether $G\alpha$·GTP binds the inhibitor subunit or $PDE\alpha\beta$. It was assumed in the above discussion that light generates a messenger. Light decomposes cGMP rather than generates it. It should be noted, therefore, that cGMP is a "negative messenger". Rhodopsin, G-protein and PDE are all associated with the disk membrane, while the channel regulator cGMP is bound to the plasma membrane. How is the message transmitted from the disk to the plasma membrane? It is pospostulated that the decreased concentration of cGMP near the disk membrane is transmitted to the plasma membrane by propagation of a negative concentration gradient. This results in the dissociation of cGMP from the channel protein and closure of the channel. Albeit it has many merits, the cGMP hypothesis is by no means complete. For example, when frog retinas or rod outer segments were irradiated and frozen immediately for cGMP assay, no appreciable decrease of cGMP was observed until several seconds after irradiation.[19,20] According to the hypothesis G-protein activation has to precede PDE activation. Yet, the rate of PDE activation (PDE turnover) is faster than the rate of G-protein activation (GTP binding).[21] Peculiarly, cGMP seems to be a messenger only in vertebrate photoreceptors; light has little effect on the cGMP level in invertebrate photoreceptors. Reception of an external signal by a variety of receptors evokes hydrolysis of phosphatidylinositol-4,5-bisphosphate (PIP_2) to diacylglycerol (DAG) and inositol-1,4,5-triphosphate (IP_3), a reaction catalyzed by phospholipase C activated by receptors[22] (Fig. 4).

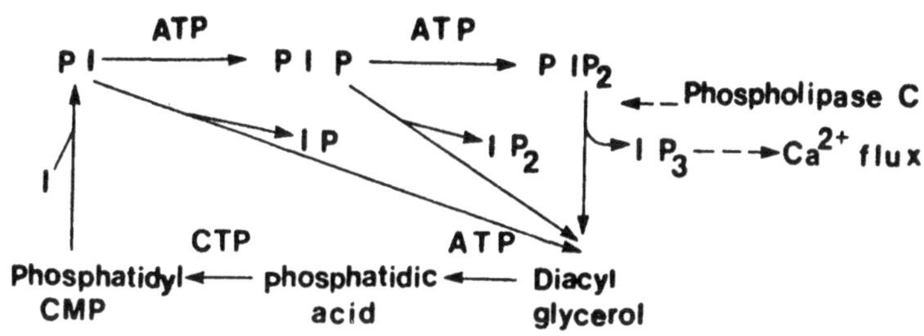

Fig. 4. Phosphatidylinositol cycle

Both IP_3 and DAG serve as signal messengers: IP_3 increases intracellular Ca^{2+} and DAG stimulates protein kinase C. Injection of IP_3 into limulus ventral photoreceptors in the dark mimics the light effect, evoking membrane depolarization without latency, and an increase in the intracellular Ca^{2+} level.[23,24] In fly photoreceptors (Musca domestica and Drosophila), light activates a PIP_2-specific phospholipase C through the activation of a GTP-binding protein.[25] Recent evidence indicates that a Drosophila mutant with impaired vision (norp A) has a defect in phospholipase C[26]. These results suggest that IP_3 may be a photosignal messenger in invertebrate photoreceptors.

Is the photoreceptor system different from other receptor systems?

Elucidation of the primary structure of rhodopsin, the muscarinic acetylcholine receptor, the β-adrenergic receptor, the α-factor receptor of the yeast, and the serotonin receptor in recent years has revealed that these receptor proteins show significant sequence homology and also structural similarities with seven hydrophobic peptide segments.[27,28,29] These proteins have a number of hydroxy amino acid residues in the C-terminal region and phosphorylation of these residues may be a general regulatory mechanism of receptor function. A β-receptor-specific kinase phosphorylates rhodopsin,[30] and both β-receptor and rhodopsin are dephosphorylated by a high-molecular membrane-bound phosphatase.[31] Similarities between rhodopsin and these receptor proteins suggest that they are evolutionarily related. A G-protein involved in PDE activation in photoreceptors belongs to a family of GTP binding proteins.[32] The Gβ subunit is common to G proteins associated with various receptors including the β-adrenergic receptor, while Gα is specific to the individual receptors. Thus, the photoreceptor system shares many properties with other receptor systems.

Acknowledgments

I thank Ms. Roxanne M. Bowman for typing of the manuscript and the National Institutes of Health (EY03807) for support.

REFERENCES

1. M. Chabre, Trigger and amplification mechanisms in visual phototransduction, Ann. Rev. Biophys. Biophys. Chem. 14:331 (1985).
2. J.I. Korenbrot, Signal mechanisms of phototransduction in retinal rod, CRC Critical Rev. Biochem. 17:223 (1985).
3. L. Stryer, Cyclic GMP cascade of vision, Ann. Rev. Neurosci. 9:87 (1986).
4. J. Nathans, Molecular biology of visual pigments, Ann. Rev. Neurosci. 10:163 (1987).
5. H. Shichi, "Biochemistry of Vision," Academic Press, New York, NY (1983).
6. J.E. Dowling, "The Retina," Harvard University Press, Cambridge, MA (1987).
7. Y.A. Ovchinnikov, Rhodopsin and bacteriorhodopsin: Structure-function relationships, FEBS Lett. 148:179 (1982).
8. P.A. Hargrave, J.H. McDowell, D.R. Curtis, J.K. Wang, E. Juszczak, S.L. Fong, J.K. Mohana Rao and P. Argos, The structure of bovine rhodopsin, Biophys. Struct. Mech. 9:235 (1983).
9. B.C. Findlay and D.J.C. Pappin, The opsin family of proteins, Biochem. J. 238:625 (1986).
10. J. Nathans and D.S. Hogness, Isolation and nucleotide sequence of the gene encoding human rhodopsin, Proc. Natl. Acad. Sci. U.S.A. 81:4851 (1984).
11. J.E. O'Tousa, W. Baehr, R.L. Martin, J. Hirsh, W.L. Pak and M.L. Applebury, The Drosophila nina E gene encodes an opsin, Cell 40:839 (1985).
12. C.S. Zucker, A.F. Cowman and G.M. Rubin, Isolation and structure of a rhodopsin gene from D. melanogaster, Cell 40:851 (1985).
13. Y.A. Ovchinnikov, N.G. Abdulaev, A.S. Zolotarev, I.D. Artamonov, I.A. Bespalov, A.E. Dergachev and M. Tsuda, Octopus rhodopsin: Amino acid sequence deduced from cDNA, FEBS Lett. 232:69 (1988).
14. J. Nathans, D. Thomas and D.S. Hogness, Molecular genetics of human color vision: The genes encoding blue, green and red pigments, Science 232:193 (1986).
15. W.A. Hagins, The visual process: Excitatory mechanisms in the primary receptor cells, Ann. Rev. Biophys. Bioeng. 1:131 (1972).
16. E. Gesenko, S.S. Kolesnikov and A.L. Lyubarsky, Induction by the cyclic GMP cationic conductance in plasma membrane of retinal rod outer segment, Nature 313:310 (1985).
17. D. Matesic and P.A. Liebman, cGMP-dependent cation channel of retinal rod outer segments, Nature 326:600 (1987).
18. N.J. Cook, W. Hanke and U.B. Kaupp, Identification, purification, and functional reconstitution of the cyclic GMP-dependent channel from rod photoreceptors, Proc. Natl. Acad. Sci. U.S.A. 84:585 (1987).
19. P. Kilbride and T.G. Ebrey, Light-initiated changes of cyclic guanosine monophosphate levels in frog retina measured with quick-freezing technique, J. Gen Physiol. 74:415 (1979).
20. V.I. Govardovski and A.L. Berman, Light-induced changes of cyclic GMP content in frog retinal rod outer segments measured with rapid freezing and microdissection, Biophys. Struct. Mech. 7:125 (1981).
21. P.A. Liebman and A. Sitaramayya, Role of G-protein-receptor interaction in amplified phosphodiesterase activation of retinal rods, Adv. Cyclic Nucleotide Protein Phosphorylation Res. 17:215 (1984).
22. "Inositol Lipids in Cellular Signalling" (R.H. Michell and J.W. Putney, eds.), Academic Press, London and New York (1988).
23. J.E. Brown, L.J. Rubin, A.J. Ghalayini, A.P. Tarver, R.F. Irvine, M.J. Berridge and R.E. Anderson, myo-inositol polyphosphate may be a messenger for visual excitation in Limulus photoreceptors, Nature 311:160 (1984).

24. A. Fein, Excitation and adaptation of Limulus photoreceptors by light and inositol 1,4,5-triphosphate, Trends NeuroSci. 9:110 (1986).
25. O. Devary, O. Heichael, A. Blumenfeld, D. Cassel, E. Suss, S. Barash, C.T. Rubinstein, B. Minke and Z. Selinger, Coupling of photoexcited rhodopsin to inositol phospholipid hydrolysis in fly photoreceptors. Proc. Natl. Acad. Sci. U.S.A. 84:6939 (1987).
26. R.D. Shortridge, B.T. Bloomquist, S. Schneuwly, M.H. Perdew and W.L. Pak, Molecular isolation and analysis of a photoreceptor-specific phospholipase C gene, norpA, of Drosophila, Abstracts of "Molecular Physiology of Retinal Proteins," L-32, (XXI Yamada Conference held in Kyoto, May 30-June 3, 1988).
27. T. Kubo, K. Fukuda, A. Mikami, A. Maeda, H. Takahashi, M. Mishima, T. Haga, K. Haga, A. Ichiyama, K. Kanagawa, M. Kojima, H. Matsuo, T. Hirose and S. Numa, Cloning, sequencing and expression of complementary DNA encoding the muscarinic acetylcholine receptor, Nature 323:411 (1986).
28. L. Marsh and I. Herskowitz, STE2 protein of Saccharomyces Kluyveri is a member of the rhodopsin/β-adrenergic receptor family and is responsible for recognition of the peptide ligand factor, Proc. Natl. Acad. Sci. U.S.A. 85:3855 (1988).
29. D. Julins, A.B. MacDermott, R. Axel and T.M. Jessell, Molecular characterization of a functional cDNA encoding the serotonin 1c receptor, Science 241:558 (1988).
30. J.L. Benovic, F. Mayor, R.L. Somers, M.G. Caron and R.J. Lefkowitz, Light-dependent phosphorylation of rhodopsin by β-adrenergic receptor kinase, Nature 321:869 (1986).
31. S.-D. Yang, Y.-L. Fong, J.L. Benovic, D.R. Sibley, M.G. Caron and R.J. Lefkowitz, Dephosphorylation of the β_2-adrenergic receptor and rhodospin by latent phosphatase 2, J. Biol. Chem. 263:8856 (1988).
32. L. Stryer and H.R. Bourne, G proteins: a family of signal transducers, Ann. Rev. Cell Biol. 2:391 (1986).

BIOLOGICAL PHOTOSENSORS: PHYTOCHROME AND STENTORIN

Pill-Soon Song

Department of Chemistry
and Section of Molecular Plant Biology
School of Biological Sciences
University of Nebraska
Lincoln, NE 68588

INTRODUCTION

Living organisms possess extremely sensitive sensors by which the organisms are capable of probing their various environmental signals such as chemicals (chemotaxis), heat (thermotaxis), gravitational force (gravitropism), magnetic field (magnetotaxis) and radiation (phototaxis, photophobic response, etc.). In this chapter, we discuss two examples of the biological radiation sensors, namely, phytochrome in plants and stentorin in an aneural unicellular protozoan ciliate *Stentor coeruleus*. The organisms perceive specific wavelength light as the stimulus signal, which is then processed through a series of transduction reactions leading to the output, i.e. response, as shown in the following scheme:

LIGHT SIGNAL ------> PERCEPTION/SENSING
↓
SIGNAL GENERATION AND AMPLIFICATION
↓
OUTPUT (RESPONSE)

Scheme. I. Photosensory transduction in light-sensitive organisms.

PHYTOCHROME AS A PHOTOSENSOR IN PLANTS

Phytochrome is a tetrapyrrolic chromoprotein which plays a primary role as the red-light sensor, regulating a number of diverse morphogenic and developmental responses to light in plants. There are over one hundred different types of responses in plants triggered by the red light signal, including expression of genes critical for the biosynthesis of photosynthetic components such as chlorophyll-binding proteins and ribulose-1,5-bisphosphate carboxylase/oxygenase subunits (for a detailed review, see Song, 1983, 1988; Furuya, 1987).

Higher plants' responses to red light are characterized by a high light sensitivity (see review by Schaefer and Briggs, 1986). The high photosensitivity is achieved by ultrafast primary photoprocesses of phytochrome upon absorption of light (660 nm) by the Pr form of the photoreceptor protein

(Holzwarth et al., 1984; Song et al., 1986). The Pr form is then phototransformed to the physiologically active Pfr form, which can be reverted to the Pr form by far red light (730 nm). The photoreversible transformation of phytochrome and its relationship to red light responses in plants are schematically shown below (Scheme II):

$$\mathrm{Pr} \underset{730\ \mathrm{nm}}{\overset{660\ \mathrm{nm}}{\rightleftharpoons}} \mathrm{Pfr} \rightarrow \rightarrow \rightarrow \text{biological response}$$

Scheme II. Photosensing (red light) by phytochrome.

Thus, phytochrome is a photochromic light sensor.

In this chapter, the structure of phytochrome and its possible functional implications will be described on the basis of results obtained from several research groups. Although phytochrome from different plant sources is now known, here we will be mainly concerned with the etiolated oat phytochrome having a molecular mass of 124 kDa. Properties of the phytochrome molecules in different plant species are believed to be similar to those of the oat phytochrome molecule.

The Phytochrome Molecule: General. The amino acid sequence of oat phytochrome has recently been deduced on the basis of the apophytochrome cDNA sequence (Hershey et al.,1985), as shown in Fig. 1. The light-absorbing chromophore is an open-tetrapyyrole. The tetrapyrrole chromophore is covalently linked via a thioether bond between the vinyl group of Ring A of the chromophore and cysteine-321. There is a significantly hydrophobic region adjacent to the chromophore-linkage site (Quail et al., 1987), which is consistent with its possible role in accommodating the chromophore in a hydrophobic pocket (Hahn and Song, 1981; Song, 1983).

As can be seen from Fig. 1, there are 10 serine residues clustered near the N-terminus. The N-terminus amino group of serine is apparently posttranslationally modified by the addition of an N-acetyl group (W. Rudiger, private communication). The serine-cluster in the N-terminus side is somewhat similar to the N-terminal transit peptides in chlorophyll a/b binding proteins (Lamppa et al., 1985). In both phytochrome and the latter, the N-terminal sequences are also rich in the basic amino acids such as arginine, as in the case in phytochrome. There are other unusual features in the amino acid sequence of phytochrome as well. For example, four lysyl residues occur in series, lys-360 through lys-363, forming a positive charge cluster, and all ten tryptophyl residues are localized in the central segment of the protein. Structural roles of these and other features of the primary structure of phytochrome await further studies with monoclonal antibodies (e.g., Chai et al., 1987), peptide mapping (e.g., Lagarias and Mercurio, 1985), and ultimately X-ray crystallography.

In solution, phytochrome exists predominantly as a dimer (Lagarias and Mercurio, 1985; Jones and Quail, 1986). *In vivo*, phytochrome is evenly distributed as a soluble cytosolic protein, but its Pfr form tends to form sequestered bodies (see Pratt, 1986, for review).

Functional Properties of Phytochrome as a Photosensor. Color discrimination by the light sensory receptors in plants is inherently determined by their absorption spectra. The absorption spectra of phytochrome have been analyzed in analogy to the free electron assignments of the electronic transitions in porphyrins (Song et al., 1979; Song, 1985). The red and far-red absorption maxima of oat Pr and Pfr species occur at 668 and 730 nm, respectively. The red absorbing band is due to an electronic transition from the

Ser-Ser-Ser-Arg-Pro-Ala-Ser-Ser-Ser-Ser-Ser-Arg-Asn-Arg-
-Gln-Ser-Ser-Gln-Ala-Arg-Val-Leu-Ala-Gln-Thr-Thr-Leu-Asp-Ala-
-Glu-Leu-Asn-Ala-Glu-Tyr-Glu-Glu-Ser-Gly-Asp-Ser-Phe-Asp-Tyr-
-Ser-Lys-Leu-Val-Glu-Ala-Gln-Arg-Asp-Gly-Pro-Pro-Val-Gln-Gln-
-Gly-Arg-Ser-Glu-Lys-Val-Ile-Ala-Tyr-Leu-Gln-His-Ile-Gln-Lys-
-Gly-Lys-Leu-Ile-Gln-Thr-Phe-Gly-Cys-Leu-Leu-Ala-Leu-Asp-Glu-
-Lys-Ser-Phe-Asn-Val-Ile-Ala-Phe-Ser-Glu-Asn-Ala-Pro-Glu-Met-
-Leu-Thr-Thr-Val-Ser-His-Ala-Val-Pro-Ser-Val-Asp-Asp-Pro-Pro-
-Arg-Leu-Gly-Ile-Gly-Thr-Asn-Val-Arg-Ser-Leu-Phe-Ser-Asp-Gln-
-Gly-Ala-Thr-Ala-Leu-His-Lys-Ala-Leu-Gly-Phe-Ala-Asp-Val-Ser-
-Leu-Leu-Asn-Pro-Ile-Leu-Val-Gln-Cys-Lys-Thr-Ser-Gly-Lys-Pro-
-Phe-Tyr-Ala-Ile-Val-His-Arg-Ala-Thr-Gly-Cys-Leu-Val-Val-Asp-
-Phe-Glu-Pro-Val-Lys-Pro-Thr-Glu-Phe-Pro-Ala-Thr-Ala-Ala-Gly-
-Ala-Leu-Gln-Ser-Tyr-Lys-Leu-Ala-Ala-Lys-Ala-Ile-Ser-Lys-Ile-
-Gln-Ser-Leu-Pro-Gly-Gly-Ser-Met-Glu-Val-Leu-Cys-Asn-Thr-Val-
-Val-Lys-Glu-Val-Phe-Asp-Leu-Thr-Gly-Tyr-Asp-Arg-Val-Met-Ala-
-Tyr-Lys-Phe-His-Glu-Asp-Asp-His-Gly-Glu-Val-Phe-Ser-Glu-Ile-
-Thr-Lys-Pro-Gly-Leu-Glu-Pro-Tyr-Leu-Gly-Leu-His-Tyr-Pro-Ala-
-Thr-Asp-Ile-Pro-Gln-Ala-Ala-Arg-Leu-Leu-Phe-Met-Lys-Asn-Lys-
-Val-Arg-Met-Ile-Cys-Asp-Cys-Arg-Ala-Arg-Ser-Ile-Lys-Val-Ile-
-Glu-Ala-Glu-Ala-Leu-Pro-Phe-Asp-Ile-Ser-Leu-Cys-Gly-Ser-Ala-
-Leu-Arg-Ala-Pro-His-Ser-cys-His-Leu-Gln-Tyr-Met-Glu-Asn-Met-
-Asn-Ser-Ile-Ala-Ser-Leu-Val-Met-Ala-Val-Val-Val-Asn-Glu-Asn-
-Glu-Glu-Asp-Asp-Glu-Ala-Glu-Ser-Glu-Gln-Pro-Ala-Gln-Gln-Gln-
-Lys-Lys-Lys-Lys-Leu-Trp-Gly-Leu-Leu-Val-Cys-His-His-Glu-Ser-
-Pro-Arg-Tyr-Val-Pro-Phe-Pro-Leu-Arg-Tyr-Ala-Cys-Glu-Phe-Leu-
-Ala-Gln-Val-Phe-Ala-Val-His-Val-Met-Arg-Glu-Phe-Glu-Leu-Glu-
-Lys-Gln-Leu-Arg-Glu-Lys-Asn-Ile-Leu-Lys-Met-Gln-Thr-Met-Leu-
-Ser-Asp-Met-Leu-Phe-Arg-Glu-Ala-Ser-Pro-Leu-Thr-Ile-Val-Ser-
-Gly-Thr-Pro-Asn-Ile-Met-Asp-Leu-Val-Lys-Cys-Asp-Gly-Ala-Ala-
-Leu-Leu-Typ-Gly-Gly-Lys-Val-Trp-Arg-Leu-Arg-Asn-Ala-Pro-Thr-
-Glu-Ser-Gln-Ile-His-Asp-Ile-Ala-Phe-Trp-Leu-Ser-Asp-Val-His-
-Arg-Asp-Ser-Thr-Gly-Leu-Ser-Thr-Asp-Ser-Leu-His-Asp-Ala-Gly-
-Tyr-Pro-Gly-Ala-Ala-Ala-Leu-Gly-Asp-Met-Ile-Cys-Gly-Met-Ala-
-Val-Ala-Lys-Ile-Asn-Ser-Lys-Asp-Ile-Leu-Phe-Trp-Phe-Arg-Ser-
-His-Thr-Ala-Ala-Glu-Ile-Arg-Trp-Gly-Gly-Ala-Lys-Asn-Asp-Pro-
-Ser-Asp-Met-Asp-Asp-Ser-Arg-Arg-Met-His-Pro-Arg-Leu-Ser-Phe-
-Lys-Ala-Phe-Leu-Glu-Val-Val-Lys-Met-Lys-Ser-Leu-Pro-Trp-Ser-
-Asp-Tyr-Glu-Met-Asp-Ala-Ile-His-Ser-Leu-Gln-Leu-Ile-Leu-Arg-
-Gly-Thr-Leu-Asn-Asp-Ala-Ser-Lys-Pro-Lys-Arg-Glu-Ala-Ser-Leu-
-Asp-Asn-Gln-Ile-Gly-Asp-Leu-Lys-Leu-Asp-Gly-Leu-Ala-Phe-Leu-
-Gln-Ala-Val-Thr-Ser-Glu-Met-Val-Arg-Leu-Met-Glu-Thr-Ala-Thr-
-Val-Pro-Ile-Leu-Ala-Val-Asp-Gly-Asn-Gly-Leu-Val-Asn-Gly-Trp-
-Asn-Gln-Lys-Ala-Ala-Glu-Leu-Thr-Gly-Leu-Arg-Val-Asp-Asp-Ala-
-Ile-Gly-Arg-His-Ile-Leu-Thr-Leu-Val-Glu-Asp-Ser-Ser-Val-Pro-
-Val-Val-Gln-Arg-Met-Leu-Tyr-Leu-Ala-Leu-Gln-Gly-Lys-Glu-Glu-
-Lys-Glu-Val-Arg-Phe-Glu-Val-Lys-Thr-His-Gly-Pro-Lys-Arg-Asp-
-Asp-Gly-Pro-Val-Ile-Leu-Val-Val-Asn-Ala-Cys-Ala-Ser-Arg-Asp-
-Leu-His-Asp-His-Val-Val-Gly-Val-Cys-Phe-Val-Ala-Gln-Asp-Met-
-Thr-Val-His-Lys-Leu-Val-Met-Asp-Lys-Phe-Thr-Arg-Val-Glu-Gly-
-Asp-Tyr-Lys-Ala-Ile-Ile-His-Asn-Pro-Asn-Pro-Leu-Ile-Pro-Pro-
-Ile-Phe-Gly-Ala-Asp-Glu-Phe-Gly-Trp-Cys-Ser-Glu-Trp-Asn-Ala-
-Ala-Met-Thr-Lys-Leu-Thr-Gly-Trp-Asn-Arg-Asp-Glu-Val-Leu-Asp-
-Lys-Met-Leu-Leu-Gly-Glu-Val-Phe-Asp-Ser-Ser-Asn-Ala-Ser-Cys-
-Pro-Leu-Lys-Asn-Arg-Asp-Ala-Phe-Val-Ser-Leu-Cys-Val-Leu-Ile-
-Asn-Ser-Ala-Leu-Ala-Gly-Glu-Glu-Thr-Glu-Lys-Ala-Pro-Phe-Gly-
-Phe-Phe-Asp-Arg-Ser-Gly-Lys-Tyr-Ile-Glu-Cys-Leu-Leu-Ser-Ala-
-Asn-Arg-Lys-Glu-Asn-Glu-Gly-Gly-Leu-Ile-Thr-Gly-Val-Phe-Cys-
-Phe-Ile-His-Val-Ala-Ser-His-Glu-Leu-Gln-His-Ala-Leu-Gln-Val-
-Gln-Gln-Ala-Ser-Glu-Gln-Thr-Ser-Leu-Lys-Arg-Leu-Lys-Ala-Phe-
-Ser-Tyr-Met-arq-His-Ala-Ile-Asn-Asn-Pro-Leu-Ser-Gly-Met-Leu-
-Tyr-Ser-Arg-Lys-Ala-Leu-Lys-Asn-Thr-Asp-Leu-Asn-Glu-Glu-Gln-
-Met-Lys-Gln-Ile-His-Val-Gly-Asp-Asn-Cys-His-His-Gln-Ile-Asn-
-Lys-Ile-Leu-Ala-Asp-Leu-Asp-Gln-Asp-Ser-Ile-Thr-Glu-Lys-Ser-
-Ser-Cys-Leu-Asp-Leu-Glu-Met-Ala-Glu-Phe-Leu-Leu-Gln-Asp-Val-
-Val-Val-Ala-Ala-Val-Ser-Gln-Val-Leu-Ile-Thr-Cys-Gln-Gly-Lys-
-Gly-Ile-Arg-Ile-Ser-Cys-Asn-Leu-Pro-Glu-Arg-Phe-Met-Lys-Gln-
-Ser-Val-Tyr-Gly-Asp-Gly-Val-Arg-Leu-Gln-Gln-Ile-Leu-Ser-Asp-
-Phe-Leu-Phe-Ile-Ser-Val-Lys-Phe-Ser-Pro-Val-Gly-Gly-Ser-Val-
-Glu-Ile-Ser-Ser-Lys-Leu-Thr-Lys-Asn-Ser-Ile-Gly-Glu-Asn-Leu-
-His-Leu-Ile-Asp-Leu-Glu-Leu-Arg-Ile-Lys-His-Gln-Gly-Leu-Gly-
-Val-Pro-Ala-Glu-Leu-Met-Ala-Gln-Met-Phe-Glu-Glu-Asp-Asn-Lys-
-Glu-Gln-Ser-Glu-Glu-Gly-Leu-Ser-Leu-Leu-Val-Ser-Arg-Asn-Leu-
-Leu-Arg-Leu-Met-Asn-Gly-Asp-Val-Arg-His-Leu-Arg-Glu-Ala-Gly-
-Val-Ser-Thr-Phe-Ile-Ile-Thr-Ala-Glu-Leu-Ala-Ser-Ala-Pro-Thr-
-Ala-Met-Gly-Gln

Fig. 1. Amino acid sequence for oat phytochrome.

ground state to the lowest excited state (Q_y) transition. There is a shoulder for each of these maximum absorbance bands located at about 608 nm and 674 nm for Pr and Pfr, respectively, corresponding to the second excited state (Q_x transition). In addition, both forms of phytochrome absorb significantly in the near UV-blue region with absorbance maxima at 379 nm and 400 nm, respectively. The near UV-blue band can be correlated with the Soret band (composed largely of higher excited state (B_x and B_y) transitions) in porphyrins and chlorophylls, and its intensity depends strongly on the chromophore conformation, as will be discussed later.

The Pr to Pfr phototransformation of phytochrome apparently occurs from the Q_y excited singlet state, with the primary reaction taking place within a few picoseconds (Holzwarth et al., 1984; Song et al., 1986) and with high efficiency (50%) (Heihoff et al., 1987), as shown in Fig. 2. A configurational isomerization about the C=C bond between ring C and D is likely to be involved in the primary photoprocess (Rudiger et al., 1983). A subsequent thermal reaction step is rate-limiting, and is likely to involve conformational changes in the apoprotein and/or chromophore. The Pfr to Pr photoreversion takes place similarly, i.e., an ultrafast primary reaction followed by thermal relaxation steps one of which is rate-limiting. However, the microscopic reversibility is not obeyed in the photoreversible photochemical pathways in phytochrome, indicating that the forward and reverse reaction paths are different. The overall forward and reverse phototransformations of oat phytochrome in phosphate buffer proceed with quantum efficiencies of 15.2 and 6%, respectively (Lagarias et al., 1987). The relatively low quantum efficiencies probably reflect highly competitive internal conversion from the Q_y excited state and/or photo- and thermal-reversibility of the photoreactions. The reader may refer to some of the recent studies on the flash photolysis of phytochrome (e.g., Pratt et al., 1984; Ruzsicska et al., 1985).

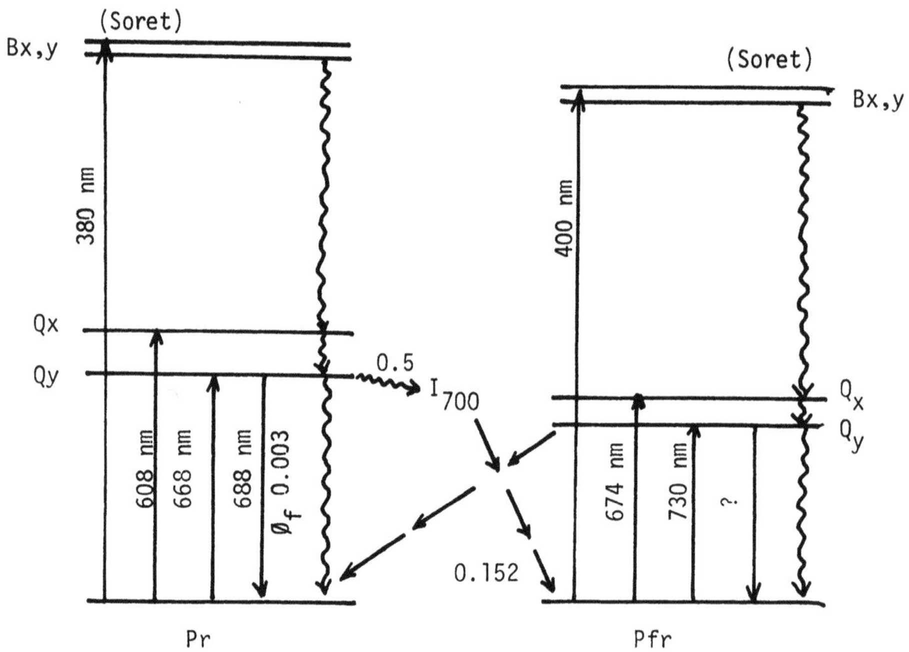

Fig. 2. Energy level diagrams for phytochrome (Pr and Pfr forms). Absorbance maxima, fluorescence emission wavelength and yields, and quantum yields for phototransformation (only the forward transformation is shown) are shown.

Figure 3 shows a simplified scheme for the overall phototransformation of phytochrome.

In Fig. 3, the chromophore is shown to be photoisomerized during the Pr → Pfr phototransformation. However, the conformation (shape) of the chromophore is preserved, in spite of the isomerization about the 15 C=C bond. It has been argued on the basis of the ratios of the oscillator strengths of $Q_{x,y}$ (visible; red) to $B_{x,y}$ (near UV-blue) absorbance bands of Pr and Pfr that the chromophore conformation of Pfr is similar to that of Pr, since the oscillator strength ratio, which is strongly dependent on the shape of tetrapyrroles, remains essentially constant in the Pr → Pfr phototransformation (Song et al., 1979). A possible conformation of the chromophore is shown in Fig. 3. Other semi-circular or -extended conformations may well be consistent with the absorption spectra of Pr and Pfr. The native conformation of the Pfr chromophore must be fixed in an overall conformation/shape similar to the Pr chromophore conformation.

In the proposed photoisomerization model shown in Fig. 3, the overall shape of the tetrapyrrole π-electron network remains approximately invariant, consistent with the absorption spectral characteristics noted above. However, the proposed model does not take into account the native chromophore structure of the Pfr form absorbing at a longer wavelength (730 nm). There are a number of proposals on the possible structure of the Pfr-phytochrome; the chemical and stereochemical structure of the Pr-chromophore has been established (Lagarias and Rapoport, 1980). However, the structure/conformations of <u>native</u> Pr and Pfr species remain to be established.

Proton transfer within the chromophore pore was suggested as a primary photoprocess of phytochrome, followed by the formation of intermediate(s) (Song, 1985). Braslavsky et al. (1985) have shown that ultrafast intramolecular proton transfer indeed occurs prior to subsequent isomerization reactions in a phytochrome model compound, biliverdin. However, it is now clear that proton transfer does not occur in the primary photoprocess (Holzwarth et al., 1984). It has yet to be elucidated at what stage of the phototransformation an intramolecular or intermolecular proton transfer takes place along the thermal pathway leading to the Pfr form of phytochrome and what is responsible for proton release/uptake in solution (Tokutomi et al., 1982, 1986).

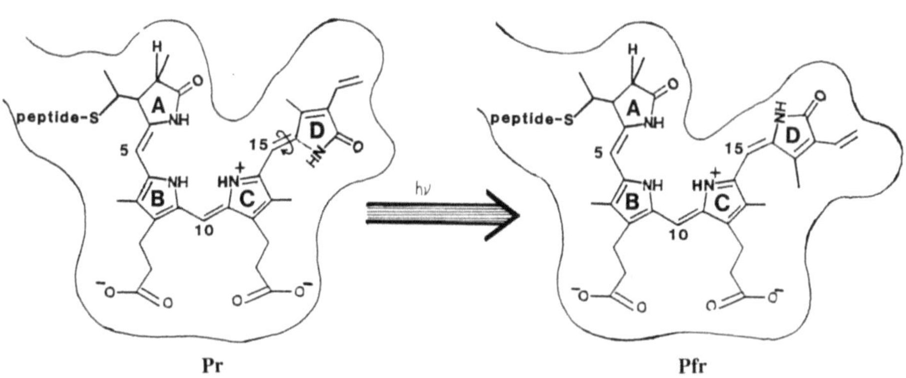

Fig. 3. A possible mechanism for the phototransformation of phytochrome based on the isomerization at 15,16-C=C bond in the primary photoprocess of Pr (left).

Proton release/uptake accompanying the phytochrome phototransformation may well be caused by a conformational change in the protein molecule, including an alteration of the chromophore topography/accessibility. According to the proposed topographic model for the phototransformation of phytochrome, the Pfr chromophore of phytochrome is more exposed than the Pr chromophore (Song et al., 1979; Hahn and Song, 1981). Recently, tetranitromethane (TNM; Hahn et al., 1984), sodium borohydride (Chai et al., 1987), and bilirubin oxidase (unpublished) have been used to probe the topography of the Pr and Pfr chromophores, all confirming the preferential exposure of the chromophore in the Pfr-phytochrome.

The degree of exposure of the Pfr-chromophore clearly depends on the size of the protein, indicating in particular that the 6000-dalton peptide chain along the N-terminus sequences shields and interacts closely with the Pfr chromophore. Since there appears to be a significant change in the topography of the Pr and Pfr chromophores accompanying phototransformation, the degree of chromophore movement/reorientation relative the fixed axis of the chromoprotein molecule can be measured by linear dichroic measurements. Ekelund et al. (1985) and Inoue (1986) carried out linear dichroic measurements of the immobilized phytochrome, yielding a reorientation angle of 32° (or 148°) and 50°, respectively. A functional implication of change in the chromophore topography during the phytochrome phototransformation has been discussed in detail elsewhere (Song, 1983, 1988).

The absorbance maxima of different molecular mass species of phytochrome are virtually identical for the Pr form, but the absorbance maximum for undegraded, 124-kDa species is at slightly longer wavelength (730-732 nm) than that for degraded phytochromes (720-725 nm) (Vierstra and Quail, 1982, 1983; Cha et al., 1983; Lagarias et al., 1987). In absolute energy units, the bathochromic shift of the Pfr absorption maximum in going from degraded species to undegraded protein corresponds to less than 200 cm^{-1} or 0.6 kcal/mol. This suggests that the difference in interaction forces between the chromophore and the apoprotein among different molecular mass species of phytochrome is quite small. The small energy difference reflects a subtle change in hydrogen bonding, hydrophobic force and/or electrostatic forces between the chromophore and its environment. In addition, a small conformational change of the chromophore as a result of interactions between the chromophore and apoprotein, for example, the N-terminus peptide chain in the 124-kDa species, can also account for the spectral shifts among different molecular mass species of phytochrome.

8-Anilinonaphthalene 1-sulfonate (ANS) and bis-ANS are well known for their fluorescence ability upon binding to a hydrophobic site on a protein. These hydrophobic fluorescence probes also bind to phytochrome. The binding causes bleaching of the chromophore absorbance bands, especially in the Pfr form of phytochrome (Hahn and Song, 1981; Song, 1988). The spectral bleaching of the Pfr species apparently arises from a competitive binding of the probe at the chromophore binding site or its vicinity. The binding of the probe disrupts the interaction between the chromophore and its binding pocket and results in the formation of a cyclic conformation of the chromophore. The susceptibility of phytochrome to ANS bleaching is a sensitive function of the apoprotein integrity, particularly the 6/10-kDa amino terminal peptide segment. This suggests that the amino terminus segment interacts and/or masks the Pfr chromophore of 124-kDa phytochrome, thus making it less susceptible to cyclize in the presence of ANS than the Pfr form of degraded proteins.

Additional evidence for the possible interaction between the chromophore and the N-terminus peptide chain has come from a circular dichroic (CD) study (Chai et al., 1987). A photoreversible CD change occurs during the phototransformation of 124-kDa phytochrome, corresponding to a 3% increase in

Table 1

Analysis of the secondary structure of 124-kDa oat phytochrome (in percent)

Phytochrome	α-helix	β-sheet	β-turn	Random coil
Pr	52.2	0.0	21.1	26.7
Pfr	55.3	0.0	18.8	26.0

α-helical folding in the Pfr form of phytochrome. However, no such photoreversible CD changes have been observed with degraded phytochromes. Secondary structure analysis of the CD spectra is summarized in Table 1. The photoreversible CD change, and the increased α-helicity in the Pfr form, are suppressed by monoclonal antibody binding near the N-terminus sequences (Chai et al., 1987). The photoreversible CD spectral change can also be inhibited by sodium borohydride, which specifically bleaches the chromophore by reducing it, and by TNM, *vide supra*.

From Table 1, the 3% increase in α-helical folding upon Pr → Pfr photoconversion arises largely at the expense of β-turns. According to the Chou-Fasman (1978) method, the N-terminal sequences do tend to assume a β-turn conformation (Hershey et al., 1985; Quail et al., 1987), as shown in Fig. 4. It is well known that prosthetic groups such as the tetrapyrrole heme of myoglobin (Beychok, 1966) exhibit a marked propensity for facilitating α-helical conformations in peptide chains; perhaps, this type of interaction is responsible for the photoreversible α-helical folding in phytochrome. Although the mode of interactions between the Pfr-chromophore and the N-terminus sequence is not known, amino acid residues near the N-terminus include serine and threonine clusters (Fig. 1), and these hydroxyl

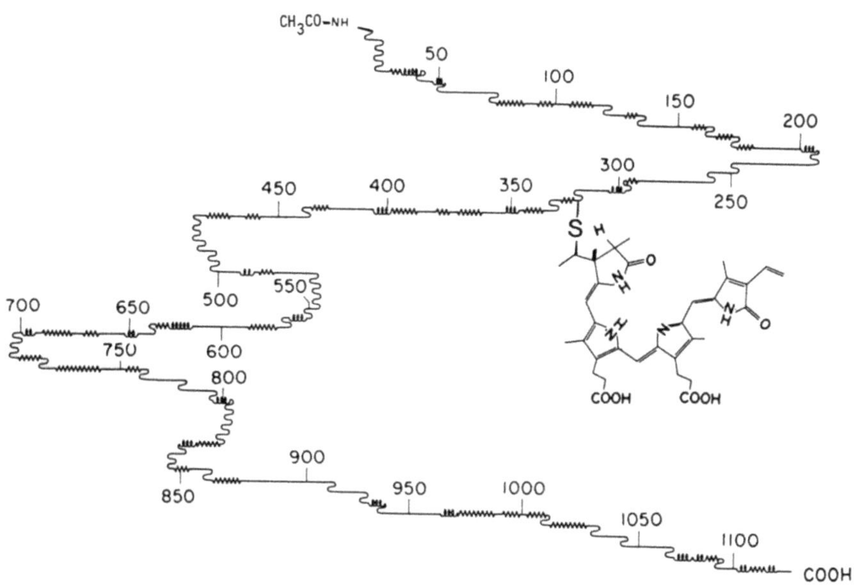

Fig. 4. The predicted secondary structure of oat phytochrome. Coiled line, α-helix; zigzagged line, β-sheet; wavy line, β-turn; smooth line, random coil; X-, unidentified amino-blocking group. (Modified from Quail, by courtesy).

amino acid residues could form hydrogen bonds with the pyrrole nitrogens, carbonyl and carboxylate functionalities of the chromophore (Fig. 4).

In a reversible conformational change, proteins cannot gain or lose significant amounts of volume. Thus, it is possible that the photoreversible phytochrome does not exhibit a gross conformational change, especially given the relatively small energy of 180 kJ/mol of red photons. Most hydrodynamic data tend to confirm this prediction. However, there are some clearly recognizable conformational changes in the phototransformation of phytochrome. For example, there is a significant difference in the number of exchangeable protons between the Pr and Pfr forms of phytochrome (Hahn et al., 1984). Peptide maps for 124-kDa oat phytochrome deduced from partial proteolysis show the protein with two domains, one for the N-terminus/chromophore domain of ca. 74 kDa and one for the C-terminus side domain of ca. 55 kDa. It turns out that the N-terminus sequence on the chromophore domain is extremely susceptible to proteolysis in the Pr form but not in the Pfr form, whereas at least one site near the central hinge region between the two major domains becomes preferentially accessible to a proteolytic enzyme (Jones et al., 1985; Lagarias and Mercurio, 1985).

Mechanism of Action. We now know much about the phytochrome molecule with respect to its structure and spectroscopic/photochemical properties. However, its mode of action largely remains to be elucidated. For example, the Pfr form of phytochrome tends to sequester without being associated with a particular organelle in plant cells (Pratt, 1986), but its physiological/molecular biological implications are still puzzling. One can also ask where the "functionally active" pool of phytochrome is localized within the cell. Is phytochrome a membrane-bound protein, at least in certain systems such as in the polarotropically sensitive *Mougeotia* (for review, see Song, 1983)? There is apparently no clear-cut, direct evidence for the membrane-bound phytochrome *in vivo*. However, there are several lines of indirect evidence that suggest the membrane as the site of phytochrome localization for certain rapid *in vitro* processes (for review, see Roux, 1983; Furuya, 1987).

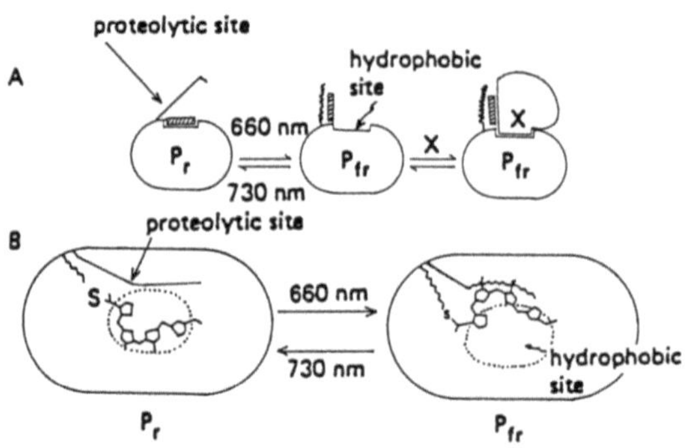

Fig. 5. A topograhic model for the phototransformation of phytochrome and its interaction with an as yet unidentified receptor (X). Proteolytic site on the N-terminus segment (6-kDa) of Pr is shown. The N-terminus segment assumes an α-helical conformation (wavy line) in the Pfr form. (From Song, 1988).

Returning to the structural aspect of the phytochrome molecule, we have focused our attention on the chromophore topography and its relationship to the secondary structure of the N-terminus sequences, particularly in the Pfr form of phytochrome. A schematic model incorporating only these features has been proposed (Song, 1983, 1988), as shown in Fig. 5. What are the possible roles of these structural features in the as yet unidentified interactions between phytochrome and its putative receptor(s)? For example, what is the functional significance of a photoreversible α-helical folding along the amino terminus sequence of phytochrome? Is the amino terminal α-helix a structure required for anchoring to the Pfr-receptor or membrane? Many more studies are called for to answer these crucial questions. It is also hoped that both Pr and Pfr forms of phytochrome can be crystallized for structure elucidation in the near future.

Molecular Electronic Aspects. As the result of the Pr → Pfr phototransformation of phytochrome, electrical signals can be generated *in vivo* as well as *in vitro*. For example, red light irradiation of mungbean seedling slices induces their attachment to the negatively charged surface (Tanada effect). In a number of cases of those fast photomorphogenic responses of plants not involving the slower processes of gene regulation, the red light-induced membrane potentials have been recorded. In particular, phytochrome modulates the ionic permeability and ion fluxes, particularly K^+ and Ca^{2+} ions, across cellular and/or organelle membranes. For example, phytochrome mediates a reversible Ca^{2+} release from the intracellular calcium ion reservoir and triggers the red light induced rotation of the chloroplast plane in algae *Mougeotia*.

STENTORIN AS A PHOTOSENSOR PIGMENT

Stentor coeruleus is a unicellular protozoan ciliate. This organism exhibits an extremely sensitive response to visible wavelength light, particularly red light of 610-620 nm (for review, see Song, 1983). In the dark, the cell swims forward by beating rows of cilia in clockwise direction, which propels the cell to rotate in the same direction. When the cell encounters visible light, it suddenly stops as the ciliary beating reverses its beating direction, and turns away from the light source. This type of photoresponse is called a "photophobic response", and it occurs whenever the cell encounters a sudden increase in light intensity, e.g. from darkness to light and from dim light to bright light. In addition, *Stentor coeruleus* is capable of sensing the direction of light propagation. Thus, the cell swims parallel to the direction of light propagation. This type of photoresponse is called "negative phototaxis". The two types of responses ensure that the ciliate can efficiently escape from the photodynamically harmful radiation (Yang et al., 1986).

The photosensing processes of *Stentor coeruleus* can be described within the general framework of the photosensory transduction chain shown in Scheme I. For example, in the photophobic response of *Stentor coeruleus*, light signal is perceived by stentorin localized in the pigment granule. The signal transduction then leads to a reversal of the direction of ciliary

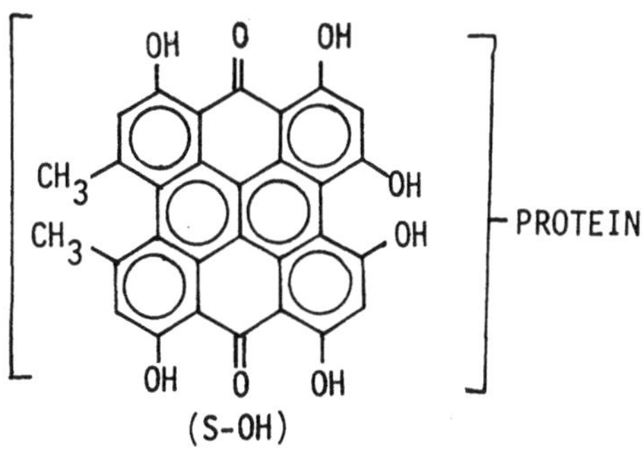

Fig. 6. The possible structure of the chromophore in stentorin. The site of linkage of apoprotein is unknown.

beating that is entailed in the stop-turning motion of the cell away from the source of light. Here, we briefly describe the photosensory transduction in the ciliate cell which utilizes stentorin as the photosensor pigment.

Stentorin Molecules. The stentorin molecules have been isolated and partially characterized (for review, see Song, 1983). However, information and data on the molecular weight, the chromophore identity and its stoichiometry, the binding site and chromophore-apoprotein linkage, the amino acid sequence, and the photochemical reactivity are still incomplete. The absorption spectrum of stentorin suggests that the chromophore of the stentorin photoreceptor is similar to hypericin (Walker et al., 1979). In the native stentorin molecule, the chromophore is presumably linked to the apoprotein via an acid-labile bond (Walker et al., 1979). Fig. 6 shows a plausible structure of the chromophore in stentorin. One of the most striking features of the stentorin chromophore is its poly hydroxyl groups.

Available evidence indicates that there are two different forms of stentorin, one small molecular weight species "stentorin I" and one large molecular mass assembly "stentorin II" (Song, 1983; unpublished). The latter appears to be a functionally active species, whereas the former is either a non-functional species or an antenna pigment. Here, we explore the mechanism of photosensing in *Stentor coeruleus* by the sensor pigment stentorin.

Signal Generation. No evidence for a photochemical cycle similar to the rhodopsin photocycle has been reported with stentorin. On the basis of indirect evidence, we proposed that the photophobic response in *Stentor coeruleus* is triggered by a signal transducing mechanism which is initiated by a light-induced proton release/dissociation from the photoreceptors in the pigment granules to the cytoplasm (Song, 1983). More recently, picosecond fluorescence kinetics of native and denatured stentorins have shown that only the native photoreceptor assembly releases protons efficiently during its excited state lifetime, with an apparent excited state pK_a^* of ca. 2.5 for the hydroxyl groups of the stentorin chromophore. Scheme III outlines the process of proton dissociation/release from the excited stentorin (S-OH*).

$\text{S-OH} + h\nu \longrightarrow \text{S-OH}^*$		(light absorption)
$\text{S-OH}^* \longrightarrow \text{S-OH} + h\nu_f$		(fluorescence emission, k_f)
$\text{S-OH}^* \longrightarrow \text{S-OH} + \text{heat}$		(heat dissipation, k_h)
$\text{S-OH}^* \longrightarrow \text{S-O}^{-*} + \text{H}^+$		(proton dissociation, k_d)
$\text{S-O}^{-*} + \text{H}^+ \longrightarrow \text{S-OH}^*$		(proton association, k_a)
$\text{S-O}^{-*} \longrightarrow \text{S-O}^- + h\nu_f$		(anion fluorescence, k_{af})
$\text{S-O}^{-*} \longrightarrow \text{S-O}^- + \text{heat}$		(heat dissipation, k_{hd})
$\text{S-O}^- + \text{H}^+ \longrightarrow \text{S-OH}$		(reassociation, k_{ra})
$\text{S-OH} \longrightarrow \text{S-O}^- + \text{H}^+$		(dissociation, k_{da})

Scheme III. Proton dissociation/release from the excited state stentorin (S-OH)

According to Scheme III, a steady state concentration of protons is determined by the following equation:

$$(H^+) = k_d\,(S\text{-}OH^*)/(k_a(S\text{-}O^{-*}) + k_{ra}(S\text{-}O^-)).$$

One can speculate that one of the functions of the protein matrix in stentorin is to facilitate proton dissociation (k_d), while retarding proton association (k_a) in the excited state and reassociation (k_{ra}) in the ground state. This possibility is illustrated in Fig. 7.

A recent picosecond time-resolved fluorescence study provides direct evidence for the proposed release of protons from the photoreceptor assembly (unpublished data). We suggest that a transient proton release induced by the light stimuli serves as an initial signal in the photosensory transduction chain for the photophobic response in *Stentor coeruleus*.

Signal Amplification: Molecular Electronic Aspects. An initial signal resulting from light perception in *Stentor coeruleus* can be a transient proton release from the excited stentorin in the pigment granules to the cytoplasm. A depolarizing effect of the intracellular and/or intraciliary pH drop may then trigger the opening of Ca^{2+} channels, as outlined in Scheme IV.

LIGHT SIGNAL ----> SIGNAL PERCEPTION ----> SIGNAL GENERATION
(stentorin) (protons)

---> SIGNAL AMPLIFICATION AND TRANSMISSION ---> MECHANOTRANSDUCTION
(opening of Ca^{2+} channels & influx) (ciliary reversal)

Scheme IV. A tentative scheme for a step-up photophobic response in *Stentor coeruleus*.

A light-induced action potential of the ciliate has been recorded, consistent with the calcium ion influx from the extracellular medium to the cytoplasm when the cell is subjected to the light stimuli. The action potential cannot be induced by light in the presence of protonophores and can be completely quenched by calcium blocking agents, in qualitative agreement with Schemes III and IV (Song, 1981).

In addition to the inhibitory effects of specific calcium blocking agents on the photophobic response of the ciliate, calimycin stimulates the photophobic sensitivity of the ciliate, apparently due to its enhancement of calcium permeability across the cellular membrane (Song, 1983). Other drugs that facilitate calcium permeability and thus enhance the photosensitivity of the *Stentor coeruleus* cell include caffeine and α-phosphatidic acid.

In Scheme IV, the link between the Ca^{2+} influx and the dynein ATPase activity in the ciliary contractile axoneme has not been established. The contractile mechanism of km fibers and myonemes is operative in the cell body contraction of *Stentor coeruleus* upon mechanical and electrical stimulation, and is based on Ca^{2+} influx as the driving force for the contractile phenomenon. However, the role of Ca^{2+} ions in the contractile axonemes of cilia and ciliary basal bodies remains to be elucidated (for review, see Song, 1981).

SUMMARY

Phytochrome is a photochromic sensor involved in the regulation of growth and development of plants. This blue pigment protein acts as a key molecule to trigger the expression of certain genes. Its inactive form (Pr) absorbs red light signal of 660 nm, and the physiologically active form (Pfr) is then formed. The latter can be converted back to the Pr form with 730 nm light which is exclusively absorbed by the Pfr form. The following diagram illustrates the mode of action of phytochrome schematically:

$$\mathrm{Pr} \underset{730\ \mathrm{nm}}{\overset{660\ \mathrm{nm}}{\rightleftharpoons}} \mathrm{Pfr} \rightarrow \rightarrow \rightarrow \text{response}$$

This photoreversible, photochromic process of phytochrome takes place in the cytoplasm of a plant cell. Once the active form of phytochrome is produced, it triggers the expression of several photosynthetically and biosynthetically important genes. How this process of light signal transmission (in the form of Pr as a photosensor) to nuclei where these genes are kept dormant under dark conditions are suddenly activated by red light is not well understood. It can be assumed that the Pfr form, once produced in the cytosol, recognizes its receptor(s) located either at the organellar membrane envelopes or inside the nuclei in initiating a cascade of biochemical events leading to the enhanced expression of the genes.

Stentorin acts as the photosensor molecule in the ciliate *Stentor*. The unicellular protozoan ciliate, *Stentor coeruleus*, is most sensitive to red light (610 nm). *Stentor* also senses the direction of light propagation, as evidenced by their light-avoiding and negative phototactic swimming behaviors. This aneural photosensory phenomenon is triggered by the photoreceptor stentorin. We will discuss the possible involvement of a light-induced proton release from the photoreceptor as a primary mechanism of light-signal processing. The primary sensory signal, in the form of proton release, triggers subsequent transduction steps that include calcium ion influx from the extracellular medium. It is proposed that the calcium ion influx causes the *Stentor* cell to reverse its ciliary beating and subsequently turn away from the light source.

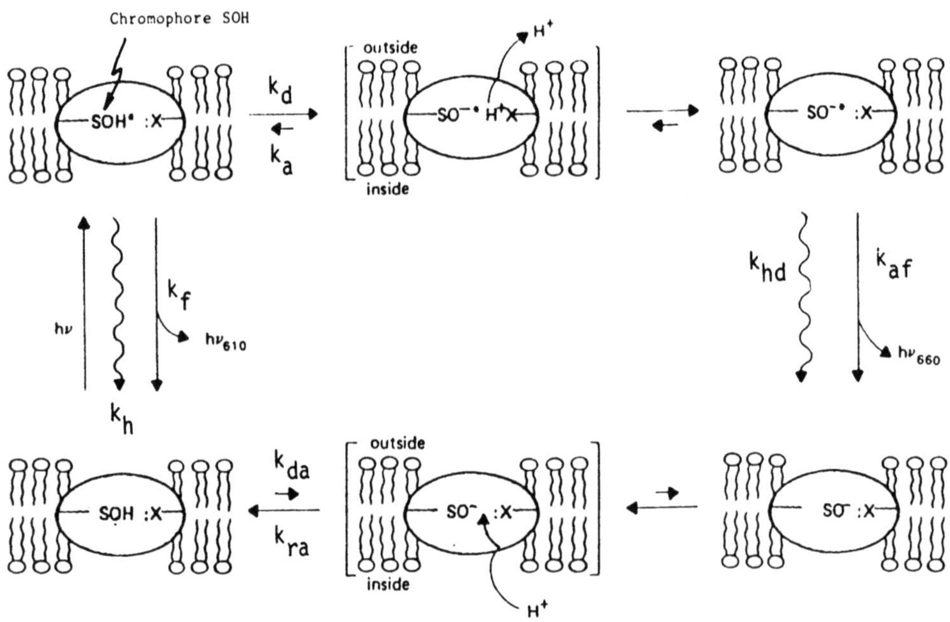

Fig. 7. The diagram representation of the proposed mechanism of proton dissociation form the excited state stentorin embedded in the pigment granule. X, a conjugate base on apoprotein. (Modified from Song, 1981).

ACKNOWLEDGEMENTS

The author's work described in this chapter has been supported by the NIH grants GM-36956 and NS-15426.

REFERENCES

Beychok, S., 1966, Circular dichroism of biological macromolecules. Science, **154**:1288.

Braslavsky, S. E., Al-Ekabi, H., Petrier, C., and Schaffner, K., 1985, Phytochrome models. 9. Conformation selectivity of the photocyclization of the biliverdin IXr and IXd dimethyl esters, Photochem. Photobiol., **41**:237.

Chai, Y. G., Song, P. S., Cordonnier, M.-M., and Pratt, L. H., 1987, A photoreversible circular dichroism spectral change in oat phytochrome is suppressed by a monoclonal antibody that binds near its N-terminus and by chromophore modification, Biochemistry, **26**:4947.

Chou, P. Y., and Fasman, G. D., 1978, Empirical predictions of protein conformation, Annu. Rev. Biochem., **47**:251.

Ekelund, N. G. A., Sundqvist, C., Quail, P. H. and Vierstra, R. D., 1985, Chromophore rotation in 124-kilodalton *Avena* phytochrome as measured by light-induced changes in linear dichroism, Photochem. Photobiol., **41**:221.

Furuya, M., ed., 1987, "Phytochrome and Photoregulation in Plants", in press, Academic Press, New York and Tokyo.

Hahn, T. R., and Song, P. S., 1981, The hydrophobic properties of phytochrome as probed by 8-anilinonaphthalene 1-sulfonate fluorescence, Biochemistry, **20**:2602.

Hahn, T. R., Song, P. S., Quail, P. H., and Vierstra, R. D., 1984, Tetranitromethane oxidation of phytochrome chromophore as a function of spectral form and molecular weight, Plant Physiol., **74**:755.

Heihoff, K., Braslavsky, S. E., and Schaffner, K., 1987, Study of 124-kilodalton oat phytochrome photoconversion *in vitro* with laser-induced optoacoustic spectroscopy, Biochemistry, **26**:1422.

Hershey, H. P., Barker, R. F., Idler, K. B., Lissemore, J. L., and Quail, P. H., 1985, Analysis of cloned cDNA and genomic sequences for phytochrome: complete amino acid sequences for two gene products expressed in etiolated *Avena*, Nucleic Acids Res., **13**:8543.

Holzwarth, A. R., Wendler, J., Ruzsicska, R. P., Braslavsky, S. E., and Schaffner, K., 1984, Picosecond time-resolved and stationary fluorescence of oat phytochrome highly enriched in the native 124 kDa protein, Biochim. Biophys. Acta, **791**:265.

Inoue, Y., 1986, Round-table discussion, Symp. on Phytochrome and Photoregulation in Plants", Okazaki, Japan.

Jones, A. M., and Quail, P. H., 1986, Quaternary structure of 124-kilodalton phytochrome from *Avena sativa* L., Biochemistry, **25**:2987.

Kendrick, R. E., 1983, The physiology of phytochrome action, in: "The Biology of Photoreception," D. Cosens and D. Vince-Prue, eds., Cambridge University Press, Cambridge, pp. 275-303.

Lagarias, J. C., and Mercurio, F. M., 1985, Structure-function studies of phytochrome. Identification light-induced conformational changes in 124-kDa *Avena* phytochrome *in vitro*, J. Biol. Chem., **260**:2415.

Lagarias, J. C., Kelly, J. M., Cyr, K. L., and Smith, W. O., 1987, Comparative photochemical analysis of highly purified 124 kilodalton oat and rye phytochromes *in vitro*, Photochem. Photobiol. **46**:5.

Lamppa, G. K., Morelli, G., and Chua, N. H., 1985, Structure and developmental regulation of a wheat gene encoding the major chlorophyll a/b-binding polypeptide, Mol. Cell. Biol., **5**:1370.

Pratt, L. H., 1986, Phytochrome: localization within the plant, in: "Photomorphogenesis in Plants," R. E. Kendrick and G. H. M. Kronenberg, eds., Martinus Nijhoff, Dordrecht, pp. 61-81.

Pratt, L. H., Inoue, Y., and Furuya, M., 1984, Photoactivity of transient intermediates in the pathway from the red-absorbing to the far-red-absorbing form of *Avena* phytochrome as observed by a double-flash transient-spectrum analyzer, Photochem. Photobiol., **39**:241.

Quail, P. H., Barker, R. F., Colber, J. T., Daniels, S. M., Hershey, H. P., Idler, K. B., Jones, A. M., and Lissemore, J. L., 1987, Structural features of the phytochrome molecule and feedback regulation of the expression of its genes in *Avena*, in: "Molecular Biology of Plant Growth Control," J. F. Fox and M. Jacobs, eds., Alan R. Liss, New York, pp. 425-439.

Roux, S. J., 1983, A possible role of Ca^{2+} in mediating phytochrome responses, in: "The Biology of Photoreception," D. Cosens and D. Vince-Prue, eds., Cambridge University Press, Cambridge, pp. 561-508.

Rudiger, W., Thuemmler, F., Cmiel, E., and Schneider, S., 1983, Chromophore structure of the physiologically active form (Pfr) of phytochrome, Proc. Natl. Acad. Sci. USA, **80**:6244.

Ruzsicska, B. P., Braslavsky, S. E., and Schaffner, K., 1985, The kinetics of the early stages of the phytochrome transformation Pr → Pfr. A comparative study of small (60 kDalton) and native (124 kDalton) phytochromes from oat, Photochem. Photobiol., **41**:681.

Schaefer, E., and Briggs, W. R., 1986, Photomorphogenesis from signal perception to gene expression, Photobiochem. Photobiophys., **12**:305.

Song, P. S., 1981, Photosensory transduction in *Stentor coeruleus* and related organisms, Biochim. Biophys. Acta, **639**:1.

Song, P. S., 1983, Protozoan and related photoreceptors: Molecular aspects, Annu. Rev. Biophys. Bioengin., **12**:35.

Song, P. S., 1985, Primary molecular events in aneural cell photoreception, in: "Sensory Perception and Transduction in Organisms," G. Colombetti, F. Lenci and P. S. Song, eds., Plenum, London, pp. 47-59.

Song, P. S., 1988, The molecular topography of phytochrome: Chromophore and apoprotein, J. Photochem. Photobiol., **Part B. 2**:43.

Song, P. S., Chae, Q., and Gardner, J. G., 1979, Spectroscopic properties and chromophore conformations of the photomorphogenic receptor: Phytochrome, Biochim. Biophys. Acta, **576**:479.

Song, P. S., Tamai, N., and Yamazaki, I., 1986, Viscosity dependence of pri mary photoprocesses of 124 kdalton phytochrome, Biophys. J., **49**:645.

Tokutomi, S., Inoue, Y., Sato, N., Yamamoto, K. T., and Furuya, M., 1986, Effect of pH on absorption spectra of pea 114 and 121 kilodalton phytochromes during and after red-light irradiation, Plant Cell. Physiol., **27**:765.

Tokutomi, S., Yamamoto, K. T., Miyoshi, Y., and Furuya, M., 1982, Photoreversible changes in pH of pea phytochrome solution, Photochem. Photobiol., **35**:431.

Walker, E. B., Lee, T. Y., and Song, P. S., 1979, Spectroscopic characterization of the *Stentor* photoreceptor, Biochim. Biophys. Acta, **587**:129.

Yang, K. C., Prusti, R. K., Walker, E. B., Song, P. S., Watanabe, M., and Furuya, M., 1986, Photodynamic action in *Stentor coeruleus* sensitized by endogenous pigment stentorin, Photochem. Photobiol., **43**:305.

ANIMAL RHODOPSIN AS A PHOTOELECTRIC GENERATOR

M. A. Ostrovsky

Institute of Chemical Physics
U. S. S. R. Academy of Sciences
Moscow, U. S. S. R.

INTRODUCTION

In 1964, Brown and Murakami[1] recorded an electric potential from monkey retina by stimulating it with an intense light flash. This potential differed from other then known membrane potentials; the risetime was so fast that it exceeded the time resolution of existing recording devices. This electric signal was named the early receptor potential (ERP). The subsequent study of the light-induced events in the photoreceptor cells led to the understanding that the ERP develops too fast to be associated with post-rhodopsin steps of the visual process (for review see Ref. 2). The amplitude of the intracellularly recorded ERP was no more than a few millivolts. Such a small magnitude of the photosignal made one regard the ERP as an epiphenomenon - a side-effect that cannot play a fundamental role in visual excitation.

In 1971, Stoeckenius, Oesterhelt and Blaurock[3,4] discovered a new retinal-containing protein called bacteriorhodopsin (bacterial rhodopsin). Subsequently, it proved to be a photoelectric generator in *Halobacterium halobium*, which transduces photon energy into a convertible form, i.e., protonic potential. When a light flash is used to stimulate a reconstituted bacteriorhodopsin membrane, a photoelectric signal similar to the ERP can be recorded.[5-8] In fact, analogous terminology was used to describe the components of the two potentials; R1 in visual membranes and B1 in bacteriorhodopsin membranes were found to be temperature insensitive whereas the corresponding slower components (R2 and B2) were found to be reversibly inhibited by low temperature. In our own laboratory, we reconstituted bacteriorhodopsin and rhodopsin with collodion films.[6,9] We found phase I, II, and III in both types of membranes:[10,11] Phase I being opposite in polarity to II and III and formally corresponding to B1 and R1.

The obvious correlation between the proton movement and the photoelectric signal in bacteriorhodopsin prompted us to study the role of the proton in the generation of the ERP. Earlier, investigators have reached the consensus that the ERP was generated by a light-induced conformational change and the associated internal charge displacement. However, internal charge displacement associated with the light-induced conformational change is not the only possible mechanism. Proton uptake by rhodopsin has been previously suggested to be a possible explanation of the ERP.[12-15]

The effect of alkalinification of the incubation medium during the bleaching of digitonin extracts of rhodopsin[16] or a suspension of fragments of rod outer segments[17,18] has been interpreted as the protonation of rhodopsin at the stage of metarhodopsin II formation. It is at this functionally important stage of visual pigment photolysis that the most significant conformational restructuring in the protein moiety of the pigment occurs. In addition to protonation, also observed is: (a) an increase in the mobility of the hydrophilic parts of the rhodopsin polypeptide chain on the cytoplasmic surfaces of the photoreceptor disc membrane,[19,20] (b) the freeing of additional titratable sulfhydryl groups,[21,22] and (c) the generation of a fast photoelectric potential across the membrane.[9,23]

The ERP R2 component coincides with the formation of metarhodopsin II.[2,24] However, the concurrence of a major conformational change and proton uptake at this stage complicates the issue as to whether the R2 component is due to the conformational change or due to proton uptake or both (see, for example, Refs. 15,25).

In analyzing ERP-like signal in reconstituted rhodopsin membranes, Hong and his colleagues[26,27] have presented evidence that the B2 component is generated by proton uptake by bacteriorhodopsin at the cytoplasmic surface. In our analysis based on photometric pH measurements of normal and inverted vesicles formed from purified photoreceptor discs,[28] we believe a similar conclusion can be reached about the R2 component, i.e., the R2 component is a result of proton binding on the cytoplasmic surface of photoreceptor discs.

It is known that the R2 component has such a polarity that the disc interior is positive, using a penetrating synthetic anion as a probe.[23] The validity of proton uptake in the generation of the R2 component requires that the proton uptake occur at the cytoplasmic side. We thus determine the side of proton uptake using normal and inverted vesicles. Furthermore, there has been controversy with regard to whether there is proton transport across the photoreceptor membrane.[29] Our experiment also serves to test the latter idea. For if there is proton transport, acidification of the medium should be detected upon illumination of inverted vesicles. Our determination of the light-induced pH change in the medium bathing a suspension of enriched inverted discs reveals a major difference between the two rhodopsin systems.

While there is obvious proton release at the extracellular surface of the purple (bacteriorhodopsin) membrane, our result indicated that there is no corresponding proton release at the intradiscal space. In other words, there is no transmembrane proton transfer in the visual membrane. This result thus settled the controversy about the presence of a proton gradient. While our finding seems consistent with the role of rhodopsin as photoinformational rather than photobioenergetic, the question of the ERP as an epiphenomenon remains unsettled. Nevertheless, our results suggests several possible roles of the ERP. Further future investigation is required before one can ascertain that the ERP really is an epiphenomenon. The problems involved here are also key issues for the research and development of molecular photosensors.

MATERIALS AND METHODS

A method of membrane reconstitution previously developed by Drachev *et al.*[6] allow us to demonstrated rhodopsin as a photoelectric generator by direct electrical measurements. Photoreceptor discs were isolated from cattle retinas as described previously.[23,30] The isolated discs were added to one of the two compartments of a Teflon chamber, separated by a collodion film impregnated with a decane solution of azolectin (100 mg/ml) or of egg lecithin (70 mg/ml) and octadecylamine (1 mg/ml). The final concentration of rhodopsin was 0.25 mg/ml. The bathing solution contained 0.1 M NaCl and 5 mM Mes-KOH pH 6.0, if lecithin and octadecylamine were used to impregnated the film. If it was azolectin, the bathing medium was supplemented with 0.025 M $CaCl_2$. During 2 - 4 hours of incubation, spontaneous incorporation of discs into the film took place. The measurement method for the open-circuit photovoltage was previously described.[10,11]

Photoreceptor disc membranes with normal orientation were prepared from cattle retina as described previously.[31] Inverted discs were obtained by freeze-thawing. A disc suspension in 1 M sucrose with a rhodopsin concentration of 30-40 μM was applied dropwise to an aluminum support previously cooled to the temperature of dry ice. The sample was then warmed to 4°C temperature and again frozen. The free-thawing operation was repeated three times.

It is known that the N-terminal part of the rhodopsin polypeptide chain contains two oligosaccharide chains exposed to the intradiscal space.[32] During successive freeze-thawing they face the outer medium as a result of disc inversion. This permits the preparation of inverted discs using affinity chromatography on concanavalin A-Sepharose.[33]

The separation of discs with differing orientations of the photoreceptor membrane was done using concanavalin A (Pharmacea) immobilized on Sepharose 4B. Applied to a chromatographic column of 10 ml were 2 ml concanavalin A, followed by rinsing with 50 ml of solution containing 115 mM NaCl, 2.7 mM KCl, 1 mM $CaCl_2$, and 0.5 mM $MnCl_2$, pH 6.0 (Solution 1). Then 6 ml of Solution 1 and 1 ml of disc suspension with a rhodopsin concentration of 30-40 μM were applied to the column, the contents of the column were mixed, and after 5 min of settling the discs unbound with concanavalin A were eluted.

The inverted discs bound with concanavalin A were eluted using Solution 2, containing all components of Solution 1 and 0.5 mM methyl-α-D-mannopyranoside. Five ml of Solution 2 were applied to the column and the discs were eluted with 10 ml of Solution 2 at the same rate.

The pH of the medium with disc membranes was monitored photometrically by recording the 585-nm absorbance of a pH indicator bromocresyl purple. An increase in the 585-nm absorbance corresponds to alkalinification while a decrease means acidification. The system is calibrated by addition of 10 μM HCl.

RESULTS

Flash-Induced Electrogenesis Mediated by Rhodopsin

A 15-ns laser flash (λ_{max} = 530 nm) induced a photoresponse with an initial negative component (Phase I, Fig. 1a) followed by a positive phase with at least two exponents indicating two additional electrogenic phases with half-time of about 0.5 ms and several millisecond (Fig. 1b). The magnitude of the initial negative phase was approximately 100 times smaller than that of the positive phase. The positive phase is correlated with the formation of metarhodopsin II, measured as an increase in the 390-nm absorbance. The difference in the magnitudes of the two components was found to decrease when the collodion film was shunted with an external resistance (Fig. 2). As the membrane resistance was reduced from 50 MΩ-cm^2 to a level (2 KΩ-cm^2) that is more comparable to that of the outer segment membrane, the negative/positive ratio changes from 0.014 to 0.16.

The shunted photosignal proved to be very similar to the ERP measured *in vivo* (Cf: Ref. 12). The action spectrum of the flash-induced photoelectric

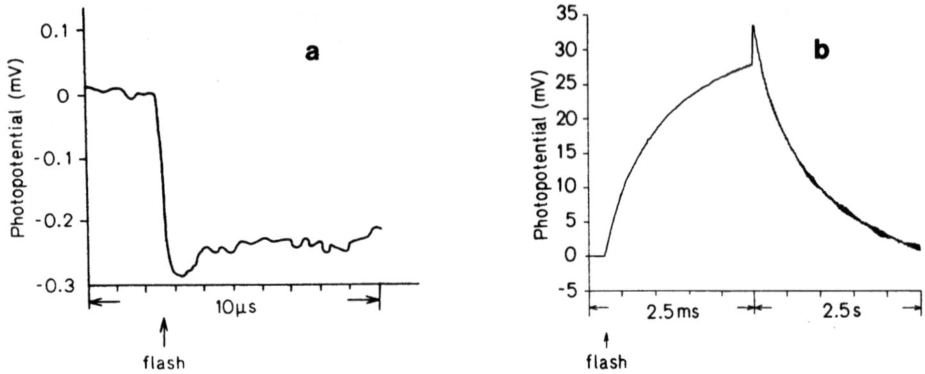

Fig. 1. Kinetics of the rhodopsin electrogenesis. The record was shown with three different time scales to reveal the faster negative phase I (a) and the slower positive phases II and III (b). A 530-nm laser flash was used (Reproduced from Ref. 11)

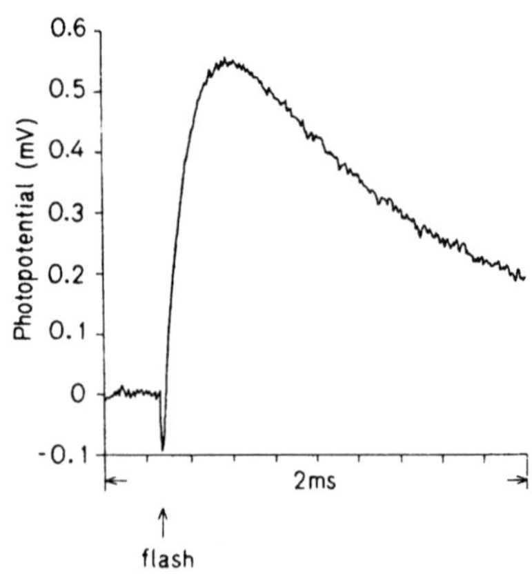

Fig. 2. Electric response induced by a 530-nm flash in the disc-collodion film system shunted with an external resistance of 2KΩ-cm^2. (Reproduced from Ref. 11)

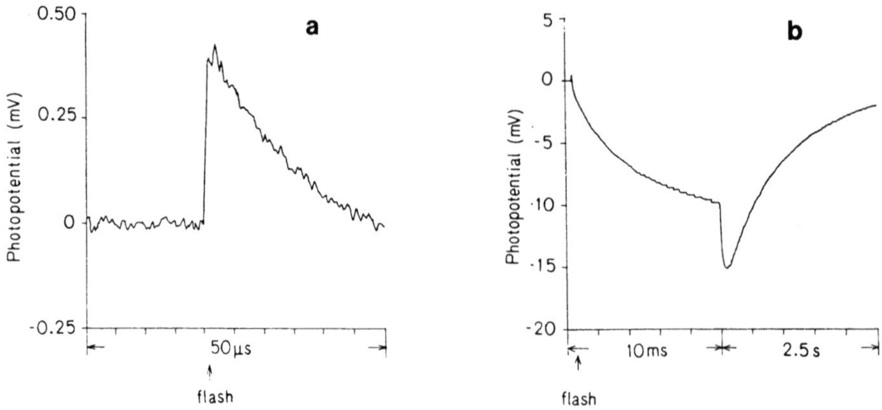

Fig. 3. Kinetics of the metarhodopsin II electrogenesis. The system was first illuminated with 31 consecutive 530-nm flashes followed by a 347-nm flash: (a) fast phase, (b) slow phase. (Reproduced from Ref. 11)

effect was in good agreement with the absorption spectrum of rhodopsin.[11] The formal identification of the negative phase with R1 and the positive phase with R2 is justified.

Flash-Induced Electric Responses of Metarhodopsins

Illumination with a 347-nm laser flash given after a 530-nm flash was found to generate a photoelectric response with a polarity opposite to that mediated by rhodopsin (Figs. 3a and 3b). For such a response, the interval between the 530-nm and 347-nm flashes proved to be critical. It must be longer than 1 ms and shorter than 5 min. The study of both spectral and time characteristics of the response at 347 nm suggests that it was mediated by metarhodopsin II. Thus, the photosignal so measured is identical with what is referred to as the photoreversal potential in the ERP literature.[12] This response can be eliminated by hydroxylamine treatment which prevents accumulation of metarhodopsin II while the normal ERP was unaffected by such a treatment.

Flash-Induced Acceleration of the Photopotential Decay and Increase in Membrane Conductivity

In Fig. 4, the amplitudes and decay half-times by a 530-nm laser flash are plotted against the number of successive flashes. It was found that both parameters of the animal rhodopsin response decrease when the number of flashes increases. In the case of bacteriorhodopsin, no changes in the decay time and the amplitude were found. In the animal rhodopsin system, the amplitude of the photopotential previously decreased by a train of flashes was found to be restored by addition of 11-*cis* retinal which however did not restore the decay time. The decrease of the photopotential amplitude by a train of flashes might be attributed to rhodopsin bleaching. As to the acceleration of the photopotential decay, it might be ascribed to a decrease in the membrane ionic resistance. This can be shown by an experiment in which photoreceptors discs were adsorbed on both sides of the collodion film.

In this system, the two photogenerators facing opposite directions (right-side orienting vs. left-side orienting rhodopsin molecules) cancel each other and therefore no photopotential can be observed. If then an external potential of either polarity is exerted on the membrane, any light-induced changes of the recorded signal must be ascribed to the change of membrane conductance alone (Fig. 5). We found there is indeed a membrane conductance increase with increased flashes (Fig. 6). We also found that the flash-induced conductance increase partially recovered in the minute time scale. The light-induced conductance increase depends on the availability of rhodopsin. The effect becomes less prominent as bleaching proceeds. This bleaching related change can be reversed by adding 11-*cis* retinal or by apply a 347-nm flash, which presumably regenerates rhodopsin from metarhodopsin II.

From the flash-induced acceleration of the decay time, we further concluded that the light-induced ion permeability change of the disc membrane occurs in the millisecond time scale or faster.[11] The effect of acceleration of photopotential decay survives the treatment of 10 mM hydroxylamine and anaerobic treatment, indicating that the effect is mediated by rhodopsin rather than metarhodopsin II and no oxidation of photoproducts is involved.

Changes in pH During Illumination of Normally Oriented and Inverted Discs

Our experiment with normally oriented discs shows an increase of 585-nm absorbance of the mixture, indicating an alkalinification of the medium (Fig. 7). The result is consistent with a proton uptake at the cytoplasmic surface. A slow partial restoration of pH, indicating proton release to the same side, was then observed in the darkness. The control using a suspension

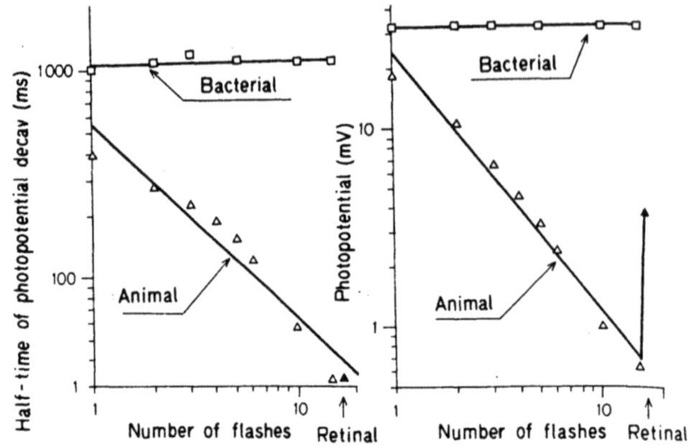

Fig. 4. Half-time of decay of a flash-induced electric potential and its amplitude as functions of the number of successive 530-nm laser flashes. (Δ,▲) Before and after addition of 20 μM 11-*cis* retinal. (Δ)'Animal', photoreceptor disc-collodion film system; (□) 'bacterial', bacteriorhodopsin sheet collodion film system. (Reproduced from Ref. 11)

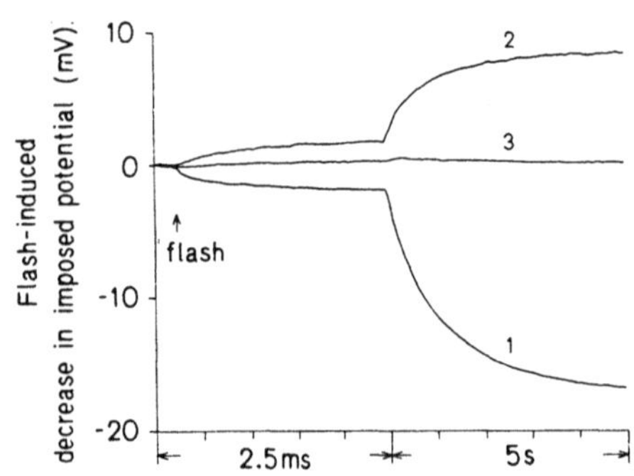

Fig. 5. Photoelectric responses in a phospholipid-impregnated collodion film covered on both sides with photoreceptor discs. The system was exposed to three successive flashes (flash numbers indicated above the curves). Curves 1 and 2, an electric potential difference was imposed across the film, +350 mV, respectively. Curve 3, no potential difference was imposed. A 530-nm laser flash was used. (Reproduced from Ref. 11)

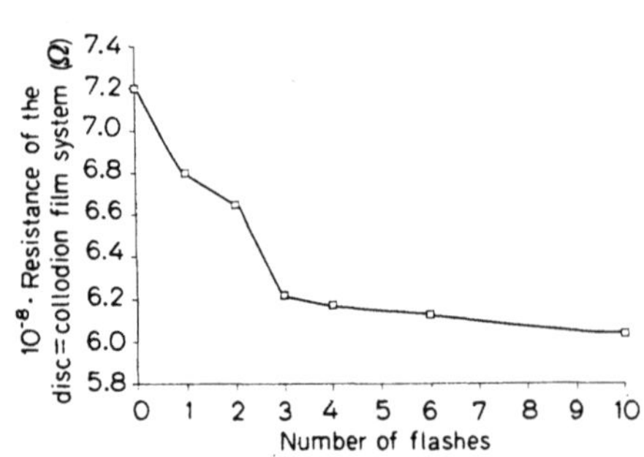

Fig. 6. Effect of successive 530-nm flashes on electric resistance of the system of the collodion film with discs on both sides. A 100 mV potential difference was imposed on the film. (Reproduced from Ref. 11).

without bromocresyl purple shows a decrease in the 585-nm absorbance upon illumination, caused by rhodopsin photolysis and the formation of metarhodopsin II. The true light-induced pH change was indicated by the difference of the two curves (curve a - curve b). At pH 6.0 the changes in pH comprised 1.40 ± 0.07 moles of H^+ per mole bleached rhodopsin.

Freezing of a suspension of photoreceptor discs and their subsequent thawing leads to the inversion of 40-50 % of the discs in the suspension (the intradiscal surface of the membrane in this case faces the external medium).

Figure 8 presents the light-induced change in the pH of a suspension of partially inverted discs obtained by means of three-fold freeze-thawing. A slight alkalinification of the medium was observed, corresponding to 0.27 ± 0.04 moles H^+ per mole bleached rhodopsin. This corresponds to 20 % of the effect observed in a suspension of normally oriented discs.

Thus, the acidification of the medium was not observed in the photoreceptor disc suspension obtained by repeated freeze-thawing as would be expected on the assumption that a transmembrane proton transport occurs. The slight uptake (and <u>not release</u>) of proton observed in the experiment could occur, for example, due to contamination with uninverted discs. This is indeed the case as is evident from the following observations.

Concanavalin A-Sepharose chromatography was carried out with a dark sample of (normally oriented) disc suspension in Solution 1 not subject to freeze-thawing along with aliquots of the same sample subject to freeze-thawing (inverted disc enriched).

In the control sample, 80 % is eluted by Solution 1 (normally oriented membrane) while only 42 % was eluted in the freeze-thawing samples. In contrast, almost equal (35 %) freeze-thawing sample were eluted by Solution 2, while only 9 % of control were eluted in Solution 2. However, about 10-20 % of the discs were irreversibly bound to concanavalin A in both cases. Apparently, while the majority of vesicles in the control were normally oriented, about 50 % of vesicles in the freeze-thawing sample were inverted. The corresponding pH measurement was shown to be consistent with this interpretation.

The fraction of freeze-thawing sample eluted by Solution 2 presumably containing "pure" inverted discs caused alkalinification of the medium, corresponding to 0.61 moles of H^+ per mole bleached rhodopsin. This may be explained by re-inversion to normal orientation, a conclusion to follow from the following observations. The presumably "pure" inverted disc sample was sedimented and a repeated chromatography on concanavalin A-Sepharose was performed. The distribution of the discs to fractions was the same as during the first separation: 46 % comprised discs with a normal orientation of the membrane, 32 % were eluted by Solution 2 (i.e., inverted discs), and 22 % were irreversibly bound with concanavalin. Thus, in fact, the "pure" fraction of inverted discs also contained discs with a normal orientation, probably reinverted. This is further supported from the following experiments.

Samples of the suspension of photoreceptor discs after freeze-thawing were stored at 4°C for 15, 60, 90, and 120 min, after which the light-induced changes in pH were recorded in each sample (Fig. 9). As is apparent, the magnitude of the light-induced proton uptake increase with increasing storage time and after 2 h reached more than 80 % of the control. Probably the process of re-inversion begins immediately after the thawing of the suspension, since about 25 % of the discs are oriented normally after just 15 min.

The initial fraction eluted with Solution 2 from the column comprises about 40 % of the proton uptake in the control. This finding is consistent

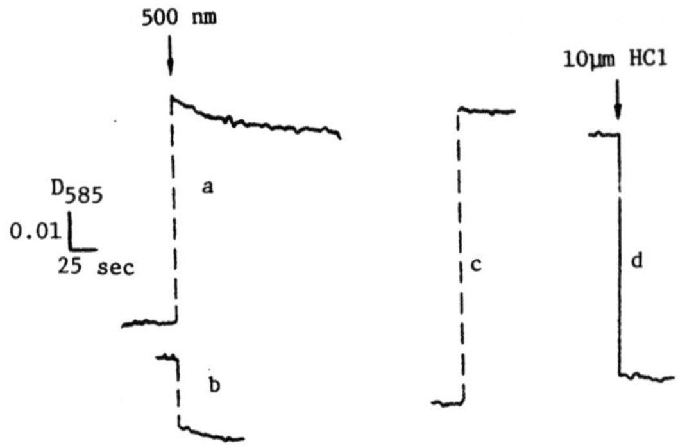

Fig. 7. Light-induced pH changes in a suspension of photoreceptor discs. (a) 100 μM bromocresyl purple, (b) without bromocresyl purple, (c) true change in pH, (d) calibrated addition of HCl. Incubation medium: 1 M sucrose, pH 6.0. Rhodopsin concentration: 14 μM. Illumination of samples with a flash of IFK-120 gas-discharge xenon lamp (4 msec pulse duration) resulted in bleaching of 40% of rhodopsin molecules. (Reproduced from Ref. 28)

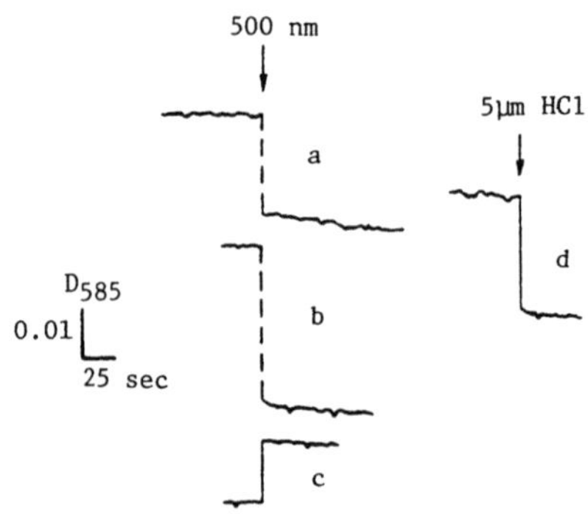

Fig. 8. Light-induced pH changes in a suspension of photoreceptor discs after freeze-thawing. (a) 100 μM bromocresyl purple, (b) without bromocresyl purple, (c) true change in pH, (d) calibrated addition of HCl. Incubation medium: 115 mM NaCl, 2.7 mM KCl, 1mM $CaCl_2$, 0.5 mM $MnCl_2$, ph 6.0. Rhodopsin concentration: 25 μM. (Reproduced from Ref. 28)

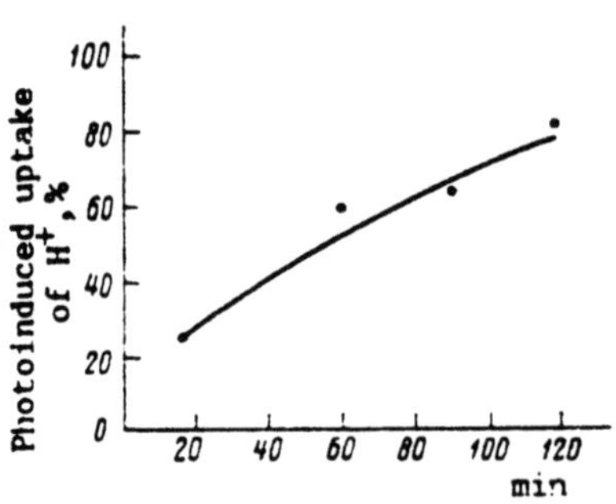

Fig. 9. Changes in the light-induced proton uptake during storage of a suspension of photoreceptor discs after freeze-thawing. Incubation medium and rhodopsin concentration: same as in Fig. 8. (Reproduced from Ref. 28)

with the above observation since a 30-40 minute interval elapses between the application of the freeze-thawed sample to the column and the time of pH recording, leading to a 35-40 % re-inversion according to the curve in Fig. 9.

Thus, a correlation is observed between the effect of proton uptake from the incubation medium and the quantity of photoreceptor discs with a normal orientation of the membrane. But if the inverted discs acidify the medium, such a correlation would evidently be absent. If the transmembrane proton transfer occurred, no changes in the pH in the medium would be observed given an equal content of discs with opposite membrane orientations. Thus, despite the complication of spontaneous re-inversion of a presumably inverted sample, the above analysis rules out the possibility of a light-induced transmembrane proton transfer in photoreceptor discs.

Light-Induced Deprotonation of Metarhodopsin II

In an earlier section, it was shown that illumination of the intermediate photoproduct, metarhodopsin II, results in the appearance of an ERP-like signal with reversed polarity (photoreversal potential). It was found that the light-induced pH change in the opposite direction occurs under analogous conditions, as shown in Fig. 10. The sample was first illuminated with 500-nm light, and the normal effect of a rapid alkalinification of the medium was recorded. Then after a minute the sample was re-illuminated, but with 365-nm light. In this case there appeared a rapid acidification of the medium (Fig. 10a). In absolute value, this effect comprised 35 % of the magnitude of the direct effect. If the sample was illuminated with 365-nm light at the end of the terminal breakdown in metarhodopsin II (30 min after the flash) no changes in pH occurred (Fig. 10b).

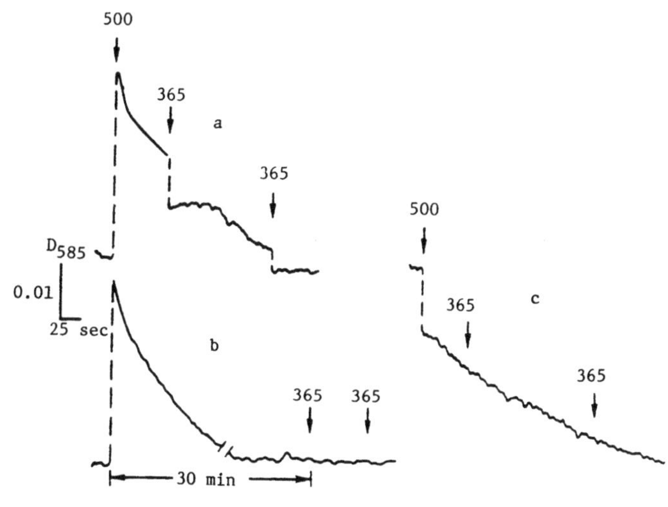

Fig. 10. Changes in pH in a suspension of fragments of photoreceptor discs during the light-induced transition of metarhodopsin II into product P_{465}. (a) 100 mM NaCl, pH. 6.0, (b) 100 mM NaCl, pH 6.0 (sample illuminated with 365-nm light 30 min after the flash), (c) 100 mM NaCl, 100 mM hydroxylamine, pH 6.0. Rhodopsin concentration: 12-25 μM; concentration of bromocresyl purple: 100 μM. Arrows indicate instants of illumination of sample with 500-nm light (IFK-120 lamp, IF-500 interference filter, 4 msec pulse duration) and 365-nm light (DRSh-250 lamp, UFS-6 filter, 10 sec illumination duration). (Reproduced from Ref. 28)

The addition to the medium of hydroxylamine, which is known to reduce the life-span of the metarhodopsin II and the formation of retinaloxime, led to the disappearance of the effect of proton release during the illumination of the sample with 365-nm light (Fig. 10c). This also indicates that the effect of rapid acidification is associated specifically with the illumination of metarhodopsin II. Apparently, this process reflects reversible changes in the rhodopsin molecule during the phototransition to the product P_{465}, which, as in metarhodopsin III, is spectrally similar to the earlier product of photolysis, metarhodopsin I.

DISCUSSION

The striking similarity of the fast photoelectric signals ascribed to the two retinal proteins, animal rhodopsin and bacteriorhodopsin, is perhaps not fortuitous. Comparing the above-reported data and those reported about the fast photoelectric responses from reconstituted bacteriorhodopsin membranes, we should note first of all the coincidence in many essential features of the fast photoelectric responses exhibited by animal and bacterial rhodopsins. In both cases: (a) a 530-nm laser flash inducing a single turnover of the pigments results in generation of a large photoelectric effect; (b) the electrogenic process consists of three main phases, the first phase being oppositely directed to the second and the third; (c) the kinetics of the first (negative) phase is faster than 200 ns, whereas the second phase develops in the microsecond and the third in the millisecond time scales; (d) if we take into account that the amino and carboxy termini of the polypeptide chain of both animal and bacterial rhodopsins are located on the opposite membrane sides, we can conclude that both rhodopsins generate photopotentials of the same direction (the N-terminal side positive); (e) the amplitudes of the three phases are: first < second < third, if they are measured under open-circuit conditions; (f) the decay of the photoresponse to the first laser flash occurs in the second time scale; (g) a blue flash added after a green one produces an opposite photoelectric effect; (h) if we identify the second and the third phase as the R2 component of the ERP, then both the R2 and the corresponding B2 component of the bacterial origin are correlated with proton uptake from the cytoplasmic surface (Cf: Refs. 26,27). Despite the apparent conflict with earlier reports of the lack of pH dependence of that component (reviewed in Ref. 15), our direct assay of alkalinification/acidification of the medium bathing the normally oriented or the inverted disc suspension seems to offer the first direct experimental support to the molecular mechanism for the R2 generation based on proton uptake. The remaining difficulty associated with the latter mechanism can be further reconciled.

Animal and bacterial rhodopsins after all are different pigments serving entirely different physiological functions. In addition to the well known lack of amino acid sequence homology,[34,35] the differences in the behavior of the two rhodopsins appear when they are exposed to repeated flashes or to continuous illumination. This is due to bleaching of animal rhodopsin and a light-induced conductance increase of disc membranes. Bacterial rhodopsin does not share these properties. Our present report further demonstrated a major difference between the two rhodopsins. While both rhodopsins take up protons from the cytoplasmic aqueous phase, only bacterial rhodopsin actually moves protons across the membrane and release them into the opposite aqueous phase to form a transmembrane proton gradient. The absence of such a gradient in the photoreceptor disc membrane is further supported by the lack of effect of proton ionophore 2,4-nitrophenol in accelerating the deprotonation (acidification) in a suspension of normally oriented disc membranes.[28] In our subsequent discussion, the bearing of our findings to the molecular mechanisms of fast photoresponses will be explored.

The fast photoresponse recorded from our disc-collodion film system is strikingly similar to the early receptor potential recorded from the outer (plasma) photoreceptor membrane *in vivo*. The only difference between the two systems is revealed when one compares the ratio of the amplitudes of the fast and slow phases. In our system this ratio is much lower than in the ERP. This is not surprising, since the conductance of the photoreceptor cell membrane must be much higher than that of discs due to the operation of sodium channels in the dark.[36,37]

In agreement with the above reasoning, shunting of the photoreceptor disc-collodion film system with an external conductance comparable to the conductance of the photoreceptor cell membrane was shown to result in such a change in the shape of the photoresponse that it proved to be practically the same as the ERP. We further notice that the acceleration of the decay and the change of the amplitude ratio as a result of shunting is similar to what has been observed in shunting or short-circuiting reconstituted bacteriorhodopsin membranes[8] and is consistent with the concept of chemical capacitance, which seems generally applicable to several pigment systems.[38]

Despite the striking similarity of the fast photosignals from the two rhodopsin systems, there is a conspicuous difference, namely, there is no proton release at the N-terminal side of the photoreceptor membrane and therefore no transmembrane proton transport. It must be pointed out that the proton release after the initial uptake occurs at the same (cytoplasmic) side. An analogous situation was observed in the corresponding stage of photoelectric process in bacteriorhodopsin reconstituted on a Teflon substrate.[5,8] Since the Teflon film precludes measurement of proton release at the extracellular side, any observed proton release must occur at the same (cytoplasmic) side. We have noticed the slow proton release in the dark and have linked the event to a possible change in the pK of the amino acid groups of rhodopsin at a late stage of photolysis, for example, during the formation of metarhodopsin III. Alternatively, this proton release has been linked to the reverse reaction of the initial proton uptake at the cytoplasmic surface.[26,27] The latter notion is consistent with many indications of reversible protonation and deprotonation processes accompanying the bleaching sequence of rhodopsin (for review, see Ref. 37).

The absence of a transmembrane proton transport in the photoreceptor membrane deserves additional comments. The lack of a sustained photovoltage under continuous illumination is not just a consequence of rhodopsin bleaching but also an expected consequence of rhodopsin's inability to establish a proton gradient. Our biochemical assay thus provided the direct evidence against a possible proton release at the intradiscal space (N-terminal side). For an obvious reason, such a proton release at the N-terminal side is mandatory for bacteriorhodopsin. This latter proton release has been suggested to contribute to an additional component of the fast photoelectric signal in the bacteriorhodopsin system.[39,40]

The ERP has long been considered a manifestation of light-induced conformational change of rhodopsin, resulting in the formation of a transient electric dipole within the rhodopsin molecule. Since the formation of metarhodopsin I is a microsecond process, the first (R1) electrogenic phase in the rhodopsin photoresponse ($t<200$ ns) must be associated with the preceding step of the photoreactions, i.e., rhodopsin → bathorhodopsin and/or bathorhodopsin → lumirhodopsin transitions. Electrogenesis may be a result of movement of positively charged protonated Schiff base inside the membrane due to retinal isomerization and, thus, a result of intramolecular dipole formation.

The second electrogenic phase may be coupled with the protonation of a group other than the Schiff base by protons from the cytosol. Because of the lack of proton release into the intradiscal space, a possible association of this process with the third electrogenic phase is no longer tenable. An explanation which links both the second and the third phase to the proton uptake at the cytoplasmic surface of a reconstituted bacteriorhodopsin membrane has been proposed.[26,27] In any case, the main electrogenic phases are coupled with the conversion of metarhodopsin I → metarhodopsin II. It is this step that is known to be accompanied by a major conformational change in the rhodopsin molecule.[37]

The R2 component has been linked to the proton uptake at the metarhodopsin stage.[12-15] Our biochemical assay lends support to the validity of the latter mechanism. Although a major conformational change indeed occur at the stage of metarhodopsin II formation, it has been demonstrated in a model system that reversible charge transfer at the membrane surface alone (without a conformational change and associated dipole formation) is sufficient to generate an ERP-like photoelectric signal.[14] The accessibility to an aqueous phase has been shown to be an essential requirement for the generation of the B2 component[41] but not for the generation of the B1 component[42] in the bacteriorhodopsin system. In this regard, it is of interest to note that the metarhodopsin I to II transition also requires the accessibility to aqueous protons.[16]

An obvious difficulty with the proton uptake mechanism for the R2 component is the lack of reported evidence of a pH dependence of the R2 component. Thus, despite previous suggestions of proton uptake as the R2 mechanism, the dipole mechanism has gained a wide acceptance. The apparent lack of pH dependence for the R2 and the B2 components has been attributed to the distortion of the photosignal by the membrane RC interaction under open-circuit conditions.[14,38] Incidentally, the variation of the ratio of the negative to the positive phase with external shunting observed in our present experiment is consistent with the same analysis, and therefore can be similarly explained.

In discussing the possible physiological role of rhodopsin-generated photoelectric effect, one should take into account its similarity to the photoelectric effect of bacteriorhodopsin, of which the role as a photoelectric generator is well documented. Therefore a minimal hypothesis seems to be the assumption that animal rhodopsin functions as a photoelectric generator too.

Assuming this point, one may suggest that electric potential produced by rhodopsin is used to change insulating properties of the disc membrane just as the electric potential across an electrically excitable membrane regulates its permeability.

Whatever the functional role the ERP may be, it is hardly probable that the completely delocalized electric field can performed any work in the disc since absorption of one photon should produce, according to our calculation, no more than a 0.1 mV delocalized membrane potential. So, a local, rather than delocalized, field should be operative.

One of the possible mechanisms of the light-induced increase in the membrane permeability might be an electrical break-down[43,44] of the disc membrane, occurring in the vicinity of that rhodopsin molecule which absorbed a photon. There are several factors that favor the electrical break-down of membranes: (a) the disc membrane contains proteins besides phospholipids, (b) the disc membrane contains a high percentage of extremely unsaturated fatty acids, (c) there is a Ca^{2+} asymmetry across the disc membranes with a higher Ca^{2+} concentration inside the disc, i.e., on that membrane side which charges positively under illumination. All these factors were found to be conditions in favor of an electrical break-down.[45,46] Artificially imposed electric field across the photoreceptor disc-phospholipid-impregnated film system changes the decay of the photopotential in a fashion which is quite similar to the effect of light.[47] In both cases, the photopotential decay is accelerated and can be observed at least several minutes after treatment. Calculation showed that such effects can be due to the formation of stable pores.

In addition to generating a (localized) electric field inside the membrane, the proton uptake process and the associated surface potential may also electrify the aqueous phase in the adjacent electrical double layer. The possible existence of such a surface potential has been suggested by Hong[15] and experimentally demonstrated by Cafiso and Hubbell.[48,49] Our experimental demonstration of the proton uptake at the cytoplasmic surface during the slow phase (R2 stage) provides the molecular basis of this surface potential. Hong[50] further suggested an electrostatic switching mechanism for transducin activation in the cyclic GMP cascade:[51] the light-induced (positive) surface potential and the subsequent negative surface potential caused by photophosphorylation of rhodopsin may be the "on-switch" and the "off-switch", respectively, of the visual phototransduction. It should be noted that this latter mechanism is consistent with our previous suggestion that a local electric field produced by rhodopsin may be used to change the activity of an enzyme affecting the cGMP level.[11]

It is not excluded that the rhodopsin function consists in a H^+ transfer *per se* rather than in the accompanying formation of an electric potential. However, in this case, it is difficult to understand why an artificially imposed electric field reproduces the effect of light on the photopotential decay.

CONCLUSION

The possibility of directly measuring the fast photoelectric signal from a collodion film reconstituted with photoreceptor discs and to correlate it with photometric pH measurements from normally oriented and inverted photoreceptor discs permit a new look at the primary mechanism of the visual process to be introduced.

A striking coincidence of all the essential features of the photoelectrogenesis mediated by bacterial and animal rhodopsins indicated that the primary function of animal rhodopsin may be similar to that of bacteriorhodopsin, namely the light-induced generation of an electric field due to proton movement. Several possible roles of the ERP in the visual phototransduction are discussed. The relevance of these possibilities to the design of molecular photosensors is apparent. Additional future investigations may prove fruitful in both furthering our understanding of the visual process as well as advancing the development of molecular photosensors.

Certainly, it would be premature to state that the above data are sufficient to accept the hypothesis on the electrogenic function of rhodopsin as verified. The possibility of rhodopsin electrogenesis as an epiphenomenon cannot be completely excluded. In the latter case, however, we ought to assume that the ability of rhodopsin to transduce light energy into electric energy is a senseless atavism occasionally preserved during the long process of evolution from bacteria to higher animals.

ACKNOWLEDGMENTS

The author wishes to thank his colleagues V. I. Bolshakov, L. A. Drachev, G. R. Kalamkarov, A. D. Kaulen, T. F. Shevchenko, L. E. Skokan and V. P. Skulachev, with whom the collaborative work is reviewed in this article. This study was carried out within the framework of the research program

'Rhodopsin', organized by the USSR Academy of Sciences and Moscow State University, and supervised by the late Vice President of the U.S.S.R. Academy of Sciences Professor Yu. A. Ovchinnikov.

REFERENCES

1. K. T. Brown and M. Murakami, Nature (Lond.) 201:626-628 (1964).
2. R. A. Cone and W. L. Pak, in: Handbook of Sensory Physiology, Vol. I. Principles of Receptor Physiology," W. R. Loewenstein, ed., pp. 345-365, Springer-Verlag, Berlin (1971).
3. D. Oesterhelt and W. Stoeckenius, Nature New Biol. 233:149-152 (1971).
4. A. E. Blaurock and W. Stoeckenius, Nature New Biol. 233:152-155 (1971).
5. H.-W. Trissl and M. Montal, Nature (Lond.) 266:655-657 (1977).
6. L. A. Drachev, A. D. Kaulen, and V. P. Skulachev, FEBS Lett. 87:161-167 (1978).
7. S.-B. Hwang, J. I. Korenbrot, and W. Stoeckenius, Biochim. Biophys. Acta 509:300-317 (1978).
8. F. T. Hong and M. Montal, Biophys. J. 25:465-472 (1979).
9. V. I. Bolshakov, G. R. Kalamkarov, and M. A. Ostrovsky, Dokl. Akad. Nauk USSR 249:1485-1488 (1979).
10. L. A. Drachev, A. D. Kaulen, L. V. Khitrina, and V. P. Skulachev, Eur. J. Biochem. 117:461-470 (1981).
11. L. A. Drachev, G. R. Kalamkarov, A. D. Kaulen, M. A. Ostrovsky, and V. P. Skulachev, Eur. J. Biochem. 117:471-481 (1981).
12. R. A. Cone, Science 155:1128-1131 (1967).
13. S. E. Ostroy, Photochem. Photobiol. 14:747-751 (1971).
14. F. T. Hong, Photochem. Photobiol. 24:155-189 (1976).
15. F. T. Hong, Bioelectrochem. Bioenerg. 5:425-455 (1978).
16. C. M. Radding and G. Wald, J. Gen. Physiol. 39:909-922 (1956).
17. M. A. Ostrovsky, I. B. Fedorovich, and S. E. Polyak, Biofizika 13(2):338-339 (1968).
18. D. G. McConnell, C. N. Rafferty, and R. A. Dilley, J. Biol. Chem. 243:5820-5826 (1968).
19. I. D. Pogozheva, V. A. Kuznetsov, V. A. Livshits, I. B. Fedorovich, and M. A. Ostrovsky, Biol. Membrany 2(9):880-896 (1985).
20. G. M. Regan, W. J. De Grip, F. G. M. Daemen, and S. L. Bonting, Biochim. Biophys. Acta 537:145-152 (1978).
21. G. R. Kalamkarov, L. E. Skokan, and M. A. Ostrovsky, Biofizika 25(4):634-637 (1980).
22. W. J. De Grip, G. L. M. Van de Larr, F. J. M. Daemen, and S. L. Bonting, Biochim. Biophys. Acta 325:315-322 (1973).
23. V. I. Bolshakov, G. R. Kalamkarov, and M. A. Ostrovsky, Dokl. Akad. Nauk USSR 240(5):1241-1244 (1978).
24. C. Gedney, J. Ward, and S. E. Ostroy, Am. J. Physiol. 221:1754-1759 (1971).
25. B. Honig, W. L. Hubbell, and R. F. Flewelling, Annu. Rev. Biophys. Biophys. Chem. 15:163-193 (1986).
26. T. L. Okajima and F. T. Hong, Biophys. J. 50:901-912 (1986).
27. F. T. Hong and T. L. Okajima, in: "Electrical Double Layers in Biology,", M. Blank, ed., pp. 129-147, Plenum Press, New York (1986).
28. T. F. Shevchenko, G. R. Kalamkarov, and M. A. Ostrovsky, Sensory Systems (USSR Acad. Sci.) 1:117-126 (1987).
29. P. P. M. Schnetkamp and U. B. Kaupp, Biochemistry 24:723-727 (1985).
30. M. A. Ostrovsky, in: "Membrane Transport Processes", D. C. Tosteson, Yu. A. Ovchinnikov, and R. Lattorre, eds., pp. 217-243, Raven Press, New York (1978).
31. T. F. Shevchenko, G. R. Kalamkarov, and M. A. Ostrovsky, Biofizika 25(3):462-468 (1980).
32. E. A. Dratz, G. P. Miljanich, P. P. Nemes, J. E. Gaw, and S. Schwartz, Photochem. Photobiol. 29:661-670 (1979).

33. A. J. Adams, M. Tanaka, and H. Shichi, Exp. Eye Res. 27:595-605 (1978).
34. Yu. A. Ovchinnikov, FEBS Lett 148:179-191 (1982).
35. R. L. Martin, C. Wood, W. Baehr, and M. L. Applebury, Science 232:1266-1269 (1986).
36. J. Toyoda, J., H. Nosaki, and T. Tomita, Vision Res. 9:453-463 (1969).
37. S. E. Ostroy, Biochim. Biophys. Acta 463:91-125 (1977).
38. F. T. Hong, in: Bioelectrochemistry: Ions, Surfaces, Membranes,", M. Blank, ed., pp. 211-237, American Chemical Society, Washington, DC (1980).
39. F. T. Hong and T. L. Okajima, in: "Biophysical Studies of Retinal Proteins,", T. G. Ebrey, H. Frauenfelder, B. Honig, and K. Nakanishi, eds., pp. 188-198, University of Illinois Press, Urbana, IL (1987).
40. F. T. Hong, J. Electrochem. Soc. 134:3044-3052 (1987).
41. S. Michaile and F. T. Hong, in: "Proc. 11th International Conference of IEEE Engineering in Medicine and Biology Society, November 9-12, 1989, Seattle, WA", F. A. Spelman and Y. Kim, eds., Institute of Electric and Electronic Engineers, Washington, DC, in press.
42. F. T. Hong, BioSystems 19:223-236 (1986).
43. R. Benz and U. Zimmermann, Biochim. Biophys. Acta 597:637-642 (1980).
44. R. Benz and U. Zimmermann, Biochim. Biophys. Acta 640:169-178 (1981).
45. G. Pilwat, R. Hampp, and U. Zimmermann, Planta (Berl.) 147:396-404 (1980).
46. L. V. Chernomordik, Thesis. Electrochem. Inst. of USSR Acad. Sci., Moscow (1979).
47. L. A. Drachev, G. R. Kalamkarov, G. R. Kaulen, M. A. Ostrovsky, and V. P. Skulachev, FEBS Lett. 119:125-131 (1980).
48. D. S. Cafiso and W. L. Hubbell, Biophys. J. 30:243-264 (1980).
49. D. S. Cafiso and W. L. Hubbell, Photochem. Photobiol. 32:461-468 (1980).
50. F. T. Hong, J. Molec. Electron. in press.
51. L. Stryer, Annu. Rev. Neurosci. 9:87-119 (1986).

J. Adams, N. Lucas, and R. Shiolli, Exp. Eye Res. [illegible] (1978).

[illegible]

[illegible] 40 [illegible]

A CORRELATION BETWEEN PHOTOCYCLE AND PHOTOELECTRIC RESPONSE OF BACTERIORHODOPSIN MONOMERS

Man-Qi Tan, Xiangming Gu*, and Kun-Sheng Hu

Institute of Biophysics
Academia Sinica
Beijing, China

ABSTRACT

The bacteriorhodopsin (BR) molecules in the purple membrane (PM) are organized into a two-dimensional hexagonal lattice of trimers which are surrounded by about 30 lipid molecules. The protein-protein and protein-lipid interactions may affect their structure and function. It is therefore interesting to solubilize the PM into BR monomers and to investigate its functional characteristics. The BR monomers were reconstituted with dimyristoylphosphatidylcholine (DMPC) to form vesicles. We studied the pH dependence of the photoelectric response in a bilayer lipid membrane (BLM) system and of the kinetics of the M412 and the O640 photocycle intermediates both in PM fragments and in BR monomer vesicles. The results show that the conformation of BR in monomeric state is more readily affected by the pH of the medium. Especially in pH higher than 9.4 the M412 slow-decaying component becomes very slow and even the polarity of the photovoltage signal is reversed. It appears that BR has two different conformational forms and their distribution depends on the pH of the medium. The two forms have different photocycle intermediates with different lifetimes and even the direction of proton pumping is reversed.

INTRODUCTION

Bacteriorhodopsin (BR) is the only protein found in the purple membrane (PM) of <u>Halobacterium halobium</u> that converts visible light energy into chemical energy through light-driven proton pumping from the inside to the outside of the membrane (see reviews in Refs. 1 and 2). BR contains a single polypeptide chain of 248 amino acids and one molecule of retinal covalently bound to the ε-amino group of Lys-216 via a protonated Schiff base linkage. Henderson and Unwin[3] suggested that BR is present as seven transmembrane α helical segments and the molecules obviously cluster to form trimers and exist in well ordered two-dimensional crystals in the PM. The lipid molecules apparently fill the space within the trimer between the monomeric units and also between trimers. But the trimer can be solubilized with nonionic detergent, Triton X-100, to form BR monomers as shown by Dencher and Heyn.[4] However, the important question of whether the aggregation of BR into the

* Present address: National Synchrotron Radiation Lab, Institute of High Energy Physics, Academia Sinica, Beijing, China

crystalline lattice of PM patches is of functional significance remains unanswered.

When BR absorbs light it undergoes a photochemical cycle identified by spectroscopically distinct intermediates: BR570 → K630 → L550 → O640 → BR570. Concurrently, protons are bound and released during the photocycle.[1,2] The exact mechanism of the proton pump remains unclear. However, evidence indicates that the M412 intermediate plays a key role in the photochemical cycle and that the Schiff base deprotonation is closely associated with the proton pump mechanism. Several studies have examined the correlation between proton movement and the photocycle.[5,6] A complication arises. While both proton release and uptake are single exponential processes, the M412 rise and decay are the sum of at least two exponential processes: M-slow and M-fast. Both Lozier et al.[7] and Govindjee et al.[8] have found, that for PM sheets under a variety of conditions, the half-time for the decay of the M-slow form of M412 is almost the same as the half-time for proton uptake. Hanamoto et al.[6] proposed that the protonated Schiff base exists at two different sites within the protein. There are two components for the deprotonation of the Schiff base that might best be described in terms of two parallel paths. Li et al.[5] showed a possible correlation between proton movements and a particular intermediate of the photocycle. At acidic pH, most of the photocycle goes through the M-slow path and this process is correlated with proton release. At alkaline pH, most of the photocycle goes through the M-fast path, but no protons are pumped.

Concurrently with the photocycle, a bilayer lipid membrane (BLM) containing oriented BR produces an electrical signal when illuminated. This photoelectric signal is probably due to internal charge movements associated with light-induced conformational changes of the BR and external charge movements such as the binding and release of protons and other ions from the PM during the photocycle.[9-11] These processes are relevant to the design of membrane-based molecular optoelectronic devices. Hong[12] did a survey of the model membrane systems that incorporate the protein pigment-BR for the design of a prototype molecular computing element. We studied some basic aspects of the structure and function relationship of BR, aiming at the development of experimental prototypes. We found that the photoelectric signal of BR monomers is different from that of BR trimers.[13] In this work we attempt to correlate the photovoltage signal with the kinetics of photocycle intermediates M412 and O640 of BR in monomeric and trimeric states.

MATERIALS AND METHODS

Preparation of PM and BR monomer vesicles

The procedure for the growth of *Halobacterium halobium* R1M1 and purification of the PM was a combination of those outlined by Oesterhelt and Stoeckenius[14] and Becher and Cassium.[15] BR monomer vesicles were prepared according to the method used by Heyn[16] and Cherry.[17] Dimyristoylphosphatidylcholine (DMPC) was purchased from Sigma (ca. 98%) without further purification. Triton X-100 was obtained from Carl Roth KG Chemische Fabrik Karisrube (especially purified). All other chemicals used were AR grade. In order to check whether the monomer sample remained in the monomeric state or not, it was characterized by the following means:

Electron microscopy. Samples were mounted on gold discs and frozen in liquid nitrogen. Freeze-fracturing was carried out in a Balzer BaF-400D apparatus. Specimens were replicated with Pt/C and were examined in a JEM-100CX electron microscope. The DMPC vesicle containing BR monomer was shown in Fig. 1.

Circular dichroism and absorption spectroscopy. To monitor the stability of the BR monomer state in the reconstituted DMPC vesicles, circular dichroism measurements were carried out with a Jasco J-500A spectrophotometer. Absorption spectra were recorded with Perkin-Elmer Lambda 7 UV/VIS spectrophotometer. Because the BR monomer vesicles exhibit considerable light-scattering, CD measurements were made with apo-vesicles in the reference cuvette. Apo-vesicles were prepared by treating vesicles with hydroxylamine in the presence of light, causing the loss of the chromophore.[14] Fig. 2 shows that the BR monomer vesicles were devoid of the exciton band couple near 570 nm in CD spectra. Fig. 3 shows a typical blue shift of the λ_{max} at 550 nm of BR monomer vesicles in absorption spectra. This demonstrated that our BR sample was in the monomeric state in the vesicles at 25°C.

Photovoltage measurements in Bilayer lipid membranes

The apparatus for measuring the photovoltage signal of DMPC vesicles containing BR monomers in the BLM system was according to Tien[18] and described previously.[19] The membrane forming solution was a mixture of BR monomer vesicles and soybean lipid/decane; the concentrations of lipid and BR were 100 mg/ml and 0.6×10^{-5} M, respectively. The planar bilayer lipid membranes (BLM) were formed by spreading the membrane-forming solution on a hole between two aqueous compartments. A 150 W high voltage xenon-lamp with a heat protection filter was used as the actinic light. Photovoltage signals were measured through a pair of calomel electrodes connect to a 10^{13} Ω high-input impedance amplifier and the output was then recorded on an X-Y recorder. The pH of the electrolyte solution of 0.1 M KCl in the inner and outer bath were adjusted using 0.1 M HCl and 0.1 M NaOH. Soybean lipid was purified with chloroform.[20] All other reagents were analytical grade.

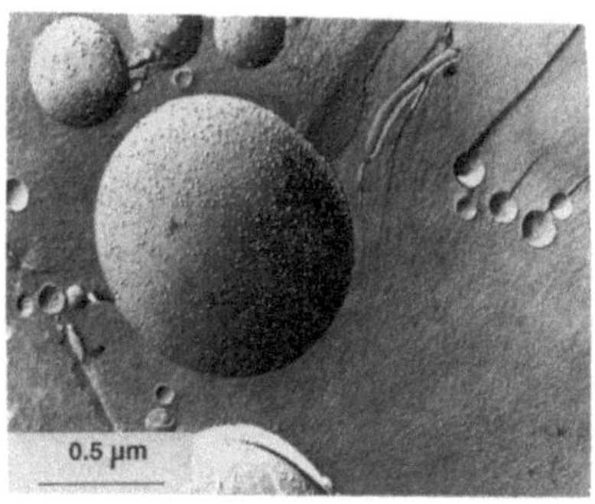

Fig. 1. Freeze-fracture electron micrograph of DMPC vesicles containing BR monomers.

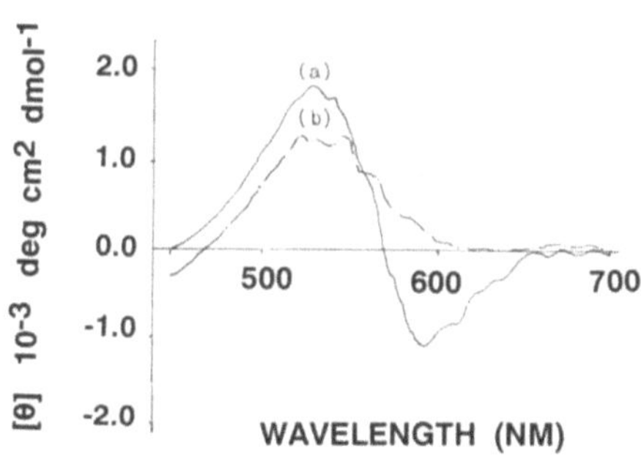

Fig. 2. CD spectra at 25°C. (a) PM fragment; (b) DMPC vesicle containing BR monomers.

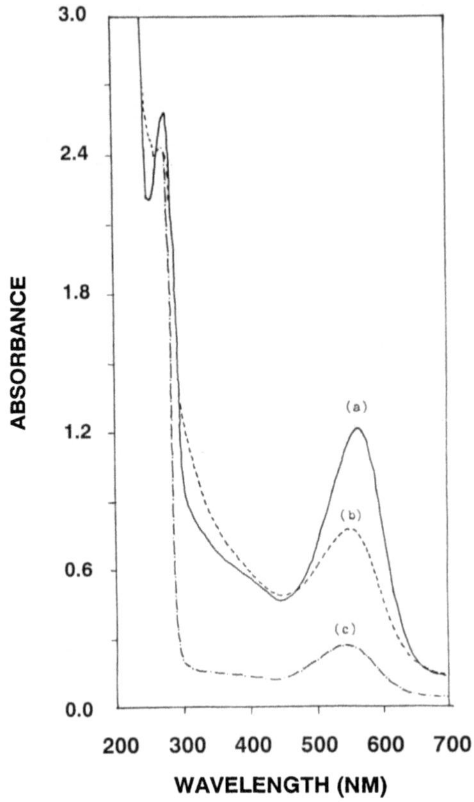

Fig. 3. Absorption spectra (a) purple membrane fragment; (b) monomeric bacteriorhodopsin; (c) DMPC vesicles containing bacteriorhodopsin monomers.

Measurements of photocycle intermediates M412 and O640

The relative amount of the photocycle intermediates M412 and O640 was determined by measuring the amplitude of flash-induced absorbance changes at 412 and 640 nm, respectively, with a single-beam kinetic spectrophotometer (assembled in our Laboratory[21]). Light-induced absorbance changes were initiated by a camera photoflash lamp (0.2 ms half-bandwidth; $\lambda > 500$ nm) with a Zeiss BG25 glass cutoff filter placed in front of the flash lamp. The signals from an IP 28A photomultiplier tube were fed through an amplifier which was connected to a digital waveform recorder (EPIC instrument, Inc.). The latter was interfaced to a Zenith microcomputer (Z-100PC, IBM-PC compatible). The computer was used to control the experiment, and to analyze and to plot data using a special computer program.[21] The sample was light-adapted before measurements and was suspended in 0.2 M KCl. The measurements were performed at 25°C. The pH was adjusted by using 0.1 M HCl and 0.1 M NaOH.

RESULTS

The photovoltage signals of BR in the BLM system

As shown in Fig. 4 there were different forms of photovoltage signals with different states of BR in the BLM system. In particular, the photovoltage signals of DMPC vesicles containing BR monomers were pH dependent (Fig. 5).

The photovoltage signals of PM fragment rose monotonically and exponentially upon illumination and decayed monotonically and exponentially upon cessation of illumination (Fig. 4a).

In contrast, the signals of BR monomers in DMPC vesicles rose transiently upon illumination but then decayed exponentially toward a steady state level, exhibiting a "spike-like" signal, even under continuous constant illumination. Upon cessation of illumination another spike with opposite polarity could be observed. The exponential decay of the spikes is faster at low pH; the decay time prolongs with increasing pH (Fig. 5). When the pH was higher than 9.4, even the polarity of the signal was reversed (Fig. 5). This behavior was not observed with PM fragments.

The pH dependence of M412 and O640 photocycle intermediates

To see if there is a correlation between the photoelectric response and the BR photocycle, we measured the pH dependence of the flash-induced absorbance change of the intermediates M412 and O640. Figs. 6 and 7 show the relaxation of M412 in PM fragments as BR trimers, and in BR as monomers in DMPC vesicles, respectively. The total amount of M412 varied with the pH as shown in Fig. 8. In the acidic range, the amount of M412 increases as the pH value increases but decreases significantly when pH exceeds 7, especially in the monomeric state. The M412 decay is the sum of two exponential components: the M fast-decaying and M slow-decaying forms. Fig. 9 showed the pH dependent decays of these two components. Both in BR trimers and monomers the M fast-decaying component was pH independent, but the M slow-decaying

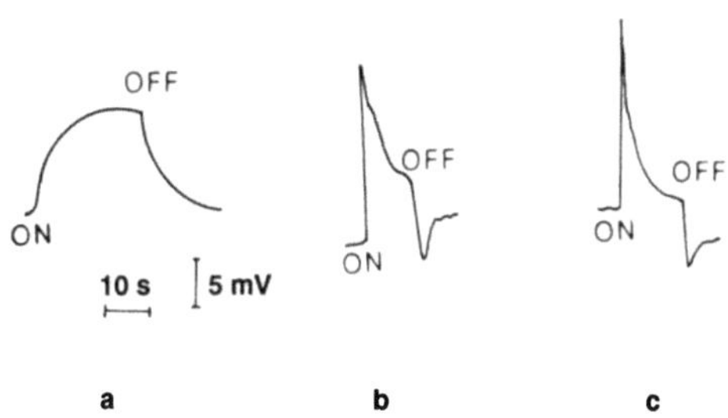

Fig. 4. The photovoltage signal of BR in BLM system at pH 6.0. (a) PM fragment; (b) PM fragment in DMPC vesicles; (c) BR monomers in DMPC vesicles.

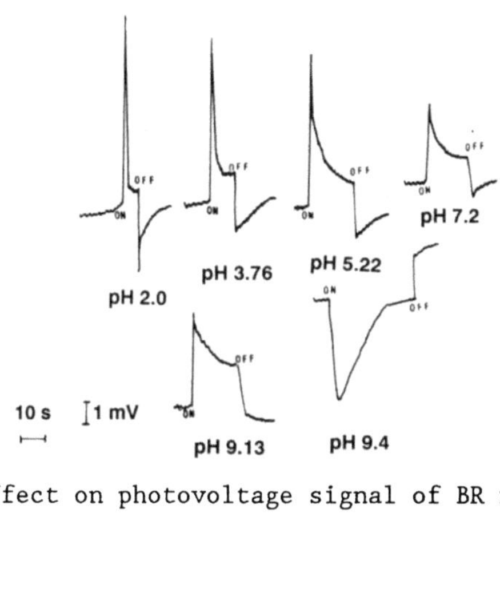

Fig. 5. The pH effect on photovoltage signal of BR monomers in DMPC vesicles.

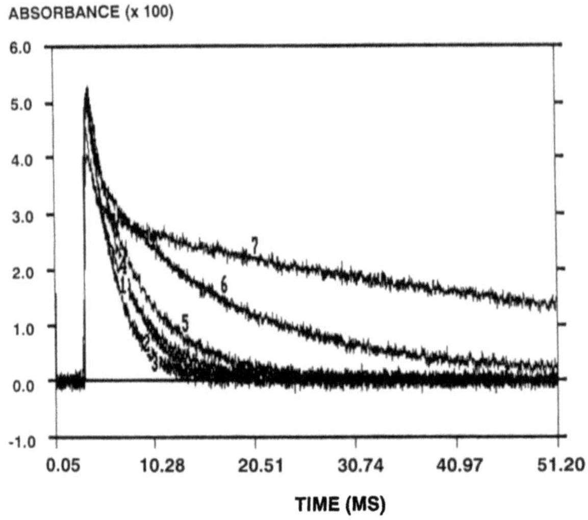

Fig. 6. Flash-induced absorbance change of the intermediate M412 of PM fragment in 0.2 M KCl at various pH. Each trace is an average of five flashes; actinic light > 500 nm; probing light 410 nm; temperature 25°C; OD_{568} of the sample = 0.2; the trace 1, 2, 3, 4, 5, 6, and 7 represent samples at pH = 3.8, 5.0, 6.0, 7.0, 7.9, 8.9 and 9.9 (±.05), respectively.

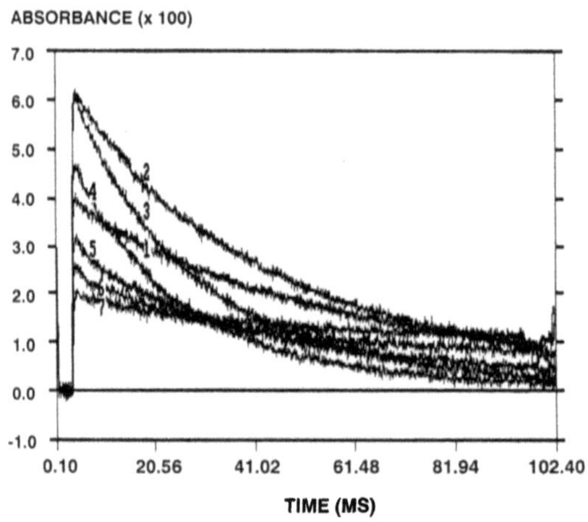

Fig. 7. Flash-induced absorbance change of the intermediate M412 of BR monomers in DMPC vesicles in 0.2 M KCl at various pH. OD_{550} of the sample = 0.2; the trace 1, 2, 3, 4, 5, 6, and 7 represent samples at pH = 4.1, 5.1, 6.1, 7.0, 8.0, 9.0, and 10.0 (±.05), respectively. The others are the same as in Fig. 6.

component was significantly dependent on the pH of the medium. The M412 decay rate ($\tau_{1/2}$) of BR trimers was 3-10 times faster than that of BR monomers in both M fast-decaying and M slow-decaying components.

The pH dependence of O640 in PM fragment was shown in Fig 10. The amount of O640 declined more rapidly as the pH of the medium increased. As shown in the traces 5 - 7 of Fig. 10, the O640 absorbance vanishes at pH higher than 8.0.

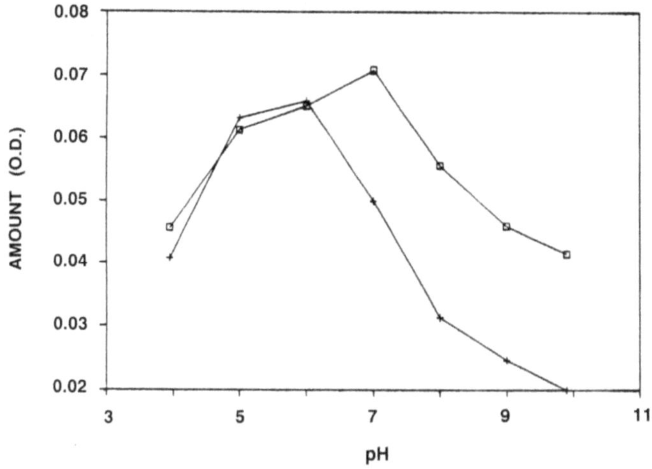

Fig. 8. The pH dependence of the intermediate M412 amount (OD value) in 0.2 M KCl. (□) the PM fragment; (+) BR monomer in DMPC vesicles.

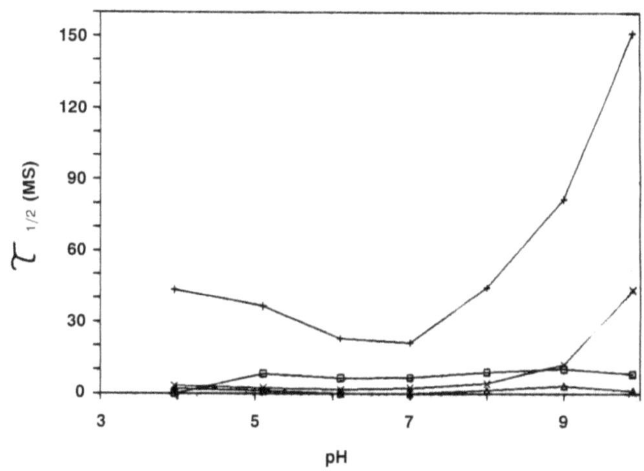

Fig 9. The pH dependent decay rate ($\tau_{1/2}$) of the intermediate M412.
(Δ) the M fast-decaying component of PM fragment;
(×) the M slow-decaying component of PM fragment;
(□) the M fast-decaying component of BR monomers;
(+) the M slow-decaying component of BR monomers.

Between pH 4-10, the O640 absorbance is rather weak in BR monomer DMPC vesicles, similar to the situation of BR fragments at pH higher than 8.0. This may be due to the cumulative effect of the slowing of the decay rate of the M slow-decaying component ($\tau_{1/2}$ > 20 ms, Fig. 9). However, the trace 7 in Fig. 10 suggests the presence of an additional intermediate with a faster rise time and decay time than the O640 signal at other pH values. Similar behaviors were observed in the BR monomer in DMPC vesicles, and may be correlated with the polarity reversal of the photoelectric signal (Fig. 5). It is possible that the pH induced change of BR at higher pH may lead to the change of photocycle intermediates as well as the reversal of the H^+ pump.

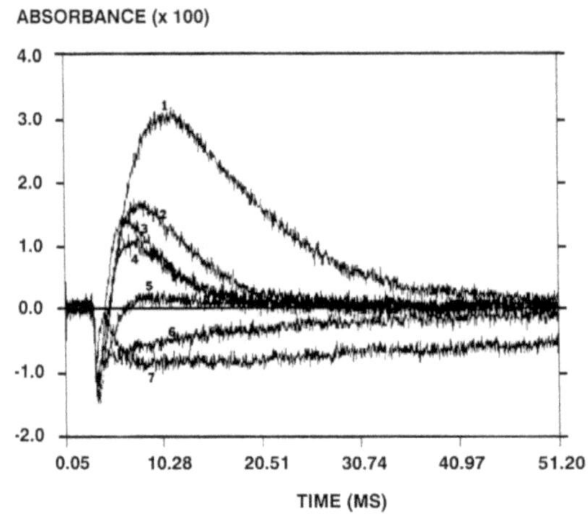

Fig 10. Flash-induced absorbance change of the intermediate O640 of PM fragment in 0.2 M KCl at different pH. Each trace is an average of five flashes; actinic λ 450-600 nm (photoflash plus VG8 Zeiss filter); measuring at 640 nm; temperature 25°; samples OD568=0.31; the trace 1, 2, 3, 4, 5, 6, and 7 represent samples at pH=4.1, 5.0, 6.0, 7.0, 8.0, 8.9 and 9.8 (±.05), respectively.

DISCUSSION

It has been shown[9,22] that there are at least three components in the photoelectric signals which are related to the direction of proton transport during the photocycle in BR model membranes. Our results show that the polarity of the photovoltage signal of DMPC vesicles containing BR monomers is reversed at pH higher than 9.4 (Fig 5). In order to correlate the pH induced changes of the photovoltage response with the corresponding changes of the photocycle, we examined the pH effect on the amount of photocycle intermediates M412 and O640 and their decay rate both in BR trimeric and monomeric states.

The results show that the decay rate of M412 slow-decaying component decreases considerably when the pH of the medium is higher than 7 (Fig 9). This probably accounts for why it is difficult to measure the O640 signals of BR monomers and BR trimers at pH higher than 8 (Fig 10). If so, our results support the suggestion that only M slow-decaying component is correlated with proton pumping and goes through O640 intermediate back to BR568.[5] It is consistent with our result that there is no pH dependence of the M fast-decaying rate (Fig. 9).

However, we suggest an alternative explanation. BR may have two different forms and their distribution depends on the pH of the medium. The two forms have different photocycle intermediates with different lifetime and even the direction of proton pumping may also be reversed. Hanamoto et al.[6] proposed that the photocycle might be best described in terms of two parallel paths. This is supported by our results. (1) Figs. 8 and 9 show that the amount of M412 and the decay rate ($\tau_{1/2}$) of M412 slow-decaying component depend on the pH of the medium both in trimeric and in monomeric states. Especially in the alkaline range, the pH effect becomes more and more prominent as the pH increases. It is consistent with the results obtained by Dancshazy et al.[23] They found a new intermediate R350 at pH 10.5. The basic changes in the two BR states are similar, but the monomeric BR seems more vulnerable. Maybe the monomeric BR molecules lack the protein-protein and protein-lipid interactions, so its conformation is easily affected by the pH of the medium. (2) In DMPC vesicles containing BR monomers the absorbance at 550 nm changes reversibly from 0.22 to 0.15 when the pH of the medium is varied from 6 to 10 (data not shown). But this is not seen in PM fragments. (3) The pH dependence of the absorbance change is correlated with that of the photovoltage signals, especially at pH 9.4 where the polarity of the signal is reversed (see Fig 5). It is consistent with the results of Li et al.[24] They suggested that the light-induced proton release at high pH is due to protons released and rebound on the carboxy terminal side of BR. We think that the change of the photoelectric signal at pH higher than 9.4 is correlated to its photocycle and conformational changes. It is more apparent in the monomeric BR state. Further detailed research is needed.

Groma et al.[25] recently reported that the risetime of the photoelectric signal from a reconstituted bacteriorhodopsin membrane is in the picosecond range. From the point of view of designing a molecular device, the fast rising photovoltage signals of monomeric BR may be configured to fabricate an ultrafast molecular photoswitch device. The very slow decay of M412 of BR monomers at approximately pH 10 (see Fig 9) may be exploited as a molecular storage device.

ACKNOWLEDGEMENTS

The authors wish to thank Dong Bao-Zhong and Shi Zhi-Yuan for helping us to assemble the single-beam flash kinetic spectrophotometer, Dong Renjie for doing the freeze-fracture electron microscopy observation, Hua Qing-Xin for measuring the CD spectra and Wang Ao-Jin for recording the absorption spectra. This work has been supported by National Natural Science Foundation of China Grant NO. 3870158 and Academia Sinica Research Fund NO. 874801-3.

REFERENCES

1. W. Stoeckenius, R. H. Lozier and R. A. Bogomolni, Bacteriorhodopsin and the purple membrane of *Halobacteria*, Biochim. Biophys. Acta 505:215-278 (1979).
2. H. G. Khorana, Bacteriorhodopsin, a membrane protein that uses light to translocate protons, J. Biol. Chem. 263:7439-7422 (1988).
3. R. Henderson, and P, N. T. Unwin, Three-dimensional model of purple membrane obtained by electron microscopy, Nature (London) 257:28-32 (1975).
4. N. A. Dencher, and M. P. Heyn, Preparation and properties of monomeric Bacteriorhodopsin, Meth. Enzymol. 88:5-10 (1982).
5. Q.-G. Li, R. Govindjee, and T. G. Ebrey, A correlation between proton pumping and the bacteriorhodopsin photocycle, Proc. Natl. Acad. Sci. USA 81:7079-7082 (1984).
6. J. H. Hanamoto, P. Dupuis, and M. A. El-Sayed, On the protein (tyrosine)-chromophore (protonated Schiff base) coupling in bacteriorhodopsin, Proc. Natl. Acad. Sci. USA 81:7083-7087 (1984).
7. R. H. Lozier, W. Niederberger, R. A. Bogomolni, S.-B. Hwang, and W. Stoeckenius. Kinetics and stoichiometry of light-induced proton release and uptake from purple membrane fragments, *Halobacterium Halobium* cell envelopes, and phospholipid vesicles containing oriented purple membrane, Biochim. Biophys. Acta 440:545-556 (1976).
8. R. Govindjee, T. G. Ebrey, and A. R. Crofts, The quantum efficiency of proton pumping by the purple membrane of *Halobacterium Halobium*, Biophys. J. 30:231-242 (1980).
9. L. Keszthelyi, Intramolecular charge shifts during the photoreaction cycle of bacteriorhodopsin, in: "Information and Energy Transduction

in Biological Membranes," C.L. Bolis, E. J. M. Helmreich, and H. Passow, eds., pp. 51-71, Alan R. Liss, Inc., New York (1984).
10. H.-W. Trissl, Primary electrogenic processes in bacteriorhodopsin probed by photoelectric measurements with capacitative metal electrodes, Biochim. Biophys. Acta 806:124-135 (1985).
11. F. T. Hong, and T. L. Okajima, Electrical double layers in pigment-containing biomembranes, In:"Electrical Double Layers in Biology," M. Blank, ed., pp. 129-147, Plenum, New York (1986).
12. F. T. Hong, The bacteriorhodopsin model membrane system as prototype molecular computing element, BioSystems, 19:223-236 (1986).
13. K.-S. Hu, L.-L. Shen, W.-Q. Sun, and M.-Q. Tan, The effect of pH on photoelectric response signal in different states of bacteriorhodopsin, Kexue Tongbao (Science Bulletin) (in Chinese) 32:1895-1898 (1987).
14. D. Oesterhelt, and W. Stoeckenius, Isolation of the cell membrane of Halobacterium halobium and its fractionation into red and purple membrane, Meth. Enzymol. 31:667-678 (1974).
15. B. Becher, and H. Y. Cassim, Improved isolation procedures for the purple membrane of *Halobacterium Halobium*, Prep. Biochemistry 5:161-178 (1975).
16. M. P. Heyn, and N. A. Dencher, Reconstitution of monomeric bacteriorhodopsin into phospholipid vesicles, Meth. Enzymol. 88:31-35 (1982).
17. R. J. Cherry, U. Muller, R. Henderson, and M. P. Heyn, Temperature-dependent aggregation of bacteriorhodopsin in Dipalmitoyl- and Dimyristoylphosphatidylcholine vesicles, J. Mol. Biol. 121:283-298 (1978).
18. H. T. Tien, "Bilayer Lipid Membranes (BLM); Theory and practice," Marcel Dekker (1974).
19. K.-S. Hu, M.-Q. Tan, D.-H. Wang, W.-Q. Sun, and T.-Z. Zhang, The effect of pH on the rotational diffusion and photoelectrical response of bacteriorhodopsin, Acta Biophysica Sinica (in Chinese) 4:221-226 (1988).
20. K.-S. Hu, and M.-Q. Tan, The effect of temperature on the structure and photoelectric response signal in bacteriorhodopsin, Acta Biophysica Sinica, in press.
21. X.-M. Gu, The pH dependence of photocycle intermediates in two states of bacteriorhodopsin, Thesis (in Chinese) (1988).
22. T. L. Okajima, and F. T. Hong, Kinetic analysis of displacement photocurrents elicited in two types of bacteriorhodopsin model membranes, Biophys. J. 50:901-912 (1986).
23. D. Dancshazy, R. Govindjee, B. Nelson, and T. G. Ebrey, A new intermediate in the photocycle of bacteriorhodopsin, FEBS Lett. 209:44-48 (1986).
24. Q.-G. Li, R. Govindjee, and T. G. Ebrey, A proton release site on the C-terminal side of bacteriorhodopsin, Photochem. Photobiol. 44:515-518 (1986).
25. G. I. Groma, F. Raski, G. Szabo, and G. Varo, Picosecond and nanosecond components in bacteriorhodopsin light-induced electric response signal, Biophys. J. 54:77-80 (1988).

NEURONS AS MICROPROCESSORS WITH

A KIND OF MEMORY FUNCTION

Gen Matsumoto and Toshio Iijima

Electrotechnical Laboratory
Tsukuba City, Ibaraki 305
Japan

INTRODUCTION

Studies on neurocomputing (Amari, 1982; Rumelhart *et al.*, 1986a) should be directed in two ways which in turn influence each other (Matsumoto, 1988): In one direction concrete neural-network solutions for specific important problems should be applied to substantiate its practical significance and, at the same time, the theoretical potentialities and limitations for computation should be explored on such network models. The alternative way is to seek a more profound understanding of the algorithms used by the CNS (central nervous system) to process information and to learn more about the molecular and cellular mechanisms underlying specific computations and memory processes in neurons and neural networks.

We will first give a brief review of current research on neurocomputing. The potentialities and limitations of neurocomputing will be discussed. We stress that these limitations primarily originate from the oversimplified model of a neuron (an artificial neuron model). We note that electrical events occurring in a natural neuron are part of the manifestations associated with neural activities and are regulated by the chemical and conformational processes inside the neuron. Neurons could correspond to LSI (large-scale integrated) microprocessors with a kind of memory function in the electronic digital-computing system. A concrete example of the neural information processing at the synaptic sites will be given for the primarily-cultured neurons of both rat hippocampus CA1 pyramidal cells and rabbit superior cervical ganglion cells. The information processing performed at the synaptic sites will be discussed from the standpoint of computational science. Finally we will discuss how we should proceed in the neuroscientific study of neurons and neural networks for information science.

NEUROCOMPUTING -COMPUTATIONS PERFORMED ON ARTIFICIAL NEURON NETWORKS-

The Artificial Neuron Model

The current models of artificial neuron networks are neurally inspired mathematical models that use a large number of simple, homogeneous but highly-interconnected processing elements, "artificial neurons". An artificial neuron collects inputs, using its internal summer after each of them is multiplied by its proper weight, and produces an output only if the sum of the inputs exceeds an internal threshold value. We shall call the computing performed on these neural networks "neurocomputing". The operation of the current models of the neural network is basically nearest-neighbour search. The efforts of research on neurocomputing have been concentrated on searching the efficiencies of these neural networks in performing this operation (Amari, 1982; Kohonen, 1984; Rumelhart *et al.*, 1986a) and on applying concrete neural network solutions for specific problems to substantiate the practical significance (Hopfield and Tank, 1985; Sejnowski and Rosenberg, 1986).

The Error Back-Propagation Model

Recently Hecht-Nielsen (1986a) has shown that a three or more-layer neural network system can generally represent the connection strength that will give the appropriate output for all inputs. In other words any continuous mapping from input to output can be done in a three or more-layer system (Fig.1). The problem now is how to implement a learning procedure for a multi (three or more)-layer system. If the input units are directly connected to the output units, it is relatively easy to find learning rules that iteratively adjust the strengths of the connections so as to progressively reduce the difference between the actual and desired output vectors (Minisky and Papert, 1969). Learning becomes more interesting and more difficult when we introduce a multi-layer network with hidden units whose actual or desired states are not specified by the task. The reason the learning becomes more difficult arises from the requirement that the learning procedure must decide under what circumstances the hidden units should be active in order to help achieve the desired input-output behavior. The reason the learning becomes more interesting is because the learning procedure itself is a means to construct a new representation. Rumelhart, Hinton and Williams (1986b) have constructed a novel algorithm which gives an effective learning procedure for the neural network with hidden units. In the three-layer system, the weights are adjusted for the output layer according to an error function that calculates a weight adjustment based on the difference between the actual and desired outputs. In other words, each error value for each actual output unit is propagated backward to the hidden layer and used to adjust the values of the weights to the hidden units. The procedure repeatedly adjusts the weights of the

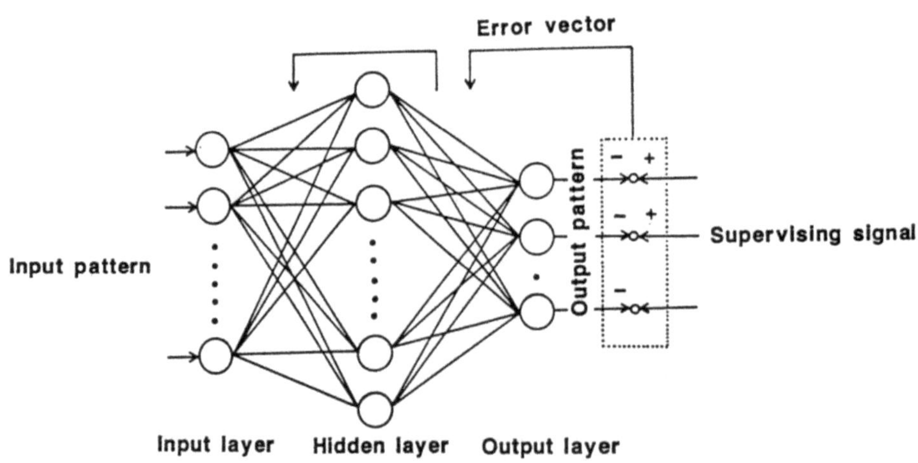

Fig.1. A three-layer artificial neuron network and the back-propagation model.

connections in the network so as to minimize the error function (a measure of the difference between the actual and desired output states). As a result of the weight adjustments, internal hidden units come to represent important features of the task domain, and the regularities in the task are captured by the interactions of these units. Practical usefulness of the learning procedure of the back-propagation model was demonstrated by Sejnowski and Rosenberg (1986). They constructed a neural network application of NETalk that did text-to-speech conversion of English. The network consists of three layers of artificial neurons; the input, the middle (or hidden) and the output layers contain 203, 80 and 26 neurons, respectively. After the learning, the network captured the text-to-speech rule and successfully operated for the new text that was different from the one used for the learning. The network has achieved 95 % accuracy. The network performance was limited to 82 % accuracy without hidden units but was enhanced to 98 % by increasing the number of hidden units to 120. This example illustrates the usefulness of the error back-propagation learning algorithm and the value of the hidden units.

However, one of the disadvantages inherent in the error back-propagation learning procedure is that it cannot generally escape from local minimum states characterized by a set of quasi-stable states in the neural network system. The other disadvantage is that the time necessary for the learning becomes tremendously longer for large size problems. According to Hinton (1987), the time of the N-interconnected network necessary for the learning process should be on the order of N^3 when the operation is executed on current serial digital computers. The time is reduced by the order of N when each weight of connection is separately calculated with a parallel computer. Therefore, one solution to overcome the disadvantage is to build a computer which is architecturally suited to handle the neural network system. Because the neural network is massively parallel in nature, a computer built with several ten or hundred thousands of processors, where each processor takes the place of an artificial neuron, would overcome the disadvantage. Many attempts are being made at building parallel computers for neural network systems (Hecht-Nielsen, 1986b). The other solution to overcome the difficulty of necessitating tremendously longer time in the learning for large size problems is to improve the learning algorithm of the error back-propagation model or to create a new learning algorithm to reduce the time. Most research efforts in neurocomputing are currently being devoted to the development of improved or new algorithms.

INFORMATION PROCESSINGS IN A NEURON

A neuron is a basic element for transmitting, processing and storing information in neural systems and the brain. The CNS is not composed of neurons with the homologous characteristics as has been assumed in the artificial neuron. Rather, the variety is one of the most prominent characteristics of the neurons. Over 50 kinds of functionally different neurons are reported in the cerebrum. This variety may be tremendously increased when the functions possessed in neurons are analyzed in more detail. The versatility is another peculiar characteristic of neurons. Indeed, dynamic molecular transformations within a neuron take place by environmental stimuli and conditions, as shown below. Both the variety and versatility are inherently associated with the neuron. As a result, a neuron could be regarded as a kind of microprocessor with multiple information-processing (synaptic) sites and with a kind of memory function.

Here we will exemplify how the information is processed and stored in a neuron for both rat hippocampus pyramidal CA1 cells and rabbit superior cervical ganglion cells.

Rat Hippocampus Pyramidal CA1 Neurons

Pyramidal CA1 Neurons of the rat hippocampus all contain glutamate receptive sites which respond excitatorily. Besides the glutamate receptors, some of the neurons contain muscarinic acetylcholine (ACh) receptors. In muscarinic responses, ACh usually works excitatorily (Swanson *et al.*, 1982) but in some cases inhibitorily (Brown and Adams, 1980; Egan and North, 1986). These two opposite effects of ACh result from interactions of ACh with two different subtypes of ACh receptors (Egan and North, 1986); M_1 and M_2 subtypes (Fig.2). The subtype is characterized by its dissociation equilibrium with pirenzepine. The dissociation constants

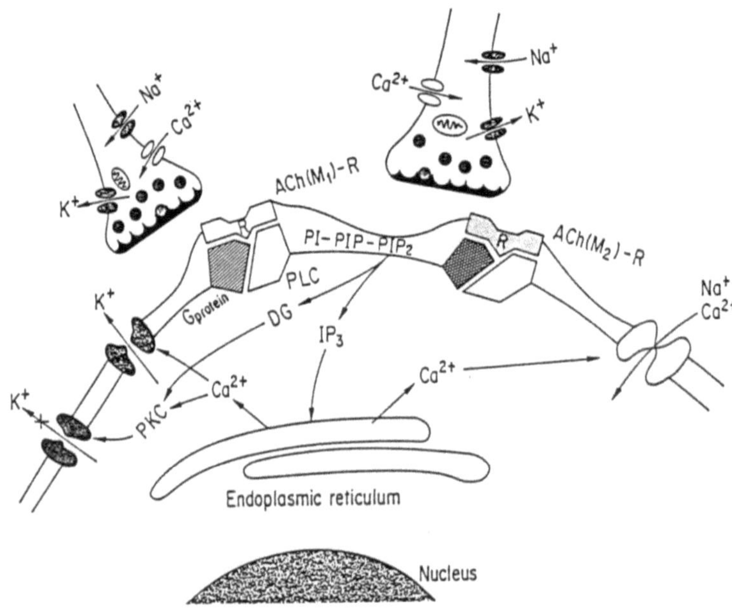

Fig.2. A schematic drawing of molecular events at the synaptic sites of muscarinic acetylcholine receptors.

of the subtypes, M_1 and M_2, are ~10 and ~600 nM, respectively, for slices of rat pons containing the nucleus parabrachialis (Egan and North, 1986). ACh bound to the M_2 receptor causes membrane hyperpolarization as a consequence to an increase in the membrane conductance to potassium ions. The Ca^{2+}-dependent potassium channel couples to the M_2 receptor. On the other hand, ACh bound to the M_1 receptor brings about membrane depolarization as a result of a decrease in the potassium conductance. The potassium M channel is coupled to the M_1 receptor. Thus, presence of multiple receptive sites specific for a neurotransmitter augments the variety of the information processing at the synaptic site. It was recently confirmed that there were three subtypes of glutamate receptors, and that gamma aminobutyric acid (GABA) could specifically bind to two subtypes of its receptor (one with high and another with low affinity).

What kinds of molecular mechanisms regulate the switching between depolarization (the pathway from the M_1 receptor to the K_M channel) and hyperpolarization (the pathway from the M_2 receptor to the K_{Ca} channel) on ACh application? The pathway to the K_{Ca} channel is delineated by the following molecular events which take place sequentially (Fig.2); binding of ACh to the M_2 receptor → release of inositol-1, 4, 5-triphosphate (IP_3), a hydrolysis product of the membrane lipid phosphatidylinositol biphosphate, into the cytoplasm from the plasma membrane → IP_3-induced release of Ca^{2+} from endoplasmic reticulum → opening of Ca^{2+}-dependent potassium channels. In the pathway from the M_1 receptor to the K_M channel, an enzyme termed protein kinase C (Takai *et al.*, 1982) is involved. Protein kinase C (PKC) is in the activated state only when sufficient levels of Ca^{2+} ions, phospholipid and diacylglycerol are present. Activated PKC phosphorylates and closes the K_M channel, resulting in the membrane depolarization.

In cultured CA1 pyramidal neurons of the rat hippocampus, we have found that some neurons are depolarized and some are hyperpolarized when we expose them to the same concentration of ACh extracellularly (Iijima and Matsumoto, 1988). This cannot be explained by the above-mentioned difference of the ACh dissociation between the M_1 and M_2 receptors and of the fixed circuits between M_1 and K_M or between M_2 and K_{Ca}, since all the neurons tested are stimulated with the same amount (10 μM) of ACh. We have concluded that there are three types of connections between the ACh receptive site and the ionic channel in the cultured CA1 pyramidal neuron, resulting in three types of differentiated neurons; the first type is the M-to-K_M type of the neuron, the second type is the M-to-K_{Ca} type, and the third type is the M-to-cation selective channel type of neuron. In the last type of neuron, ACh causes the membrane depolarization by opening the (Ca^{2+}-dependent) cation-selective channel (Fig.2).

Opening and closing of the ionic channels are regulated, directly or indirectly, by intracellular Ca^{2+} ion concentration (Fig.2). Therefore, measurement of the regional change in Ca^{2+} ion concentration within a

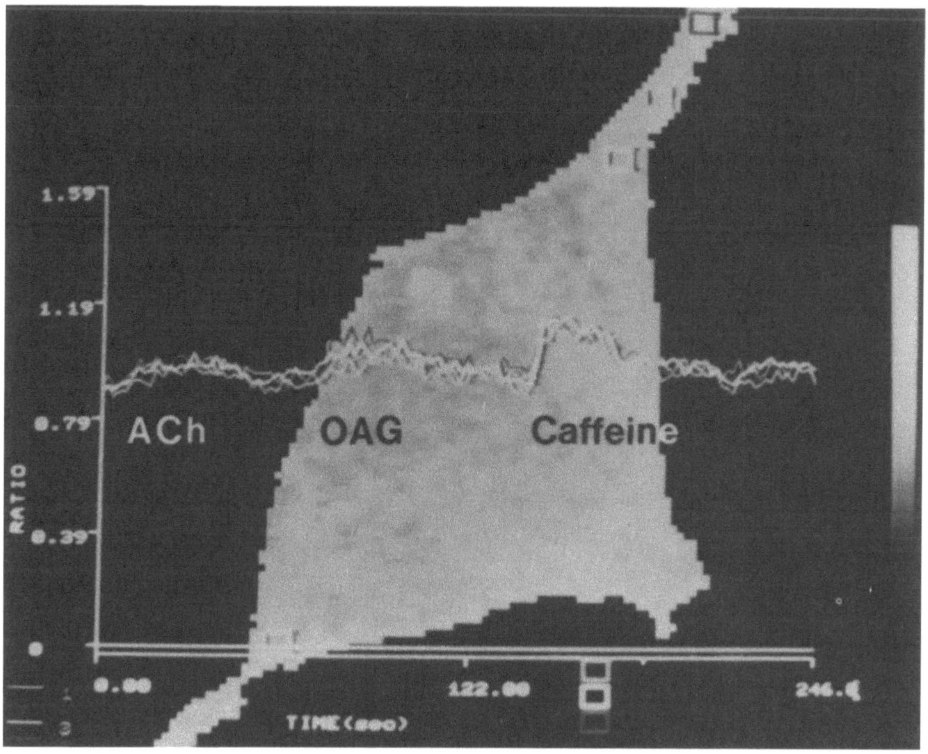

Fig.3. Regional and temporal Ca^{2+} concentration change in a cultured CA1 pyramidal cell of the rat hippocampus on sequential applications of ACh, OAG (oleoyl acetyl glycerol; this is functionally a synthesized substitute for diacylglycerol, but permeable to the membrane.) and caffein.

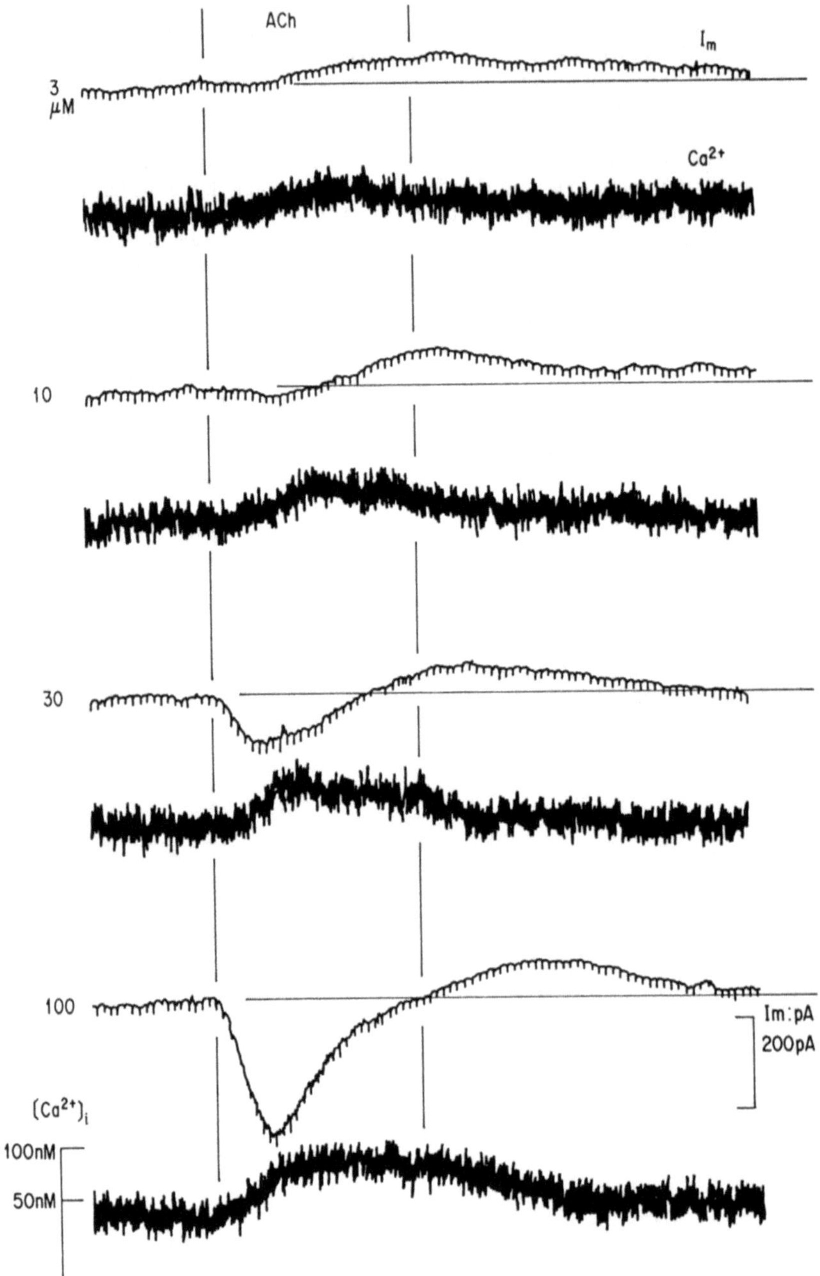

Fig.4. Simultaneous Recording of both intracellular calcium concentrations (Ca^{2+}) and the membrane currents (Im). A cultured superior cervical ganglion cell is voltage-clamped at -60 mV. Four pair of records (Ca^{2+} and Im) are obtained on ACh application of 3, 10, 30 and 100 μm, respectively.

neuron is crucial for understanding neural information processing. This can be realized for the neuron loaded with Fura-2 (a fluorescent Ca^{2+} sensitive dye developed by Tsien *et al.*, 1985) with a SIT (silicon-intensified target) camera. A typical example of the experiment is illustrated in Fig.3, where the cultured CA1 pyramidal cell of the rat hippocampus is stimulated with ACh homogeneously for a period of time after it is loaded with 50 μM Fura-2 (Iijima and Matsumoto, 1988).

Rabbit Superior Cervical Ganglion Cells

Rabbit superior cervical ganglion cells all possess both nicotinic and muscarinic ACh receptors. In these cultured cells, muscarinic ACh responses are found to be dose-dependent (Iijima *et al.*, 1988), as shown in Fig.4. In the low concentration range (<10 μM), ACh interacts specifically with the M_1 receptor and causes Ca^{2+} release from endoplasmic reticulum. Calcium ions then activate the K_{Ca} channel and cause membrane hyperpolarization (activation of the outward K current). On the other hand, in the high concentration range (>10 μM), ACh also interacts with the M_2 receptor and activates the cation selective channel, which causes membrane depolarization (or the inwardly-flowing Na/Ca current). In other words, the muscarinic ACh receptors respond to ACh bimodally (Fig.4); first excitatorily and then inhibitorily. These suggest that the sympathetic neuron becomes inactivated for a while after the neuron is exposed to high concentrations of ACh.

In the sympathetic ganglion cells intracellular Ca^{2+} ions are again crucial in regulating the ionic channel activities. In these cells intracellular Ca^{2+} ions are released from endoplasmic reticulum on exposure to ACh. Very recently, it has been observed that the calcium releasing sites are not homogeneously distributed but localized (Iijima *et al.*, 1988), as shown in SIT imaging of a Fura-2 loaded neuron (Fig.5). On the contrary, the intracellular Ca^{2+} ions homogeneously increase over the neuron on caffein application (Fig.5). Electron microscopic observation has indicated that the spatial distribution of rough endoplasmic reticulum in a neuron is well correlated to the local sites where the Ca^{2+} ion concentration augments on exposure to ACh (Iijima *et al.*, 1988). Calcium ions may be released from rough endoplasmic reticulum on which membrane surface IP_3 receptors are concentrated.

INFORMATION STORAGE IN A NEURON

Synaptic Plasticity

Synaptic information processing undergoes constant modification by the events which take place in neighbouring synapses (heterosynaptic potentiation and potentiation induced by associative inputs; associative learning) or by the events which have taken place previously in the identical synaptic site (homosynaptic potentiation; non-associative learning). How long the effect persists is synapse-dependent; it can last for several hundred milli-seconds to several months. The synaptic modulations are shown to be temporally divided into two kinds; (1) when the effect persists for relatively short periods of time (several 100 msec - several min) and (2) when it persists for long periods of time (several 10 min - several 10 months). The short-term modulation (plasticity) takes place in all synapses by a common mechanism characterized by an increase of transmitters released from the pre-synaptic side. It can be phenomenologically classified into three classes; facilitation, augmentation and potentiation. Long-term modulation, on the other hand, has been so far observed in limited cases in cells such as the hippocampal pyramidal cells (LTP; long-term potentiation), cerebellar Purkinje cells (LTD; long-term depression) (Ito, 1984) and Aplysia neurons (Goelet *et al.*, 1986).

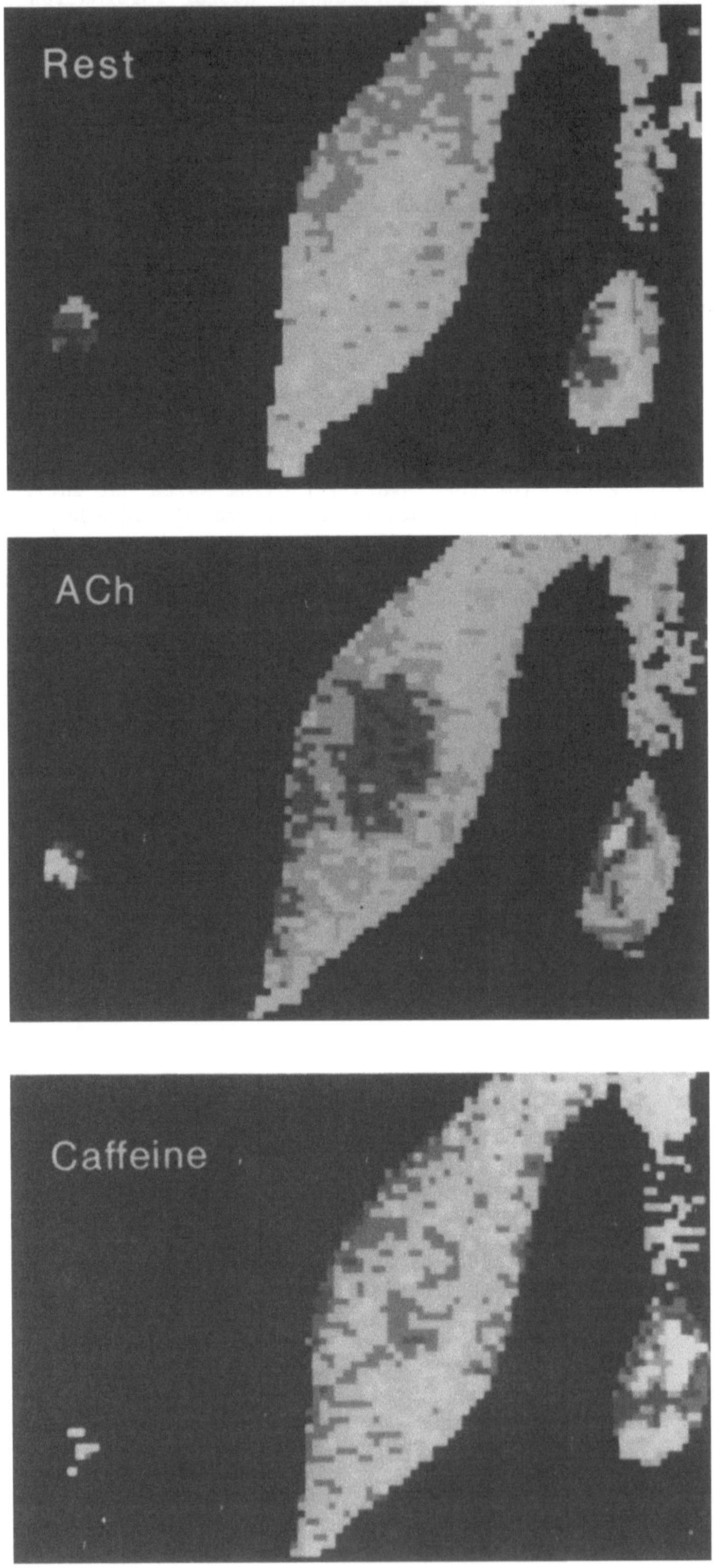

Fig.5.
Regional Ca concentration change in a cultured superior cervical ganglion cell. *The upper;* control. *The middle;* on ACh application. *The lower;* on caffein application.

Long-term Modifications

What molecular mechanisms govern the translation of the effect of brief stimuli into relatively long-lasting molecular changes of high specificity? Recently it has been clarified that environmental stimuli can alter steady state levels of messenger RNA species encoding neurotransmitters by altering gene transcription, thereby altering synaptic, neuronal, and network function over time (Black *et al.*, 1987). Thus, brief external events can elicit long-term modifications in neuronal function. Information about the external environment is stored through the very mechanism that serves neuronal intercommunication. Multiple molecular species are available in the neuron to code for environmental events over time. As a result, the basic elements for the information storage and retrieval in a neuron are proteins, RNA and DNA while those for the processing are proteins and ions (Fig.6). Current opinion suggests that proto-oncogenes in the brain may represent genetic elements mediating both short- and long-term modulations in neuronal function (Hanley, 1988). Proto-oncogenes are the cellular counterparts of viral transforming genes. Therefore, we can expect that the study of carinogenesis, a process related to proliferation and tumor formation, would provide insights into the functions of neurons.

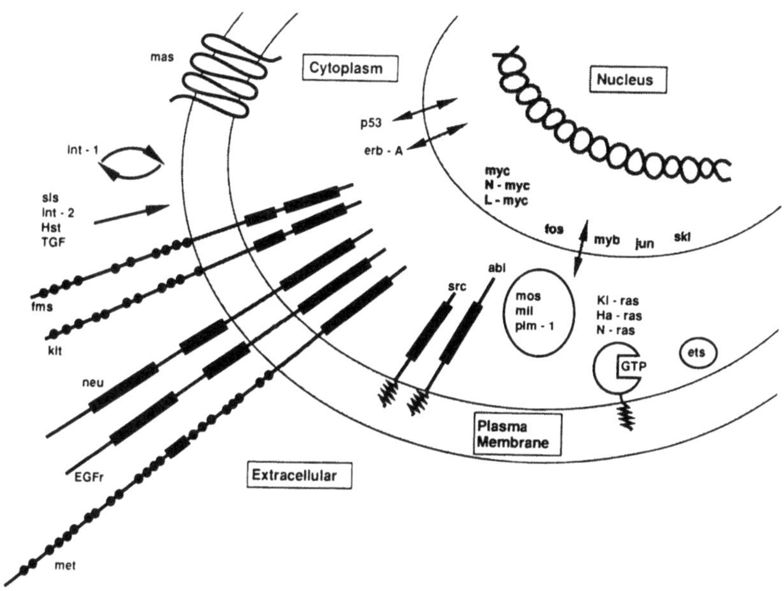

Fig.6. A schematic drawing of the predicted cellular localization of proto-oncogenes (Hanley, 1988).

CONCLUSIONS

Neurocomputing exploits the algorithm and architecture for the information processing of neurons and neural networks to construct a computer. The basic assumption of artificial neurons on which the artificial neural networks are modelled are currently studied. We have pointed out that the assumption is oversimplified in view of the function of a real neuron. We have exemplified that the recent application of molecular and genetic approaches to the study of neural function has yielded a number of notable advances. For example, a synaptic site is a molecular processing unit that can possess a kind of memory function. The characteristics of the information processing are versatile and variable through the molecular and/or structural changes at the site. Further, contrary to traditional views, it is now well understood that single neurons use multiple transmitter signals, that different transmitters are independently expressed and regulated in the same neuron, and that experience alters gene expression in the nervous system (Black *et al.*, 1987).

These and related insights suggest that neurons should be regarded as LSI microprocessors with a kind of memory function similar to a digital computing system. Therefore, studies on the neurocomputing should be ideally performed to address the problems as to what kind of characteristic processors are used in the brain and what kind of molecules and structures support the functioning. Then, it will become essential to determine quantitatively what the neural circuits are and what their associated neurobiological processes can do.

REFERENCES

Amari, S., 1982, Competitive and cooperative aspects in dynamics of neural excitation and self-organization, in: "Competition and Cooperation in Neural Nets", S. Amari and M.A. Arbib, eds., Springer Lecture Notes in Biomath. 45: 1.

Black, I.B., Adler, J.E., Dreyfus, C.F., Friedman, W.F., LaGamma, E.F. and Roach, A.H., 1987, Science 236: 1263.

Brown, D.A. and Adams, P.R., 1980, Nature (London) 283: 673.

Egan, T.M. and North, R.A., 1986, Nature (London) 319: 405.

Goelet, P., Castellucci, V.F., Schacher, S. and Kandel, E.R., 1986, Nature (London) 322: 419.

Hanley, M.R., 1988, Neuron 1: 175-182.

Hecht-Nielsen, R., 1986a, UCSD Extention Class Notes.

Hecht-Nielsen, R., 1986b, TRW Rancho Carmel Artificial Intelligence Center Technical Report.

Hinton, G.E., 1987, Science (Tokyo) 57: 228 (Translated into Japanese by T. Ohmori and S. Amari).

Hopfield, J.J. and Tank, D.W., 1985, Biol. Cybern. 52: 141.

Iijima, T. and Matsumoto, G., 1988, in preparation.

Iijima, T., Sekiguchi, T. and Matsumoto, G., 1988, in preparation.

Ito, M., 1984, "The Cerebellum and Neural Control", Raven Press, N.Y.

Kohonen, T., 1984, "Self-Organization and Associative Memory", Springer-Verlag, Berlin.

Matsumoto, G., 1988, Future Generations Computer Systems (in press).

Minsky, M.L. and Papert, S., 1969, "Perceptrons", MIT Press, Cambridge.

Rumelhart, D., Hinton, G. and Williams, R., 1986a, "Parallel Distributed Processing : Explorations in Microstructure of Cognition, Vol. 1: Foundations", Bradford Books/MIT Press, Cambridge.

Rumelhart, D.E., Hinton, G.E. and Williams, F.J., 1986b, Nature (London) 323: 533.

Sejnowski, T.J. and Rosenberg, C.R., 1986, Johns Hopkins Univ.Elect.Eng. and Comp.Sci.Tech.Rep. 86: 1.

Swanson, L.W., Teyler, T.J. and Thompson, R.F., 1982, Neurosci.Res.Prog. Bull. 26: 611.

Takai, Y., Kishimoto, A. and Nishizuka, Y., 1982, Calcium and phospholipid turnover as transmembrane signaling for protein phosphorylation, in: "Calcium and Cell Function, Vol. 2", W.T. Cheung, ed., Academic Press, NY, p.385.

Tsien, R.Y., Rink, T.J. and Poenie, M., 1985, Cell Calcium 6: 145.

PHASE WAVE AND PERCEPTION OF AMOEBOID CELLS

Yonosuke Kobatake, Tetsuo Ueda, and Kenji Matsumoto

Faculty of Pharmaceutical Sciences
Hokkaido University
Sapporo, Japan

INTRODUCTION

Cell behavior is a good example for studying how the machinery of information processing is working in a single cell. Let us see, for example, how an amoeba catches a prey organism. When a ciliate comes close to or touches the amoeba, the amoeba detects it through chemical signals which the prey organism secretes and/or through the sound of beating cilia and/or through a sense of touch. The sensed information should then be judged as to whether it is edible or hazardous. When the amoeba judges it to be edible, the predator tries to catch it by extending pseudopodia just enough to hold the prey, whereas the pseudopodia extending in the other direction retract. The prey organism enclosed by the membrane is then incorporated into the cell and digested. When the amoeba senses an alarm signal, the organism tries to avoid it.

In this way, although primitive, the amoeba has sensors to detect the nature, position and size of the prey organism, and in the protoplasm there are networks which make possible the information transmission and processing which leads ultimately to an organized and integrated cell behavior. In higher animals all these functions are attributed to a nervous system and especially to the brain when "judging" is concerned. Our aim here is to elucidate the mechanism of information processing in the protoplasm where chemical as opposed to electrical processes are fundamental. We show that phase waves in a population of coupled oscillators are related to perception of an amoeboid cell.

AMOEBOID LOCOMOTION IS OSCILLATORY

The shape of an amoeboid cell changes during locomotion. A series of photographs taken at 20-sec intervals are shown in Fig. 1 where Amoeba proteus moves freely in a petri dish. The organism takes an elongated shape when it moves fast, and becomes rather round when it stops moving. These processes repeat. The direction of pseudopodium extension is mostly stochastic, and hence the trajectory of the locomotion looks like that of a Brownian particle. The cell shape is quantified by a parameter 'complexity' defined by (periphery)*(periphery)/(area). The time series of the complexity varies rather periodically, and the power spectra indicate that oscillations with a period of several min are dominant.

Other amoeboid cells show similar oscillatory locomotion (Satoh et al, 1985). This includes leukocytes, macrophages, fibroblasts, cellular slime molds, the true slime molds, etc. This ubiquity, in turn, seems to suggest that the sustained oscillation plays important roles in cell behavior.

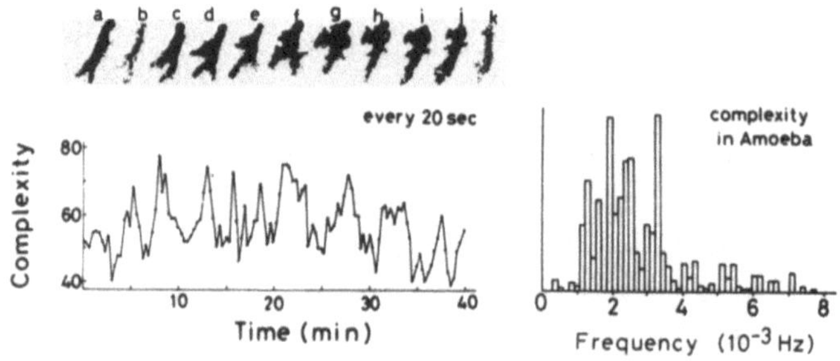

.Fig.1 Temporal order in the locomotion of Amoeba proteus. (Ueda and Kobatake, 1983)

USE OF A GIANT AMOEBOID CELL, THE PLASMODIUM OF PHYSARUM POLYCEPHALUM

The plasmodium of Physarum polycephalum is a multi-nuclear organism. The protoplasm extends over several cms like a sheet at the frontal part, and, toward the rear, intricate networks of the protoplasmic veins develop in which the protoplasmic sol moves to and fro accompanying the rhythmic contraction. Using this large 2-D extension of the protoplasm, one can analyze the variation of intracellular chemical components with time. This is accomplished simply by cutting off successively the small area of the plasmodium at the frontal part where oscillations are synchronous. Each piece is frozen by liquid nitrogen, and then subjected to chemical analysis.

OSCILLATION OF INTRACELLULAR CHEMICAL COMPONENTS IN A PLASMODIUM

Time courses of intracellular concentration of NADH and cyclic AMP are shown in Fig. 2 where dotted lines show the contractile activity. NADH oscillates with the same period as the rhythmic contraction, and the phase advances about 1/3 of the period (A). Cyclic AMP also oscillates, and the phases retard about 1/3 of the period (B). It is known that ATP, Ca ion and pH also oscillate with respective phase relationship with the rhythmic contraction. Thus, metabolic oscillations may form a basis for rhythmic contractility.

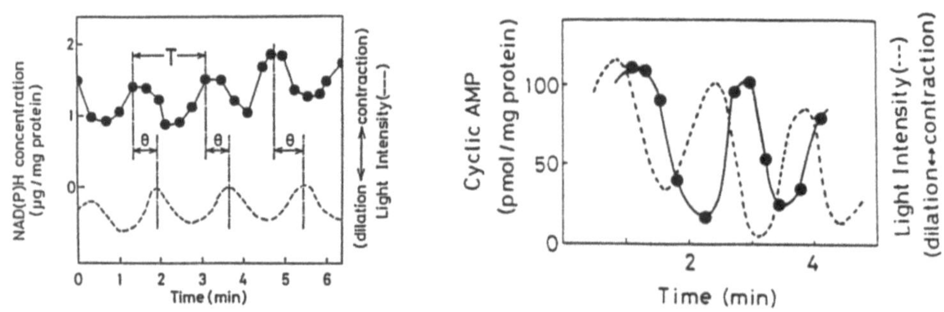

Fig. 2. Oscillations in NAD(P)H and cAMP concentration accompanying the rythmic contraction in the plasmodia. (Mori et al, 1987: Ueda et al, 1986)

The chemical oscillation of the slime mold shows multi-periodicity. When the organism is exposed to low temperatures or anaerobic conditions, the ATP concentration fluctuates with long periods (about 30 min) as compared to about a few minutes normally. When the organism is exposed to an ethanol solution, longer period oscillations occur with faster oscillations superimposed. Thus, the organism takes on versatile dynamic states depending on the environment.

SPATIAL-DISTRIBUTION OF INTRACELLULAR ATP CONCENTRATION

Spatial distributions of ATP concentration are shown in Fig. 3 where the organism takes various shapes. The plasmodium extends like a fan and advances by keeping semi-circular shape after passing through a narrow channel. The ATP concentration becomes high toward the periphery in the radial direction upon which the undulation around the periphery superimposes (A). When the plasmodium is passing through a narrow channel, the frontal part becomes rich in protoplasm. In this condition, the ATP is high at the frontal part (B). When the organism is confined in a square wall, the protoplasm distributes uniformly by developing intricate meshworks of veins. Here the ATP distribution undulates and shows three-fold rotational symmetry (C).

In a unidirectionally migrating plasmodium, the AMP concentration is uniform, whereas ATP and ADP are high at the frontal region and decrease gradually toward the rear. Thus, cell polarity seems related to a polar ATP distribution.

The polar ATP pattern turns into a wavy one upon local stimulation with blue light (Mori et al, 1986). At first, the perturbation to the ATP distribution is limited to within the stimulated region. But eventually this expands over the whole organism, and a wavy ATP pattern results. Accompanying this variation in the ATP distribution, the plasmodium stops and moves backward. Thus, propagation of waves in the intracellular ATP concentration seems to result in different ATP patterns, which also seem to govern the cell behavior.

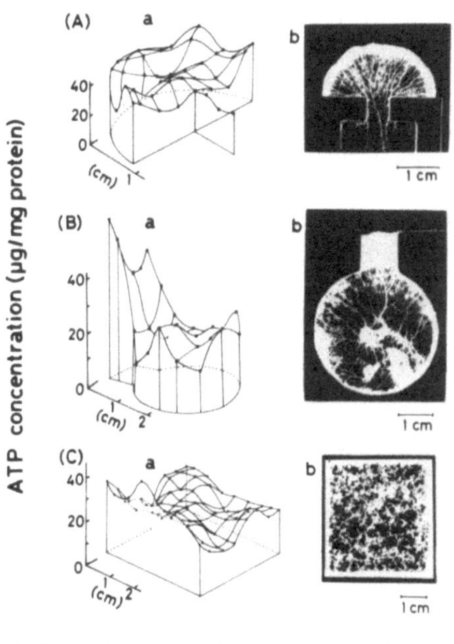

Fig. 3 Various ATP patterns induced by constraining the motility of the plasmodia. (Ueda et al, 1987)

The organism moves away from low wavelength radiation. The action spectrum for this photoavoidance shows peaks, in order of the stronger response, around 260 nm, 370 nm and 460 nm. As for photochemical reactions, we found that superoxide is generated upon irradiation and that the action spectrum for superoxide generation agrees with that for the photoavoidance. Thus, we speculate that flavins are photoreceptors for the photoavoidance.

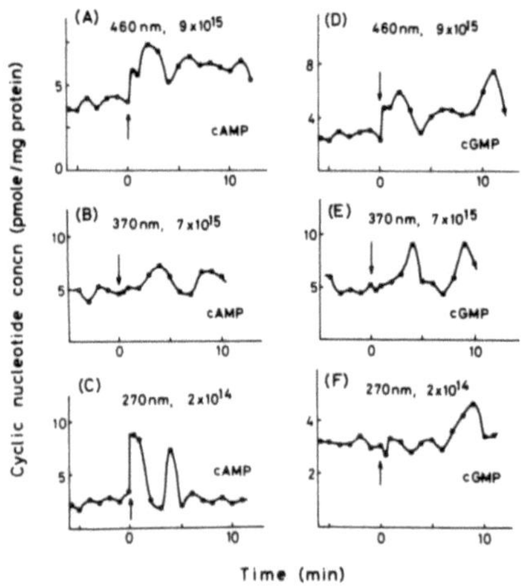

Fig. 4 Response of intracellular cAMP and cGMP concentrations to radiation. (Ueda et al, 1988)

Time courses of the intracellular cAMP and cGMP concentrations before and after irradiation are shown in Fig. 4. Upon irradiation with UV and blue light, the two cyclic nucleotides show oscillatory variations. The mode of variation, however, differs for different wavelengths. For irradiation in the far UV, cAMP increases soon after irradiation, while cGMP begins to change only at later times. With near UV irradiation, both cAMP and cGMP begin to change with a time lag, while with blue light the two cyclic nucleotides vary soon after irradiation. On the average, both cAMP and cGMP levels increase after irradiation. Microinjection of the cyclic nucleotides induces contraction in this organism, as does irradiation. Therefore, cyclic nucleotides seem to participate in the photosensory transduction.

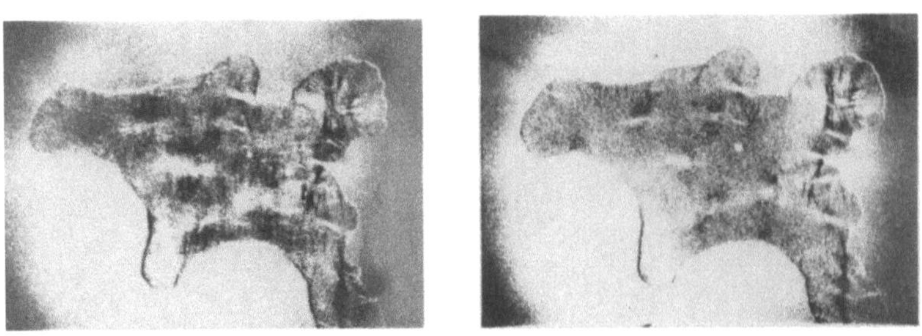

Fig. 5 Dynamic changes of cytoskeletons as viewed by birefringent fibrils in the plasmodia. (Ueda et al, 1988)

Cytoskeletons of the plasmodia are observed as birefringent fibrils under a polarizing microscope (Fig. 5). Soon after UV irradiation, the birefringent fibrils become thick and numerous almost everywhere in the cell. At later stages, the thick fibrils become re-distributed non-uniformly accompanying the oscillatory variation. The organism moves toward less fibrous regions as if the fibrils squeeze the protoplasm.

A MODEL FOR THE CELL BEHAVIOR IN TERMS OF THE SELF-ORGANIZATION OF CHEMICAL PATTERNS

Oscillations and pattern formation of intracellular chemical components such as ATP are explained in terms of a reaction-diffusion coupling model. Let us take, e.g., a well-known model of the Brusselator (Fig.6). In this model, the system is regarded as a collection of chemical oscillators which are coupled to each other spatially with diffusion. As a computer simulation demonstrates, this chemical system takes on various dynamic states such as oscillations and spatial non-uniform distributions, known as dissipative structures. As shown above, this notion also holds for the cellular system. If a chemical component, as denoted by X or Y in the reaction scheme, is able to regulate cytoskeletons such as G-F transformation of actin, cell behavior is then governed by the chemical patterns. This mechanism of self-organization for cell behavior explains how the cell behavior occurs as an integrated unit.

Brusselator (a theoretical chemical oscillator)

$$A \rightarrow X$$
$$B + X \rightarrow Y + D$$
$$2X + Y \rightarrow 3X$$
$$X \rightarrow E$$

$$\frac{\partial X}{\partial t} = A-(B+1)X+X^2Y+D_1\nabla^2X$$

$$\frac{\partial Y}{\partial t} = BX-X^2Y+D_2\nabla^2Y$$

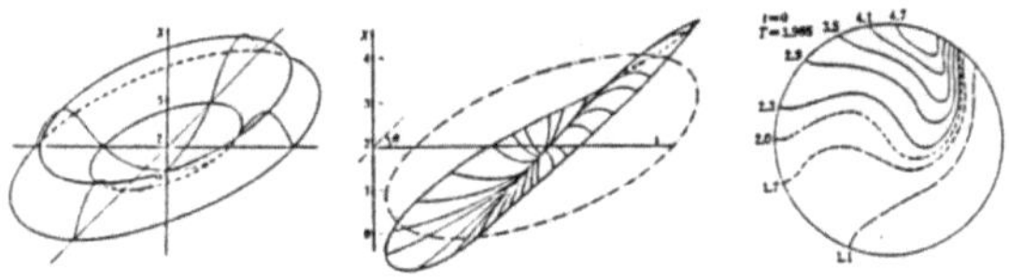

Fig. 6 Self-organization of chemical patterns. (Nicolis and Prigogine, 1977)

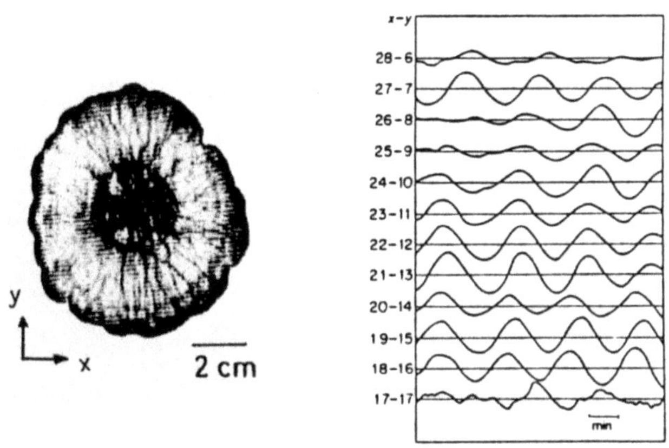

Fig. 7 Computer image processing of the collective behavior of oscillators in the plasmodium. (Matsumoto et al, 1986)

COLLECTIVE BEHAVIOR OF OSCILLATORS IN THE PLASMODIUM

The coupled oscillator model seems to be realized experimentally, and can be visualized by computer image processing (Fig. 7). The organism is illuminated from below with weak IR light (the organism does not react to IR light) and monitored by an IR sensitive video camera. So the brightness level reflects the thickness of the plasmodium. To pick up the oscillatory components, the difference in two brightness levels at a given time interval is calculated. The entire organism oscillates with the same period, but the phase differs for different positions. The spatial relationship among the oscillators is visualized by expressing the amplitude of the oscillation at a given point as a square with differing shades.

PROPAGATION OF PHASE WAVE IN RESPONSE TO EXTERNAL STIMULI

In a concentrically expanding plasmodium, oscillation patterns are radially symmetric, and waves propagate from periphery to inside in most cases. Upon local stimulation, a new wave begins to propagate around the stimulated region. Response to hot stimulation is shown in Fig. 8. The wave propagates from the stimulated region to the surroundings.

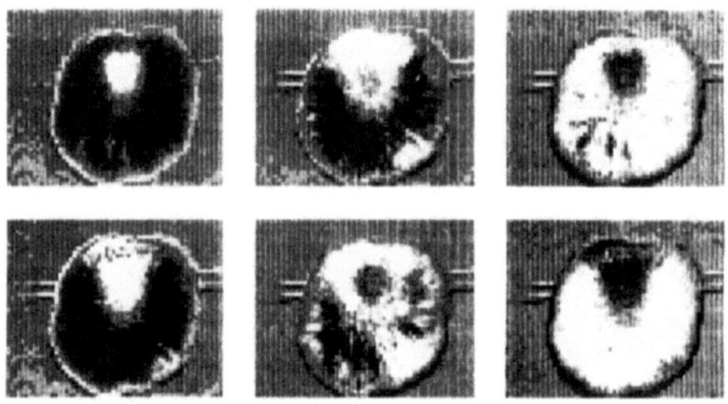

Fig. 8 Oscillation patterns in response to local attractive stimulation. (Matsumoto et al, 1986)

Thus, the brightness level at point x and at time t, I(x, t) is expressed as I(x, t) = I(k x + wt), where k is the phase gradient vector, w the frequency of the oscillation. The phase gradient vector k is determined by k = -w*grad(I)/(dI/dt). The distribution of k is shown in Fig. 9, where the organism is stimulated locally. When stimulated with attractants such as high temperature, oatflakes, glucose, the k vectors orient to point away from the stimulated region. On the other

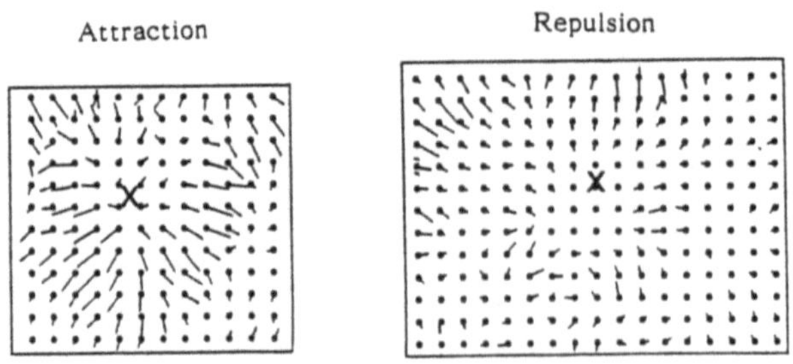

Fig. 9 Responses of phase gradient vectors to local stimuli. (Matsumoto et al, 1986)

hand, the k vectors orient to point toward the region when stimulated with repellents such as low temperature, UV and blue light, and high salts.

When the organism is stimulated uniformly, the frequency of the oscillation is modified. The oscillation becomes faster and slower for attractants and repellents, respectively. It seems common in a system of non-linear oscillators that the faster oscillator entrains the slower one, and that the phase gradient vectors point from the entraining to the entrained regions. The phase wave and phase gradient vectors observed in the slime mold suggest that a similar mechanism is working in plasmodia. The fact that phase gradient vectors classify attraction and repulsion uniquely indicates that the sensed information is integrated at the level of interacting oscillators.

REVERSAL OF THERMOTAXIS WITH OSCILLATORY STIMULATION

By oscillating the external stimulation, we can modify the orientation of phase gradient vectors and tactic behaviors as well. The response to oscillatory high temperature is shown in Fig. 10. When stimulated with stationary high temperature, the phase at the stimulated region advances and the thickness there increases. When oscillated slightly slower than the intrinsic rhythm, the phase at the stimulated region becomes retarded and the thickness there decreases. Thus, the positive thermotaxis turns into a negative one. This reversal of thermotaxis is maintained as long as the intrinsic rhythm entrains to the external oscillation. Similarly, negative thermotaxis is reversed by oscillating the low temperature stimulation faster than the intrinsic rhythm. In this way, the metachronal phase relation is determined by the distribution of oscillators with different periods, the phase on the side of intrinsically faster rhythm being advanced. The net movement of the protoplasm occurs along the direction of phase gradient toward the region where the phase advances.

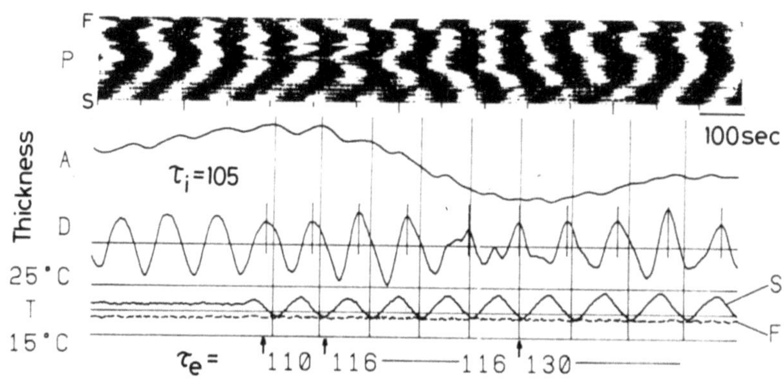

Fig. 10 Responses of the organism to slowly oscillating high temperature. (Matsumoto et al, 1988)

Based on a coupled oscillator model, we can outline the mechanism of information processing in the Physarum plasmodium: (1) both the kind and the strength of stimulation are encoded into the period of local oscillators (frequency modulation); (2) a change in the period of these oscillators induces an overall change in the phase relation throughout the plasmodium, whereby either the faster oscillators entrain the slower ones, or they entrain mutually; and (3) the resulting metachronal phase relation in a population of oscillators governs the net movement of the protoplasm.

CONCLUSION

Information processing in the chemical system of the amoeboid cell of the Physarum plasmodium is shown to proceed along the following hierarchy toward integration. External and internal conditions are sensed by receptors, which modify the dynamic states of chemical oscillators. The latter in turn generate new patterns in a population of coupled oscillators. The cell then behaves according to these patterns.

REFERENCES

Matsumoto, K., Ueda,T. and Kobatake, Y. (1988) Reversal of thermotaxis with oscillatory stimulation in the plasmodium of Physarum polycephalum. J. Theor. Biol. 131, 175-182.

ibid. (1986) Propagation of phase wave in relation to tactic response by the plasmodium of Physarum polycephalum. J. Theor. Biol. 122, 339-345.

Mori, Y., Ueda, T. and Kobatake, Y. (1987) NAD(P)H oscillation in relation to the rhythmic contraction in the Physarum plasmodium. Protoplasma 139, 141-144.

Mori, Y., Matsumoto, K., Ueda, T. and Kobatake, Y. (1986) Spatio-temporal organization of intracellular ATP content and oscillation patterns in response to blue light by Physarum polycephalum. Protoplasma 135, 31-37.

Nicolis, G. and Prigogine, I. (1977) Self-Organization in Nonequilibrium Systems, John Wiley & Sons, Inc., New York.

Satoh, H., Ueda, T. and Kobatake, Y. (1985) Oscillations in cell shape and size during locomotion and in contractile activities of Physarum polycephalum, Dictyostelium discoideum, Amoeba proteus and macrophages. Exp. Cell Res. 156, 79-90.

Ueda, T., and Kobatake, Y. (1983). Quantitative analysis of changes in cell shape of Amoeba proteus during locomotion and upon responses to salt stimuli. Exp. Cell Res. 147, 466-471.

Ueda, T., Mori, Y., Nakagaki, T. and Kobatake, Y. (1988). Changes in cAMP, cGMP concentration, birefringent fibrils and contractile activity accompanying UV and blue light photoavoidance in plasmodia of an albino strain of Physarum polycephalum. Photochem. Photobiol. 47, 271-275.

Ueda, T., Mori, Y., and Kobatake, Y. (1987) Patterns in the distribution of intracellular ATP concentration in relation to coordination of amoeboid cell behavior in Physarum polycephalum. Exp. Cell Res. 169, 191-201.

Ueda, T., Matsumoto, K., Akitaya, T. and Kobatake, Y. (1986) Spatial and temporal organization of intracellular adenine nucleotides and cyclic nucleotides in relation to rhythmic motility in Physarum plasmodium. Exp. Cell Res. 162, 486-494.

MICROTUBULES: POSSIBLE APPLICATION TO COMPUTER TECHNOLOGIES

Djuro Koruga

Molecular Machines Research Center
Faculty of Machine Engineering, University of Belgrade
27. Marta 80, 11000 Belgrade, Yugoslavia.

INTRODUCTION

The three fields of technology, microelectronics, biotechnology and computer technology, are prime candidates for establishing a new field of science and technology called *information biotechnology*. The first step is done in the field of computer architecture. Today there are a few types of *neurocomputers* based on neural networks (Table 1).[1] The second step has already been accomplished in the research and development of *neurochips*, under laboratory conditions, on the basis of VLSI technology.[2] The first research results in the field of molecular electronics point out that the application of organic and biological polymers will be possible in the beginning of the 21st century. In this way the third step will be accomplished with the construction of *biomolecular information machines*.

Of all the biological structures that participate in bioinformation processing only a small number work on such a principle that can be applied to the architecture of computer chips and hardware chips. One of these rare biological structures are microtubules (MT), self-organized organelles usually consisting of 13 subunits. These cytological structures create a network of protofilaments in the cell (neuron), similar to the way in which the neurons create a network in the brain (Fig. 1).

MICROTUBULES

Microtubules (MT) are organelles present in nearly all eukariotic cells, composed of equimolar amounts of the two globular (~50,000 Dalton) subunits, α- and β-tubulin, each having a similar amino acid composition and a similar overall shape. The subunits of tubulin molecules are assembled into long tubular structures with an average exterior diameter of ~24 nm, capable of changing length by self-assembly or disassembly of their subunits; sensitive to cold, high hydrostatic pressure and several specific chemicals, such as colchicine, vinblastine, etc. The stable form of tubulin, both *in vivo* and *in vitro*, is a dimer, and the functional unit in MT assembly is the α-β heterodimer. Some proteins complex assemblies, such as mitotic spindles, centrioles, cilia and flagella, axonemes and neurotubules, are built from MT. They modulate cell shape, cell motility and mitosis, and some part of functional information processing in cells.

Table 1. Types of Neurocomputers.

	Year		Capacity			Speed
Neurocomputer	introduced	No. of Technology	No. of P.E.*	No. of connect.	Connect. networks	per sec
Perceptron	1957	Electromech. & electronic	8	512	1	10^3
Adaline/ Madaline	1960 -62	Electroch. (now elect)	1/8	16/128	1	10^4
Electro-optic crossbar	1984	Electro-optic	32	10^3	1	10^5
Mark III	1985	Electronic	8×10^3	4×10^5	1	3×10^5
Neural emul. processor	1985	Electronic	4×10^3	1.6×10^4	1	4.9×10^5
Optical resonator	1985	Optical	6.4×10^3	1.6×10^7	1	1.6×10^5
Mark IV	1986	Electronic	2.5×10^5	5×10^5	1	5×10^6
Odyssey	1986	Electronic	8×10^3	2.5×10^5	1	2×10^6
Crossbar chip	1986	Electronic	256	6.4×10^4	1	6×10^9
Optical novelty filter	1987	Optical	1.6×10^4	2×10^6	1	2×10^7
Anza	1987	Electronic	3×10^4	5×10^5	No limit	2.5×10^4
Parallon 2	1987	Electronic	10^4	5.2×10^4	No limit	1.5×10^4
Parallon 2x	1987	Electronic	9.1×10^4	3×10^5	No limit	1.5×10^4
Delta floating point process.	1987	Electronic	10^6	10^6	No limit	2×10^6
Anzy plus	1988	Electronic	10^6	1.5×10^6	No limit	1.5×10^6

*P.E. - Processing Elements

Microtubules participate in intracellular transport, addressing, growth form and many other dynamic activities.[3-5] Some experimental results link tubulin and microtubules to bioinformation processes such as memory and learning.[6,7] Because their subunits are arranged by sphere packing symmetry derived from information coding laws, tubulin and microtubules are primary information processors.[8,9] They are remarkable because they play two distinct and separable roles: as structural materials (cytoskeleton) and as dynamic machines that operate on the molecular level. They are carbon based three-dimensional intelligent machines which function on the nanometer scale with picosecond speed.

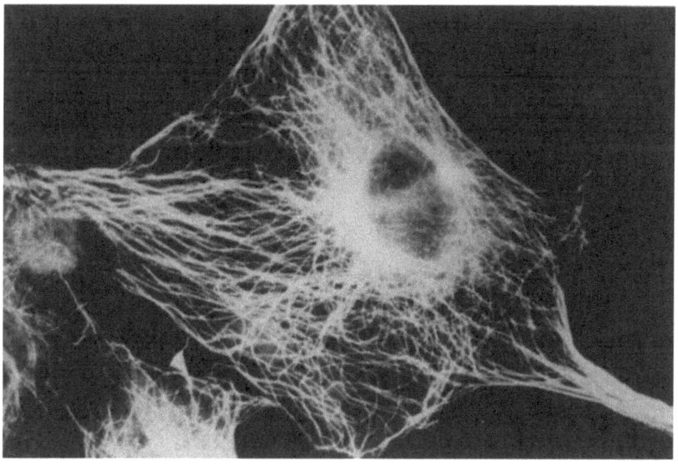

Fig. 1. Microtubules as the subcellular "neural network".

MICROTUBULES AND NEURAL NETWORKS

During the past thirty years interest in neural networks has generated a number of different models: binary (McCulloch-Pitts, Cianiello, Rosenblatt), linear (Widrow, Kohonen) and non-linear (Hartline-Ratliff, Hodgkin-Huxley, Grossberg, Wilson-Cowan) (Table 2).[1] Here we consider such nonlinear control networks in which the neuronal response depends on input signal, subneural activities, and its interactions with other neurons.[10,11]

We have found that the structure of microtubules is optimal for information processing on the subneural level. Having in mind its connection with the membrane, we believe that microtubules play an important role in neural networks as a subneuronal factor of neuronal activities, such as memory and learning. Having this in mind the neural control networks model based on microtubules may exist in the form:

$$a_{io}^{-1} X_i(t) + X_i(t) = \mathrm{NL}\{f_i + \sum_{j=1}^{n} C_{ij} X_j(t-\sigma_{ij}) + \sum_{k=1}^{m} b_{ij} \int_0^t X_i(\tau) e^{-a_{ik}(t-\tau)} d\tau\}$$

where

t	time.
n	the number of neurons in the network.
m	the number of MT in the subneuronal network.
f_i	the input to the i-th neuron at time t.
$X_i(t)$	the activity function.
a_{io}	the constant rate characterized by the fact that the stop charge input to the i-th neuron produced an exponential approach from the initial value $X_i(0)$ to a steady-state firing rate X_i with a rate constant a_{io}.
b_{ik}	inhibition factor for the i-th MT in case $b_{ik} < 0$. excitation factor for the i-th MT in case $b_{ik} < 0$.
C_{ij}	the interaction coefficient between the i-th neuron and the j-th neuron.
σ_{ij}	the time lag occurring in the transfer of the activity from the j-th neuron to the i-th neuron.
NL	nonlinear element $1/(1+e^{-u})$.

Table 2. Types of Neural Networks.

Name of network	Inventors and developers	Years introduced	Primary applications	Limitations	Comments
Adaptive resonance theory	Gail Carpenter, Northeastern U.; Stephen Grossberg, Boston U.	1978-86	Pattern recognition, especially when pattern is complicated or unfamiliar to humans (radar or sonar readouts, voice prints).	Sensitive to translation, distortion, changes in scale.	Very sophisticated; not yet applied to many problems.
Avalanche	Stephen Grossberg, Boston, U.	1967	Continuous speech recognition; recognition; teaching motor commands to robotic arms.	Literal playback or motor sequences - no simple way to alter speed or interpolate movements.	Class of networks -- no single net work can do all these tasks.
Back propagation	Paul Werbos, Harvard U.; David Parker, Stanford U.; David Rumelhart, Stanford U.	1974-85	Speech synthesis from text; adaptive control of robotic arms; scoring of bank loan applications.	Supervised training only - correct input-output examples must be abundant.	The most popular network today - works well; simple to learn.
Bidirectional associative memory	Bart Kosko, U. of Southern California.	1985	Content-addressable associative memory.	Low storage density; data must be properly coded.	Easiest network to learn - good educational tool; associates fragmented pairs of objects with complete pairs.
Boltzmann and Cauchy machines	Jeffrey Hinton, U. of Toronto; Terry Seijnowsky, Johns Hopkins U., Harold Szu, Naval Research Lab.	1985-6	Pattern recognition for images, sonars, radar.	Boltzmann machine: long training time. Cauchy machine: generating noise in proper statistical distribution.	Simple networks in which noise function is used to find a global minimum.
Drain state in a box	James Anderson, Brown U.	1977	Extraction of knowledge from data bases.	One-shot decision making - no iterative reasoning.	Similar to bidirectional associative memory in completing fragmented inputs.

Cerebellatron	David Mar, MIT; James Albus, NBS; Andres Pellionez, NYU.	1969-82	Controlling motor action of robotic arms.	Requires complicated control input.	Similar to avalanche network; can blend several command sequences with different weights to interpolate motions smoothly as needed.
Counter-propagation	Robert Hecht-Nielsen, Hecht-Nielsen Neuro-computer Corp.	1986	Image compression; statistical analysis; load application scoring.	Large number of processing elements and connections required for high accuracy for any size of problem.	Functions as a self-programming look-up table; similar to back propagation only simpler, although also less powerful.
Hopfield	John Hopfield, Caltech and AT&T Bell Labs.	1982	Retrieval of complete data or images from fragments.	Does not learn - weights must be set in advance.	Can be implemented on a large scale.
Madaline	Bernard Widrow, Stanford U.	1960-62	Adaptive nulling of radar jammers; adaptive modems; adaptive equalizers; (echo cancelers) in telephone lines.	Assumes a linear relationship between input and output.	Acronym stands for multiple adaptive linear elements; powerful learning low; in commercial use for more than 20 years.
Neocognitron	Kunihiko Fukushima, NHK Labs.	1978-84	Handprinted character recognition.	Requires unusually large number of processing elements and connections.	Most complicated network ever developed; insensitive to differences in scale, translation, rotation; able to identify complex characters (such as Chinese).
Perceptron	Frank Rosenblatt, Cornell U.	1957	Typed character recognition.	Cannot recognize complex characters (such as Chinese); sensitive to difference in scale, translation, distortion.	The oldest neural network known; was built in hardware; rarely used today.
Self-organizing map	Teuvo Kohonen, Helsinki U. of Technology.	1980	Maps one geometrical region (such as a rectangular grid) onto another (such as an aircraft.	Requires extensive training.	More effective than many algorithmic techniques for numerical aerodynamic flow calculations.
Sub-neural	Djuoro Koruga, U. of Belgrade.	1987	Adaptive control, simulation of neuron and brain activities.	Cannot recognize distortion of time.	Can be implemented in both chip and computer technology.

The outer shell of any particular atom holds no more than 8 electrons. When an atom in its outer shell has:

from 1 to 4 electrons	——→ conductor
exactly 4 electrons	——→ *semiconductor*
from 4 to 8 electrons	——→ insulator
exactly 8 electrons	——→ inert gases

The outermost shell of a particular atom is called the valence shell and the electrons that orbit within this shell are referred to as valence electrons. When an atom contains exactly 4 electrons in its outer shell, it does not readily give up or accept electrons. Elements which contain atoms of this type do not make good insulators or conductors, and are referred to as semiconductors. The elements *germanium*, *silicon* and *carbon* are typical examples of semiconductor materials. There are more than 50 commonly used semiconductor compounds like: cadmium sulfide (CdS), cadmium selenide (CdSe), gallium arsenide (GaAs), gallium phosphide (GaP) and so on.

Each atom of germanium, silicon or carbon shares its four valence electrons with four neighboring atoms. This sharing of electrons creates a bond which holds the atoms together, commonly referred to as a covalent bond. When an electron breaks away from a covalent bond, an open space or vacancy exists in the bond. The space previously occupied by an electron is generally referred to as a hole. The electrical characteristics of semiconductor materials are highly dependent upon temperature. Since at room temperature a small number of electron-hole pairs exist, the pure semiconductor materials are doped by adding other materials called impurities. Basically, there are pentavalent and trivalent types of impurities.

When a pure semiconductor material such as germanium or silicon is doped with a pentavalent material such as arsenic (As), some of the atoms in the semiconductor lattice are replaced with arsenic atoms. By replacing one of the germanium or silicon atoms, an arsenic atom will share four of its valence electrons with adjacent atoms in a covalent bond. Since arsenic has five electrons in its outer shell, the fifth valence electron is closely attached to the nucleus of the arsenic atom and can be easily set free. Because there are a lot of donor atoms within the semiconductor material, there are many free electrons. Due to the presence of these extra negatively charged electrons in the doped germanium or silicon structure, these materials are commonly referred to as N-type (negative type) semiconductors. The pentavalent atom (arsenic or phosphorus) produces a free electron available for conduction.

When a pure semiconductor material is doped with trivalent material (gallium or aluminum), some of the atoms in the semiconductor are replaced with doped atoms. The gallium atom can share its electrons with three adjacent atoms in the structure but cannot share with the fourth adjacent atom because it has only three valence electrons. The missing electron creates a hole in the covalent bond. Since many gallium or aluminum atoms are added to the semiconductor material, a large number of these holes will be present which are free to drift about in the lattice. Due to the presence of these additional holes, the doped material is referred to as a P-type (positive type) semiconductor.[12,13]

It is important to note that N-type and P-type semiconductors have a much higher conductivity than pure semiconductors based on germanium or silicon.

In biomolecular structures composed of proteins, the main atom is *carbon*. From a semiconductor point of view, nitrogen, hydrogen, oxygen and sulfur are "impurity atoms". Since proteins are three-dimensional materials, P-type and N-type semiconductor solutions are different.

Proteins fall into the general class of polymers which are linear molecules built up from simple repeating units, amino acids. There are 20 different amino acids, but some of them (i.e. aspartic and glutamic acid) possess negative charges ("N-type") in solution, and some of them (i.e. lysine, arginine and histidine) possess positive charges ("P-type").

For α- and β-tubulin, the subunit proteins of microtubules, the relative amounts of these "P-type" and "N-type" amino acids at physiological pH is:

Amino acids	Type of Carbon impurities	α-tubulin	β-tubulin
Asp.	"N-type"	10.3	11.1
Glu.		13.5	12.7
		23.8	23.8
Lys.	"P-type"	3.9	3.9
Arg.		4.7	4.7
		8.6	8.6

This means that there are three times more "N-type" than "P-type" amino acids based on *carbon impurities*. In that case tubulin and microtubules have a surplus of electrons whose mobility may be important for microtubules lattice soliton machine-like and soliton automata-like functions within cells.

MICROTUBULES AS A MOLECULAR INFORMATION MACHINE

If microtubules are primary information processors, their subunits should be arranged by sphere packing symmetry derived from information coding laws. Here we investigate the packing symmetry of microtubules relevant to the information coding laws for molecular information machines.

The symmetry $Oh(\overline{6}/4)$ group possesses both cubic and hexagonal packing characteristics and appears to apply to tubulin packing in microtubules. From this symmetry group information coding laws are derived for a 48-dimensional space for face-centered-cubic packing.[14] Hexagonal packing of protein monomers independent of the $Oh(\overline{6}/4)$ symmetry group has been used to explain the form patterns of viruses, flagella, and MT.[15] Since hexagonal packing and face-centered-cubic packing have equal density, we use both to explain MT organization.

Let face-centered-cubic packing of spheres be in the volume of a 48-dimensional sphere of unit radius, and let the normalized packing density depend on the packing dimension.[16]

In this case the normalized density is $\overline{D} = \mathrm{Log}_2 D + n(24-n)/96$. Figure 2 shows the value of $\overline{D}$ for laminated lattice packing. Hexagonal packing may be used by fixing conditions. This is possible if the centers of the spheres lie on the surface of a cylinder (with the radius equal to that of the 48-dimensional unit sphere) and if (and only if) the sphere values in the axial direction (lattice) of the cylinder, by order of sphere packing, is the same as the dimension of the face-centered-cubic packing. Thus each sphere on the circumference of the cylindrical circle will represent one dimension of space of the face-centered-cubic packing.

With these conditions, packing on the cylindrical surface is hexagonal. Because the $\overline{6}$-fold symmetry axis of the *Oh* group is inverse, there must be two kinds of spheres ("white" and "black") on the cylinder surface, but

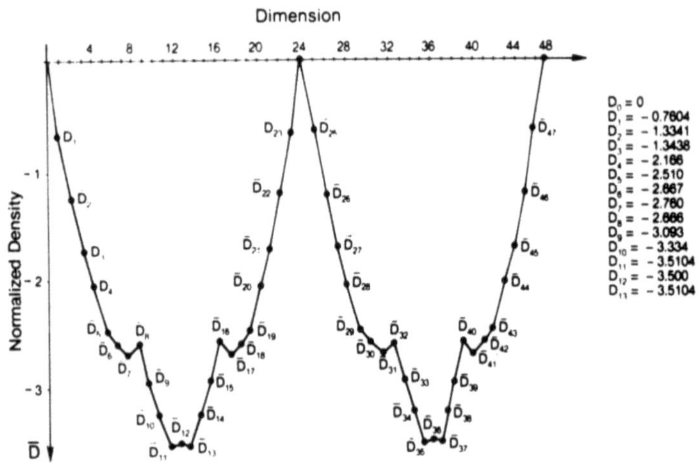

Fig. 2. The value of normalized density ($\overline{D}$) of the packing depends on the dimension in which the packing is done. Since packing is constructed from codes for the digital transmission of information, dimensions 11,12 and 13 are optimal for information processing.

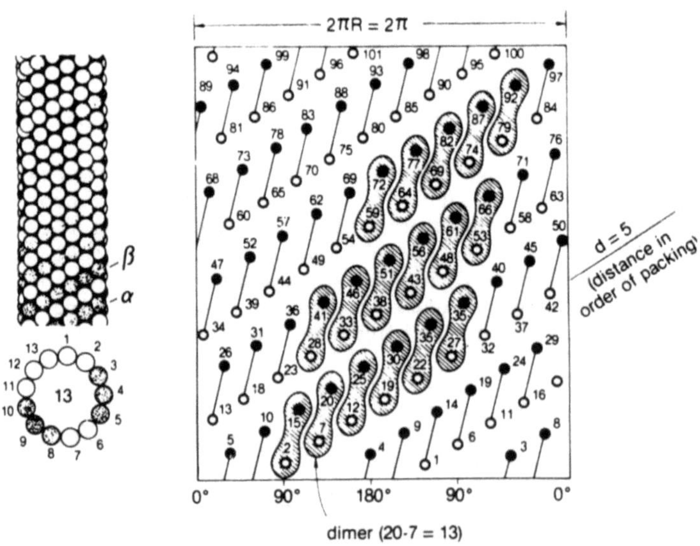

Fig. 3. The sphere indices as results of hexagonal packing of spheres on the surface of the circular cylinder.

linked so that they have the dimension value ("b" - "w" = n) of the face-centered cubic packing. Hexagonal packing on the cylinder surface leads to "screw symmetry". Figure 3 shows the solution for packing the 13 spheres.

Figure 2 shows that maximal normalized density ($\bar{D}$) has 11, 12 or 13 spheres. Since these packings are constructed from codes for the digital transmission of information, the number of spheres whose centers lie on the circumference of a circle is optimal for information processing.

Furthermore, Figure 3 shows that if the 13 spheres are on the surface of a circular cylinder, the distance between the spheres in order of packing will be 5. Sphere indices in the axial direction of dimers ("white" - "black") must be 13, because each dimer represents the dimension of face-centered-cubic packing. From the information theory, the arrangement of "white" and "black" spheres corresponds to a binary code of length n=13 and distance d=5.

Microtubules are comprised of α- ("white") and β- ("black") tubulin subunits which constitute 11, 12 or 13 protofilaments with screw symmetry. Screw symmetry in hexagonal packing determines distances between spheres and implies that MT possess all elements of a coding system. From the coding theory it is known that a code of length n=13 and distance d=5 is among the best known binary error-correcting codes with 64 codewords.[17] Consequently a latent MT bioinformation code $K_1(13,2^6,5)$ emerges.

Code K_1 is the result of the hexagonal property of the $Oh(\bar{6}/4)$ symmetry group, but also uses its cubic property. Since the density of packing in a

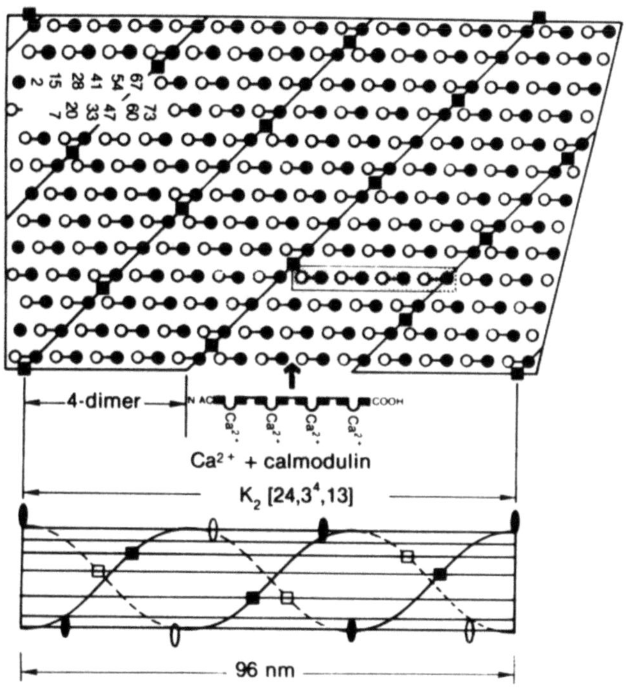

Fig. 4. K_1* code as a result of interaction between 12 dimers and high molecular weight proteins (K_1* = K_2). Under the influence of Ca^{2+}-calmodulin binary dimers may give dimer-ternary sequence due to Ca^{2+} attraction of surplus electrons in tubulin.

Table 3. K functions codes

Digital Sum	Code		
	$K_1[13,2^6,5]$	$K_2[24,3^4,13]$	
0	18	19	
1	32	16^+	16^-
2	10	10^+	10^-
3	4	4^+	4^-
4	-	1^+	1^-
All	$2^6 = 64$	$3^4 = 81$	

48-dimensional space is symmetrical (Figs. 1 and 2), the analogous symmetry group which derives the same law of a dense packing of spheres is 24-dimensional. This means that on the surface of a circular cylinder axial direction, there must be a code of length of 24 monomer subunits (12 dimers). The coding theory states that binary information cannot normally be transmitted over long distances; the information must be transformed into a new code more suitable for transmission. If the coding efficiency is used as a criterion of transmission, then 6-binary dimers of the K_1 code must be coded to give a 4-dimer ternary sequence of the K_2 code.[18] As the K_2 code should have length n=24 in the axial direction of a circular cylinder, from Figure 4 it is possible to write K_2 $(24,3^4,13)$. This code may result from interaction among 24 tubulin monomers and high molecular weight MAP (microtubule associated proteins). Under the influence of calcium-calmodulin, binary dimers of the K_1 code give dimer ternary sequences of the K_2 code. In this way the K_1 and K_2 code which result from the property of the $Oh(\overline{6}/4)$ symmetry group, leads to a $K(B^6T^4)$ transmission code, which may combine with energy sources, such as GTP, for intracellular dynamic and organizational activities, like soliton automata based on "P-type" and "N-type" carbon properties.

Coding relations between the K_1 and K_2 codes are given in Table 3. Eighteen binary words of the K_1 code yields 19 tertiary words of the K_2 code: there exists one "nonsense" word. Furthermore, we see that 14 binary words are ambiguous with tertiary words. Finally, two more "nonsense" ("stop") and ambiguities are seen in the coding of amino acids in the genetic code. Thus MT may code for dynamic information processing, as DNA codes for genetic structure.

CONCLUSION

1. If we use the subneuronal network based on MT as part of the neural network, it is a more realistic approach to functions such as learning and intelligence than if we use today's concept of neural networks.

2. If we put into the neural network model a subneuronal factor of neuronal activities based on MT, the neural activity profile is similar to the profile of action potentials.

3. From these investigations it emerges that MT activities are strongly cooperative with actions potentials in a neuron.

4. Subneuronal networks based on MT may be linked with neurocomputers and biocomputers in the future.

5. Although comparisons may be made between proteins and electronic devices through semiconductor concepts, significant differences exist between biomolecules and semiconductor electronic devices. Biomolecules are three-dimensional devices on the nanometer scale; electronic devices are two-dimensional devices on the micrometer scale. Viewing proteins as special states of semiconductor properties of carbon and its impurities based on electrons transport by soliton may be useful. Microtubules may be used as a basic material to build molecular electronic devices, or their structural architecture may be duplicated with a different material. Since microtubules possess a code system for memory and dynamic activities, they may be "Nature's computers" in living cells.

6. One of the fast self-assembly processes which has been thoroughly studied is the formation of sheets of tubulin in the presence of Zn^{2+} ions. The principle of self-assembly of MT and Zn-sheets provides an excellent example for possible future applications of microtubules to the development of molecular electronic devices.

REFERENCES

1. R. Hecht-Nielsen, Neurocomputing: picking the human brain, IEEE SPECTRUM 25(3):36, March (1988).
2. C. A. Mead and M. A. Mahowald, A Silicon Model of Early Visual Processing, Neural Networks, Vol. 1, 91:97 (1988).
3. P. Dustin, "Microtubules", Springer-Verlag, Berlin - New York (1978).
4. J. D. Cronly-Dillon and C. Birks, The possible involvement of brain microtubules in memory fixation, J. Exp. Biol., 61:443 (1974).
5. S. R. Hameroff, "Ultimate Computing: Biomolecular Consciousness and Nanotechnology", North-Holland, Amsterdam (1987).
6. R. Mileusnić et al., Learning and Chick Brain Tubulin, J. Neurochemistry, Vol. 34, 4:1007 (1980).
7. S. P. R. Rose, Early visual experience, learning and neurochemical plasticity in the rat and the chick, Phil. Trans. R. Soc. London B. 278:307 (1977).
8. D. Koruga, Microtubular Screw Symmetry: Packing of Spheres as a Latent Bioinformation Code, Ann. New York Academy of Sciences, 466:953 (1986).
9. D. Koruga, S. R. Hameroff, Microtubules as a Soliton Automata, Automatika, 28: 3 (1987).
10. K. Ogata, "Modern Control Engineering", Prentice-Hall, Englewood Cliffs, (1970).
11. M. N. Oguztoreli, G. M. Steil and T. M. Caelli, Control Mechanisms of a Neural Network, Biological Cybernetics, 54:21 (1986).
12. B. Streetman, "Solid State Electronic Devices", Prentice-Hall, Englewood Cliffs (1980).
13. S. M. Sze, "Semiconductor Devices, Physics and Technology", John Wiley & Sons, New York (1985).
14. J. Leech and N. J. A. Sloane, Sphere packings and error-correcting codes, Can. J. Math., Vol XXIII, 4:718 (1971).
15. R. D. Erickson, Tubular Packing of Spheres in Biological Fine Structure, Science, Vol. 181, 4101:705 (1973).
16. N. J. A. Sloane, The Packing of Spheres, Scientific American, Vol. 250, 1:116 (1984).
17. F. J. MacWilliams and N. J. A. Sloane, "The Theory of Error-Correcting Codes", North-Holland, Amsterdam (1977).
18. R. J. Catchpole, Efficient Ternary Transmission Codes, Electronics Letters, Vol. 11, 20:482 (1975).

INFORMATION PROCESSING IN MICROTUBULES:

BIOMOLECULAR AUTOMATA AND NANOCOMPUTERS

Stuart R. Hameroff

Advanced Biotechnology Laboratory
Department of Anesthesiology
University of Arizona Health Sciences Center
Tucson, Arizona 85724

Steen Rasmussen

Center for Nonlinear Studies and
Theoretical Division
Los Alamos National Laboratories, MS-B258
Los Alamos, New Mexico 87545

INTRODUCTION

To approach the cognitive capabilities of the human brain, computing technology must emulate brain structure at the intra-neuronal molecular level as well as the level of inter-neuronal networks. Strategies described by Forrest L. Carter (1984) and others at his historic meetings on Molecular Electronic Devices (in 1981, 1983, 1986) propose to implement nanometer scale ("nano" = 10^{-9}; nanometer = 10^{-9} meter) computing through switching in arrays of polyacetylenes, Langmuir-Blodgett films, electro-optical molecules, proteins and/or a number of other materials. Carter envisioned three dimensional, molecular scale memory and switching densities of 10^{15} to 10^{18} elements per cubic centimeter, near the theoretical limit for charge separation and the density of subunit dipoles in protein assemblies such as microtubules (Hameroff, 1987).

The potential for protein based molecular computing has been described by Michael Conrad (1986) who portrays input processed as dynamical physical structures, patterns, or analog symbols. Multidimensional conditions determine the conformational state of any one protein: temperature, pH, ionic concentrations, voltage, dipole moment, electroacoustical vibration, phosphorylation state, conformation of bound neighbor proteins, etc. Proteins integrate all this information to determine output, thus each protein is a rudimentary computer and converts a complex analog input to an output state or conformation. Organized protein assemblies, therefore, may be particularly suited to complex information processing.

A 1984 conference, "Chemically Based Computer Designs" (Yates, 1984), considered fundamental limits on computing due to quantum mechanical or thermodynamic effects, the nature of serial processing, or the ability to further miniaturize computer elements. The potential contributions of biological systems (DNA, protein antibodies, receptors, enzymes, cytoskeleton, etc.) were reviewed and a major conclusion was:

> None of these materials is as rich in chemoelectric physical phenomena as are (cytoskeletal) microscopic biological objects. Microtubules offer the most possibilities for inspiring chemically based computation! (Yates, 1984)

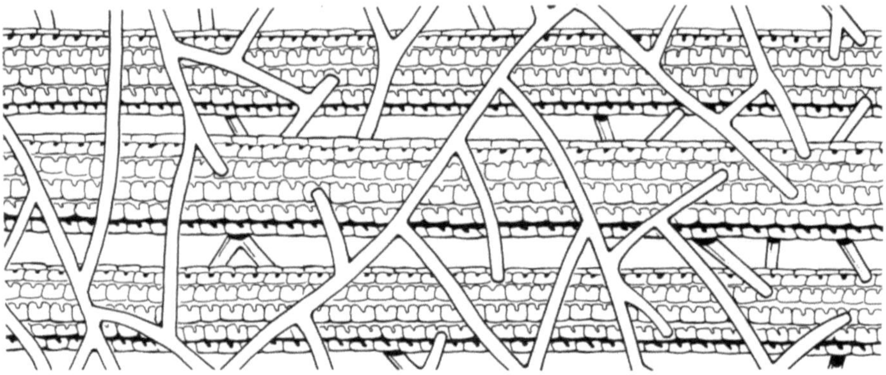

Fig. 1. Connectionist network of parallel arrayed microtubules (MT) interconnected by lateral crosslink filaments. Illustrated by Fred Anderson.

MICROTUBULES AND THE CYTOSKELETON

Interiors of living cells contain networks of filamentous proteins called microtubules which are arrayed in parallel and interconnected by smaller filamentous strands to form a dynamic gel (Fig. 1) whose intracellular activities (i.e. mitosis, growth, differentiation, locomotion, synapse modulation, neurotransmitter release, etc.) are essential to the living state (DeBrabander and DeMey, 1985).

Microtubules (MT) are hollow cylinders 25 nanometers ("nm") in diameter whose lengths may span meters in some mammalian neurons. MT cylinder walls are assemblies of 13 longitudinal protofilaments which are each a series of subunit proteins known as tubulin. Tubulin is a polar, 8 nm "dimer" which consists of two slightly different classes of 4 nm, 55 kilodalton "monomers" known as α and β tubulin. Tubulin dimer subunits in MT are arranged in a hexagonal lattice which is slightly "twisted", resulting in differing neighbor relationships among each subunit and its six nearest neighbors. MT assemble and disassemble by dynamic, complex processes which depend on various factors including temperature and calcium ion concentration. Oriented by MT based organelles called centrioles, MT polymerization determines the architecture and form of cells which can quickly change by MT depolymerization and reassembly in another direction (Kirschner and Mitchison, 1986). GTP, an energy providing analog of ATP, binds to polymerizing tubulin; GTP hydrolysis energy is subsequently delivered to assembled MT. The consumption of GTP energy in MT is not understood, although one possible utilization is the production of coherent lattice excitations as proposed by Fröhlich (1970, 1975, 1986; see next Section).

When viewed in cross section by electron microscopy, MT outer surfaces are surrounded by a "clear zone" of several nm which apparently represents an electronegative field due to excess electrons in tubulin and may also serve to organize cytoplasmic water and enzymes (Stebbings and Hunt, 1982). MT, as

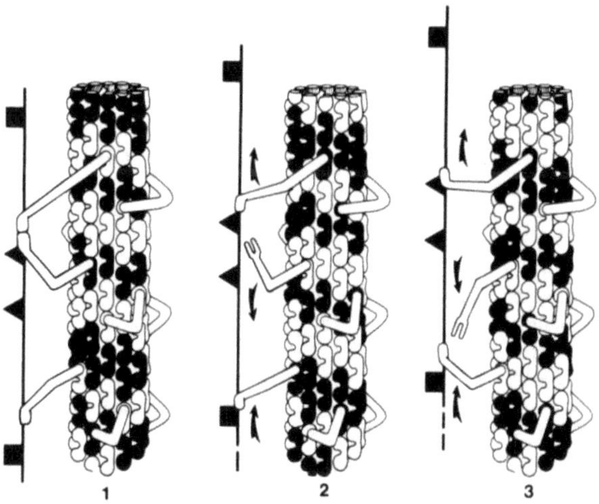

Fig. 2. Axoplasmic transport occurs by coordinated activities of microtubule attached contractile proteins ("dynein") which cooperatively pass material in a "bucket brigade". The orchestration mechanism is unknown, but shown here as the consequence of traveling conformational pattern waves such as arise from automaton behavior. Illustrated by Fred Anderson.

well as their individual dimers, have dipoles with negative charges localized toward α monomers (DeBrabander, 1982). Thus MT are "electrets": oriented assemblies of dipoles which are predicted to have piezoelectric properties (Athenstaedt, 1974; Mascarenhas, 1974). Contractile or enzymatic proteins (microtubule associated proteins: "MAPs") may be attached to MT at specific dimer sites and MAP attachments result in various helical patterns on MT surface lattices (Burns, 1978). Proteins, organelles, calcium ions and other materials are transported by MT and their contractile MAPs in a "bucket brigade". In neuronal axons and dendrites, neurotransmitter precursors, enzymes, membrane components and other materials necessary for synaptic function are conveyed by MT dependent axoplasmic transport at rates from one to 400 millimeters per day [Fig. 2 (Lasek, 1981; Ochs, 1982)]. The contractile MAPs use ATP as their energy source, but the controlled orchestration of these tiny "arms" is not understood.

MT and cytoskeletal activities are essential to organized, intracellular behavior which may be described as "intelligent". These activities and the lattice polymer structure of MT have suggested "computer-like" capabilities for information processing. At least a dozen author groups have published theoretical models of MT/cytoskeletal information processing which liken cytoskeletal function to computer related technologies (summarized in: Hameroff, 1987). These models remain untested because nanometer scale technologies (i.e. scanning tunneling microscopy: Schneiker and Hameroff, 1988) capable of recognizing (and interfacing) information processing in cytoskeletal polymers are only beginning to emerge. Direct evidence for signal propagation in MT has been generated by Vassilev et al. (1985) who suspended parallel, excitable membranes in ionic solution. Only when two membranes were connected by MT, excitation in one membrane provoked excitation in the other. The authors suggested that similar communication signals occurred routinely within the cytoskeleton. Becker et al. (1975) studied energy resonance transfer among fluorescent groups separately attached to different MT subunits or to membranes. They showed energy resonance transfer both among MT subunits and among MT subunits and membrane proteins. Also, Matsumoto and

Sakai (1979) and Alvarez and Ramirez (1979) have shown how cytoskeletal processes may be coupled to nerve membrane depolarization.

An approach to MT signaling and/or information processing requires an understanding of protein conformational state switching. Presently, a precise understanding is lacking, however one relevant theory will be discussed.

COHERENT PROTEIN DIPOLE EXCITATIONS

Proteins are vibrant, dynamic structures in physiological conditions. Proteins and their component parts undergo conformational motions over a range of time scales from femtoseconds (10^{-15} sec) to many minutes. Functionally significant conformational vibrations with switching capabilities appear to be in the range of 10^{-9} to 10^{-11} sec, or from nanoseconds (10^{-9}) to ten picoseconds ("pico" = 10^{-12}). Biologically relevant protein motions in this time scale are thought to include "collective elastic body modes, coupled atom fluctuations, solitons and other nonlinear motions, and coherent excitations" (Karplus and McCammon, 1979; 1983).

Coherent conformational excitations near the nanosecond time domain have been proposed by physicist Herbert Fröhlich. Fröhlich (1975) proposed that changes in protein conformation are triggered by charge redistributions such as dipole oscillations within specific hydrophobic regions of proteins. Another Fröhlich (1970; 1988) concept is that a set of proteins connected in a common physical structure and voltage gradient field such as within a polarized membrane or polymer electret like a microtubule may be excited coherently if biochemical energy such as ATP or GTP were supplied. Coherent excitation frequencies on the order of 10^9 to 10^{11} Hz are deduced by Fröhlich, who cites as evidence sharp windows of sensitivity to electromagnetic energy in this region by a variety of biological systems (Grundler and Keilmann, 1983). Other aspects of Fröhlich's model include "metastable states" (longer-lived conformational state patterns stabilized by local factors) and polarization waves, traveling regions of conformational states out of phase with the majority of coherently excited states. "Polarization waves", "massless bosons" (Del Giudice et al., 1982; 1986), "solitons" (Davydov, 1977; Scott, 1984) and "automata gliders" may be synonymous descriptions of "self-trapped" states which can propagate through protein lattices and process information.

Fröhlich's theory suggests that random supply of energy to a system of nonlinearly coupled dipoles in a common voltage field can lead to coherent excitation of a single vibrational mode, provided the energy exceeds a critical threshold. Important biological consequences may be expected from such coherent excitations and long range cooperativity. These include a capacity for intelligence and information processing due to "molecular automata" behavior in microtubules.

MOLECULAR AUTOMATA

Complex behavior resulting from collective activities of simple subunits occurs in dynamic entities known as "cellular automata." Von Neumann's (1966) original cellular automaton consisted of a large number of identical "cells" connected in a uniform matrix. The term "cell" was chosen by Von Neumann and others as the indivisible subunit in "cellular automata" based on biological "cells" as indivisible subunits of life. It is now apparent that biological cells are complex entities whose actions depend on collective functions of many intracellular structures including the cytoskeleton. Nevertheless, "cellular" in cellular automata jargon came to imply an indivisible grain, a discrete subunit with a finite number of states. The essential features of "cellular" automata are (Wolfram, 1984a; 1984b): 1) At a given time, each cell is in one of a finite number of states. 2) The cells are organized according to a fixed geometry. 3) Each cell communicates only with other cells in its neighborhood; the size and shape of the neighborhood are the same for all cells. 4) There is a universal clock; each cell may change to a new state at each tick of the clock depending on its present state, and states of its neighbors. The rules by which neighbor states influence each cell's state are the transition rules of the cellular automaton. At each clock tick (or "generation") the behavior of each cell depends only on the states of its neighbors and its own state. Simple neighbor transition rules can lead to complex, dynamic patterns.

Von Neumann showed how cellular automata could perform useful computations. By placing a number of contiguous cells in a specific pattern and running the clock through a sequence of generations, an output could be obtained by the patterns of states at a later time. Von Neumann showed that for any solvable problem, there existed proper transition rules and an initial configuration pattern which would evolve to a configuration containing the solution. Thus cellular automata are essential general purpose computers. With the right rules, cellular automata can simulate partial differential equations, or formation of snowflakes, mollusk shells, or galaxies. The dynamics of cellular automata can be sensitive to small perturbations in initial configurations; with a given set of rules, changing a single initial site value can lead to markedly different outcome patterns. Michael Conrad and colleagues (1973) introduced the idea of "molecular" automata <u>within</u> neurons as a basis for intracellular information processing related to cognition. We believe molecular automata activities in cytoskeletal structures such as microtubules could explain their organizational activities and provide a possible coding mechanism for interfacing and controlling their functions as molecular devices.

MOLECULAR AUTOMATA IN MICROTUBULES

Fröhlich's (1970) model of coherent excitations and cooperative coupling among proteins arrayed in a common voltage field may be applied to tubulin subunit dimers within MT to define discrete "time steps" or "generations" for automaton behavior. Fröhlich's model predicts coherent excitations within a range of 10^{-9} to 10^{-11} seconds, correlating with the conclusions of Karplus and McCammon (1979; 1983) that biologically relevant protein motions occur in this time domain. We will nominally refer to the clocking period, or generation time, as "nanosecond" (10^{-9} sec), although shorter times (i.e. 10^{-11} sec) are possible and would imply faster automata. Orientation, or phase of any tubulin subunit dimer relative to other coherently oscillating subunits at any "nanosecond" excitation period would depend on many factors such as initial conformational states, binding of water, ions, or microtubule associated proteins (MAPs), bridges to other MT, tyrosinated/glutamated state, energy providing phosphate nucleotides (i.e. GTP) and associated proteins, genetically determined subunit factors [seventeen different isozymes of mammalian brain tubulin have been determined (Lee et al., 1986)], and electrostatic dipole interactions among neighboring subunits. Many of these factors may be considered programming modes (either hereditary or environmental) whose net effects would serve to alter phases of particular subunits in coherently excited biomolecular arrays. To simulate MT molecular automata, we will focus solely on electrostatic dipole interactions among coherently excited MT subunits.

Original MT automaton simulations (Smith, Watt and Hameroff, 1984; Hameroff, Watt, and Smith, 1984; 1986) demonstrated patterns of dipole coupled tubulin conformational states which included traveling wave fronts, oscillatory blinkers, standing waves, gliders, kinks, and traveling diamonds

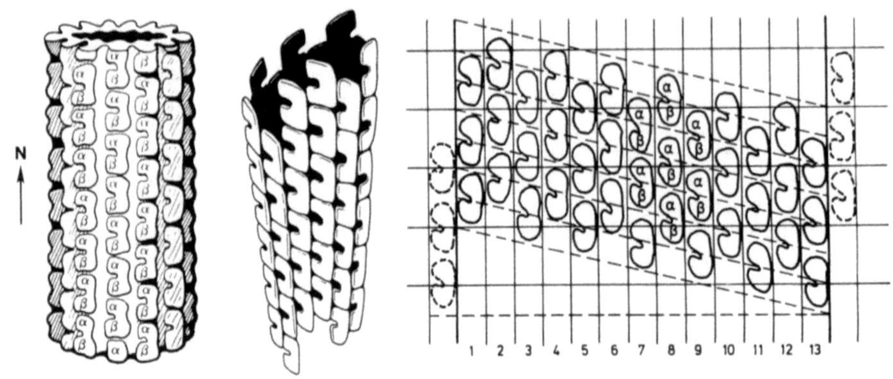

Fig. 3. Microtubule (MT) structure and computer simulation unfolding of MT lattice. Left: MT structure by X-ray diffraction analysis (Amos and Klug, 1974). Each subunit is dimer with an α and β monomer. Middle: stylized tubulin showing spiral offset of 1½ dimers (3 monomers) per row. Right: MT lattice flattened for computer simulation. α and β monomers within a single neighborhood of central dimer and surrounding hexagonal dimers are labeled.

and triangles. More recent simulations (Hameroff, Rasmussen and Mansson, 1988) and the current study sought to investigate different boundary conditions, catalogue effects of altered neighbor coupling thresholds, describe behavior of "virtual" MT automata [the dynamical features enabling automata to perform computations (Langton, 1987)], and consider MT molecular automata as an information medium in molecular devices.

The structure of MT and our derivation of automaton neighborhoods and neighbor dynamics for computer simulation are illustrated in Figs. 3-5. The MT cylinder (25 nanometers in diameter) has a circumference of 13 tubulin subunit dimers and the pitch pattern of the leftward helix is one and one-half dimers (three tubulin monomers - Fig. 3). For automaton simulation we

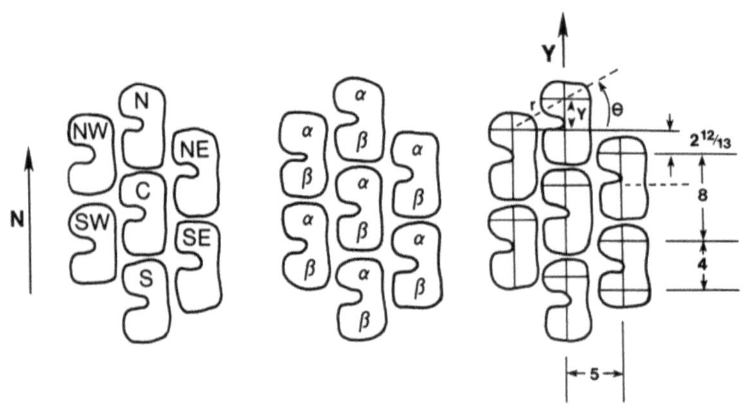

Fig. 4. Microtubule automata (MTA) neighborhood. Left: Neighborhood dimers oriented to longitudinal MT axis with α ends to the north (C=center, N=north, NE=northeast, SE=southeast, S=south, SW=southwest, NW=northwest). Center: α and β monomers within each dimer are labeled. Right: Distances (nm) and orientation among lattice neighbors. Interaction forces are calculated using y = r sin θ.

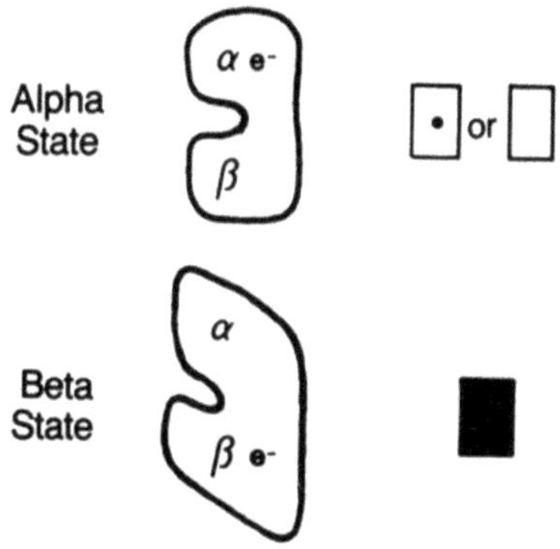

Fig. 5. Switching in microtubule (MT) dimers as function of Fröhlich coherent dipole excitations coupled to conformational states and influence by neighbor dimer dipole states. Top row: α states, bottom row: β states. Left: dimer conformation coupled to mobile electron localization, Right: representation of states in computer simulation - α states are white dots, β states are black.

consider 7 member neighborhoods of tubulin dimers: a central dimer (C) surrounded by a tilted hexagon of 6 neighbor dimers (N, NE, SE, S, SW, NW, see Legend Fig. 4).

To calculate the neighbor dipole interaction forces as a basis for automaton transition rules, each dimer may be viewed as having a mobile electron shared by the two monomers (Fig. 5). At each time-step the electron's average position is considered to be oriented either more toward the α monomer ("α state") or more toward the β monomer ("β state"), and the dimer conformation changes accordingly. Because MT net dipoles are negative toward the α ends, oscillations need not necessarily change direction totally to effect a conformational change. For example a "relative" dipole reorientation toward the β monomer may be sufficient to induce a tubulin dimer conformational state. For our calculations, however, we have assumed that electrons localize alternately in the centers of the two monomers. The electrostatic forces exerted on each dimer's mobile electron by the electrons of its nearest neighbor dimers would depend on the neighbors' dipole coupled conformational states. Consequently the location of neighbor dimer mobile electrons in either α or β monomer centers may serve as the basis for automaton transition "rules".

Considering the net forces due to MT lattice geometry, the resulting electrostatic force acting on an electron in a central dimer can then be calculated as:

$$f_{net} = \frac{e^2}{4\pi\epsilon} \sum_{i=1}^{6} \frac{y}{r_i^3} \qquad (1)$$

where ϵ is the average permittivity in MT [determined by the permittivity of free space times the MT dielectric constant-typically 10 for proteins, (Lindeburg, 1982)], e is the charge on an electron and r is the distance between the electrons. The distance r between electrons differs among neighbor pairs because of the MT lattice screw symmetry and dipole state of each neighbor at any particular nanosecond. Net positive (in "y" direction) neighbor forces will induce center dimer α states, and net negative forces will induce center dimer β states at each time step. We have assumed that only the y-component of the interaction forces are effective and have neglected any net force around the MT circumference. From equation (1) it is seen that only the <u>relative</u> magnitudes of neighbor forces are necessary for MT automaton simulations. These relative forces, which we use as transition "rules", may be multiplied times 2.3×10^{-14} Newtons to obtain absolute values of inter-subunit forces. Accordingly, the typical absolute values for dipole coupling interactions among MT dimer neighbors are of the order of 10^{-13} Newtons.

To simulate MT automata, we represent MT structure as a 2-dimensional grid in which the cylindrical MT has been fileted open and flattened (Figs. 6-11). The grid consists of MT subunit dimer loci which can exist in either an "α state" (blank with dot-Fig. 5) or "β state" (solid black-Fig. 5) at each time step or generation. To run the simulation, each dimer in an MT grid is treated as the "C" central dimer for each time step and net neighbor forces calculated. Net forces which are positive induce α states, and net negative forces induce β states, provided "threshold" values are exceeded. The extent to which each dimer is influenced by net neighbor forces acting upon its mobile electron may be represented by a "threshold" parameter. The higher the threshold, the greater are the summated neighbor forces necessary to induce a transition. For example, a threshold of ±9.000 means that net neighbor forces greater than $9.000 \times 2.3 \times 10^{-14}$ Newtons will induce an α state, and negative forces of less than $-9.000 \times 2.3 \times 10^{-14}$ Newtons will induce a β state. Biological factors which might be represented by "threshold" include temperature, acid/base status, voltage gradients, ionic concentration, genetically determined variability in individual dimers, and binding of molecules including MAPs and/or drugs to dimer subunits. In nerve cells, traveling membrane depolarizations could induce transient waves of lowered threshold along parallel arrayed MT. Consequently, the frequency of depolarization for a particular nerve cell would directly influence the elaboration of patterns within that nerve cell's MT automata. Such coupling could provide a hierarchical nesting mechanism between neuronal level activities ("neural nets") and intra-neuronal cytoskeletal automata. For boundary conditions in the present study, we have chosen a torus (periodic boundary conditions) in which the two ends of an MT segment which is 40 dimers in

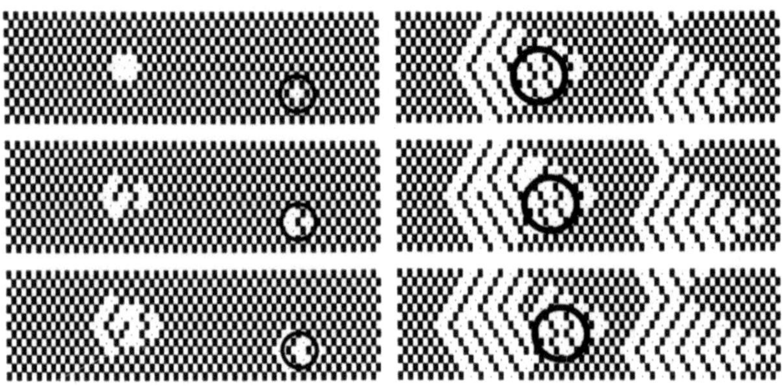

Fig. 6. Microtubule (MT) automaton simulation with threshold ±1.000, beginning with α seeds on β background on toroidal boundary conditions. Left column from top to bottom shows first 3 "nanosecond" time steps with a stationary oscillator ("blinker") encircled. Right column shows time steps 6 through 8 with a traveling "spider" glider encircled. α seeds grow in all four directions to ripple patterns of traveling waves.

length are connected to each other and information can flow around the closed system toroidal surface. Using toroidal boundary conditions, we have investigated effects of varying thresholds on MT automata behavior.

In Fig. 6, "nanosecond" time steps 1-3 and 6-8 of a toroidal MTA at a symmetric threshold ±1.000 develop from an initial configuration with a β background and a few α "seeds". This means that at the initiation of a sequence of time steps, most MT subunits are in β states with a few randomly selected α states. The α seeds grow in all four directions and an asymmetric ripple pattern is generated which evolves to standing and traveling waves on the torus. Also observed are blinkers, stationary objects which alternate between two patterns (small circles, steps 1-3), and gliders (i.e. "spider" glider - large circles, steps 6-8; also see Fig. 9), objects which travel at a velocity of one dimer per time step. Gliders are very important from a computational point of view; they can "transport" data in a very localized way from one area of the computer to another. Both α and β gliders exist simultaneously in the system: the α gliders always traveling south and the β gliders always traveling north. Generally, interactions among the traveling gliders and stationary patterns may be viewed as calculations and endow MT automata with potential computational capabilities.

Increasing the threshold leads to new features. In Fig. 7 (threshold ±5.750), the initial configuration is again α seed on β background. The wave pattern now develops only in three directions forming wedge shaped structures. The wavefronts are all in the SW-NE orientation, a preference which appears to be an effect of the "left-handed" helical structure of MT.

Figure 8 shows a virtual automaton structure, the "bus glider", for thresholds around ±9.000. For a very broad class of initial configurations the system organizes the β seeds into α "bus", which then goes South (to get back, you have to take the α bus on a β background). All β seeds move South after a transient "wiggle", but only a limited number preserve their shape. The bus glider is one of several structures which are extremely stable to perturbations and develop regardless of initial conditions; these gliders are examples of virtual automata (Langton, 1987). Other, similar structures are shown at two different thresholds in Fig. 9. The upper left portion of Fig. 9 shows a dot glider and three other β seed gliders at threshold ±9.000. The next two generations are shown below and demonstrate movement of these gliders southward. The upper right portion of Fig. 9 shows 4 objects at threshold ±1.000. We refer to these objects (from left to right) as: dot glider,

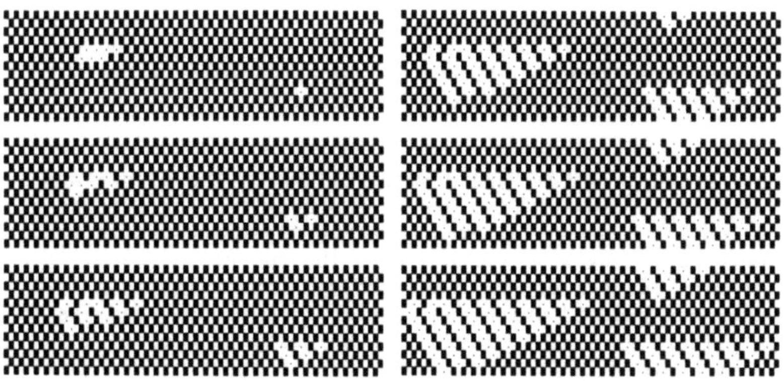

Fig. 7. Microtubule automaton toroidal simulation with threshold ±5.750 beginning with α seeds on a β background. SW-NE wave patterns develop in 3 directions forming wedge shaped structures. Time steps 1-3 and 6-8 are shown.

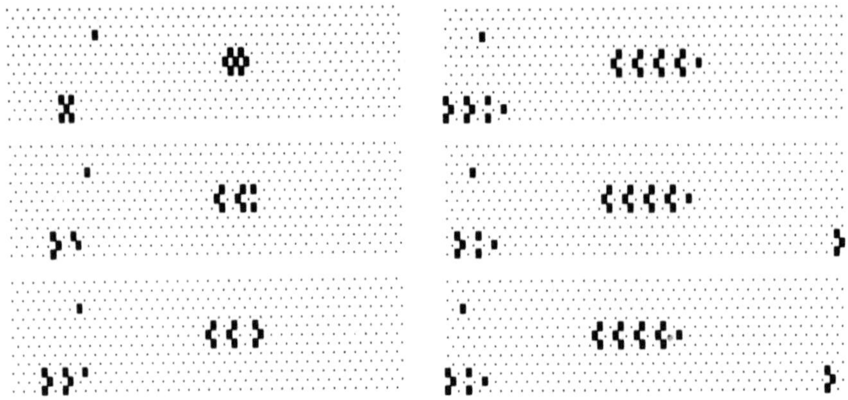

Fig. 8. Microtubule automaton toroidal simulation at threshold ±9.000 showing a "bus glider", an example of a virtual automaton. After transient wiggle, the bus glider heads South. Time steps 1-3 and 6-8 are shown.

Fig. 9. Some important virtual microtubule automata existing for two different threshold values. Upper left: dot glider and three other gliders with threshold ±9.000. Middle and lower left: next two generations showing gliders moving south without "wake". Upper right: (from left to right) dot glider, spider glider, triangle glider and diamond blinker with threshold ±1.000. Middle and lower right: next two generations showing gliders moving and leaving wakes which develop to traveling wave structures.

spider glider, triangle glider and diamond blinker. The middle and lower sections of the right side of Fig. 9 show the next two generations of these β seed objects leaving "wakes", altered background patterns which lead to traveling and/or standing wave structures. The dot glider behavior exemplifies the differing effects of the 2 thresholds. The triangle glider is an element of the "bus glider" which can develop from random seeds as in Fig. 8. MT automaton gliders generally travel at a velocity of one dimer per generation. Since each dimer is 8 nm in length, over a range of time-steps from 10^{-9} to 10^{-11} seconds this glider velocity translates to 8 to 800 meters per second. Perhaps coincidentally, this velocity range correlates with that of traveling nerve action potentials! Correlation between nerve membrane depolarization and MT automata gliders may signify coupling between "neural networks" and cytoskeletal information processing. Sequences of gliders could be used to transmit "messages" with MT automata always ready to read a new message soon after the previous one. Encoding of such messages may be both spatial (location of glider on MTA) and temporal (time sequence of gliders). Isolated gliders existing for thresholds around ±9.000 also seem ideal for regulation of certain biological functions such as sequential contraction of MAPs and axoplasmic transport of molecules along microtubules (Fig. 2: Hameroff, 1987).

At higher symmetric threshold values (±37.580) activity is inhibited and even the dot glider cannot exist any more. Typical dynamics for MT automata in the range up to ±46.880 include "bean sprout growth" from any induced seed. Bean sprouts are linear patterns which add one dimer each time step. As the threshold is increased further, all patterns freeze. At thresholds which are not too high, (i.e. ±47.000) blinkers may exist in the frozen structures. For symmetric thresholds at ±62.020, no dynamics exist any more. The capability to freeze a pattern may usefully function as a memory mechanism for MT automata, however the frozen configuration heavily depends on the dynamics of the freezing process. Relations between pattern updating time (clocking frequency), velocity of the freezing process and threshold will determine the stored pattern, a situation analogous in many ways to "spin-glass" systems. For example, a slow "freezing" (i.e. increase in threshold) always results in a north-south striped pattern due to the bean sprout growth interval, but a more rapid freezing process allows for conservation of a variety of memory structures. For details see Hameroff, Rasmussen and Mansson (1988).

Thus far we have only discussed MT automata dynamics for symmetrical threshold values. However MT are electrets with net dipoles on the entire macromolecule and like-oriented dipoles on each subunit; the α ends are negative relative to the β plus ends (Athenstaedt, 1974; DeBrabander, 1982; Mascarenhas, 1974). Thus electrons are more likely to be found in the α end of the tubulin dimers, suggesting an asymmetrical threshold for MT automata in which escape from the α state is more difficult. Asymmetrical thresholds offer new dynamical properties: Figure 10 shows a toroidal MT automata sequence with an α to β threshold of -20.000, and a β to α threshold of +2.000. Random seeds develop to form α and β linear gliders which travel in opposite directions simultaneously (ovals). Many initial MT encodings with these

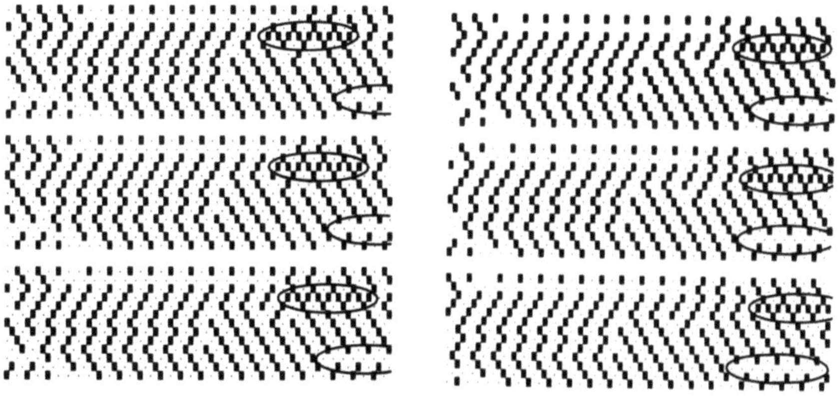

Fig. 10. Microtubule automata toroidal simulations with asymmetric thresholds. For α to β, the threshold is -20.000. For β to α, the threshold is +2.000. Linear gliders travel in opposite directions (ovals) amidst other interesting dynamics.

threshold values compute into linear gliders traveling in opposite directions. Such a phenomenon could account for observations of concurrent bilateral transport along single MT (Schnapp et al., 1986), as well as bilateral information exchange.

Figure 11 summarizes the dependence of toroidal MT automata behavior on threshold. The upper portion of each sector of the diagram schematically indicates the dominant global behavior for (a) β seeds on an α background, and (b) α seeds on a β background. Below the schematic behaviors are shown the corresponding threshold intervals followed by our qualitative description of the behavior. At the bottom of each sector, the "dimensionality" (number of directions) of the growth behavior is indicated. Thus we have demonstrated specific MT automata behaviors dependent on threshold conditions which could have relevant biological correlates.

DISCUSSION

Our simulations demonstrate automaton behavior in MT of both a general nature (i.e. waves, gliders and blinkers) similar to behavior in other automaton systems, and of a specific nature (i.e. wedge patterns, stable gliders and NE-SW orientations) derived from unique MT geometry. Further, we have catalogued a variety of MT automata behaviors dependent on "threshold", a parameter of dipole coupling sensitivity among coherently oscillating MT subunits. Threshold dependent behaviors include gliders (dot, spider and bus gliders), traveling and standing wave patterns, blinkers, linearly growing patterns ("bean sprouts"), memory wakes, bidirectional gliders, and frozen patterns. Subject to hereditary and environmental "programming" of MT subunits (genetic tubulin isozymes, binding of GTP, MAPs, inter-MT links, tyrosination/glutamation, nerve depolarization frequency, etc.), these behaviors appear capable of useful biological information representation and computation. For example, MT automata may be utilized for guidance and movement of

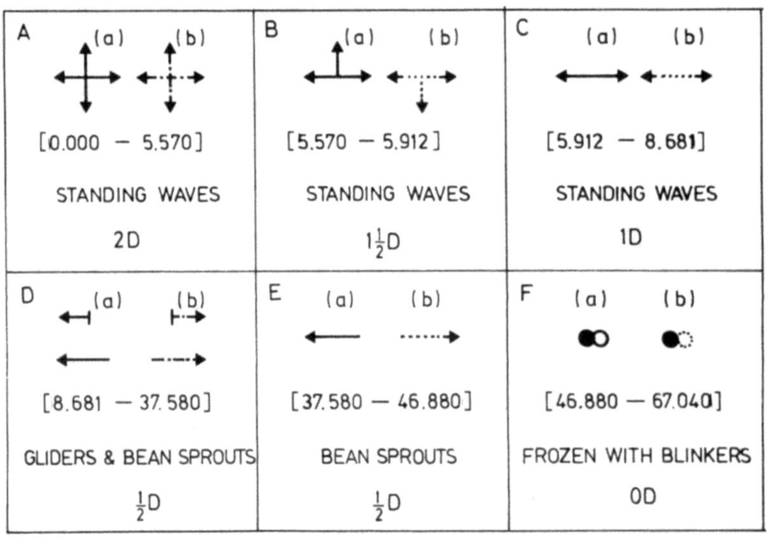

Fig. 11. Dependence of toroidal microtubule automata dynamics on symmetrical dipole coupling threshold. The upper items in each sector (A-F) indicate the dominant global behavior for (a) β seeds on an α background, and (b) α seeds induced on a β background. Below the dominant behaviors are the threshold intervals, what we call this type of dynamic phenomena, and the dimensionality of the growth patterns.

cells, transport of molecules within cells, and functions related to cognition in the human brain.

In computer based artificial intelligence ("AI"), data is represented by virtual pointers which connect symbols (DeCallatay, 1986). In computers, these virtual relations are actual wires with potential gate connections, and among brain neurons they appear to be synaptic connections. Cross-bridge filaments within neurons which connect cytoskeletal MT, neurofilaments and membrane proteins may be virtual pointers connecting symbolic MT automaton representations. To take maximal advantage of parallelism, Hewitt (1985) has described parallel elements connected by "arbiters" which "weigh" and reconcile differing content with continuous input and output. Similarly, Reeke and Edelman (1984) have modelled parallel recognition automata which communicate laterally and have distinct and complementary personalities. Lateral communications between the parallel automata resolve conflicting output and form an associative memory. Filamentous bridges among parallel MT and/or neurofilaments could serve comparable functions.

Connectionist networks within single elements of larger connectionist networks suggest a self similar, "fractal-like" relationship. If so, the information capacity of a finer scale fractal basement should be far greater than larger scales. Roboticist Hans Moravec (1986) has attempted to compare the "computing power" of a computer and of the human brain. Considering the number of "next states" available per time in binary digits ("bits"), a microcomputer has a capacity of about 10^6 bits per second. Moravec calculates brain "computing" power by assuming 40 billion neurons which can change states hundred of times per second, resulting in 4×10^{12} bits per second. Densely packed MT contain more than 10^{18} tubulin subunits per cubic centimeter (equivalent to the limit for charge separation). We estimate approximately 10^{14} MT tubulin subunits in a human brain, based on MT spaced 1000 nanometers apart and 40 percent of brain volume being neuronal cytoplasm. In our model of MT automata these 10^{14} brain MT subunits can change states 10^9 to 10^{11} times per second leading to a total brain capacity of 10^{23} to 10^{25} bits per second! Such a huge capacity could provide a fine resolution to the grain of cognition, permit massive parallelism and redundancy, and account for cytoplasmic "housekeeping" chores not directly related to brain/mind function. Such a huge capacity might be co-opted for molecular devices. Tubulin may be produced by genetic engineering techniques and polymerized in large arrays whose parallel, collective computing properties may approach or far surpass the cognitive capacity of the human brain (assuming interfacing could be accomplished). The dream of "downloading" human consciousness could be realized. Properly programmed MT may also be utilized as nanoscale machines to assemble structures, or seek and destroy lethal infectious agents, cancer cells, or other pathology in the human body. As a less ambitious goal, the architecture and organization of MT may be emulated in molecular devices of different materials. However, to take full advantage of self assembling, regenerative capabilities, MT themselves may be recruited as molecular biocomputers.

REFERENCES

Amos, L. A., and A. Klug, 1974, Arrangement of Subunits in Flagellar Microtubules, J. Cell Sci., 14:523-550.

Athenstaedt, H., 1974, Pyroelectric and Piezoelectric Properties of Vertebrates, Ann. NY Acad. Sci., 238:68-93.

Becker, J. S., J. M. Oliver and R. D. Berlin, 1975, Fluorescence Techniques for Following Interactions of Microtubule Subunits and Membranes, Nature (London), 254:152-154.

Burns, R. B., 1978, Spatial Organization of the Microtubule Associated Proteins of Reassembled Brain Microtubules, J. Ultrastruct. Res., 65:73-82.

Carter, F. L., 1984, The Molecular Device Computer: Point of Departure for Large Scale Cellular Automata, Physica, 10D:175-194.
Conrad, M., 1986, The Lure of Molecular Computing, IEEE Spectrum, 23(10):55-60.
Conrad, M., W. Guttinger and M. Dal Cin, 1973, Physics and Mathematics of the Nervous System, Proceedings of a Summer School Organized by the International Centre for Theoretical Physics, Trieste, and the Institute for Information Sciences, University of Tübingen.
Davydov, A. S., 1977, Solitons and Energy Transfer Along Protein Molecules, J. Theor. Biol., 66:379-387.
DeBrabander, M., 1982, A Model for the Microtubule Organizing Activity of the Centrosomes and Kinetochores in Mammalian Cells, Cell. Biol. Intern. Rep., 6 (10):901-915.
DeBrabander, M., and J. DeMey, 1985, "Microtubules and Microtubule Inhibitors," Amsterdam, Elsevier.
De Callatay, A. M., 1986, "Natural and Artificial Intelligence: Processor Systems Compared to the Human Brain," Amsterdam, Elsevier North-Holland.
Del Giudice, E., S. Doglia and M. Milani,, 1982, Self Focusing of Fröhlich Waves and Cytoskeleton Dynamics, Phys. Lett. A, 90:104-106.
Del Giudice, E., S. Doglia and M. Milani, 1986, Collective Properties of Biological Systems - Solitons and Coherent Electric Waves in a Quantum Field Theoretical Approach, in: "Modern Biochemistry," F. Gutmann and H. Keyzer, eds., New York, London, Plenum Press, pp. 263-287.
Fröhlich, H., 1975, The Extraordinary Dielectric Properties of Biological Materials and the Action of Enzymes, Proc. Natl. Acad. Sci USA., 72 (11):4211-4215.
Fröhlich, H., 1970, Long Range Coherence and the Actions of Enzymes, Nature (London), 228:1093.
Fröhlich, H., 1986, Coherent Excitations in Active Biological Systems, in: "Modern Bioelectrochemistry," F. Gutmann and H. Keyzer, eds., New York, Plenum Press, pp. 241-261.
Fröhlich, H., 1988, "Biological Coherence and Response to External Stimuli," Heidelberg, Germany, Springer-Verlag.
Grundler, W., and F. Keilmann, 1983, Sharp Resonances in Yeast Growth Prove Nonthermal Sensitivity to Microwaves, Phys. Rev. Lett., 51:1214-1216.
Hameroff, S. R., 1987, "Ultimate Computing: Biomolecular Consciousness and Nanotechnology," Amsterdam, Elsevier North-Holland.
Hameroff, S. R., and R. C. Watt, 1982, Information Processing in Microtubules, J. Theor. Biol., 98:549-561.
Hameroff, S. R., S. A. Smith, and R. C. Watt, 1984, Nonlinear Electrodynamics in Cytoskeletal Protein Lattices, in: "Nonlinear Electrodynamics in Biological Systems," W. R. Adey and A. F. Lawrence, eds., New York, Plenum Press, pp. 567-583.
Hameroff, S. R., S. A. Smith, and R. C. Watt, 1986, Automaton Model of Dynamic Organization in Microtubules, Ann. NY Acad. Sci., 466:949-952.
Hameroff, S. R., S. Rasmussen and B. Mansson, 1988, Molecular Automata in Microtubules: Basic Computational Logic of the Living State?, in: "Artificial Life: Proceedings of a Los Alamos International Workshop on the Simulation and Synthesis of Living Systems," C. Langton, ed., Santa Fe Institute, Addison Wesley, pp. 521-553.
Hewitt, C., 1985, The Challenge of Open Systems, Byte, 16(4):223-242.
Karplus, M., and J. A. McCammon, 1983, Dynamics of Proteins: Elements and Function, Ann. Rev. Biochem., 52:263-300.
Karplus, M., and J. A. McCammon, 1979, Protein Structural Fluctuations During a Period of 100 ps, Nature (London), 277:578.
Kirschner, M., and T. Mitchison, 1986, Beyond Self Assembly: From Microtubules to Morphogenesis, Cell, 45:329-342.
Langton, C. G., 1986, Studying Artificial Life with Cellular Automata, Physica, 22D:120-149.

Langton, C. G., 1987, Virtual State Machines and Cellular Automata, "Complex Systems", 1:257-271.
Lasek, R. J., 1981, The Dynamic Ordering of Neuronal Cytoskeletons, Neurosci. Res. Prog. Bull., 19 (1):7-31.
Lee, J. C., D. J. Field, H. J. George and J. Head, 1986, Biochemical and Chemical Properties of Tubulin Subspecies, Ann. NY Acad. Sci., 466:111-128.
Lindeburg, M. R., 1982, "Engineer In Training Review Manual," Professional Publications, San Carlos, California.
Mascarenhas, S., 1974, The Electret Effect in Bone and Biopolymers and the Bound Water Problem, Ann. NY Acad. Sci., 238:36-52.
Moravec, H., 1987, "Mind Children," San Francisco, University Press.
Ochs, S., 1982, "Axoplasmic TranSport and Its Relation to Other Nerve Functions," New York, Wiley Interscience.
Reeke, G. R. and G. M. Edelman, 1984, Selective Networks and Recognition Automata, in: "Computer Culture: The Scientific, Intellectual, and Social Impact of the Computer," Heinz R. Pagels, ed., Ann. NY Acad. Sci., 426:181-201.
Schnapp, B. J., R. D. Vale, M. P. Sheetz and T. S. Reese, 1986, Microtubules and the Mechanism of Directed Organelle Movement, Ann. NY Acad. Sci., 466:909-918.
Schneiker, C. W., and S. R. Hameroff, 1988, Nanotechnology Workstations Based on Scanning Tunneling/Optical Microscopy: Applications to Molecular Scale Devices, in: "Proceedings of the Third International Symposium on Molecular Electronic Devices, Oct. 6-8, 1986, Arlington, VA.," F. L. Carter, R. E. Siatkowski, and H. Wohltjen, eds., Elsevier, North-Holland, pp. 69-90.
Scott, A. C., 1984, Solitons and Bioenergetics, in: "Nonlinear Electrodynamics in Biological Systems," W. R. Adey, and A. F. Lawrence, eds., New York, Plenum Press, pp. 133-142.
Smith, S. A., R. C. Watt, and S. R. Hameroff, 1984, Cellular Automata in Cytoskeletal Lattices, Physica, 10D:168-174.
Stebbings, H., and C. Hunt, 1982, The Nature of the Clear Zone Around Microtubules, Cell Tissue Res., 227:609-617.
Vassilev, P., M. Kanazirska and H. T. Tien, 1985, Intermembrane Linkage Mediated by Tubulin, Biochem. Biophys. Res. Comm., 126(1):559-565.
Von Neumann, J., 1966, "Theory of Self-Reproducing Automata," A. W. Burks, ed., Urbana, University of Illinois Press.
Waltz, D. L., 1982, Artificial Intelligence, Sci. Am., 247(4):118.
Wolfram, S., 1984a, Cellular Automata as Models of Complexity, Nature (London), 311:419-424.
Wolfram, S., 1984b, Universality and Complexity in Cellular Automata, Physica, 10D:1-35.
Yates, F. E., 1984, "Report on Conference on Chemically-Based Computer Designs," Crump Institute for Medical Engineering Report CIME TR/84/1, U.C.L.A.

BLM-BASED BIOSENSORS AND DEVICES: APPLICATIONS AND FUTURE PROSPECTS

H.T. Tien, Z. Salamon[1], V. Kochev[2], A. Ottova[3] and M. Zviman

Membrane Biophysics Laboratory (Giltner Hall)
Department of Physiology, Michigan State University
East Lansing, Michigan 48824 (USA)

Life is nothing but a movement of electrons!

Albert Szent-Gyorgyi, c. 1960

Life is molecular electronics! Sym. on MEBC

Budapest, Hungary, August 27, 1987

INTRODUCTION

Biomembranes are composed of two main classes of compounds: lipids and proteins. Insofar as can be determined, the lipids of biomembranes are organized in the form of a bilayer which is an universal element in all biomembranes. This is so because of one unique property of the lipid molecule. One end of the molecule is hydrophilic with a great affinity towards a medium of high dielectric constant. The other end of the lipid molecule is hydrophobic, which is sequestered away from the aqueous solution. With hindsight, we now know that the remarkable stability of a lipid bilayer is due to the combination of hydrophilic and hydrophobic forces, which makes the lipid bilayer a thermodynamically favored structure. Thus, lipids and other amphipathic compounds, when placed in aqueous media, will spontaneously arranged themselves in the form of a bilayer. This outstanding characteristic has contributed to the discoveries of two experimental bilayer lipid membrane systems, namely, planar BLMs and spherical liposomes. The thoughts and experiments leading to the planar bilayer lipid membrane (BLM) formation, however, owed much to 'black' soap film studies and the Langmuir-Blodgett monolayer technique, as described in the experimental section of this paper.

Since the early 1960s, reconstituted bilayer lipid membranes have been extensively investigated as models of biological membranes (Burton, 1968; Tien, 1974; Davison, 1985; Pethig and Kell, 1987; Ringsdorf et al., 1988; Ivanov, 1988). However, until recently, relatively few attempts have been made to exploit their potential in practical applications such as sensors

Permanent Address:
[1]Institute of Physics, Technical University of Poznan, Poland
[2]Faculty of Physics, University of Sofia, Bulgaria
[3]Institute of Biotechnology, Slovak Technical University, Bratislava, Czechoslovakia

and molecular devices. From the viewpoint of membrane biophysics and physiology, biological membranes are essentially the basic structure of Nature's sensors and devices. For examples, the thylakoid membrane of green plants functions as an energy transducer converting sunlight into electrical/chemical energy, the photoreceptor membrane of a rod's outer segment detects photons as the initial step in visual perception, and the plasma membranes of cells and organelles possess ability for ion sensing, for instance, differentiating Na^+ and K^+ with great specificity. Further, the plasma membrane provides sites for a host of ligand-receptor contact interactions such as antigen-antibody formation (Dryhurst and Niki, 1988). In view of these and numerous other vital functions associated with cell membranes, it is not surprising that the past two decades or so have witnessed an enormous research effort spent on membranes in which reconstituted planar BLMs and spherical liposomes have played a superlative role. Today, our interest in biomembranes and their experimental models is still increasing, as evidenced by the papers presented at this Symposium, of which more than one-half are directly concerned with membranes. In this paper we will first of all give a brief description of the BLM system. This will be followed by a summary of our work and that of others on BLMs in relation to biosensors and molecular electronics. In the last section, prospects of BLMs and their modifications are discussed with a view of potential applications in the future.

THE BLM SYSTEM

The history of the BLM system and its development as a model for biomembranes has been recounted elsewhere (Burton, 1968). Together with 'black' soap films (Exerowa and Kashchiev, 1986), the Langmuir-Blodgett (L-B) technique provided the crucial insight for its realization. It should be mentioned, however, there is one major difference between the L-B layers on rigid substrate and the BLMs. Apart from its bimolecular thickness, a BLM is a liquid-like, dynamic structure in a metastable state (Mittal, 1988). On the other hand, a layered structure made by the L-B technique is rigid and in a solid state. The so-called 'solvent-free' BLMs made by the L-B technique requires the pretreatment of the aperture with petrolatum gel or squalene. From a self-assembling point of view, we conclude that it is difficult to envision how a BLM separating two aqueous solutions can be made from two rigid, solid-like monolayers of lipids without a Plateau-Gibbs border (Tien, 1974). For biosensor development, it is our opinion that a fluid bilayer is of crucial importance. The aim of this section is to describe in sufficient detail how to set up a simple BLM system using the self-assembling technique. For more details, a number of articles should be consulted (Ivanov, 1988; Davison, 1985).

BLM cell. A BLM cell comprises of two compartments, a Teflon cup and an outer container with two chambers, into one of which the Teflon cup is placed. The outer container is usually made of plexiglass for viewing the BLM in formation. The two chambers are connected via a 2 mm hole in the wall of the cup, as shown in Fig. 1(A).

BLM-forming solution. Depending upon the interest of the investigator, many different lipid solutions for BLMs have been formulated. The recipes for three easily prepared lipid solutions are given here. The first one is 1% lecithin (phosphatidylcholine) in n-decane. The second one is oxidized cholesterol, which is readily obtained by refluxing a 4% cholesterol in n-octane at its boiling point (about 126 ° C) for 5.5-6.0 hrs, stirring with air or oxygen at a rate of 100-125 ml/min through a gas dispersing tube. The cooled supernatant is used for BLM formation. The third BLM-forming solution is simply prepared by dissolving 0.1 ml glycerol monooleate (GMO) or glycerol dioleate (GDO) in 10 ml squalene. Analogously, GDO and squalene may be replaced by lecithin and paraffin (Nujol) oil, respectively. BLMs formed according to the last recipe are known to be

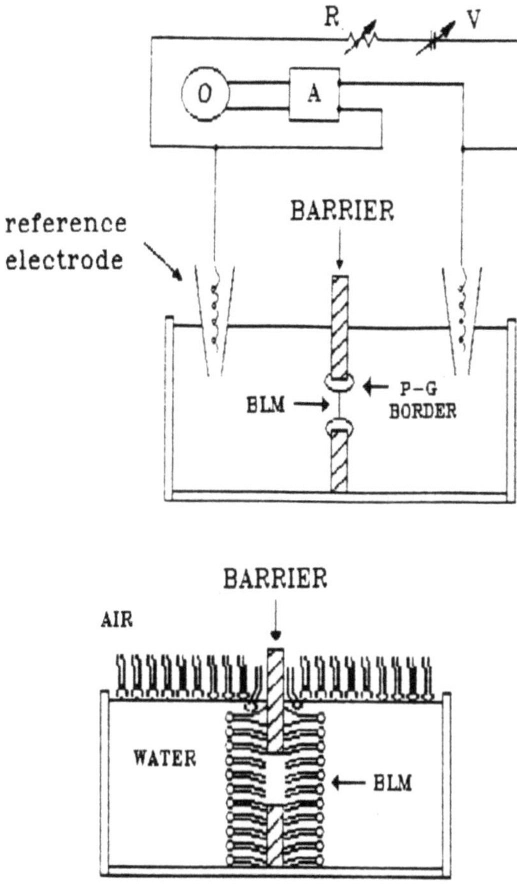

Figure 1. Experimental cells for investigating the electrical properties of bilayer lipid membranes (BLM). Upper: A conventional setup with BLM supported by Plateau-Gibbs (P-G) border. Lower: A BLM is formed by a modified Langmuir-Blodgett technique according to Takagi et al. (see Tien, 1974, Fig. 11-3, p. 477).

"solvent-free" with C_m = 0.78 μF/cm^2 and t_m (the thickness of hydrocarbon core only) = 2.5 nm (White, 1978; Dilger et al., 1982). It is interesting to note that both squalene and paraffin oil were used in the seminal studies on BLMs (Mueller et al., 1963; Danielli et al., 1964).

BLM formation. BLMs with diameters up to 10 mm can be formed by a variety of methods. Fig. 1 shows two such methods: (A) after both chambers have been filled with, for example, 0.1 M KCl, a drop of BLM-forming solution is spread over the aperture, which will undergo spontaneous thinning to a 'black' or bilayer lipid membrane under appropriate conditions. If the thinning process is too slow, the 'black' hole formation may be initiated by applying a voltage (about 50-100 mV) of short duration. The second method shown in Fig. 1(B) is based on the Langmuir-Blodgett technique introduced by Takagi et al. (1965) and starts with monolayer formation at the air/water interface in the cell containing two chambers separated by a Teflon septum with an aperture (1 mm or less). By lowering and raising the liquid level of the bathing solution across the aperture, a BLM may be formed in the process, provided that the aperture has been conditioned with squalene or petrolatum gel (Ivanov, 1988). Alternatively, a new type of BLM may be formed by a method being developed in our laboratory in which one side of a solvent-free BLM is anchored to a hydrophilic (either conducting or non-conducting) support while the other side is in contact with aqueous solution in the usual manner. Such a system could be useful in the development of biosensors and molecular electronic devices.

Electrical measurements. BLM/solution interfaces are heterogeneous and as such they are usually electrified as a result of charge transfer, separation, and/or redox reactions. Thus, the charges on BLM/solution interfaces create electrical fields which greatly influence the movement of ions, electrons, holes, and redox species (Blank, 1986; Blank and Findl, 1987). The current through the BLM and the potential across it as well as other electrical parameters of BLMs can readily be measured via a pair of Ag/AgCl or calomel electrodes (with or without salt bridges) placed in the bathing solution. These parameters include membrane potential (E_m), resistance (R_m), current (I_m), capacitance (C_m), breakdown voltage (V_b), and I/V curves (voltammograms). Both dc and ac methods have been used. For a basic BLM system, a high impedance ($>10^{14}$ ohms) electrometer and a low impedance picoammeter are essential. For measuring C_m, either a dc transient technique or ac bridge method may be used (e.g., with a RLC bridge). In the dc technique a potential step is applied across the series combination of the BLM and a known resistor (R_s), and the voltage across the R_s is

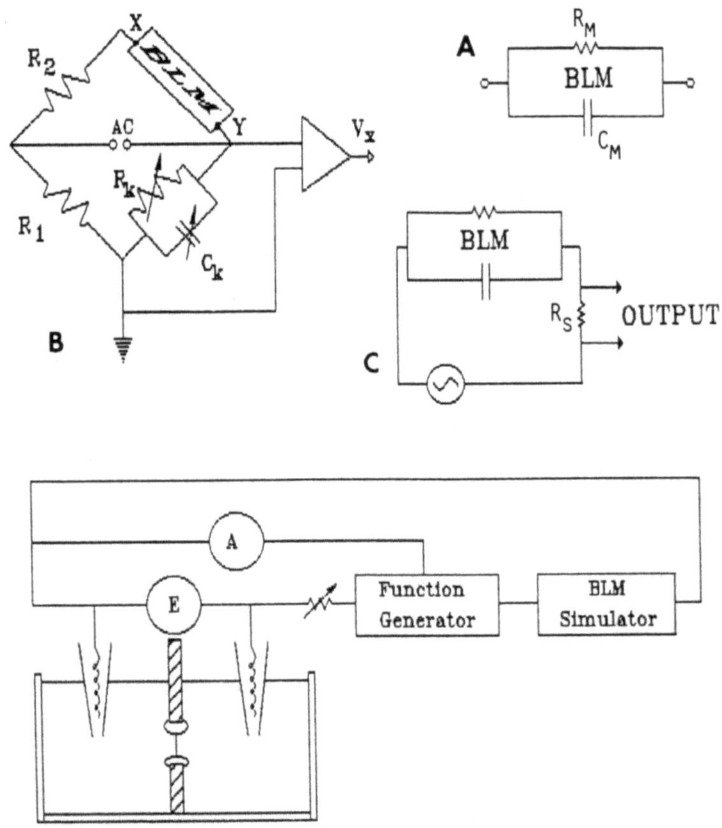

Figure 2. Equivalent circuit diagrams of BLM for the measurement of membrane electrical properties. A: Equivalent circuit of a BLM (R_m = membrane resistance, C_m = membrane capacitance), B and C: Circuits for membrane capacitance measurement. Lower figure is a block diagram of an overall setup.

monitored by means of an oscilloscope. In the ac bridge method, shown in Fig. 2, the C_m is evaluated using the equation

$$C_m = (R_1/R_2)\ C_k \tag{1}$$

where R_1 and R_2 are, respectively, 1 K Ω and 100 K Ω, and C_k is the value at the balance, at which terminal X is at a virtual ground so that at terminal Y the amplifier measures the ac signal appearing across the BLM. Alternatively, the C_m can be measured by a charge injection technique using commercially available instruments (pulse response analyzer, Taylor and MacDonald, 1987). A variable voltage source made of a 1.5 V dry cell and a precision 10-turn potentiometer (Helipot) is adequate to provide a selection of voltage to the BLM.

EXPERIMENTS RELEVANT TO DEVICE APPLICATIONS

In the past ten years or so, planar BLMs have been suggested to hold promise in the field of chemical/bio sensors for potential applications in medicine, industry, and clinical laboratories (Ivanov, 1988; Janata, 1987; Krull, 1987; Reichert, 1987; Tien et al., 1988; Carter and Wohtjen, 1988). In this section, past experiments relevant in the research and development of sensors and devices using BLMs will be summarized.

Ion-selective BLM. BLMs doped with 2,4-dinitrophenol (DNP) have been found to be sensitive to H^+ ions, where the DNP molecule acts as a carrier of H^+. Also BLMs formed from chloroplast extracts are highly specific to H^+ below pH 8, and in the pH range 4 to 7, a potential difference (E_m) as a function of $[H^+]$ obeys the Nernst equation. For ions other than H^+, polypeptide antibiotics such as valinomycin, enniatin, gramicidin, and others act at the 10^{-6} M level to decrease the BLM resistance by 10^4 to 10^8 fold. Further, these antibiotic-doped BLMs exhibit K^+ specificity with respect to other ions. Thompson et al. (1983) found a relationship between valinomycin concentration and membrane conductance with the limit of detection at 3×10^{-11} M (Krull, 1987). Another example is iodine-containing BLM specific to I^- ions (Krysinski and Tien, 1986; 1988). Along this line, it is evident that a variety of compounds could be incorporated into BLMs to make them useful in electrochemical/bio sensor development. In this connection, attempts have been made to fabricate BLM-based sensors (Thompson et al., 1983; Krull et al., 1986; 1987). These authors have deposited BLMs onto a polyacrylamide gel using the L-B technique. The use of BLMs as a sensing element for measurement of receptor activity and for drug response assay has been described by examples employing auxin receptor from maize, valinomycin, phloretin, and concanavalin. The stability of BLM can be improved by means of a microporous polycarbonate filter (i.e., Nuclepore membranes, see O'Boyle et al., 1982; Yoshikawa et al., 1987). Membranes deposited into these microporous filters exhibited switching and negative resistance phenomena (Yoshikawa et al., 1986). BLMs formed from synthetic lipids deposited on quartz oscillators as a model of chemical receptor have been also reported (Okahata et al., 1987). Mention should also be made of experiments of Kurihara et al. (1987), who have been investigating the transduction mechanisms of gustation and olfaction using lipid bilayers. Recently, use of the patch-clamp technique for study of the properties of single-ion channels of membranes is in vogue (Vassilev and Tien, 1988). Thus, it has been proposed that the function of single-ion channels such as their various conformation states may find applications in biosensor development (Williams et al., 1987; Cunnane et al., 1988).

BLM-based electrochemical/bio sensors for immunology. The ability of BLM to substitute for biomembranes in immune cytolytic phenomena have been first demonstrated by del Castillo and his colleagues in 1966 and later by others (for references and reviews, see Mountz and Tien, 1978; O'Boyle et al., 1982). The latter workers formed BLMs in microporous filters to enhance stability (Mountz and Tien, 1978). The basic principle is that any immunological reactions taking place at the BLM will alter the electrical properties which in turn will be converted into an electrical signal useful for assay purposes. Clearly, the successful development of voltametric immunoassay requires first of all, a BLM-based sensor which should be specific, sensitive, and stable. In this connection, it should be mentioned that BLM-based electrochemical/bio sensors could be also useful in cancer research (Ivanov, 1988; Vassilev et al., 1986; 1987).

BLM as bipolar electrodes. Unmodified BLMs are excellent insulators. However, upon suitable modification by incorporation of molecular iodine or π-electron rich complexes (TCNQ, tetracyanoquinodimethane; TTF, tetrathiafulvalene), BLMs have been shown to behave as a bipolar redox electrode (Blank, 1986; Ivanov, 1988; Marino, 1988). The very high electrical resistivity of BLM can be reduced dramatically by iodine and iodide. As already mentioned, the BLM can be made into an iodide-specific electrode by incorporating iodine into the BLM-forming solution. More recently, it has been found that the current is not due to movement of I^- across the BLM, but due to electrons (Krysinski and Tien, 1986; 1988). With TCNQ (or TTF) modified BLMs, oxidation takes place on the side of the membrane facing the negative electrode, with reduction occurring on the side facing the anode. A transverse movement of electrons occurs across the membrane; the BLM functions essentially as a bipolar redox electrode (Tien, 1974; Marino, 1988; Bockris and Diniz, 1988). In this regard semiconductors containing BLMs exhibit similarly interesting properties (Kutnik and Tien, 1987).

Long-lasting BLMs. For device development and applications, long-lasting BLMs have been formed by pyrrole polymerization (Kotowski et al., 1988). The polypyrrole-BLM exhibits a marked resistance against mechanical shock and a long-term stability. The electrical strength of pyrrole BLM has been determined by measurements of membrane breakdown voltage, which is over 600,000 volts per cm and about three times the value in comparison with a usual lecithin BLM. Cyclic voltammograms of pyrrole-BLM show two symmetric current peaks at applied voltages of ±356 mV, which are consistent with those reported for polypyrrole-coated gold electrodes.

Photoelectric BLMs. Photoactive BLMs in the last decade have found their place in fields other than membrane biophysics and physiology (Davison, 1985; Ivanov, 1988). Insights gained in studying photoactive BLMs have helped in designing more practical systems for light transduction (Worthy, 1988). We will describe our most recent efforts in relation to device applications (Kutnik and Tien, 1987; Salamon and Tien, 1988).

With BLMs formed from lecithin solution containing Mg-phthalocyanine, dyes or tetraphenylporphyrins (TPP) together with colloidal AgBr, considerable photoeffects were observed. The photovoltage/time curves are dependent upon the kind of redox compounds present (electron donors such as ascorbic acid and EDTA, and electron acceptors such as MV^{2+} and Fe^{3+}) in the bathing solutions and the photosensitizers in the BLM. Photovoltage/time curves for several combinations of photosensitizers and redox agents have been investigated. The combination of Mg-TPP in the lipid bilayer phase, photosensitizing dyes (malachite green and crystal violet) and redox agents (EDTA and MV^{2+}) in the aqueous solutions yielded the largest photoresponse. However, when only one of these substances was present alone in the system, there was no photoresponse. Certain dyes can act as photosensitizers or redox agents or both. The results indicate that there is a great latitude in experimentation. Further, experimental findings on pigmented BLMs show that an electric field across the membrane causes both a compression of the bilayer and an increase of its area. Both parameters affected by the electric field are connected with a capacitance of BLM giving its dependence on voltage across the membrane, which in turn is related to the difference in surface potentials on the two sides of the BLM (Blank, 1987; Ivanov, 1988). Therefore, the capacitance of pigmented BLMs (PBLM) should also be sensitive to the photovoltage changes. Thus, a model of the PBLM which takes into consideration light-induced changes of all basic electrical parameters (E_m, R_m and C_m) would be very useful not only from the theoretical point of view but would facilitate the experimental approach towards the application of BLMs in the basic research area of biomolecular electronic devices (BED). We have formed BLMs using thermotropic liquid crystals such as 4,4'-n- hexyl-cyanobiphenyl (6CB), 4,4'-n-octylcyanobiphenyl (8CB) or cholesteryl oleyl carbonate (ChoC) together with TCNQ as the photosensitizer. The BLM capacitance has been enhanced as much as five times in comparison with conventional BLMs (Salamon and Tien, 1988).

FUTURE PROSPECTS

The principal idea behind the development of BLM-based biosensors is quite simple. We envision that, for detection in biological environments, the sensing element should be biocompatible and be biomembrane-like. Thus, the bilayer lipid membrane (BLM) has great potential and is an ideal choice upon which to develop a new class of electrochemical/bio sensors. To illustrate our approach, the following case is used as an example. In immunochemical reactions, the antigen (Ag) that corresponds to antibody (Ab) to be detected is incorporated into a BLM. This modified BLM then becomes a sensing element specific for the antibody. Upon interaction between Ag and Ab, a change of some of the BLM's electrical properties is expected. With techniques now available, the electrical properties of the BLM such as the membrane potential, capacitance, resistance, dielectric breakdown voltage and other electrical parameters can be readily measured. With electrical methods, detection sensitivities of 10^{-9} M or better can be achieved. Biosensors based on BLMs, once successfully developed, would be very cheap with the added advantage of ease of measurements.

For biosensing purposes, BLM formed by the conventional method, however, has a number of problems, the most serious of which is the extreme fragility of the BLM. This shortcoming greatly limits its usefulness as a practical device since it cannot be easily fabricated and will not sustain rugged laboratory use. One successful attempt to overcome this fragility problem is being achieved in our laboratory by forming BLMs on either smooth substrates or freshly created metallic surfaces. We have found that the electrical properties of these BLMs are consistent with those BLMs formed by conventional techniques.

There are reports in the literature (Janata, 1987; Kell, 1987; Reichert et al., 1987; Van Roggen and Meijer, 1986; Ottova, 1988) in which field effect transistors (FET) in conjunction with lipid layers have been either proposed or employed in sensor devices development. Generally, FET carries current ("holes" or positive charges) through a silicon substrate from a "source" to a "sink". The flow of charges can be altered by a voltage applied to a third terminal (the so-called "gate") situated between the source and sink but insulated from the substrate, since the field alters the distribution of charge carriers. In principle, if the gate is replaced by an electroactive BLM (or an equivalent), voltage changes caused by charge accumulation by carriers in the membrane will also change the current which may be proportional to species concentration, thus allowing its quantification. Fig. 3 illustrates one of our approaches in developing BLM-based/FET devices. Also shown in Fig. 3 is the simplest possible arrangement for testing our ideas, namely, ligand-receptor contact interactions which are the principal mechanism underlying almost all physiological activities. Taking these into consideration, the heart of a BLM-based electrochemical device should be a modified lipid bilayer. For signal generation and transduction, the electrical properties of BLM are easily measurable. Since some of the modified BLMs are photoactive, a battery of spectroscopic techniques, therefore, can be coupled to the electrical measurements. From the device reliability standpoint, the ability to monitor simultaneously or sequentially a number of parameters in order to verify a positive test reaction will be our objective (Tien, 1988).

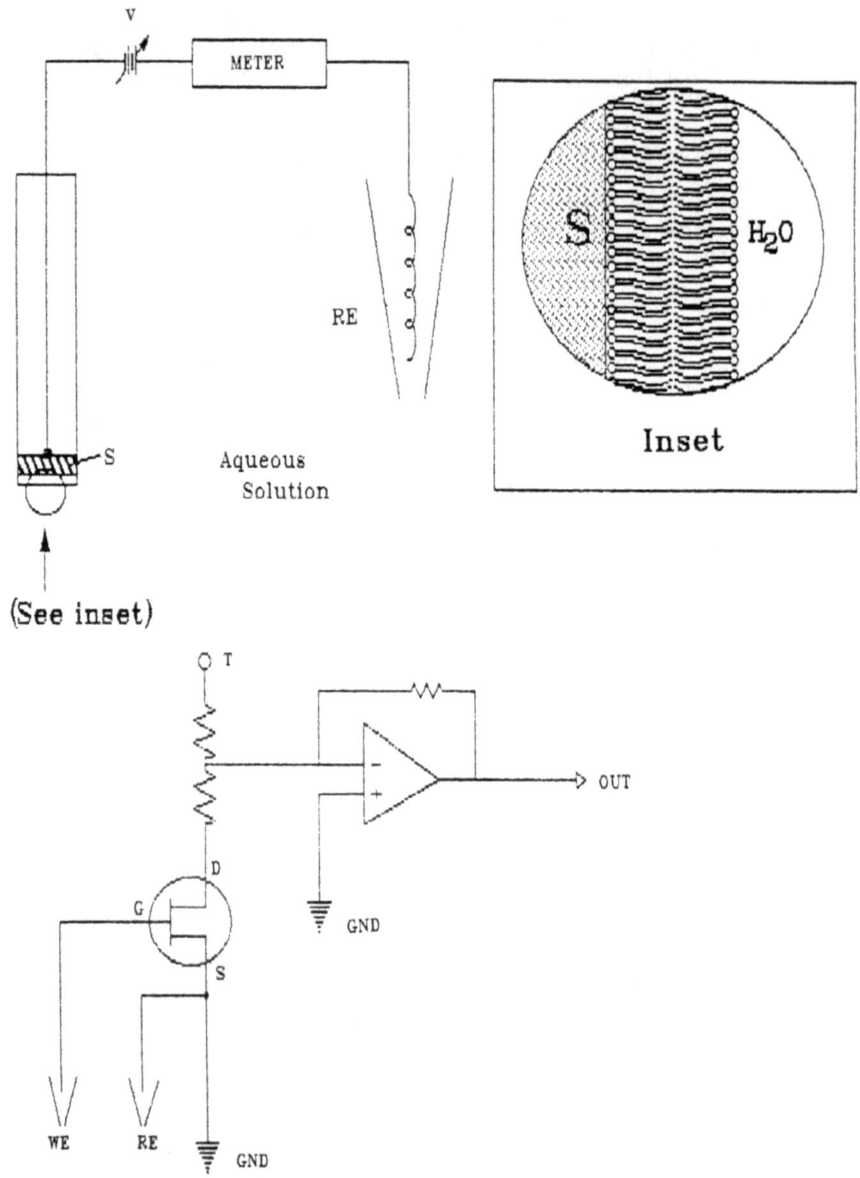

Figure 3. Schematic diagrams of BLM-based devices. Upper: A basic experimental arrangement. Lower: Representation of a BLM-based sensor with associated field effect transistor where the input (WE = working electrode) is connected directly to the gate electrode (G), D = drain (sink), S = source, RE = reference electrode, T: terminal to voltage source. Inset: Enlarged view of a bilayer lipid membrane (BLM) showing asymmetrical contacts (S, substrate and aqueous solution).

CONCLUDING REMARKS

As described in the preceding sections, a large number of modifiers which drastically alter the BLM properties have been incorporated into BLMs. For example, synthetic organic metals (TCNQ, TTF), finely divided semiconductor particles (CdSe, AgBr), and photoactive pigments (TPP, methylene blue), when incorporated into the lipid bilayer, endow non-linear electronic characteristics to the BLM system, some of which exhibit light-sensitivity. Moreover, metallic and/or semiconductive layers deposited on to the BLM surface can serve as electrodes directly contacting the membrane. Thus, in summary, the following statements can be made: (1) BLM formation is a self-assembling process; the structure is dynamic, fluid-like and self-sealing, (2) the BLM system mimics aspects of biomembranes in terms of structure and functions, (3) a BLM is a nanostructure of less than 6 nm thick and may be considered as a molecular junction free from the usual rigid, bulk phase effects, (4) both electronic and photoelectrical processes have been demonstrated in BLMs, and (5) the confined space of a BLM may display quantum size effects.

In concluding, it seems certain that the original BLM system together with its modification described here can provide unique experimental opportunities in research and development of bilayer-based biosensors and molecular electronic devices. To accomplish these goals, interdisciplinary collaboration is essential. It is in this spirit that this paper has been written.

ACKNOWLEDGMENTS

The authors are grateful for financial support from the ONR and the NIH. Thanks are due Debbie Benedict for her superb secretarial assistance.

REFERENCES

Blank, M., ed., 1986, "Electrical Double Layers in Biology," Plenum Press, New York, pp. 129-166.
Blank, M., and Findl, E., eds., 1987, "Mechanistic Approaches to Interaction of Electric and Electromagnetic Fields with Living Systems," Plenum Press, New York, pp. 301-324.
Burton, R. M., ed., 1968, J. Am. Oil Chemists' Soc., 45:201.
Carter, F. L., and Wohtjen, H., eds., 1988, "Proc. Third Int'l. Symposium on Molecular Electronic Devices," North-Holland, Amsterdam.
Cunnane, V. J., Schiffri, D. J., Fleischmann, M., Geblewic, G., J. Elec. Chem., 243(2):455.
Danielli, J. F., Pankhurst, K. G. A., and Riddiford, A. C., eds., 1964, "Recent Progress in Surface Science," Vol. 1, Academic Press, New York, pp. 379-393.
Davison, S. G., ed., 1985, "Progress in Surface Science," Vol. 19, No. 3, Pergamon Press, New York, pp. 169-274.
Dilger, J. P., Fisher, I. R., and Haydon, D. A., 1982, Chem. Phys. Lipids, 30:159.
Dryhurst, G., and Niki, K., eds., 1988, "Redox Chemistry and Interfacial Behavior of Biological Molecules," Plenum, New York.
Exerowa, D., and Kashchiev, D., 1986, Contemp. Phys., 27(5):429.
Ivanov, I., ed., 1988, "Thin Liquid Films," Marcel Dekker, Inc., New York.
Janata, J., 1987, in: "Proc. Sym. Chemical Sensors," D. R. Turner, ed., The Electrochemical Society, Inc., Pennington, N. J., pp. 258.
Kell, D. B., 1987, in: "Biosensor Fundamentals and Applications," A. P. F. Turner, I. Karube, G. S. Wilson, eds., Oxford Press, Oxford, p. 427.
Kotowski, J., Janas, T., and Tien, H. T., 1988, J. Electroanal. Chem., 19:277, 283.
Kurihara, K., Nomura, T., Kashiwayanagi, and Kumazawa, T., 1987, "Conf. on Solid State Sensors," Tokyo, p. 569.
Krull, U. J., 1987, Anal. Chim. Acta, 197:203.
Krull, U. J., Thompson, M., and Wong, H. E., 1986, in: "Fundamentals and Applications of Chemical Sensors," D. Schuetzle, R. Hammerle, J. W. Butler, eds., Am. Chem. Soc., Washington D. C., p. 351.
Krysinski, P., and Tien, H. T., 1986, Bioelectrochem. Bioenerg., 16:185; 1988, 19:227.
Kutnik, J., and Tien, H. T., 1987, Photochem. Photobiol., 46:413; 46:1009.

Marino, A. A., ed., 1988, "Modern Bioelectricity," Marcel Dekker, Inc., New York, p. 181-241.
Mittal, K. L., ed., 1988, "Surfactants in Solution," Plenum, New York.
Mountz, J. D., and Tien, H. T., 1978, J. Bioenerg. Biomembr., 10:139.
Mountz, J. M., and Tien, H. T., 1978, J. Photochem. Photobiol., 28:395.
Mueller, P., Rudin, D. O., Tien, H. T., and Wescott, W. C., 1963, J. Phys. Chem., 67:534.
O'Boyle, K., Siddiqi, F. A., and Tien, H. T., 1984, Immunological. Commun., 13:85.
Okahata, Y., Ebato, H., Taguchi, K., Shimizu, O., and Shioji, G., 1987, "Ann. Meet. Chem. Soc. Jap.," 26.
Ottova, A., 1988, Proc. Second Bionka Conf., Bratislava, May 24-26. Abst., p. 54.
Pethig, R., and Kell, D. B., 1987, Phys. Med. Bi., 32:933.
Reichert, W. M., Bruckner, C. J., and Joseph, J., 1987, Thin Solid Films, 152:345.
Ringsdorf, H., Schlarb, B., and Venzmer, J., 1988, Angen. Chem., 27:113.
Salamon, Z., and Tien, H. T., 1988, Molecular Cryst. Liq. Cryst., 154:195; 1988, Liquid Crystals, 3:169; 1988, Photochem. Photobiol., 48:281.
Takagi, M., Azuma, K., and Kishimoto, U., 1965, Ann. Rpt. Biol. Works Fac. Sci., Osaka Univ., 13:107; see also Tien, H. T., p. 477.
Taylor, D. M., and MacDonald, A. G., 1987, J. Phys. D: Appl. Phys., 20:1277.
Thompson, M., Krull, U. J., and Bendell-Young, L. I., 1983, in: "Proc. Int'l. Meeting on Chemical Sensors," T. Seiyama, K. Fueki, J. Shiokawa, and S. Suzuki, eds., Elsevier-Kodansha, New York and Tokyo, pp. 576-581.
Tien, H. T., 1974, "Bilayer Lipid Membranes (BLM): Theory and Practice," Marcel Dekker, Inc., New York.
Tien, H. T., 1988, J. Surface Sci. Technol., 4:1.
Tien, H. T., Salamon, Z., Kutnik, J., Krysinski, P., Kotowski, J., Ledermann, D., and Janas, T., in press, 1988, J. Mol. Electronics, 4.
Van Roggen, A., and Meijer, P. H. E., 1986, "Abst. 3rd. International Sym. on Molecular Electronic Devices, October 6-8, 1986, Arlington, VA," p. 9.
Vassilev, P. M., and Tien, H. T., 1988, in: "Subcellular Biochemistry," Vol. 14, J. R. Harris and A.-H. Etemadi, eds., Plenum, New York.
Vassilev, P. M., Kanazirska, M., Charamella, L. E., Dimitrov, N. V., and Tien, H. T., 1986, Cancer: Biochem. Biophys., 9:85; 1987, Cancer Research, 47:519.
White, S. H., 1978, Biophys. J., 23:337.
Williams, D. E., Walker, A., Chadwick, A., and Fleischmann, M., in press, 1988, "Sym. Biosensor Tech.," (private communication, 1987).
Worthy, W., 1988, Chem. Eng. News, 66:32.
Yoshikawa, K., Omochi, T., Fujmoto, T., and Terada, H., 1986, J. Coll. Int. Sci., 113:585.
Yoshikawa, K., Hayashi, H., Shyimooka, T., Terada, H., and Ishii, T., 1987, Biochem. Biophys. Res. Commun., 145:1092.

BIOMOLECULAR INTERFACE

Masuo Aizawa, So-ichi Yabuki, and Hiroaki Shinohara

Department of Bioengineering, Tokyo Institute of Technology

O-okayama, Meguro-ku, Tokyo 152

INTODUCTION

Bioelectronic devices (BED) may be classified into three categories;
(1) Molecular electronic devices processed by biochemical methods,
(2) Molecular electronic devices composed of biomolecules or biomimetic molecules,
(3) Molecular electronic devices modeled on information-processing in the living systems.
Of these bioelectronic devices, there has been intensive research on the second category of bioelectronic devices.

A variety of biomolecules may be candidates for the components of bioelectronic devices. These include chlorophylls, rhodopsins, enzymes, and other functional proteins. The biomolecules may be incorporated into the bioelectronic devices to transduce and transfer molecular, photonic, and electronic information. Due to the limit of electronic device technology, the biomolecules should be primarily or semiconductive materials, which might be advantageous in signal processing. The biomolecules, especially proteins, find difficulties in communicating with electronic conductive materials such as metal and semiconductor solid surfaces[1-5]. Protein molecules might be designed to make an intermolecular communication through a specific site of the molecule mostly situated within the concaved structure.

The concept of biomolecular interface has been proposed to make a smooth communication between such biomolecules as proteins and the electronic conductive solid surface (Fig. 1). There are several possible biomolecular interfaces, which may grouped into; (1) Electron mediators, (2) Electron promoters, and (3) Molecular wires (Fig. 2). The electron mediator transfers an electron by a redox reaction of the biomolecule in association with a redox reaction at the electronic conductive solid surface[6-9]. On the other hand, the electron promoter transfers no electron by itself, but promotes the electron transfer between the biomolecule and the electronic conductive solid surface primarily by maintaining the proper conformation of the biomolecule[10,11]. The molecular wire electronically connects the redox center of the biomolecule and the electronic conductive solid surface through such conductive molecular chains as a conjugated double bond polymer.

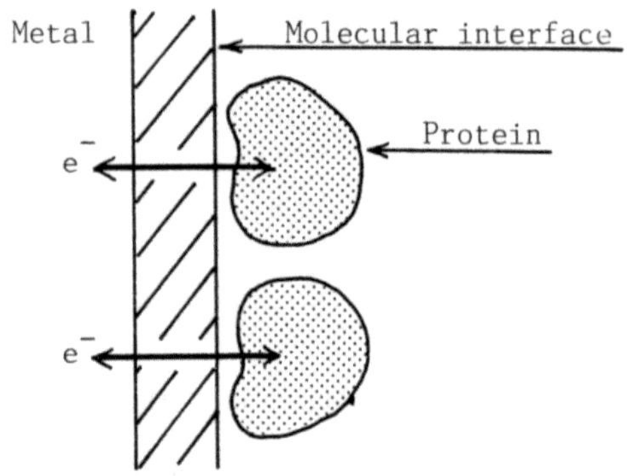

Fig. 1 Schematic illustration of biomolecular interface.

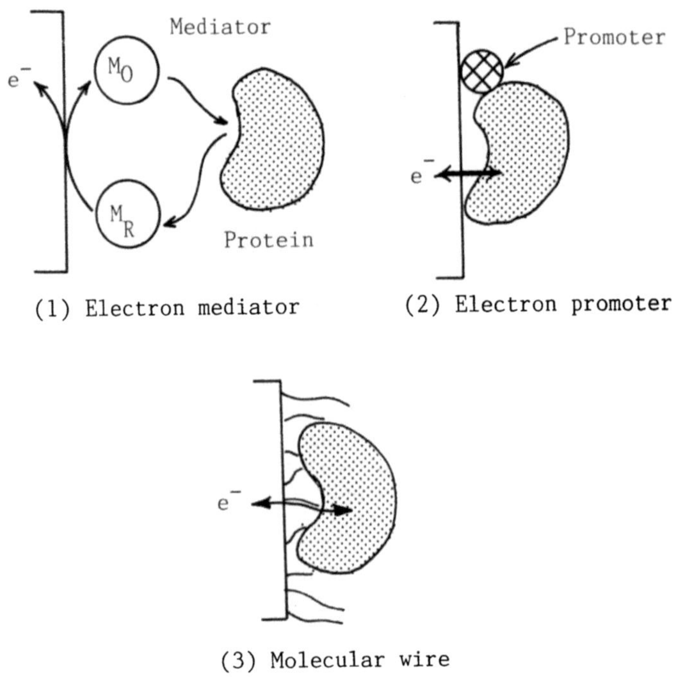

Fig. 2 Several possible biomolecular interface. Biomolecular interface may be grouped into three types.

In this paper a molecular wire of biomolecular interface has been fabricated in a monomolecular level of thickness to demonstrate the electron transfer between redox enzyme molecules and the metal electrode surface. Glucose oxidase was used as a model of redox enzyme molecules. It has two flavin adenine dinucleotides (FAD) per molecule. Since these FADs of glucose oxidase find difficulties in transferring electrons to the metal electrode surface, these have long been intensive investigations on the electron transfer of glucose oxidase in association with the electron mediators and the electron promoters. The electron mediator has covalently been conjugated with glucose oxidase by Heller et al[12]. These investigations have indicated that glucose oxidase in solution can communicate with the electrode surface through electron transfer.

Our strategy has been directed to fabricate a conductive enzyme membrane using a conductive polymer as a molecular wire. The membrane should be at a monomolecular protein level in thickness with retained enzyme activity. Glucose oxidase was adsorbed on the platinum electrode surface in monolayer by the potential-controlled method. The electrode was then contacted with a pyrrole solution, which was followed by potential-controlled electrochemical polymerization of pyrrole. Polypyrrole deposited on the platinum electrode surface to form a thin membrane incorporating adsorbed glucose oxidase. It is noted that the electrochemically synthesized glucose oxidase/polypyrrole membrane shows electric conductivity with retained enzyme activity, and polypyrrole works as a molecular wire between the redox center of glucose oxidase and the electrode surface.

EXPERIMENTAL

A platinum plate electrode (5 x 10 mm^2) was electrochemically characterized by neat and clean surface. The electrode was soaked in a citrate buffer solution (0.1 M, pH 5.5) containing glucose oxidase. The electrode potential was controlled at +0.5 V vs. Ag/AgCl and 25°C for 30 min to achieve a monomolecular layer of adsorption.

The electrode was then rinsed and transferred into a potassium chloride solution (0.1 M) containing pyrrole. The electrode potential was controlled at +0.7 V vs. Ag/AgCl to initiate electropolymerization of pyrrole. The electrolysis was stopped at a passed charge of 200 $\mu C \cdot cm^{-2}$. The membrane thickness was estimated to be 0.8 nm, which might correspond to the one fourth of the monolayer of glucose oxidase. The electrode was thoroughly washed with a citrate buffer solution.

The electrochemical measurements were made using a glass cell equipped with a working, counter, and reference electrodes. A potentiostat (Hokuto Denko, Model HA-301) pulse voltammetric analyzer (Yanagimoto, Model P-1100), and function generator (Hokuto Denko, Model HB-105) were used.

RESULTS AND DISCUSSION

Potential-controlled Adsorption of Enzyme

Glucose oxidase is negatively charged in a pH 5.5 citrate buffer solution because it has an isoelectric point of 4.2. It is presumed that glucose oxidase should adsorb on the positively charged electrode surface, although such polyelectrolytes as proteins might behave in a sophisticated manner of adsorption.

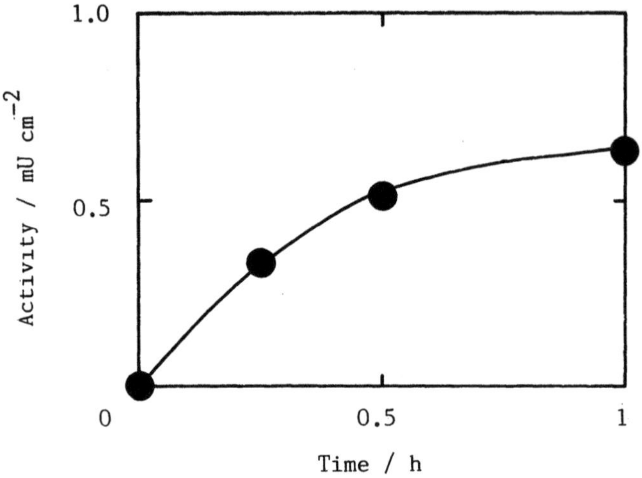

Fig. 3 The enzyme activity of the adsorbed glucose oxidase on the platinum plate suraface.

Glucose oxidase was adsorbed on the platinum electrode surface of which potential was controlled at +0.5 V vs. Ag/AgCl. Adsorbed glucose oxidase was colorimetrically assayed for enzyme activity using glucose, peroxidase, 3,3'-dimethoxybenzidine in a pH 5.5 citrate buffer solution. Adsorbed glucose oxidase increased as was shown in Fig. 3, reaching a saturated state in 30 min.

On the assumption that adsorbed glucose oxidase retained full activity of negative enzyme in solution, glucose oxidase molecules might be adsorbed with 60 % surface covered at the saturated state. It indicates that a monolayer protein membrane should be accomplished by the potential-controlled method.

Enzyme Activity of Glucose Oxidase/Polypyrrole Membrane

Polypyrrole was electrochemically synthesized in the intermolecular spaces of adsorbed glucose oxidase. Polymerization should be originated on the electrode surface, which might be followed by elongation of polypyrrole chains into the vicinity of the redox center and the surroundings of glucose oxidase. The cleaves of the redox center is supposedly charged in negative so that the positively charged polypyrrole chain has a great possibility to penetrate into the vicinity of the redox center.

Polymerization was stopped when the intermolecular spaces of adsorbed glucose oxidase was packed. A glucose oxidase/polypyrrole membrane is schematically illustrated in Fig. 4.

The glucose oxidase/polypyrrole membrane was assayed for enzyme activity in the similar manner as described for electrode-adsorbed glucose oxidase, which indicated that polypyrrole-entrapped glucose oxidase retained enzyme activity.

Electron-transfer of Glucose Oxidase/Polypyrrole Membrane

Differential pulse voltammetry was carried out with the glucose oxidase/polypyrrole membrane adhered on the platinum electrode surface to

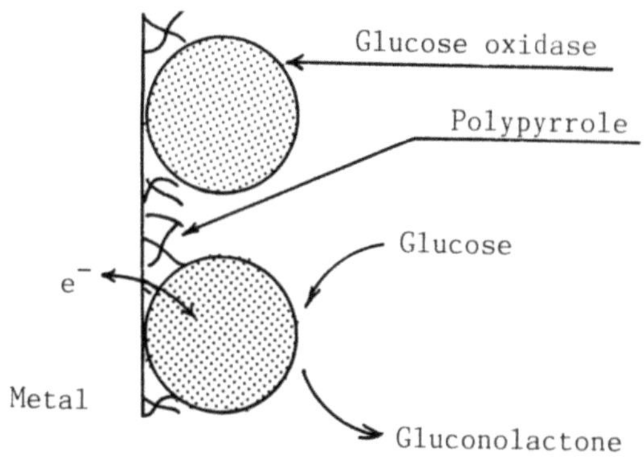

Fig. 4 Schematically illustration of polypyrrole/glucose oxidase membrane.

elucidate the electron transfer from the electrode to the redox center of glucose oxidase via polypyrrole. The glucose oxidase/polypyrrole membrane electrode was set in a pH 5.5 citrate buffer solution containing potassium chloride. Potential was scanned from -0.2 V to -0.4 V vs. Ag/AgCl with a pulse interval of 1.0 s.

Figure 5 represents the differential pulse voltammogram, indicating that a cathodic current peak appears at -0.3 V vs. Ag/AgCl. The peak potential was attributed to the electrochemical reduction of the FAD of glucose oxidase entrapped in polypyrrole with reference to that of polypyrrole-entrapped FAD.

These results clearly show that it is possible to transfer electrons between glucose oxidase/polypyrrole membrane in an conductive enzyme membrane.

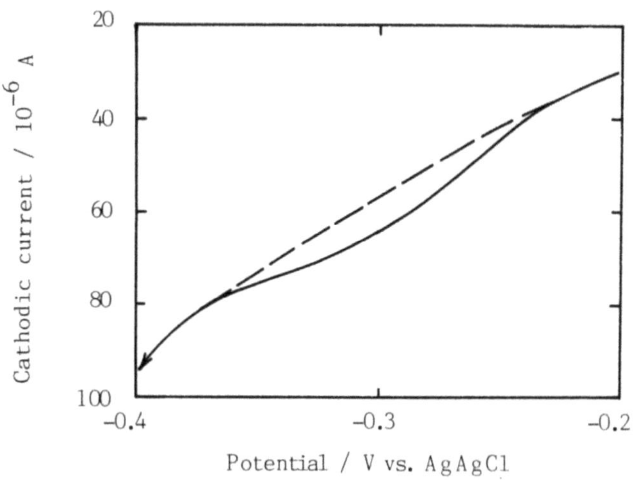

Fig. 5 Differential pulse voltammograms of polypyrrole/glucose oxidase membrane (———) and platinum plate(— — —).

On the basis of the differential pulse voltammetry, the electronic communication between glucose oxidase and the electrode was demonstrated. Glucose oxidase is endowed with a molecular function of highly selective molecular recognition. In case that polypyrrole-entrapped glucose oxidase recognizes β-D-glucose in solution, its oxidation catalytically proceeds with consumption of dissolved oxygen. No further oxidation of glucose, however, takes place without dissolved oxygen unless oxidized FAD is regenerated form reduced FAD of glucose oxidase.

The glucose oxidase/polypyrrole membrane electrode was immersed in a pH 5.5 citrate buffer solution containing potassium chloride with a counter electrode and a Ag/AgCl reference electrode. The solution was bubbled with N_2 gas to be deoxygenated prior to the experiment. The electrode potential was controlled at 0.0 V vs. Ag/AgCl which might be sufficient to oxidize the reduced form of FAD to the oxidized form.

After a steady anodic current was measured, 3.0 ml of 1.0 M glucose solution was added to the solution (30 ml). The anodic current immediately increased with addition of glucose, which resulted from the oxidation of reduced form of FAD due to enzymatic oxidation of glucose in solution. The anodic current reached a steady state in 6 min. The current increase at a steady state was linearly corrected with the concentration of glucose.

These results show that the molecular function of glucose oxidase is transduced to the electron flow of the electrode with the aide of polypyrrole.

REFERENCES

1 K. Niki, T. Yagi, H. Inokuchi, K. Kimura, Electrode Reaction of Cytochrome c_3 of Desulfovibrio vulgaris, Miyazaki, J. Electrochem. Soc., 124: 1889 (1977).
2 B. A. Kuznetsov, N. M. Mestechkina, G. P. Shumakovich, Electrochemical Behavior of Proteins Containing Coenzyme Groups and Metals, Bioelectrochem. Bioenerg., 4: 1 (1977).
3 H. Durliant, M. Comtat, Amperometric Enzyme Electrode for Determination of Glucose Oxidase, Anal. Chem., 56: 148 (1984).
4 F. A. Armstrong, H. A. O. Hill, B. Nigel Oliver, D. Whitford, Direct Electrochemistry of the Photosynthetic Blue Copper Protein Plastcyanin. Electrostatic Promotion of Rapid Charge Transfer at an Edge-Oriented Pyrolytic Graphite Electrode, J. Am. Chem. Soc., 107: 1473 (1985).
5 O. Miyawaki, L. B. Wingard,Jr, Electrochemical and Enzymatic Activity of Flavin Adenine Dinucleotide and Glucose Oxidase Immobilized by Adsorption on Carbon, Biotechnol. Bioeng., 26: 1364 (1984).
6 M. Senda, T. Ikeda, T. Kakutani, K. Kano, H. Kinoshita, Polarographic Studies on Proteins, Bioelectrochem. Bioenerg., 8: 151 (1981).
7 B. A. Parkinson, P. F. Weaver, Photoelectrochemical Pumping of Enzymatic CO2 Reduction, Nature, 309: 148 (1984).
8 A. E. G. Cass, G. Davis, D. Francis, H. A. O. Hill, W. J. Aston, I. J. Higgins, E. V. Plotkin, L. D. L. Scott, A. P. F. Turner, Ferrocene-Mediated Enzyme Electrode for Amperometric Determination of Glucose, Anal. Chem., 56: 667 (1984).
9 J. M. Johnson, H. Brian Halsall, W. R. Heineman, Redox Activation of

Galactose Oxidase : Thin-Layer Electrochemical Study, Biochemistry, 24: 1579 (1985).
10 I. Taniguchi, T. Funatsu, K. Umekita, H. Yamaguchi, K. Yasukouchi, Effect of Poly-L-Lysine Addition of the Redox Behavior of Horse Heart Cytochrome c at Functional Electrode, J. Electroanal. Chem., 199: 455 (1986).
11 H. A. O. Hill, D. J. Page, N. J. Walton, Surface Substitution Reactions at Modified Gold Electrodes and Their Effect on the Electrochemistry of Horse Heart Cytochrome c, J. Electroanal. Chem., 217: 141 (1987).
12 Y. Degani, A. Heller, Direct Electrical Communication between Chemically Modified Enzymes and Metal Electrodes, J. Phys. Chem., 91: 1285 (1987); Y. Degani, A. Heller, Direct Electrical Communication between Chemically Modified Enzyme and Metal Electrode. 2. Method for Bonding Electron-Transfer Relays to Glucose Oxidase and D-Amino Acid Oxidase, J. Am. Chem. Soc., 110: 2615 (1988).

ON THE VOLT-AMPERE CHARACTERISTICS OF MOLECULAR MONOELECTRONIC ELEMENTS

Oleg V. Gritsenko and Pavel I. Lazarev

Research Computing Center
U.S.S.R. Academy of Sciences
Pushchino, Moscow Region 142292 U.S.S.R.

INTRODUCTION

The development of monoelectronic circuitry[1,2] suggests that molecular materials could replace traditional materials and production means. That should allow circuit elements as small as a molecule (10 Å) to be made.

The central element of a monoelectronic circuit is a tunnel transition (TT) which is an ensemble of ultrasmall donor-acceptor systems (DAS) separated by dielectric layers. In those ensembles, a large enough external voltage applied to the terminal DAS (electrodes) draws a current of single electrons through the TTs as a result of the tunneling of electrons through dielectric barriers. Two types of TTs were used in the first experimental monoelectronic devices made of traditional materials.

Kuzmin and Likharev[3] described a TT that consisted of a layer of oxide between two metal macroelectrodes that contained a metal granule the size of 1,000 Å.

The monoelectronic transistor, made simultaneously by Fulton and Dolan,[4] was based on electron-beam nanolithographs with the estimated level of miniaturization at 10^4 Å^2 per circuit.[5]

New prospects were opened in monoelectronics that made it possible to create TTs made of molecular systems used as DASes and dielectric fragments.

The charge transfer processes in such molecular systems can in principle be made compatible with quasi-continuous current in common conductors. The two newly built monoelectronic devices mentioned above illustrate the possibility of such a connection.

We consider such a contact between a molecular ensemble of any complexity that fulfills logical functions with monoelectronic signal and common conductor with quasi-continuous current a necessary part of a computing device based on molecular-scale elements. In our previous papers[6-8] we suggested that proteins from electron transport chains could be an appropriate molecular material for TTs. Their donor-acceptor centers (DAC) contain either single metal ions or whole clusters of them and can be regarded as analogs of the granule in Ref. 3. In this case, the "granule" is much smaller than in those in Ref. 3 and consists of 1-4 metal atoms instead of several thousand. The tunnel contact is made by the protein "envelope" of DAC which adjoins another protein molecule or electrode. That the mechanism of electron transfer between the DACs in proteins is tunneling is proved beyond doubt.[9]

The aim of this paper is to study the similarities and differences of charge transfer processes in the simplest traditional and molecular monoelectronic elements.

To compare the two kinds of element we consider, for the sake of simplicity, a system consisting of two electrodes (the carriers of quasi-continuous current, e.g. metal plates), and a DAS positioned between them that could be either a metal granule of the 10^4 Å size or a molecule the size of 10 Å. The DAS is connected with the electrodes by tunnel interaction. The element under study can be denoted

M1 - (x) - M2,

where M1, M2 are the electrodes, x is the DAS, and the parentheses denote a tunnel contact.

In the first section of this paper, we will examine features of the electron transfer process typical of the two systems above which is due to a set of discrete DAS states that take part in the tunneling of an electron. There we will introduce an equation for volt-ampere characteristic (VAC) of a TT.

In the second section, we will analyze the differences in electron-transfer processes caused by differences of the corresponding discrete electron states.

In the third section, we will discuss prospective properties of molecular monoelectronic devices. They arise because of the aforementioned considerable difference in size between molecules and metal granules of the existing TT specimens.

SIMILARITIES IN CHARGE TRANSFER PROCESSES IN TUNNEL CONTACTS OF GRANULES AND MOLECULAR CLUSTERS

One common feature of all systems M1-(x)-M2 where x is either a metal granule or a molecular cluster is the existence of discrete physical parameters z_i which determine the state of x in the course of tunnel transfers of an electron M1 $\rightleftharpoons$ x and x $\rightleftharpoons$ M2. Here the energy of x that corresponds to the values of z_i forms a discrete spectrum $\{E_{zi}\}$.

Those parameters include the charge z_i of a granule or a molecular cluster which can be changed only by an integer number of electron charges in the process of charge transfer by single electron tunneling. Other discrete parameters for molecular clusters with the charge z_i are quantum numbers $\{i\}$ of the ground state and the lowest excited electron-vibrational states of x that, together with z_i, determine the energy spectrum, $\{E_{zi}\}$.

When the size of the granule is about 1,000 Å the boundary electronic states $\{i\}$ for a definite z_i form a quasi-continuous spectrum[10] so that the granule charge is the only parameter determining the energy spectrum of the granule states, $\{E_z\}$.

More detailed characteristics of E_z and E_{zi} along with the analysis of their differences are given in the next section. Here we only want to dwell on common features or charge transfer in various monoelectronic TTs which allow us to treat them within the framework of a single approach.

Such an approach for the case of the state of x in a TT being determined by discrete parameters z was developed in Refs. 10-12. Here we are only presenting the basic expressions used to deduce the VAC of TT M1-(x)-M2. The final expression for VAC somewhat differs from that in Ref. 11 since this form is more convenient for further applications to molecular clusters.

The current I through the TT with a constant voltage applied to it was calculated in Refs. 10 and 11 as a sum over all possible values of partial rates $G^{\pm}(E_z^{\pm}[x\text{-}M2])$ of direct and reverse single-electron transfer for the state of x with the charge z weighed by the states populations, W_z:

$$I = e \ {}_1\sum_{z} W_z \{G^-(E_z^-[x\text{-}M2]) - G^+(E_z^+[x\text{-}M2])\}. \qquad (1)$$

In Eq. 1, W_z is the probability density for x to be in the state with the charge z, $G^+(E_z^+[x\text{-}M2])$ is the rate of direct electron transfer from x to M2 with z increasing by unity, and $G^-(E_z^-[X\text{-}M2])$ is the rate of electron transfer from M2 to x with z decreasing by unity. The symbol 1Σ means that the consecutive values of z in the sum in Eq. 1 differ by 1.

Partial transfer rates $G^{\pm}(E_z^{\pm}[x\text{-}M2])$ are functions of the energy of the single-electron transfers x ← M2 and x → M2 which can be presented as

$$E_z^+[x\text{-}M2] = E_z - E_{z+1} + I(M2) + eV_{xM2} = -I_{z+1}(x) + I(M2) + eV_{xM2} \qquad (2)$$

$$E_z^-[x\text{-}M2] = E_z - E_{z-1} - I(M2) - eV_{xM2} = I_z(x) - I(M2) - eV_{xM2}. \qquad (3)$$

In Eqs. 2 and 3, I_z is the ionization potential of x with the charge z-1 (if z is an integer number, I_z is the consecutive z-s ionization potential of x), I(M2) is the electronic work function from the metal M2, and V_{xM2} the voltage drop in the sub-circuit x--M2.

For partial rates $G^{\pm}(eU)$ as a function of the single-electron transfer energy, eU, the following expressions were used:

$$G^{\pm}(eU) = 1/e \ \ I(U) \exp(\pm eU/2kT)/\{\exp(eU/2kT)+\exp(-eU/2kT)\}, \qquad (4)$$

where I(U) is a classical VAC of a transfer that does not account for the transferred charge being discrete

$$I(U) = U/R_{xM2}, \qquad (5)$$

and the ratio of direct and reverse currents that give I(U) on summation is determined by the Boltzmann factor

$$G^+(eU)/G^-(eU) = \exp(eU/kT). \qquad (6)$$

When x is a metal granule, the formula Eq. 4 was derived by a quantum mechanical method of tunnel Hamiltonian[13] in which the following formula is valid for the resistance R_{xM2}:

$$1/R_{xM2} = 4\pi e^2 \sum_{pq} T_{pq}^2 \ \delta(\varepsilon_p - \mu_x) \ \delta(\varepsilon_q - \mu_{M2}) = 4\pi e^2 N_x N_{M2} \bar{T}_{xM2}^2. \qquad (7)$$

In Eq. 7, $\{\varepsilon_p\}$ and $\{\varepsilon_q\}$ are the spectra of the single-electron transfer energy x and M2; μ_x, μ_{M2}, and N_x, N_{M2} the corresponding chemical potentials and states densities on the Fermi surface, T_{pq} the matrix element of interaction between the states p and q. However, the formula Eq. 7 was not used in Ref. 11 and an empirical R_{xM2} was used for the numerical estimates of VAC.

For the case of x being a molecular cluster, the formula Eq. 4 can be used for calculation of partial transfer rates from the states z_i as a quasi-classical approximation.

The populations W_z were found from the condition of stationarity of charge transfer that in the accepted notation is

$$F_{z-1} - F_z = 0 \tag{8}$$

$$F_z = W_z \{G^+(E_z[x\text{-}M1]) + G^+(E_z[x\text{-}M2])\} -$$

$$- W_{z+1} \{G^-(E_{z+1}[x\text{-}M1]) + G^-(E_{z+1}[x\text{-}M2])\}. \tag{9}$$

The solution of Eq. 8 is sought in the form

$$F_z = K, \tag{10}$$

where K as a constant determined from the normalization equation

$$1 \sum_z W_z = 1. \tag{11}$$

As an exact solution cannot be found in the general case, we consider the physically realistic case of an asymmetric TT M1-(x)-M2 with the rates of tunneling x ⇄ M1 and x ⇄ M2 differ greatly, that is, the resistances R_{xM1} and R_{xM2} in Eq. 7 are related as $R_{xM1} << R_{xM2}$.

In this case, we can solve Eq. 10 by iteration writing W_z as follows:

$$W_z = W_z^o + (R_{xM1}/R_{xM2})\, W_{z1} + \dots . \tag{12}$$

The populations W_z^o that satisfy Eq. 10 with the additional condition Eq. 11 when M1 and M2 are identical metals and I(M1)=I(M2)=I(M), have the form

$$W_z^o = \exp(-\tilde{E}_z/kT) / \sum_y \exp(-\tilde{E}_y/kT), \quad y = z-1,\ z,\ z+1. \tag{13}$$

$$\tilde{E}_z = E_z - E_o + z[eV_{xM1} - I(M)]. \tag{14}$$

$\tilde{E}_z$ in Eqs. 13 and 14 is the energy of the system x+M1 with the charge z on x, and is measured relative to a system with a zero charge on x.

Thus, the set $\{W_{zi}\}$ in the zero approximation corresponds to the Gibbs distribution for solids x and M1. Hence a physically evident conclusion follows that for an asymmetric contact the electronic distribution of the system x is controlled by the equilibrium of x with the "nearest" of the two metals (meaning R_{xM1} is smaller than R_{xM2}), and the relatively rare transfers of electron to the "farther" metal M2 upset the equilibrium thus generating a current.

If we determine W_{z1} from Eqs. 10 and 11 and substitute into them the expressions for W_z obtained by solving Eq. 12, we get the following expressions for current I through the asymmetric TT M1-(x)-M2 at fixed voltage V:

$$I = sh(eV/2kT)/[eR_{xM2} \ 1\sum_y \exp(-\tilde{E}_y/2kT)]$$

$$1\sum_z \exp\left((-\tilde{E}_z/2-\tilde{E}_{z+1}/2)/kT\right) \{(-I_z(x) + I(M2) + eV_{xM2})/$$

$$/sh[(-I_z(x) + I(M2) + eV_{xM2})/2kT]\}. \qquad (15)$$

The VAC derived with discrete effect taken into account differs from the classical VAC by a nonmonotonous behavior of the current I dependence of the voltage applied, V. Such a VAC is said to have negative differential conductivity (Fig. 1). This is due to the term in the brackets in Eq. 15 which reaches local maxima at voltage values $V_{xM2,max}$ derived from the conditions that the energy of single-electron transfers to M from x state with a definite z be equal to zero:

$$E_z^+[x\text{-}M1] = -I_{z+1}(x) + I(M2) + eV_{xM1} = 0. \qquad (16)$$

In other words, the local maxima of VAC correspond to the resonance tunneling from states with definite values of z, the term resonance denoting equality of the energy $E_z{}^+[x\text{--}M2]$ to zero. Such VAC is a characteristic feature of all monoelectronic TTs with both granules and molecular clusters. In the next Section, we examine their differences and introduce a special case of Eq. 15 for x being a molecular cluster.

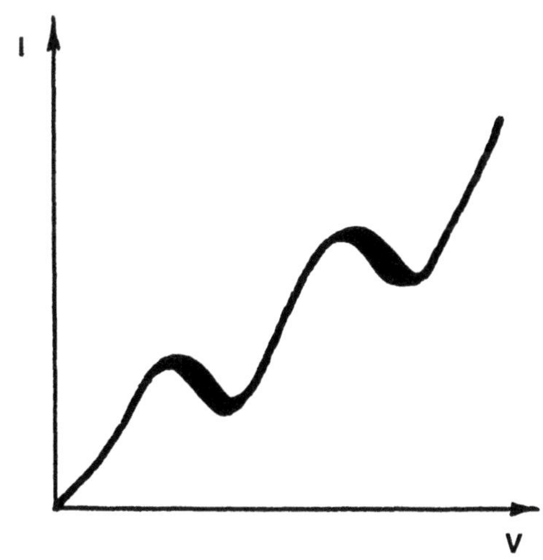

Fig. 1. VAC of a TT with the parts (bold) of negative differential conductivity.

DIFFERENCES OF THE CHARGE TRANSFERS IN TUNNEL CONTACTS OF GRANULES AND METAL ELECTRODES

The different nature of discrete states of x is responsible for the differences of charge transfer processes in the monoelectronic TT M1--(X)--M2 where x is a metal granule or molecular cluster. This is caused by the difference in sizes of a granule (10^3 Å) and a cluster (10 Å).

As we have mentioned in the Introduction section, individual features of charge transfer of monoelectronic TT are explained by different characteristics of discrete electronic states of x depending on whether x is a granule or a cluster. The sizes of regions of electron localization differ too greatly in the two cases (10^3 Å for a granule, 10 Å for a cluster) to be left without consequences.

We have shown in the previous section that the non-monotonous VAC of a TT may be a result of states of x with different charges being involved in electron tunneling. In this case, as follows from Eq. 16, the interval $V^{z+1} - V^z$ between the two neighboring maxima of VAC is equal to the difference in consecutive ionization potentials of x differing by an electron charge:

$$V^{z+1} - V^z = (I_{z+2} - I_{z+1})/e. \qquad (17)$$

The dependence between the difference Eq. 17 and the size of x is convenient to be estimated using the "jelly" model in which a neutral system can be represented by a sphere of radius d with uniform distributions of electron and nuclear densities. The uncompensated charge is dispersed over the surface of the sphere and the ionization potential can be approximated by

$$I_{z+1} \approx (z+½)e^2/\varepsilon d + W_x. \qquad (18)$$

The first term in Eq. 18 is the change in electrostatic energy of x upon removal of an electron and W_x is the electron work function from x without account of electrostatic ionization energy. It follows from Eq. 18 that the difference Eq. 17 can be estimated as follows:

$$V^{z+1} - V^z \approx e/2\varepsilon d. \qquad (19)$$

The "jelly" model neglecting inhomogeneity of electron distribution and discrete character of single-electron energy bounds for a given z is a good enough approximation for a metal granule.

Since the granule is relatively large and its dielectric permeability is high, the difference $V^{z+1} - V^z$ is small. This leads us to conclude that the isolated maxima of VAC that correspond to resonance tunneling from different charge states can be observed at comparatively small changes of voltage.

In the monoelectronic TT with metal granules, the capacity of a granule can be substituted into expression Eq. 18 instead of the energy, which gives Eq. 16 the form

$$I_{z+1} \approx (z+½)\ e^2/c + W_x. \qquad (20)$$

Using Eq. 20 and the condition that if M1 and M2 are large bodies compared to the granule x, the ratio of voltage drops over x-M1 and x-M2 is proportional to their sizes

$$V_{xM1} = d1/d\,V;\; V_{xM2} = d2/d\,V, \tag{21}$$

where d1, d2, and d_+ are the distances x-M1, x-M2, and M1-M2 respectively, the formula Eq. 2 for $E_z^+[x\text{-}M2]$ and the formula Eq. 14 for $\tilde{E}_z$ can be rewritten as follows:

$$\tilde{E}_z' = az^2 - 2azs, \tag{22}$$

$$E_z^+[x\text{-}M2] = 2a\,(s-z-\tfrac{1}{2}) + beV, \tag{23}$$

$$b = 1/kT,\; a = be^2/2c,\quad s = c(W_x - I(M) - d1/dV)/e. \tag{24}$$

By substituting Eqs. 22-24 into Eq. 15 we get the VAC of metal granule TT derived in Ref. 11:

$$I = \mathrm{sh}(ebV/2) \sum_{z=-\infty}^{z=+\infty} \exp[(-a(z-s+\tfrac{1}{2})^2]\ [a(z-s+\tfrac{1}{2})-beV/2]/$$

$$/\mathrm{sh}[-ebV/2+a(z-s+\tfrac{1}{2})]\ /\ \{eR_{xM2} \sum_{z=-\infty}^{z=+\infty} \exp[-a(z-s)^2]\}. \tag{25}$$

Figure 2 shows the VAC and the diagram of energy balance of the resonance tunnel transfer corresponding to the maxima of the current for the granule with its charge z varying from 0 to 2. It follows from Eq. 25 that the distance between VAC maxima is a constant value that can be derived from Eq. 19:

$$V^{z+1} - V^z = e/2c. \tag{26}$$

This shape of VAC was actually observed in the first experimental samples of monoelectronic devices.[3] The observed distance between maxima was of the order of 1 mV.

When x is a donor-acceptor cluster of a protein, the uncompensated charge is localized within a limited volume much smaller than that of a granule. Its dielectric permittivity is also much smaller. It follows then from the "jelly" model estimate that the difference for protein DAC

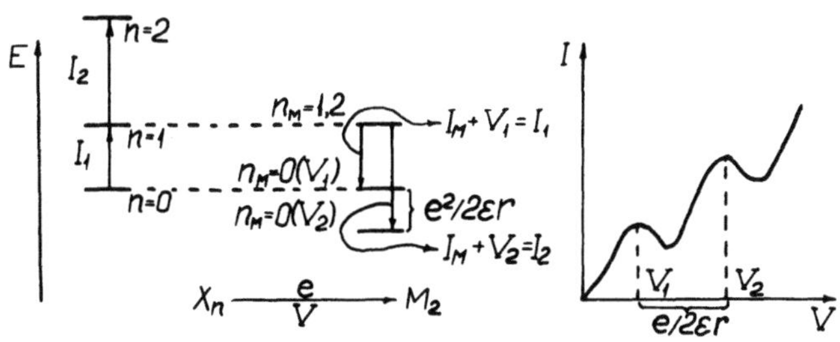

Fig. 2. Energy balance diagram of resonance tunnel transfers for various charge states x of granule x and the corresponding VAC of tunnel transfer in monoelectronics.

$$I_{z+2} - I_{z+1} = e^2/20\,\varepsilon_x \tag{27}$$

should be by two or three orders of magnitude greater than the same in metal granules, up to several eV. The distance between the corresponding maxima is of the order of several volts.

One feature of the electronic structure of clusters, the large difference between consecutive ionization potentials, leads to a natural restriction of number of electronic states involved in charge transfer. For better understanding of this restriction in biological systems, one should recall that in biological electron transport chains the electrons travel over large distances without substantial energy loss. The explanation is that the electron goes always from a DAS with lower oxidation-reduction potential to one with higher potential, the differences between consecutive DACs along the electron's path being very small.

The following conditions must be satisfied for an arbitrary DAS x2 and its two neighbors in the chain, x1 and x3:

$$I_{z1+1} < I_{z2+1} < I_{z3+1}, \tag{28}$$

while the energy of single-electron transfers[9] $\Delta E = I_{zi+1} - I_{zk+1}$ must not exceed several tenths of eV. One consequence is that the electron can tunnel only from the states of x2 with the charge z_2. Actually, the ionization potentials $I_{z2+1\pm1} = I_{z2} \pm \Delta$ of states z2+1±1 differ from I_{z2+1} by a quantity Δ which, as follows from Eq. 27 is larger than the difference $I_{zi+1} - I_{zk+1}$.

Accordingly, the state z_2-1 is not involved in electron transport chain since the corresponding electron transfer with energy $I_{z1+1} - I_{z2+1} + \Delta > 0$ is energetically disadvantageous. Nor the electron will tunnel to x3 since the transfer x2 → x3 with the energy $I_{z2+1} - I_{z3+1} + \Delta > 0$ is also disadvantageous.

These considerations explain why an individual DAC of a definite protein has only one charge state from which an electron tunnels. It is called the native reduced form of the protein (in our notation, its charge z=0) that transforms into the oxidized form with z=+1 upon removal of an electron; the energy of such a transition is the first ionization potential of the DAC, I_1.

In monoelectronic devices, nonmonotonous behavior of VAC of a protein TT within several volts of V^o, the latter being determined by resonance tunneling condition from the ground state of the native reduced form of DAC with z=0, can be caused by participation of the lower excited states (i) of the same form with z=0 in tunneling. These states in fact participate in electron transfer in biological electron transport chains which is proved by biphasic temperature dependence of electron transfer rates in photosynthetic reaction centers of bacteria.[9]

In the case of a molecular cluster, these states form a discrete spectrum $\{E_{oi}\}$ (the ground state is denoted by i=0) which allows us to use the method of VAC calculation for a TT M1-(x)-M2 with a discrete set of states of x described in the previous section.

The objective of this paper is to lay the theoretical foundations of VAC calculations and parameters of the working region of a TT with protein elements starting from spectroscopic data on the electronic structure of a protein DAC. These data include the protein ionization potential, I_1, and the energy of electronic transitions from the ground state into the lower excited states {i} of the native reduced form of the protein. Substituting these values into Eqs. 2 and 14 we can find the energy of the single-electron transfers from the ith state of x to the metal M2, and the energy terms $\tilde{E}_i$

and $\tilde{E}_1$ of the system x+M2 for the case of x in the state i with z=0 or in the oxidized form with z=1 measured relative to the ground state of the native reduced form:

$$\tilde{E}_i = \Delta E_i, \tag{29}$$

$$\tilde{E}_1 = I_1(x) - I(M) + eV_{xM1}, \tag{30}$$

$$E_i^+ = -I_1(x) + I(M) + \Delta E_i + eV_{xM2}. \tag{31}$$

Those energy terms determine the VAC. The VAC of an asymmetric TT M1-(x)-M2 with a molecular cluster as x can be obtained using the approach described in the previous section if we assume that the direct transitions between the ground state and the excited states {i} happen slower than the charge transfers M1 ⇌ x and x ⇌ M2. Here, for the sake of simplicity, we disregard the fact that not only the ground state but also the lower excited states of the oxidized form can be involved in transfer. The corresponding generalization can be obtained within the same approach.

Taking into account Eqs. 29-31, we are using Eq. 1 for the current, Eq. 4 for the partial rates $G^{\pm}$ of electron transfer from individual states, and Eqs. 8-13 to determine populations of the states, W.

Then, neglecting all the states of both the reduced and oxidized forms but their native states (z=0 and z=1 respectively) the following expression can be obtained for VAC of a TT with a molecular cluster:

$$I = sh(eV/2kT/\{\sum_j \exp(-\Delta E_j/2kT)+\exp(-(I_1(x)-I(M)+eV_{xM1})/kT\}$$

$$\sum_i (1/eR^i_{xM2})\exp[-\Delta E_i+I_1(x)-I(M)+eV_{xM1})2kT]$$

$$\{(\Delta E_i-I_1(x)+I(M)+eV_{xM2})/sh[\Delta E_i-I_1(x)+I(M)+eV_{xM2})2kT]\}. \tag{32}$$

Different effective resistance values are acceptable in Eq. 32 for tunnel transitions from various stated i. Since the excited states i>0 are more delocalized in space than the ground state i=0 the matrix elements of their interactions with states of the metal can be greater and their effective resistances smaller than those of the ground state.

The formula Eq. 32 gives VAC with the principal local maximum that corresponds to resonance tunneling from the ground state i=0 of the native reduced form[15], and additional maxima corresponding to the resonance tunneling from the lower excited electron-vibrational states i>0. Figure 3 presents a chart of energy balance of tunnel transfers from the ground state and the lower excited states of x with z=0 that are transformed into the ground state with z=1 upon removal of the electron, and the resulting VAC of TT in monoelectronics.

The principal maximum is determined from Eq. 32 using the following condition:

$$V^o \approx V_{xM2}{}^o = (I_1(x) - I(M))/e. \tag{33}$$

It then follows from Eq. 33 that for a TT to work near the principal maximum at a relatively small voltage it is necessary that the electronic work function from the metal should be approximately the same as the first ionization potential of the cluster $I(m) \approx I_1(x)$. In this sense, metal conductors and protein electron-transfer elements should be compatible.

Distances between the major and the minor maxima on VAC are equal to energy ΔE_i of transition from the ground state with i=0 to the excited States with i>0 divided by electron charge:

$$V^i - V^o = \Delta E_i/e. \tag{34}$$

It follows that VAC of a TT can reflect the electronic spectrum of the molecular cluster. This novel and interesting effect, though it is only hypothetical, could be called the tunnel volt-ampere spectrum. The energy of protein electronic transitions can in their turn determine the VAC of a protein TT. The energy ΔE_i may vary broadly from hundredths of eV to several eV depending on the electronic structures of clusters so that distances Eq. 34 can vary from hundredths of volts to volts. The difference values, ΔE_i, are smaller than those of neighboring ionization potentials which, at a small change of voltage, results in a non-monotonous behavior of VAC caused by resonance tunneling from the ground state and excited states of the native reduced DAC.

Thus our analysis shows that the difference between granules and molecular clusters in tunnel charge transfer is due to different characteristics of their discrete electronic states. In the case of the granule, the nonmonotony of VAC is caused by different charge states of the granule while for a protein DAC it is because of the ground and excited electron-vibrational states of x that form a discrete spectrum.

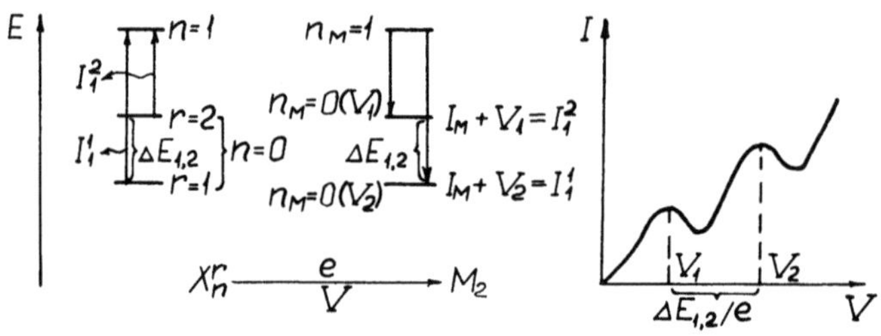

Fig. 3. Energy balance diagram of resonance tunnel transfers for the ground state and various excited states x of protein DAC x and the corresponding VAC of tunnel transfer in monoelectronics.

PROMISING POSSIBILITIES OF SINGLE ELECTRON TT WITH MOLECULAR CLUSTERS

The nonmonotonous behavior of VAC with parts of negative differential conductivity may find practical application in monoelectronics.[1] Actually, the monoelectronic TT itself is a miniature analog of a tunnel diode[16] with VAC with negative differential conductivity and in this capacity could be used for signal generation and amplification as well as in computing electronic circuits.

One advantage of protein monoelectronics is the possibility of designing protein contact sites with the required parameters: sizes and structures of protein ensembles, their constituents, etc.[6-8] It can be achieved by directed polymer system synthesis like the Langmuir-Blodgett technique or the method of Merrifield as well as genetic engineering.

The theory we have outlined is applicable to the case with a single part of the tunnel transfer path on which the voltage drops more than on other parts. Modified proteins with such parts of fixed sizes obtained by the methods above may be used in monoelectronic TT. One hypothetical example is a TT in which the electron transfer protein is separated from one metal electrode, M1, by a polymer layer of high resistance, and from the other, M2, by a molecular chain with conductivity almost that of a metal. The width of the layer and the length of the chain can be set by synthesizing modified protein systems. The asymmetric protein junction of this composition may show a nonmonotonous VAC with the local maxima corresponding to resonance tunneling from the ground and excited states of the native reduced form of the protein DAC.

Protein elements are expected to increase the working temperature of TTs as shown in Ref. 2. It is a consequence of the decrease in delocalization area for a tunneling electron upon replacement of a metal granule by a molecular cluster. Actually, the T_{work} can be estimated from the energy of thermal motion being smaller than the electrostatic energy corresponding to distance Eq. 19 between the maxima of VAC so that

$$kT_{work} \ll e^2/2\varepsilon d. \qquad (35)$$

If this condition is satisfied, as follows from Eqs. 15, 25, and 32 the VAC will preserve its specific shape. When x is a molecular cluster the right-hand side of Eq. 35 is the upper bound of the differences $(V^{l+1} - V^{l})e$ of VAC. The estimates of these difference values we have given in the previous section suggest that T_{work} may exceed 100 K for proteins which is an improvement compared to 1-4 K, the T_{work} of the existing monoelectronic devices made of conventional materials.

REFERENCES

1. K. K. Likharev, Mikroelektronika 16:189-194 (1987).
2. K. K. Likharev, V. K. Semenov, Int. Superconductivity Electronics Conference Tokyo, Rep. SP-27, pp. 182-185 (1987).
3. L. S. Kuzmin, K. K. Likharev, Letters to Z. Eksp. Teor. Fiziki 45:389-390 (1987).
4. T. A. Fulton and G. J. Dolan, Phys. Rev. Lett. 59:109-112 (1987).
5. G. Owen, Repts. Progr. Phys. 48:795-851 (1985).
6. R. I. Gilmanshin, P. I. Lazarev, Artificial Electron Transfer Protein Ensembles for Molecular Electronics, Puschino, 1987, 12 pp. (Preprint).
7. R. I. Gilmanshin, P.I. Lazarev, J. Mol. Electronics (in press).
8. R. I. Gilmanshin, P.I. Lazarev, Biotekhnologia 3(4):421-432 (1987).
9. D. De Vault, "Quantum-Mechanical Tunneling in Biological Systems," 2nd Ed., Cambridge Univ. Press, Cambridge (1984).
10. R. I. Shekhter, Z. Eksp. Teor. Fiziki 63:1410-1416 (1972).
11. I. O. Kulik and R. I. Shekhter, Z. Eksp. Teor. Fiziki 66:623-640 (1975).
12. D. V. Averin, K. K. Likharev, Z. Eksp. Teor. Fiziki 90:733-743 (1988).
13. I. O. Kulik and I. K. Yanson, "The Josephson Effect in Superconducting Tunnel Structures," Nauka, Moscow (1982).
14. S. Lundqvist and N.H. March (eds.), "Theory of the Inhomogeneous Electron Gas," p. 18, Plenum, New York (1963).
15. A. B. Rubin, "Biofizika," Vol. 1, p. 267, Vyshaya Shkola, Moscow (1987).

16. V. V. Pasynkov and L. K. Chirkin, "Semiconductor Devices," p. 177, Vyshaya Shkola, Moscow (1987).
17. F. L. Carter (ed.), "Molecular Electronic Devices," Marcel Dekker, New York (1982).

IMMOBILIZATION OF NICOTINAMIDE ADENINE DINUCLEOTIDE: IMPLICATIONS IN MOLECULAR ELECTRONICS AND BIOENGINEERING

Ratna S. Phadke, Rajesh Manchanda and Girjesh Govil

Tata Institute of Fundamental Research
Homi Bhabha Road
Bombay 400 005, India

ABSTRACT

Nicotinamide adenine dinucleotide (NAD) has been immobilised on Pt surface with the help of a prepolymer PU-3. On immobilization NAD exhibits its normal electrochemical activity and is chemically as well as electrically addressable. Switching between oxidised and reduced states can be monitored through optical methods. NAD containing enzyme alcohol dehydrogenase on immobilization on platinum electrode by means of PU-3 is active and approachable to substrates. The process of immobilization facilitates the electron transfer between the enzyme and the electrode which is potentiometrically measurable.

The potential applications of NAD immobilized surfaces in the field of molecular electronics and biochemical engineering have been discussed.

INTRODUCTION

In living organisms, enzymes bring about precise molecular operations with a high degree of efficiency and selectivity[1]. Recent awareness of utilization of enzymes for industrially important purposes has initiated a trend to study and apply various enzymatic processes under *in-vivo* conditions[2]. For bioengineering applications, the enzymes are usually tagged to a solid support to enable continuous operation. This is a process called immobilization[3].

A number of studies have been reported on immobilization of enzymes. The techniques used can be roughly classified as (i) covalent bonding (ii) physical and ionic adsorption and (iii) entrapping in insoluble matrix. Through suitable molecular design, such immobilized system can be made to mimic their biological functions under laboratory and industrial conditions.

An important class of enzymes are the oxido-reductases which are involved in providing energy to living systems. These enzymes oxidise fuels such as sugars, lipids and peptides and the free energy thus released is utilised to synthesise the units of biological energy - adenosine triphosphate (ATP). The whole set of reactions occur in a series of coupled steps so that the energy is released in small packets rather than in a single large pulse. Through proper organisation and well planned vectorial fluxes of protons and electrons released in oxidation-reduction reactions, the living systems are able to achieve a very high energy conversion efficiency. The oxido-reductase enzymes usually function in conjunction with certain small molecules called coenzymes or cofactors which act as mediators in channelling the flow of electrons and protons in suitable directions.

In earlier papers[4,5], we have shown that through appropriate strategies, it is possible to immobilize two such coenzymes flavin adenine dinucleotide (FAD) and cytochrome-c and direct electron flows from solid surfaces to the coenzymes and vice-versa. Such biomaterials can then be used as active components in molecular electronics[6]. This paper deals with immobilization and potentialities of the coenzyme nicotinamide adenine dinucleotide (NAD). This coenzyme acts as a cofactor for more than 200 kinds of enzymes involved in oxidation-reduction reactions such as conversion of alcohol to aldehydes, pyruvate to lactate, malate to oxaloacetic acid, hydroxylation of steroids, etc. It is one of the most important coenzymes in biological redox reactions. The reduction reaction involves two electrons and one proton:

$$NAD^+ \underset{}{\overset{2e, H^+}{\rightleftharpoons}} NADH,$$

and is characterised by two electrochemically distinct redox stages with

$$E(1) = -0.56 \text{ VOLTS and } E(2) = -0.32 \text{ VOLTS}$$

with respect to saturated calomel electrode. The coenzyme can be reversibly driven to donate or accept electrons by controlling the externally applied potentials. Thus the molecule can be switched between its two stable states, NAD^+ and NADH.

Earlier attempts to covalently link NAD to a solid support have indicated that chemical modifications in nicotinamide ring leads to loss of redox characteristics.[7] The electroactivity of NAD on most surfaces is rather poor. Added to this are problems on account of its extreme sensitivity to visible radiations and difficulty of reoxidizing reduced form, ($NADH \longrightarrow NAD^+$). We have used water-miscible urethane prepolymer (PU3)[8] to entrap NAD and NAD containing enzymes. The advantages in using the prepolymer are two fold. Firstly, the gels are easily formed within 30-60 min. by gentle mixing of prepolymer in water. Thus the biomolecules desired to be entrapped are subjected to minimal perturbations. Secondly, the gels formed are porous and transparent to UV-visible radiations and therefore the state of the molecules can be easily monitored.

EXPERIMENTAL

(a) Materials

The materials used are Platinized Platinum (1 cm by 1 cm), Nicotinamide adenine dinucleotide (Sigma Chemicals, USA), Alcohol dehydrogenase EC 1111 from Baker's yeast (Sigma Chemicals, USA), Propanol (Glaxo Chemicals, India), Potassium chloride. AnalaR grade reagents were used in the experiments. The preplymer polyurethane (PU3) is a gift from Prof. Tanaka, Kyoto University, Japan.

(b) Immobilization of Coenzyme

The immobilisation of NAD was done on a clear solid surface of Platinum. Initially the surface was thoroughly washed with double distilled water and vacuum dried. The prepolymer was dissolved in an aqueous solution of known concentration of the coenzyme. The gel so produced was applied on the surface of the electrode evenly. The surface was subjected to vacuum drying for 1-2 hours until a fine film of polymer was visible. The entire operation was carried out in dark as NAD is known to be extremely sensitive to light.

Immobilisation of alcohol dehydrogenase was done following a similar procedure.

(c) Physical Characterisation of Immobilised Material

Cyclic voltammetry experiments were done to assess the electrochemical characteristics of the coenzyme - polymer system. A conventional three electrode system consisting of a saturated calomel as reference electrode, Platinum wire mesh as the counter electrode and the modified Platinum as the working electrode was used[8]. The supporting electrolyte was 1.0 M KCl which is purged to remove dissolved oxygen by passing nitrogen for 1-2 hours prior to experimentations. The experiment was carried out in nitrogen atmosphere in dark. A home-built cyclic voltammeter was used[9].

Optical spectroscopy was done to monitor the optical properties of the coenzyme and the polymer. Using a quartz cuvette filled with distilled water a baseline was recorded. The spectra were obtained in 400 nm to 240 nm range. The spectra for solid surfaces were recorded when it was immersed in distilled water in the cuvette. Uncoated surface immersed in water was used to define the baseline. O.D. was maintained in all experiments to lie between 1 and 2. A commercial SPECTRONIC 1201 system was used for these experiments.

Potentiometric measurements were done in a H-type cell with a frit separating the two compartments. A highly sensitive voltammeter was used to record the changes in electric potential. The cathode was the modified electrode while a plain platinum strip was the anode in a supporting electrolyte (1.0 M KCl). The experiment was done in nitrogen atmosphere in the dark. The substrate was added and stirred for two minutes continuously. Reading was taken five minutes after stopping stirring.

RESULTS AND DISCUSSIONS

(a) Electrochemical Activity of Immobilised NAD

Fig. 1(a) depicts the cyclic voltammogram of Platinum electrode in supporting electrolyte (1.0 M KCl). Even at high currents, there is no indication of redox peaks originating from adsorbed hydrogen on the Platinum surface. NAD dissolved in supporting electrolyte exhibits characteristic peaks at - 0.56 (broad) in the forward cycle and - 0.34, - 0.18 in return cycle. The prepolymer itself when immobilised shows weak peaks in the region of - 0.5 and - 0.22. PU3 immobilised NAD gives sharp intense peaks indicating that it is electrochemically active (Fig. 1(d)).

The results of cyclic voltammetry indicate that the electrons can be made to flow from the Platinum surface to the entrapped coenzyme molecule and vice-versa by application of suitable bias voltages.

(b) Monitoring the Redox State of NAD

NAD^+ is known to exhibit a broad band at 260 nm due to overlap of absorption of adenine and pyridine rings. On reduction to NADH an additional broad peak appears around 340 nm due to the quinone structure of the reduced nicotinamide ring. Hence the reduction and oxidation of NAD can be easily monitored by observing the change in light absorption at 340 nm.

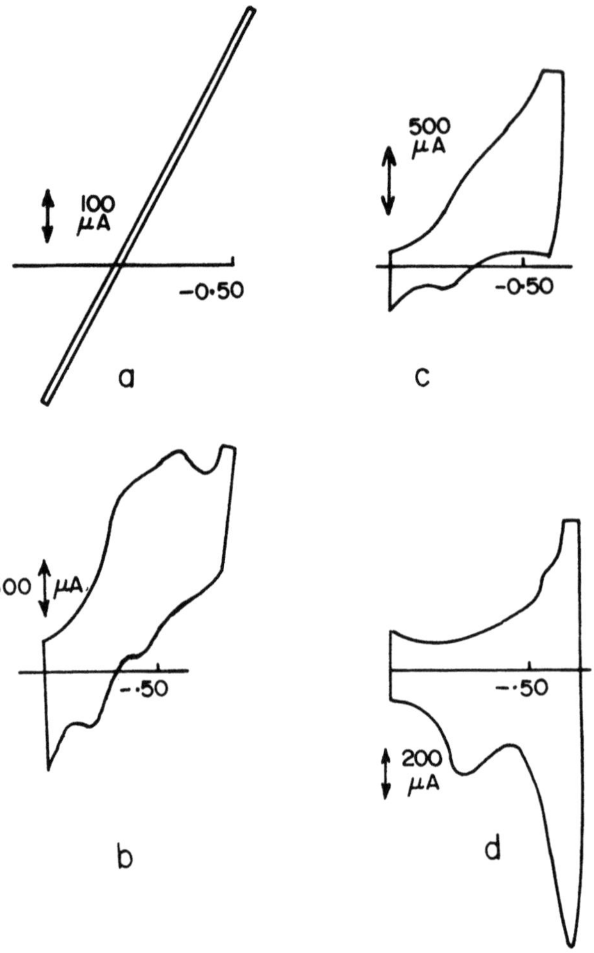

Fig. 1. Cyclic voltammograms
(a) Pt. electrode in 1.0 M KCl
(b) (a) + NAD
(c) (b) + PU3
(d) NAD^+ immobilized on Pt. surface in supporting electrolyte 1.0 M KCl.

(c) Chemical Reduction of Immobilized NAD

The switching from oxidised to reduced form can be affected with the help of appropriate reducing agent. For example, on adding freshly dissolved $Na_2S_2O_4$ in degassed aqueous solution (1.0 mg per ml final concentration) to cuvette containing 0.1 mM NAD, the peak around 340 nm builds up at the expense of the 260 nm peak. To assess the accessibility and chemical switching of immobilised NAD we have performed the following experiments.

NAD was dissolved in a dilute solution of PU3. The spectrum exhibited characteristic peak at 260 nm together with a noisy background of the polymer (Fig. 2(a)).On addition of $Na_2S_2O_4$, to this solution a

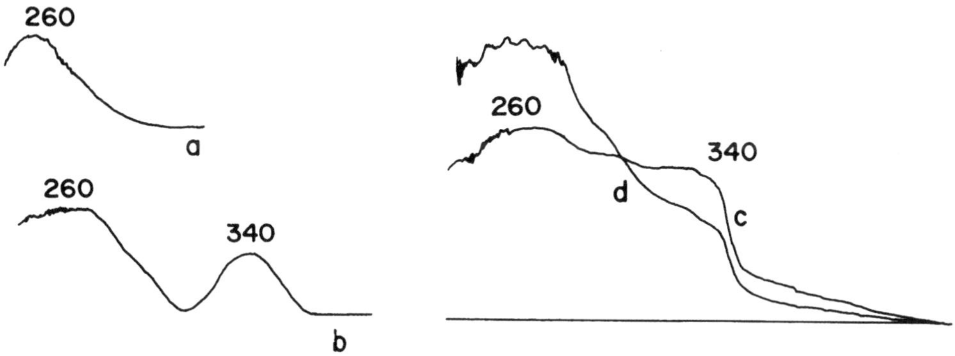

Fig. 2. UV-visible spectra
(a) NAD^+ (1 mg/ml)
(b) NAD^+ + NADH (1 mg/ml)
(c) NADH immobilized on Platinum mesh with PU3
(d) (c) after application of potential for 1/2 hr.

distinct peak appeared in the region of 340 nm (Fig. 2(b)). In another set of experiments, NAD immobilized PU3 film was formed onto a solid surface (quartz glass plate). Although the quartz surface gave a strong background absorption, it did not interfere with the peaks of NAD or NADH (figure not shown). In fact, interference from the quartz absorption can be eliminated by using difference spectrophotometer. A film of PU3 and NAD exhibited a strong peak well above the glass background absorption. It may be noted that all observations were taken when the quartz surface and that coated with PU3 and NAD were immersed in double distilled water contained in a cuvette, to which $Na_2S_2O_4$ was added for reduction purposes. The spectrum recorded after half an hour of reduction showed peaks around 260 nm and 340 nm (Fig. 2(c)). These experiments clearly indicate that NAD on immobilisation with PU3 on a solid support is optically active and can be switched chemically.

(d) Electrical Switching

Fig. 2(c) depicts optical spectra of Platinum mesh coated with NADH by means of PU3. Both the absorption maxima corresponding to adenine and reduced nicotinamide parts of the molecule are distinctly observable around 260 and 340 nm respectively. The Platinum mesh was kept at - 0.32 w.r.t. ground. The peak at 260 nm started building up at the expense of peak at 340 nm indicating oxidation of NADH to NAD (Fig. 2(d)). In a separate series of experiments NAD^+ was converted to NADH by application of appropriate electrical potential. The conductivity measurements confirm the above observation in that the surface conductivity of PU3 reduces from 3.44×10^{-3} S/cm to 3.08×10^{-3} S/cm when NAD is doped into the matrix. Thus it can be concluded that NAD immobilised by the means of PU3 is electrochemically active and forms stable system. It can be switched from oxidised to reduced form and vice-versa chemically as well as electrically.

(e) Immobilised Enzyme

Alcohol dehydrogenase has NAD as its cofactor. The enzyme was immobilised onto Platinum strip as described in Materials and Methods. The activity of the immobilised enzyme was checked by immersing the electrode in ethanol or cinnamic aldehyde overnight and assaying using standard techniques[10]. As a result of alcohol/aldehyde conversion electrons were either absorbed or liberated depending upon the direction of the reaction. If the enzyme is immobilised onto an electrode by means of a conducting polymer, then the electron transfer from the enzyme active site of the enzyme and electrode should take place unhindered. Under appropriate conditions a potential build up should be observable. The concentration of alcohol was varied and the potential build up was measured at fixed interval of time under identical conditions. The experimental points lie on a straight line. In a separate set of experiments aldehyde concentration was varied (Fig. 3). These findings can be graphically utilised in construction of a biosensor and electrode in a biobattery.

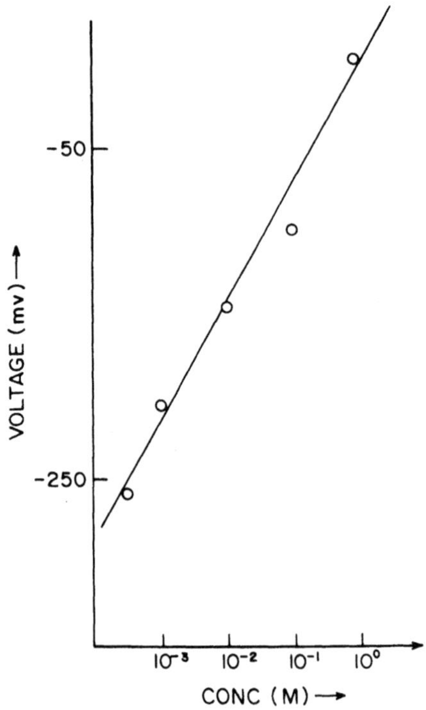

Fig. 3. Potentiometric measurements. Different concentrations of alcohol in supporting electrode vs. voltage developed.

POSSIBLE APPLICATIONS OF NAD IMMOBILISED MATERIALS

(a) Molecular Switches

The work done on NAD by immobilising it on Platinum opens new vistas to utilise NAD for making molecular switches. The electrical switching can well be utilised for various purposes in instruments used for specific tasks in the field of biomedical, bioengineering and biochemical measurements. It can also be used as a memory storage device operating between its two stable states.

(b) Biosensors

As the matrix is porous and easily accessible, it is easy to fabricate a sensor to detect specific substrates in a solution. For example, alcohol dehydrogenase is specific to alcohol in the medium and the electrode potential would depend upon the concentration of the alcohol present. In conjunction with a solid state device, the sensitivity of such a biosensor can be improved significantly. In view of the high specificities of the enzymes, such biosensors may have very little interference from other biomaterials and can prove ideal for 'in-vivo' monitoring of molecules.

(c) Bioconvertor

The reactivity of alcohol dehydrogenase can be effectively utilised for converting alcohol/aldehyde.

(d) Bioanode

It can form a bioanode for the use in batteries consuming cheap crude biofuels to generate electricity as the coenzyme-enzyme couple is selective.

REFERENCES

1. T. E. Barman, "Handbook of Enzyme", Vol. 1, Springer-Verlag, New York (1969).
2. L. B. Wingard Jr., I. Y. Berezin, and A. A. Klysov, Eds. "Enzyme Engineering," Plenum Publishing Corpn., New York (1980).
3. G. C. Guibault, in Comprehensive Analytical Chemistry (Ed.) G. Svehla (Amsterdam Pergamon Press) Vol. 8, (1977).
4. Ratna S. Phadke, H. M. Sonawat and G. Govil, Electron Delocalization during the Oxidation-Reduction Cycle of FAD and NAD: A Quantum Chemical Approach to the Design of Coenzyme-Immobilized Bioanode for Biochemical Fuel Cells, Int. J. Quantum Chem. Quantum Biology Symposium, 10: 251 (1983).
5. Ratna S. Phadke, H. M. Sonawat and G. Govil, Biomolecular Electronics Using Coenzymes Immobilized on Solid Supports, J. Molecular Electronics, New York 4: 000 (1988).
6. F. L. Carter, "Molecular Electronic Devices," Vol. I and II, Dekker, New York (1982, 86).
7. S. Fukui, and A. Tanaka, "Advances in Biochemical Engineering Bio-Technology," Vol. 29, Eds. A. Fiecher, Springer Berlin, Heidelberg, New York (1984).
8. I. J. Higgins, and H. A. O. Hill, Bioelectrochemistry, In: "Essays in Biochemistry," 21: 119 (1985).
9. V. R. Bhagat, and K. S. V. Santhanam, A Multipurpose Electrochemical Instrument, J. Scientific & Industrial Research, 30: 235 (1971).
10. N. Tamaki, and T. Hama, Aldehyde Dehydrogenase from Baker's Yeast, In:"Methods in Enzymology" Eds. W. A. Wood, Cademic Press, New York Vol. 89, pp 469, (1982).

ISFET-LIKE DEVICES COUPLED TO NEUROBLASTOMA CELLS: CYTOMETRIC AND ELECTRICAL CHARACTERIZATION

M. Grattarola (+), A. Cambiaso (+), S. Cenderelli (+), G. Parodi (+), M. Tedesco (+), B. Nie (+,*), G. Cerofolini ('), L. Meda (") and S. Solmi (#)

(+) Biophysical and Electronic Engineering Department, University of Genoa, Genoa, Italy
(*) Permanent address : Dept. of Biomedical Engineering, Sun Yat-sen University of Medical Science, Guangzhou, P.R. China
(') ENICHEM, Milan, Italy, (") SGS - Thompson, Agrate, Italy, (#) LAMEL, CNR, Bologna, Italy

INTRODUCTION

One challenging goal in cell biology is to monitor "on line" the exchange of electrochemical signals between various cells of a cell population. Especially in the field of cell neurobiology, the simultaneous recording of the electrical activities of various interconnected neurons and/or the simultaneous recording of the activities in all the arborizations of a single cell are results of major importance for the understanding of the computational properties of neuronal networks.

Several lines of research are presently being pursued in order to attain these results, including the use of fluorescent dyes sensitive to variations in cell membrane potentials (1) and the use of flat microelectrodes onto which small organs or tissues are deposited (2).

We describe here an alternative approach which has been adopted in our Department and which involves the direct coupling of living cells to the insulating layer of ion-sensitive field effect transistors (ISFETs).

This approach, which was pioneered by Bergveld (3,4), can be regarded as an extension of the ideas underlying the use of integrated biosensors, where an organic or biological component is in intimate contact with a microelectronic device of which it modulates the electrical response (5). The project has far-reaching implications for several research areas, which include, in addition to those of biosensors and electrophysiology, also clinical rehabilitation (that is, the functional coupling of severed nerves to microelectronic devices). From a more general viewpoint, the project aims to tackle the problem of biocompatibility, that is, to characterize the modifications induced by "wet" living materials in inert silicon microelectronic devices and vice versa. This is a general problem for the whole area of bio-electronics.

In this paper, we describe two cytometric techniques utilized in order to characterize the adhesion of mouse neuroblastoma cells to the insulating layer (Si_3N_4) of ISFETs. The electrical response of an electrolyte-insulator-semiconductor (EIS) system, modified by addition of biological material (laminin), is also analyzed.

MATERIALS AND METHODS

Biological materials

Laminin was purchased from Sigma. It was dissolved in a PBS buffer solution to a final concentration of 1 mg/ml. The solution was divided into small aliquots and stored at -20 °C before use. 25 μl of solution were deposited on the Si_3N_4 gates of our devices for the electrical measurements, smaller amounts were used for the cell adhesion studies (cf. next section). The devices were utilized after at least 6 h of incubation at 37 °C. Mouse neuroblastoma cells were a generous gift of E. Vigneti (CNR Center for Cellular Biology, Rome). The cells were seeded on the laminin treated devices and observed after few days of incubation (37 °C, 5% CO_2).

Microelectronic devices

Two kinds of devices were utilized. The first one (type 1) was a generous gift of SGS - Thompson. It was not originally designed to operate as an ISFET, and this can explain its unusual geometry. Before electrical measurements, the devices were stored in distilled water for 24 hours.
The second kind of device (type 2) was a generous gift of dr. Solmi (LAMEL, CNR, Bologna, Italy). It is a combination of two ISFET and two REFET sensors. In the present study, this device was utilized for cytometric purposes only. The two devices are sketched in Figure 1.

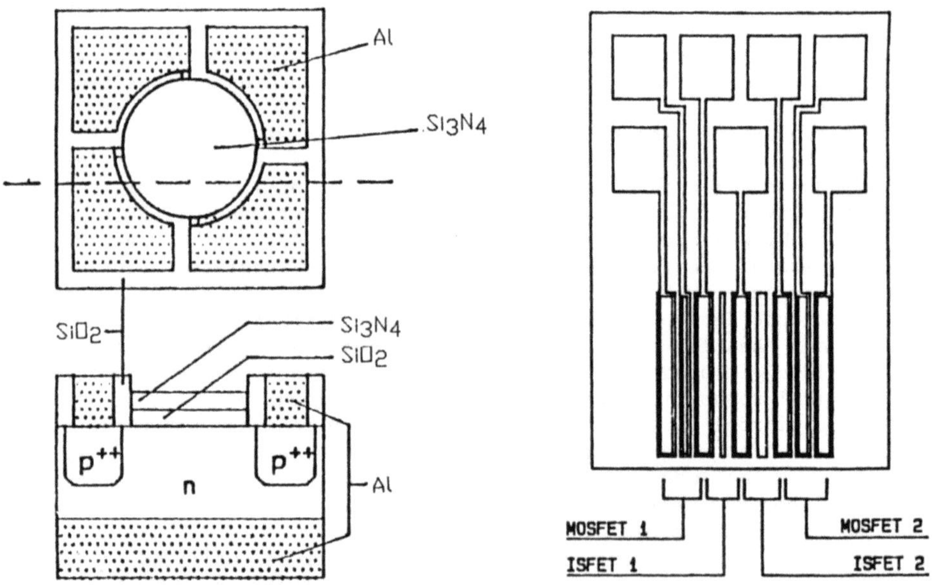

Fig. 1. Sketch of the ISFET-like devices: left type 1; right type 2.

Fluorescence digital microscopy

An epifluorescence microscope (Leitz, W. Germany) equipped with a Xenon lamp was utilized. The microscope was interfaced with a high-sensitivity ISIT TV camera (RCA, USA). The images generated by the TV camera were digitized (Data Translation DT2851 frame grabber) and saved on the floppy disks of a personal computer (AT, IBM). Mouse Neuroblastoma cells were made fluorescent by using the dye RH-421 (1). This dye is capable of binding to the cell membrane so that the cell body and arborizations can be visualized. The dye can also be used to detect possible variations in membrane potentials by utilizing matrices of photodiodes as sensors and a much more sophisticated measuring set up. In the present work , the dye was used for visualization purposes only.

High resolution scanning acoustic microscopy

Scanning reflection acoustic microscopy is quite a new technique which is very useful in imaging both microelectronic devices and biological materials, in a non-destructive way (6,7).
Acoustic images were obtained with the ELSAM scanning acoustic microscope (Leitz). This microscope can operate under reflection conditions at 100 MHz (low resolution) and in the range .8 - 2 GHz (high resolution). During the experiments described in this paper, it was operated at 1.4 GHz, with a resolution of about .9 µm. The images, generated by the electromechanical scanning of the samples, were digitized and stored in the memory of an AT personal computer. The lens was coupled to each sample by using a drop of PBS.

Automatic measurements of electrical parameters

A Model 72B Boonton Capacitance meter was employed to measure the capacitance of the type 1 device utilized as an EIS system (8). The capacitance of the EIS system was measured between the bulk metallization of the device and an Ag-AgCl reference electrode (Orion - 9005). This 1 MHz bridge furnishes as output an analog signal proportional to the susceptance value of the device under test and accepts in input a bias voltage.
In order to digitize and to store each measurement and to supply the device under test with the bias voltage, an IBM AT personal computer equipped with a DT2801A Data Translation acquisition board was used (general purpose A/D, D/A and I/O interface board). Our device may be modelled (to a first approximation) with a resistance in series with the MOS capacitance. This resistance takes also into account the reference electrode. The original data were computer-corrected in order to obtain the capacitance values according to this simplified model.

RESULTS

Fluorescence imaging of mouse neuroblastoma cells

Figure 2 shows an image of mouse neuroblastoma cells adhering to a "type 2" silicon device coated with laminin. The spreading of the cells on the various regions of the device can be evaluated at low magnification. At higher magnification details of the cell arborizations can be seen, indicating the healthy state of the cells (figure 3). Figure 4 shows the result of a digital elaboration on one of these arborizations. At the bottom of the figure, a profile of fluorescence intensity is shown, corresponding to a line travelling the arborization.

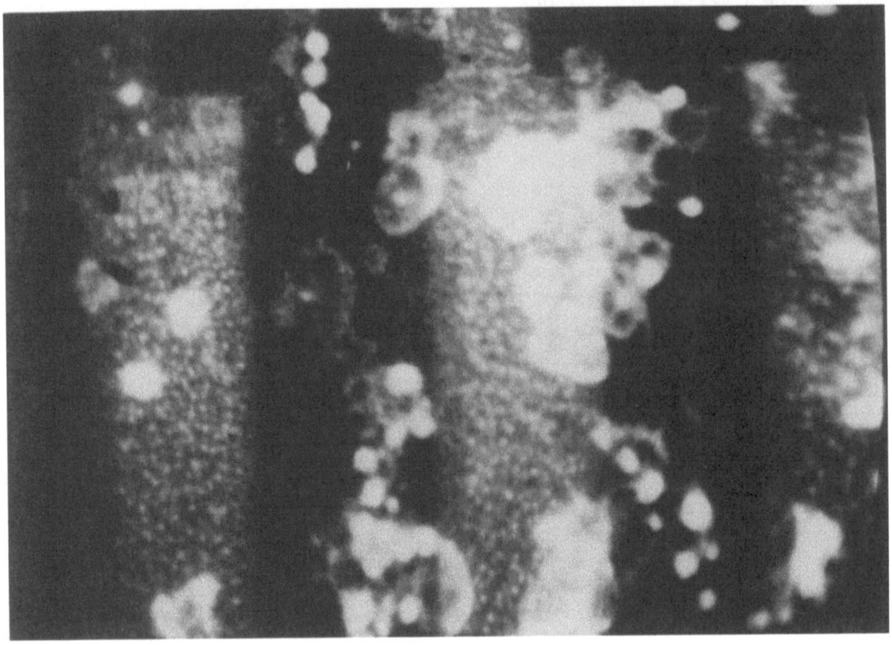

Fig. 2. Epifluorescence digital image of mouse neuroblastoma cells adhering to a laminin-coated type 2 device.

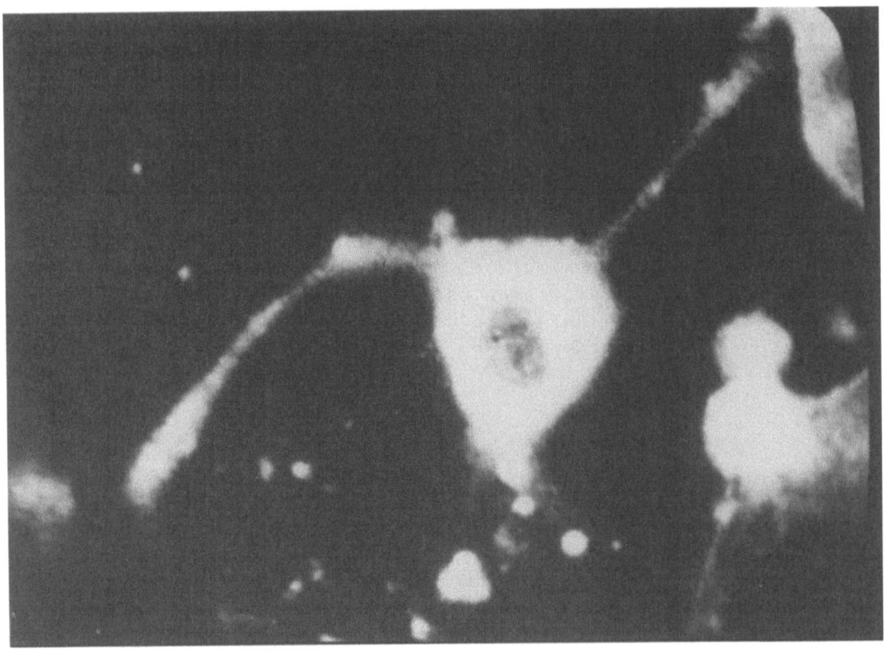

Fig. 3. High magnification epifluorescence digital image of a single mouse neuroblastoma cell adhering to a laminin-coated type 2 device.

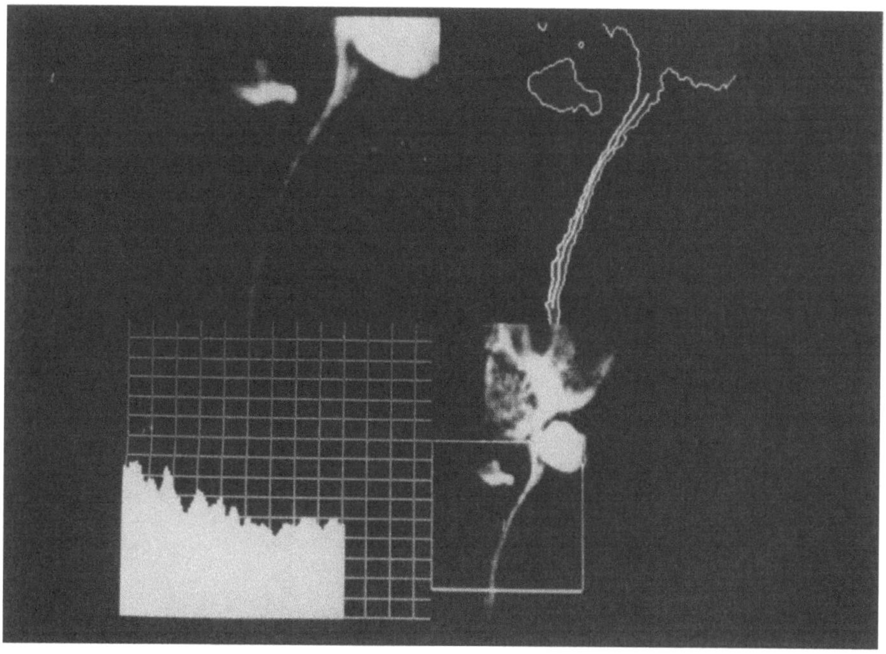

Fig. 4. Digital elaboration along an arborization of a neuroblastoma cell.

In this way, variations in fluorescence intensity along a dendrite, in response to external physico-chemical perturbations can be quantified.

High resolution acoustic imaging

Figure 5 presents an acoustic image of cells growing on a type-2 device coated with laminin. The contrast between the exposed gate (Si_3N_4) and the metallized gate is very sharp because of the different acoustic properties of the gate materials.
Under the conditions of focus utilized for this image, the cells appear as dark, structureless regions. A much more detailed acoustic image of a single neuroblastoma cell adhering to a laminin-coated insulating layer is depicted in figure 6. This image was obtained by focusing the acoustic lens on the top of the cell, instead of the substratum. Details of discrete regions of adhesion are clearly visible (small arrows). Interference fringes can also be observed (large arrows): They are generated by the interference between the acoustic waves reflected at the top of the cell and those reflected at the surface of the substratum. Such figures can be utilized to measure the thickness of the biological material (9).

High-frequency capacitance-voltage (C-V) measurements

C-V measurements were made on the large (type-1) devices coated/uncoated with laminin, as a first preliminary step towards the measurement of electrical parameters in devices functionally coupled to living neuroblastoma cells.
Laminin is a glycoprotein which is produced by the basal membrane (10). It is known to coat epithelial tissues in vivo, and is assumed to direct the migration of cells during embryogenesis. It is also routinely utilized to coat artificial surfaces in vitro, in order to improve cell adhesion. We used laminin as a kind of prototypic biological material,

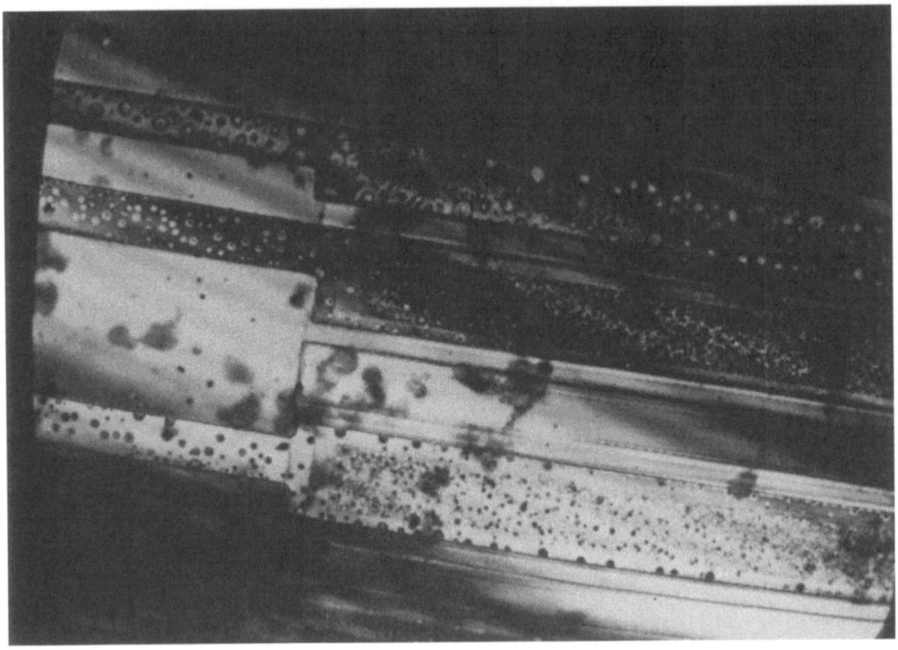

Fig. 5. Reflection acoustic digital image of mouse neuroblastoma cells adhering to a laminin-coated type 2 device.

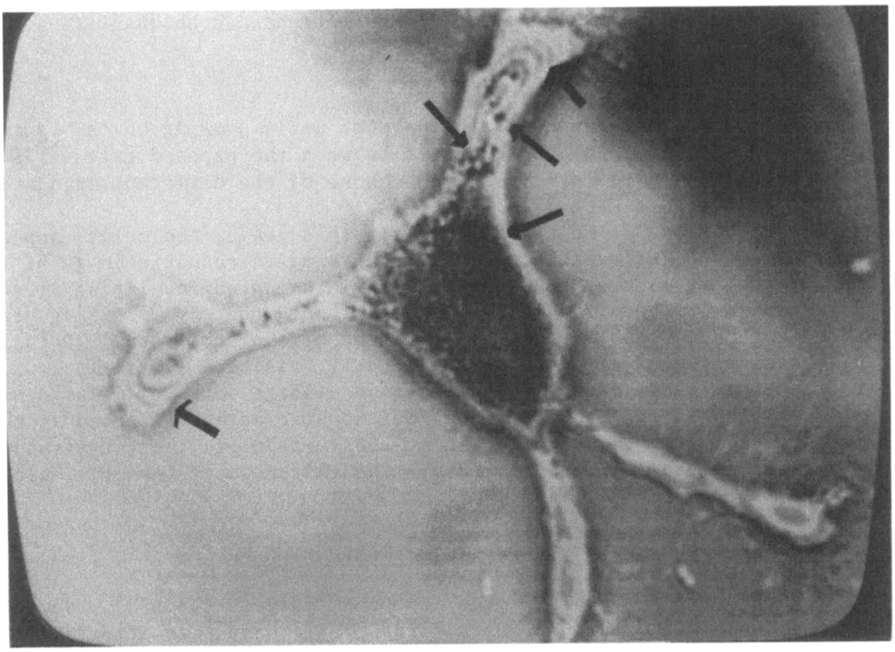

Fig. 6. High magnification reflection acoustic digital image of a single mouse neuroblastoma cell adhering to the laminin-coated Si3N4 surface of a type 1 device.

capable of invading the extraneous surfaces of microelectronic devices. Moreover, the characterization of laminin will be useful to future work, e.g., in studying the electrical properties of bio-electronic devices, where the adhesion of biological components (neurons) will be improved and guided through the patterned interposition of layers of laminin.
Figure 7 shows a set of high frequency (1 MHz) C-V curves. The response to pH of a device with a bare Si_3N_4 gate is compared to the response of the same device after coating its gate with laminin.
Each graph is the average of three full sequences of repeated automatic measurements.

DISCUSSION AND FUTURE WORK

The main results of this paper can be summarized as follows:

i) Both fluorescence digital microscopy and high resolution acoustic microscopy proved to be useful techniques for viewing bio-electronic devices obtained by coupling living neuroblastoma cells to ISFET-like devices.
Fluorescence microscopy is useful in imaging stained living cells. In the future, fluorescence microscopy will be used mainly to obtain "functional" images of cells, that is, intracellular distributions of free Ca^{++} (11) and, possibly, variations in membrane potentials.
Acoustic microscopy is particularly useful in analyzing the surfaces (and the internal regions) of microelectronic devices. In the future, it will be used to inspect the modifications induced by a biological material in the silicon devices. Moreover, the possibility of imaging

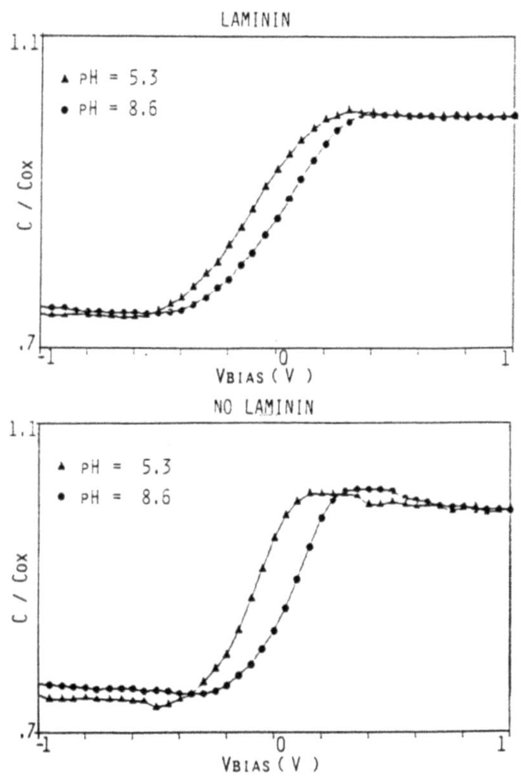

Fig. 7. High frequency capacitance-voltage curves as a function of pH, in presence/absence of laminin.

the local acoustic properties of unstained living cells will be exploited to characterize at the microscopic level, the adhesion of cells to the substrata.

ii) The coating of laminin, which simulates a natural process of biological invasion of artificial material, does not destroy the electrochemical properties of the devices. In fact, the sensitivity to pH of the Si_3N_4 EIS system is preserved, even if somehow reduced. In the future, electric measurements will be extended to EIS systems and real microelectronic ISFETs covered with living cells.

In conclusion, our results, together with the pioneering finding of Bergveld (3,4) (and with numerous data obtained with surface microelectrodes) , suggest the feasibility of a direct detection and amplification of cellular electrochemical signals by use of transistors put in intimate contact with cells.

ACKNOWLEDGEMENTS

Work supported by CNR, PF MADESS

REFERENCES

1. A. Grinvald, Real-time optical mapping of neuronal activity, Ann. Rev. Neurosci., 8:263, (1985).
2. J.L. Novak and B. C. Wheeler , Recording from the Aplysia abdominal ganglion with a planar microelectrode array , IEEE Transactions on Biomedical Engineering, BME - 33 : 196, (1986).
3. P. Bergveld, Development , operation and application of the ion-sensitive field effect transistor as a tool for electrophysiology, IEEE Transactions on Biomedical Engineering, BME-19, 5 :342, (1972).
4. P. Bergveld, J. Wiersma and H. Meertens, "Extracellular potential recording by means of a field effect transistor without gate metal, called OSFET", IEEE Transaction on Biomedical Engineering, Vol. BME-23:136,(1976).
5. A. Sibbald, Recent advances on field-effect chemical microsensors, J. of Mol. Electronics, 2 :51, (1986).
6. M. Hoppe and J. Bereiter-Hahn, Applications of scanning acoustic microscopy- Survey and new aspects, IEEE Transactions on Sonics and Ultrasonics, SU-32 :289, (1985).
7. M. Grattarola, M. Tedesco, A. Cambiaso, G. Perlo, G. Giannetti and A. Sanguineti :"Cell adhesion to silicon substrata : characterization by means of optical and acoustic cytometric techniques, Biomaterials, 9 : 101, (1988).
8. W. M. Siu and R. S. Cobbold, Basic properties of the electrolyte - SiO2 - Si - system: Physical and theoretical aspects, IEEE Transactions on Electron Devices, ED -26 :1805, (1979)
9. J. A. Hildebrand, D. Rugar, Measurements of cellular elastic properties by acoustic microscopy, J. Microscopy, 134 :245, (1984).
10. D. Barnes, in :"Mammalian Cell Culture", J. P. Mather ed, (1984), 195.
11. W. G. Wier, M. B. Cannell, J. R. Berlin, E. Marban and W. J. Lederer, Cellular and subcellular heterogeneity of Ca++ in single heart cells revealed by Fura -2, Science, 235 :325, (1987).

INCORPORATION OF ION CHANNELS IN POLYMERIZED MEMBRANES AND FABRICATION OF A BIOSENSOR

T. L. Fare, A. Singh, K. D. Seib, J. W. Smuda,
P. L. Ahl, F. S. Ligler, and J. M. Schnur

Naval Research Laboratory, Bio/Molecular Engineering Branch
Washington, DC 20375-5000

INTRODUCTION

A receptor-based biosensor uses biomolecules with specific molecule-recognition capability to detect chemicals. For this proposed receptor-based biosensor, proteins composed of binding sites and ion channels are embedded in a lipid bilayer. The protein-lipid membrane is deposited onto a transducing electrode.[1,2] When a chemical of interest binds to the receptor site, the ion channel opens to permit an ionic current to flow through the membrane. The lipid bilayer has two purposes: first, it serves as the medium to support the proteins, and, second, it acts as an insulating membrane on the transducing electrode. The presence of the chemical of interest may thus be detected by the ion current.

Receptors with ligand-gated channels have several advantages as detection elements. Receptors (such as the acetylcholine receptors, olfactory receptors, and GABA receptors) recognize families of chemicals, thus one receptor can be used to monitor a number of physiologically active compounds. Unlike antibody-based sensors, a toxic chemical does not have to be specifically identified before a sensor is constructed. Any chemical which induces the opening of the ion channel is identified as being toxic. The measurement of the physiological function can also be used to detect combinations of chemicals, each of which may not be harmful, but which act in tandem to become toxic. Since the detection mechanism is based on a physiological response rather than solely on ligand binding, the incidence of false positives is greatly reduced. The potential advantages of such systems include exquisite sensitivity, tailored specificity, and speed of response (seconds).

In the work described here, a three-pronged approach has been taken to the development of an amperometric receptor-based biosensor. First, asymmetric lipid bilayers have been investigated to produce more stable membranes. These asymmetric membranes include a polymerized lipid monolayer designed to enhance membrane stability and a nonpolymerized monolayer in which the lipids are chosen so as to optimize function of the receptor protein. Preliminary work on these asymmetric membranes was reported by Dalziel et al.[3]

Second, two moieties which form voltage-induced channels were tested in the asymmetric membranes. Alamethicin was selected because it is well characterized and commercially available, is stable for long periods at room temperature, and can be added directly to the aqueous phase without additional detergents or lipids. Most important, alamethicin does not function in membranes which are thicker than bilayers,[4,5] so its function implies that a bilayer and not a multilayer has been constructed. The calcium channel, by comparison, is a much larger, mammalian protein,[6-8] semipurified from bovine brain the week of the experiment. The calcium channels used here were selective for barium[9] and appear to be similar to the calcium channels isolated by Takahashi and Catterall[8] (J. Smuda, data not shown).

Third, an electrode has been fabricated which is: (a) porous so that there is an aqueous space below the receptor-containing membrane and (b) hydrophilic so that the lipid membrane adheres well to its surface. A more complex version of this porous electrode with the potential for increased sensitivity is also described.

METHODS

Materials

Asolectin, dipalmitoyl phosphatidylcholine, dimyristoyl phosphatidylcholine, and egg phosphatidylcholine were obtained from Avanti Polar Lipids (Birmingham, AL). Cholesterol was obtained from Calbiochem (La Jolla, CA). Soy lipid was obtained from the American Lecithin Co. (Atlanta, GA); 1,2-bis(12-methacryloyloxy-dodecanoyl)-sn-glycero-3-phosphocholine (MaPC) was synthesized by the method of Regen et al.;[10] 1,2-bis(10,12-tricosadiynoyl)-sn-glycero-3-phosphocholine ($DC_{8,9}PC$) was synthesized as described by Singh and Schnur.[11] Photopolymerizable lipids were checked by thin layer chromatography before the start of an experiment to ensure that all the lipid was in the monomer form. Alamethicin was obtained from Sigma (St. Louis, MO). Calcium channels were isolated from bovine basal ganglia by differential centrifugation and ion exchange chromatography and incorporated into asolectin/cholesterol (4:1, w/w) liposomes in the presence of 10 mM 3-[(3-cholamidopropyl)dimethylammonio]-1-propane sulfonate (CHAPS) (Sigma) by the method of Borsotto et al.[6]

Experiments involving calcium channels were performed using 10 mM $BaCl_2$, 20 mM HEPES pH 7.5 buffer. All other experiments were performed using 150 mM NaCl, 5 mM $CaCl_2$, 1 mM $MgCl_2$, 10 mM HEPES, 1 mM $MnCl_2$, pH 7.4 unless otherwise stated. All water used was double distilled. Hexane, chloroform, and methanol were chromatographic grade from Fisher Scientific (Pittsburgh, PA).

Analysis of bilayers and ion channel function

Microelectrode current measurements through lipid bilayers were performed using the patch clamp apparatus as described by Coronado and Latorre,[5] using equipment previously specified.[12] Briefly, lipid films were spread over buffer in a home-made Langmuir trough. Monolayer of asolectin were spread from a 5 mg/ml hexane stock solution by carefully adding the lipid to the air-water interface.[13] Asolectin/cholesterol (4:1 w/w) films were formed by self-assembly of monolayers from liposomes at the air/water interface[14] and used only in experiments involving calcium channels. All polymerizable lipids were spread from chloroform. Polymerization was initiated with a mercury penlamp (Ultra-Violet Products Inc., San Gabriel, CA) placed above the monolayer film. Asymmetric membranes were made by first passing the patch pipette up through the asolectin monolayer and then passing the same pipette down through the polymerized lipid film.

Alamethicin stock solutions of 10 μg/ml in $MeOH/H_2O$ (1:3) were prepared to prevent alamethicin aggregation. Alamethicin (1 μg/ml buffer) was added to the NaCl buffer in the subphase and partitioned into preformed bilayer membranes by diffusion.[15] For the incorporation of calcium channels, vesicles containing the channels were added to the $BaCl_2$ buffer (5 μg lipid/ml buffer) in the trough following the method of Schindler.[14,16] The asolectin/cholesterol film containing protein formed at the air-water interface. The patch electrode was passed upward through this film and subsequently down into the second monolayer (asolectin/cholesterol or $DC_{8,9}PC$) to form the bilayer.

Fabrication of the porous electrodes

Porous electrodes were constructed from an etched silicon substrate with either a platinum wire (backside contact) or gold coating (frontside contact) for electrical connection. Some silicon samples were anodically bonded to pyrex glass slides and modified with the formation of a porous silicon surface as described below.

To construct the device, the silicon substrate was first coated with a layer of silicon dioxide and then a layer of poly-silicon. Microlithography was used to define a strip down the center of the device. The poly-silicon and the oxide were etched away from the strip and the exposed silicon substrate is doped to a p^+ layer. A gold contact was deposited on the end of the strip (the small square apparent at the right end of the small device). The exposed substrate surface measured approximately 1 mm by 5 mm.

The porous silicon is formed in the exposed degenerately doped center strip by passing an anodic current through the exposed silicon substrate in an HF/ethanol/water bath (1:2:2) using a platinum electrode for the cathode. The current density is about 10 mA/cm^2 for a period of about 1 minute.[17] The silicon is heated to 300°C in air to stabilize the porous structure and to form a thin hydrophilic oxide on the silicon. A thin film of platinum (about 25 nm) is then evaporated or sputtered onto the porous surface.

Deposition and analysis of films on porous electrodes

Lipid films were deposited onto the porous electrodes by an LB dipping technique. Monolayers and multilayers of lipids were controllably deposited onto the surface.[13] In the LB dipping method, lipid monolayers were spread on the surface of a Joyce-Loebl Langmuir Trough 4 (Vickers Instruments, Gateshead, UK) equipped with a servo-feedback mechanism to maintain constant surface pressure independent of surface area. The trough is modified to accommodate two ultraviolet radiation lamps (Rayonet Photochemical Reaction Lamp, Cat. no. R.P.R. 2537A, The SO.N.E. Ultraviolet Company, Hamden, Connecticut). The $DC_{8,9}PC$ monolayer is spread from a chloroform or dichloromethane/hexane (70/30, v/v) solution on a distilled water or buffered subphase. After waiting twenty minutes for the solvent to evaporate, the teflon bars of the trough are moved slowly together so that a pressure-surface area curve may be drawn. The bars are drawn together at the rate of about 2-3 mm/min to a surface pressure of 30 mN/m (dyn/cm) which takes about one hour. The film is now allowed to relax for four hours, during which time the surface pressure may drop 5 to 10 mN/m. At this stage, the film is polymerized by exposure to the ultraviolet radiation source for twenty minutes. The teflon bars are held fixed during the exposure so that the area of the film is held constant. The surface pressure of the film increases by 15 to 20 mN/m during the polymerization to a final surface pressure of about 40 mN/m. A final surface pressure of 35 to 37 mN/m is reached after about an hour relaxation.

The trough was equipped with a speed-controllable dipping arm to pass substrates through the standing monolayer. A clean, hydrophilic substrate was passed up through the air-water interface to deposit the first monolayer; subsequent passes of the substrate through the film may result in the deposition of multiple layers. The surface pressure was maintained at 35 mN/m by a servo-feedback. The amount of lipid deposited onto the surface was measured from the reduction in surface area. The transfer ratio was calculated by comparing the change in the surface area of the film to the area of the substrate.

Films of DPPE/DPPA (1:1) were also investigated for controlled lipid deposition. Lipid bilayer deposition onto platinum surfaces were monitored by the transfer ratio and by measuring the impedance of the lipid-coated platinum wire in the trough. The impedance was measured with a EG&G PAR 5208 Lock-in-Amplifier. The counter electrode was a platinum wire at the end of the trough. The subphase was distilled water with 2 mM $CaCl_2$ as the electrolyte. The impedance of the transferred film was measured as a function of the film surface pressure.

RESULTS

Fabrication and analysis of asymmetric bilayers

Individual monolayer films were formed in small, side-by-side Langmuir troughs. The glass patch electrode was passed upward through the first trough and down into the second trough, so that a bilayer was formed and remained in buffer during the analysis. Three different bilayers were fabricated on the patch electrode: asolectin/asolectin, asolectin/MaPC, and asolectin/$DC_{8,9}PC$. With the asymmetric membranes, there was a 10-20% success rate in forming bilayers with 1- 10 gigaohms resistance. This resistance was considered to reflect a good seal between the membrane and the glass electrode and produced data with low noise levels.

The three membranes were analyzed for strength and integrity in the absence of protein channels (see Fig. 1). Voltage increasing to 1000 mV was applied across the membranes over a period of 30 sec. The asolectin/asolectin membranes had a voltage break-down at 386 ± 196 mV (25 tests). The asolectin/MaPC membranes broke at 851 ± 119 mV (3 tests). The asolectin/$DC_{8,9}PC$ membranes were extremely stable, maintaining their integrity at 1000 mV in all 3 tests.

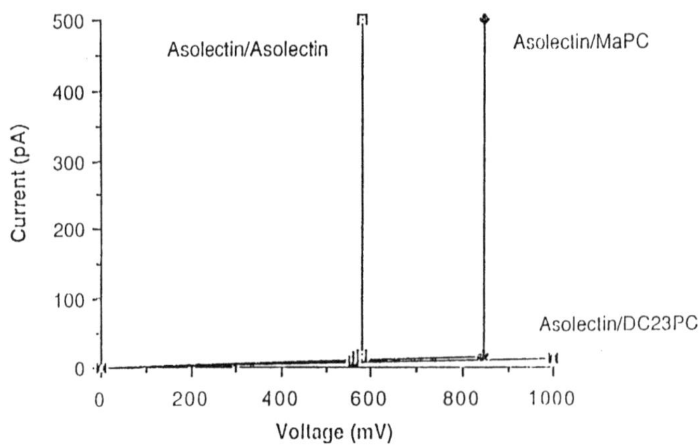

Fig. 1. Stability of lipid bilayers. Voltage across asymmetric bilayers, formed on a patch electrode as described in Methods, was increased from 0 to 1000 mV over a 30 second period. Sharp increases in current reflect breakage of the membranes in the representative tests overlayered above.

Analysis of alamethicin in the bilayer membranes

To ensure that channel-forming proteins could form a pore through the polymerized monolayer and still function, alamethicin was added to the aqueous phase in the trough below the bilayer membranes. The nature of the actual channels formed in each case was quite different (see Fig. 2). In the asolectin bilayer, the channels were open 15 msec or longer and showed the characteristic step increases in amplitude previously described.[5]. A higher voltage was required to form alamethicin channels in the asymmetric membranes. Correspondingly, a lower conductance across the membrane was observed. Once formed, the channels remained open for much shorter times. Step increases in amplitude were visible for alamethicin channels in the asolectin/$DC_{8,9}PC$ membrane, but were difficult to discern with the asolectin/MaPC membrane. Channel open times appeared slightly longer in asolectin/$DC_{8,9}PC$ than in asolectin/MaPC.

Analysis of calcium channels in asolectin/$DC_{8,9}PC$ bilayer

A second channel tested in asymmetric membranes was the calcium channel from bovine brain. In this system, the channels were first incorporated into a monomeric lipid film. The patch electrode was passed up through that

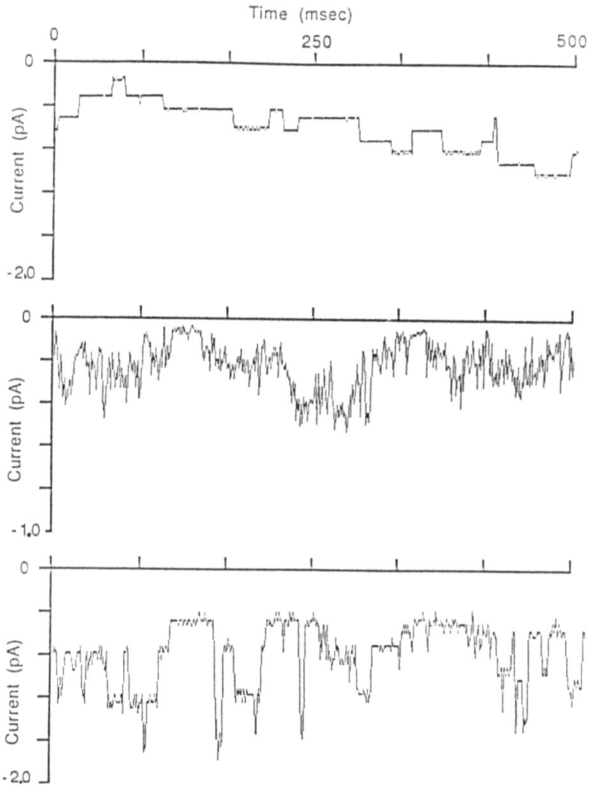

Fig. 2. Amplitude of alamethicin channels. Alamethicin was added to the NaCl buffer in the subphase below bilayers of asolectin/asolectin (upper), asolectin/MaPC (middle), and asolectin/$DC_{8,9}PC$ (lower) and allowed to diffuse into the membrane. Channel formation was measured at 60, 150, and 160 mV, respectively.

monolayer and then dipped through the second monolayer to form the bilayer membrane. Calcium channels integrated into both the asolectin/asolectin and asolectin/$DC_{8,9}PC$ bilayers were functional (see Fig. 3). Since this channel did not exhibit the step increases in amplitude characteristic of alamethicin channels,[16] the amplitude and corresponding conductance of the channels could be resolved with a fairly simple computational analysis. The mean conductance was 2.7 ± 0.8 pS for channels in the asolectin bilayer and 3.1 ± 0.6 pS in the asymmetric bilayer.

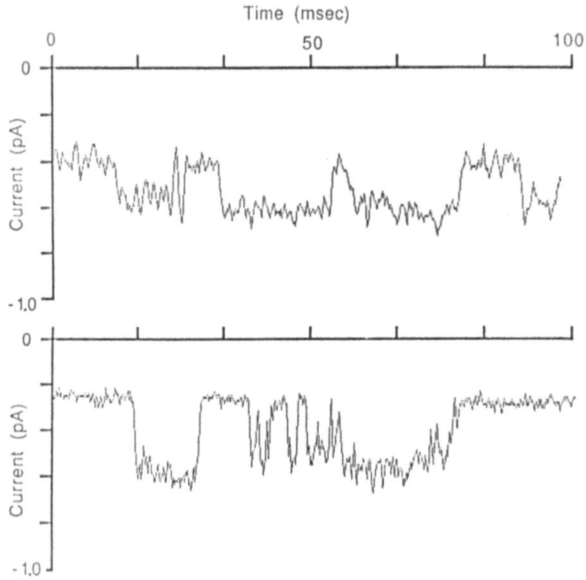

Fig. 3. Amplitude of calcium channels. Patch electrodes were passed through an asolectin/cholesterol film containing calcium channels and then through a monolayer of asolectin/cholesterol (upper panel) or $DC_{8,9}PC$ (lower panel). Channel activity was measured at 50 mV and 80 mV, respectively.

LB deposition of lipid layers

Initial testing of the device focused on the deposition of lipid monolayers and multilayers on the porous surface. Monolayers of dipalmitoyl phosphatidylcholine, dimyristoyl phosphatidylcholine, egg phosphatidylcholine, monomeric $DC_{8,9}PC$, and polymerized $DC_{8,9}PC$ were deposited by passing the device through an LB film. Deposition on the porous substrate was assessed by the transfer ratio of the lipid film calculated from the reduction in area at constant pressure.

Monolayers of polymerized $DC_{8,9}PC$ were prepared as described in the Methods section. Reproducible surface pressure-area diagrams were obtained and the transfer ratio after a single pass through the LB film was consistent with a monolayer film covering the electrode. The first withdrawal of the substrate through the LB film was sufficient to change the surface of the porous electrode from hydrophilic to hydrophobic. While one layer of lipid could be deposited on the device, attempts at depositing multiple layers of monomeric lipid films failed. The films were transferred onto the substrate, but the multilayers would peel off upon immersion or removal of the substrate from the trough during dipping. Multiple monolayers of $DC_{8,9}PC$ could be deposited as long as the film was polymerized prior to transfer.

For the DPPE/DPPA films, lipid bilayers formed much more readily on platinum wires. The impedance between a coated and an uncoated electrode was measured in the trough. The capacitance is used to measure the coverage of the film. We assume that the coverage, Θ, can be described by the following relationship:[18]

$$C_t = C_{pf}\Theta + C_e(1 - \Theta)$$

where C_t is the total capacitance, C_{pf} is the capacitance of the coated electrode (taken to be 0.1 $\mu F/cm^2$), and C_e is the capacitance of the electrode prior to coating. A plot of $\ln[(1-\Theta)/\Theta]$ v.s. $\ln[p]$ (Fig. 4), where p is the film surface pressure shows a linear relationship with a slope of 1.5 ± 0.1. A strictly Langmuirian adsorption would have had a slope of 1. Current investigations have now focused on using this technique to characterize lipid deposition onto platinum-coated porous silicon. This technique is to be developed as a means to evaluate small area film transfer where conventional film area transfer ratios would be too insensitive.

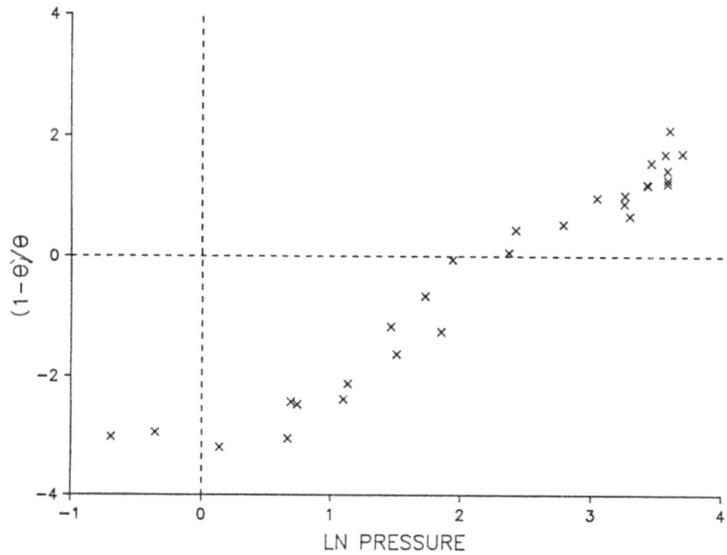

Fig. 4. Log-log plot of coverage v.s. surface pressure. A plot of the logarithmic dependence of the coverage of the platinum wire surface as a function the surface pressure shows a linear relationship over the compression range of the film.

PROPOSED BIOSENSOR CONFIGURATION

Bipolar junction transistors (BJT) will be used as the building blocks for higher order signal processing circuitry. Figure 5 shows the scheme for using the BJT and receptor-containing membranes. A protein positioned over the base of a BJT would inject current directly into the transistor circuit, allowing an amplification of the signal in the collector circuit. The intimate contact of the protein with the base would reduce extraneous signals from the environment.

Circuitry measuring the difference in signals that occur at respective base regions will greatly decrease noise. This is illustrated schematically in Figure 6 showing the base and collector of transistor 1 (at left) and transistor 2 (at right): base 1 is to be coated by an insulating layer and base 2 is to be coated by the receptor-lipid film. By blanking off base 1, the receptor signal at base 2 will be amplified without amplifying the common noise due to external sources at the inputs. More complex circuitry may be

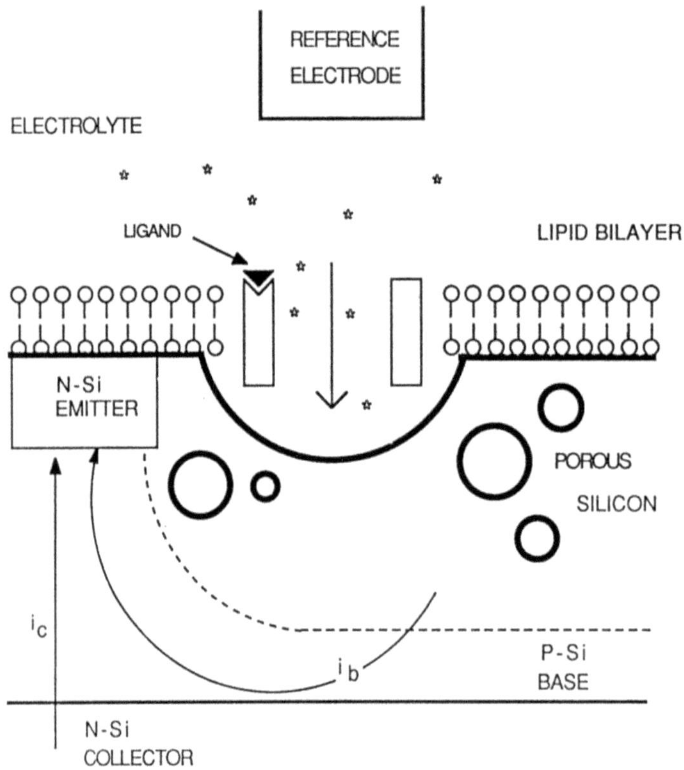

Fig. 5. Schematic of next generation biosensor. The basic unit will take advantage of the charge injection of an ion channel directly into the base of a bipolar junction transistor.

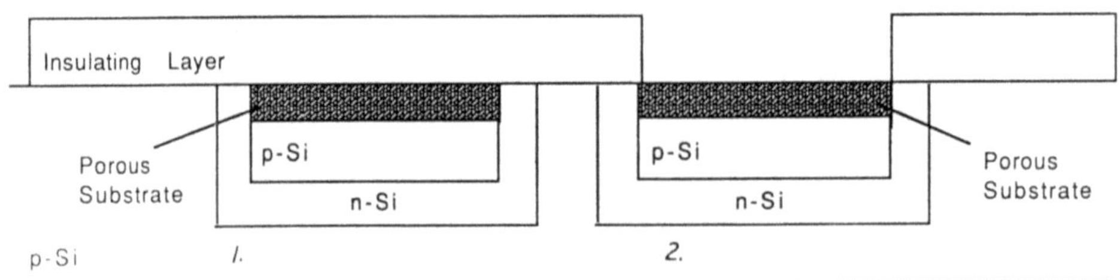

Fig. 6. Schematic of a possible circuit configuration using BJT's. The next version of the porous electrode will include a reference bilayer (base 1, left) and a receptor-containing bilayer (base 2, right) for differential signal analysis.

added to include temperature compensation. Alternatively, arrays of BJTs can be fabricated, each of which is coated with a different receptor or chemically sensitive layer.

For this circuitry, a planar BJT is fabricated in the npn configuration, as shown in Figure 5b. A pnp transistor may also be used. Various technologies now available are used to fabricate such a device and the circuitry to be developed from it.[19,20] In Figure 5b, the cut-away view of the p-type silicon substrate shows an n-type well for the collector with a p-type well for the base. The doping has been performed by alternately masking and ion-implanting into the substrate. The emitter doping well (not shown in the schematic) is to be ion-implanted as an n-type region in the base well. Typical dimensions for this configuration of planar transistors consists of square or rectangular regions which are 8-10 μm for the collector, 4-6 μm for the base, and 2-3 μm for the emitter. The depth of implantation are usually less than 1 μm.

The next step is to passivate the surface by depositing a silicon oxide. A hydrofluoric acid-resistant layer of poly-silicon is deposited on top of the insulating layer. This poly-silicon layer is doped and grounded to act as a shield for noise. A photo-resist is used to define an area over the base. The next step is to form a porous substrate surface on the exposed base region. The samples are now ready for deposition of the lipid bilayers and receptor proteins. Bilayers will be deposited by traditional LB dipping techniques or by the vesicle fusion technique.

The device is placed into a reference bath with controlled pH, temperature, and/or ionic strength. To make contacts to the transistor for biasing or for measurements, a metal film lead may be patterned on the surface, or the device may be biased through the electrolyte solution by a reference electrode. It is preferable to apply the biasing voltages with a metal lead to avoid noise induced between the reference electrode and the base. The current between a reference electrode and the substrate is monitored either in the a.c. or d.c. mode. When a channel is opened in the lipid bilayer, a pulse of current is generated and recorded at the output of the device. The biosensor is operated in a similar mode as with a patch clamp electrode,[3] with excitations to be applied by voltage stimulation or ligand gating.

DISCUSSION

This paper documents progress towards fabrication of a receptor-based biosensor. Three critical components have been fabricated: a stabilized membrane in which ion channels can function, a porous surface compatible with the channel-containing membrane and electronic components, and a small device which can be used as a test bed to measure ion channel activity for a variety of receptors.

Prior to the development of the asymmetric membranes, experiments with symmetric polymerized bilayers were conducted. Some protein, such as bacteriorhodopsin, tolerated the exposure to the ultraviolet light used to polymerize the $DC_{8,9}PC$ and continued to function in the symmetric polymerized membrane.[21] Others, such as the acetylcholine receptor, ceased to function.[6] Attempts to characterize bilayers of $DC_{8,9}PC$ or MaPC using the patch clamp apparatus failed since neither membrane, whether transformed in the monomeric or polymeric form, adhered to the bare glass electrode.

The asymmetric membranes formed from a monomeric lipid monolayer and a polymerized lipid monolayer were rugged and easily manipulated. The membranes were of high quality as judged from the low noise levels in the conductance data. Such membranes are adaptable in that the composition of the monomeric monolayer can be modified to optimize the function of whatever receptor is integrated into the membrane.

Ion channels incorporated in the asymmetric bilayer formed pores through the polymerized monolayer. Alamethicin[7-10] was selected because it is well characterized and commercially available, is stable for long periods at room temperature, and can be added directly to the aqueous phase without additional detergents or lipids. Most importantly, alamethicin does not function in membranes which are thicker than bilayers,[4,5] so its function implies that a true bilayer and not a thicker multilayer has been constructed. The calcium channel, by comparison, is a much larger, mammalian protein,[11-13] semipurified from bovine brain the week of the experiment. The calcium channels used here were selective for barium[9] and appear to be similar to the calcium channels isolated by Takahashi and Catterall[8] (J. Smuda, data not shown).

The channel functioned in the asymmetric membranes, but the function was affected by the polymerized monolayer. Alamethicin has been previously reported to be sensitive to the composition of the surrounding lipid layer.[15] In the asymmetric membranes, the channels exhibited decreased conductance and shorter open times compared to channels in asolectin bilayers. For the calcium channel, however, conductance did not change significantly in the asymmetric bilayers. Conductance values (2.7 ± 8 pS, 3.1 ± 6 pS) were remarkably consistent with the values reported for calcium channels in intact cells.[22]

Recent work at this laboratory has turned to incorporation of proteins in the polymerized lipid bilayers after U.V. radiation exposure. One means to do this is to create fluid regions in the polymerized films by adding a short chain "spacer" lipid to the polymerizable lipid. Preliminary results have shown that bacteriorhodopsin can be inserted into $DC_{8,9}PC$/dinonanoyl phosphatidylcholine vesicles by means of a detergent dialysis.

Lipid deposition onto platinum surfaces has been characterized. It has been found that PE head groups adhere more readily than PC head groups to silica or platinum surfaces. The major problem now is to develop means of depositing platinum on porous silicon surface that will adhere during cleaning and processing procedures.

CONCLUSION

The biosensor described here takes advantage of both the sensitivity, specificity, and physiological relevance of receptor function and the convenience and versatility of silicon based micro-fabrication technology. Asymmetric, polymerized lipid membranes have been fabricated in which ion channels can function. A stable, hydrophilic, porous substrate has been integrated into electrodes to support the membrane. More sophisticated versions of the biosensor will include highly sensitive arrays of these porous electrodes.

ACKNOWLEDGEMENTS

This work was supported by an Accelerated Research Initiative of the Naval Research Laboratory and the Office of Naval Research. Discussions with Paul Yager of the Bioengineering Department of the University of Washington and with Adam Dalziel are gratefully acknowledged and appreciated. The measurements for the impedance of the lipid-coated platinum wires were taken by Robert Wilson from Michigan Technical University. His work is greatly appreciated.

REFERENCES

1. M. T. Jarvis, "Biosensors: Today's Technology, Tomorrow's Products," pp. 155, SEAI Technical Publications, Madison, GA (1986).
2. A. P. F. Turner, I. Karube, and G. S. Wilson, "Biosensors: Fundamentals and Applications," pp. 770, Oxford University Press, New York (1987).
3. A. W. Dalziel, J. Georger, R. Price, A. Singh, P. Yager, Progress report on the fabrication of an acetylcholine receptor-based biosensor, in: "Membrane Proteins: Proceedings Membrane Protein Symposium", S. C. Goheen, ed., pp. 643-673, Bio-Rad Publishing Co., New York (1987).
4. R. Latorre and O. Alvarez, Voltage-dependent channels in planar lipid bilayer membranes, Physiol. Rev. 61:77-149 (1981).
5. R. Coronado and R. Latorre, Phospholipid bilayers made from monolayers on patch-clamp pipettes, Biophys. J. 43:231-236 (1983).
6. M. Borsotto, J. Barhanin, M. Fosset, and M. Lazdvnski, The 1,4 dihydropyridine receptor associated with the skeletal muscle voltage-dependent Ca^{++} channel, J. Biol. Chem. 260:14255-14263 (1985).
7. M. C. Nowycky, A. P. Fox, and R. W. Tsien, Three types of neuronal calcium channel with different calcium agonist sensitivity, Nature 316:440-443 (1985).
8. M. Takahashi and W. A. Catterall, Identification of an alpha subunit of dihydropyridine-sensitive brain calcium channels, Science 236:88-91 (1987).
9. B. Hille, "Ionic Channels of Excitable Membranes," pp. 426, Sinauer Associates, Inc., Sunderland, MA (1984).
10. S.L. Regen, A. Singh, G. Oehme, and M. Singh, Polymerized phosphatidylcholine vesicles. Synthesis and characterization, J. Amer. Chem. Soc. 104:791-795 (1982).
11. A. Singh and J. M. Schnur, A general method for the synthesis of diacetylenic acids, Synthe. Commun. 16:847-852 (1986).
12. P. Yager, Patch clamping of bacteriorhodopsin and its reconstitution in a polymerizable lecithin, Biophys. J. 47:899-906 (1985).
13. G. L. Gaines, Jr., "Insoluble Monolayers at Liquid-Gas Interfaces," pp 326-346, Interscience Publishers, New York (1966).
14. H. Schindler, Exchange and interaction between lipid layers at the surface of a liposome solution, Biochim. Biophys. Acta 555:316-336 (1979).
15. I. Vodyanoy, J. E. Hall, and T. M. Balasubramanian, Alamethicin-induced current-voltage curve asymmetry in lipid bilayers, Biophys. J. 42:71-82 (1983).
16. B.A. Suarez-Isla, K. Wan, J. Lindstrom, and M. Montal, Single-channel recordings from purified acetylcholine receptors reconstituted in bilayers formed at the tip of patch pipette, Biochemistry 22:2319-2323 (1983).
17. M. I. J. Beale, J.D. Benjamin, M. J. Urenn, N. G. Chew, and A. G. Cullis, An experimental and theoretical study of the formation and microstructure of porous silicon, J. Cryst. Growth. 73:622-636 (1985).
18. H. Arwin and I. Lundstrom, Adsorption of a tripeptide on platinum electrodes, I. AC admittance measurements, Surface Science. 140:321-338 (1984).
19. B. G. Streetman, "Solid State Electronic Devices," pp. 463, Prentice Hall, Inc., Englewood Cliff, NJ (1972).
20. J. Millman, "Microelectronics: Digital and Analog Circuits and Systems," pp. 881, McGraw-Hill, New York (1979).
21. P. Yager, Functional reconstitution of a membrane protein in diacetylenic polymerizable lecithin, Biosensors 2:363-373 (1986).
22. J. Ma and R. Coronado, Heterogeneity of conductance states in calcium channels of skeletal muscle, Biophys. J. 53:387-395 (1988).

MOLECULAR SENSOR BASED ON OLFACTORY TRANSDUCTION

Vitaly Vodyanoy*

Department of Physiology and Biophysics
University of California, Irvine
Irvine, California 92717

INTRODUCTION

The initial step in odor perception by an olfactory system is the momentary binding of odorant molecules to receptor sites of olfactory receptor cells. The interaction between odorant molecules and some receptor sites is highly specific. Humans and animals can discriminate between odorants with a high degree of molecular similarity, for example, between optical isomers. The olfactory system is capable of responding at threshold to very small concentrations of airborne chemicals.

The basic idea of the present work is the use of functional reconstitution of the olfactory receptor cell to produce an artificial system responsive to small (subnanomolar) concentrations of odorant for quantitative analysis of the odorant-receptor interaction, study this interaction, and then use this information toward the development of a chemosensitive device.

METHODS

We have used two techniques to transfer native membrane macromolecules into the model system: (1) Chemosensitive membrane fragments were incorporated into large planar lipid bimolecular membranes (BLM) by fusion of the vesicles derived from the olfactory epithelial homogenates with BLM. (2) Chemosensitive membrane fragments were transferred into bilayers formed on the tip of patch electrodes.

The first method of functional reconstitution is based on the transfer of the chemosensitive membrane from vesicles to large planar bilayers. We prepared essentially solvent-free planar BLMs of area 1 mm^2 using a simple modification of the Montal-Mueller method by passing an hydrostatically closed chamber successively through a single monolayer interface.[2,3] The chamber (BLMC-2, Forward Technology Research Laboratory, NJ) consisted of a small hydrostatically closed teflon container with a frontal aperture (Fig. 1). The chamber was attached to a micromanipulator for the dipping operation, using a small trough for monolayer formation. The membrane vesicles were added to the *cis*-compartment of the chamber. Fusion of channel-contain-

*Present Address: Department of Physiology and Pharmacology, Auburn University, Auburn, Alabama 36849

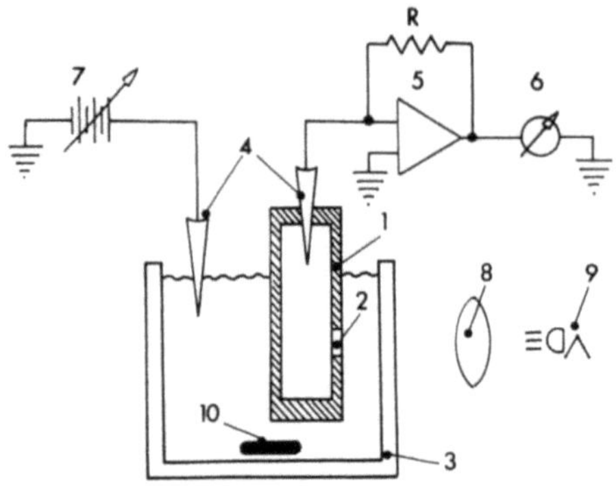

Fig. 1. Experimental apparatus used. (1) cylindrical Teflon chamber. (2) round aperture of about 1 mm^2. (3) glass 10 ml beaker. (4) Ag/AgCl electrodes. (5) operational amplifier. (6) recorder. (7) voltage source. (8) focusing lens. (9) light source. (10) magnetic stirring bar.

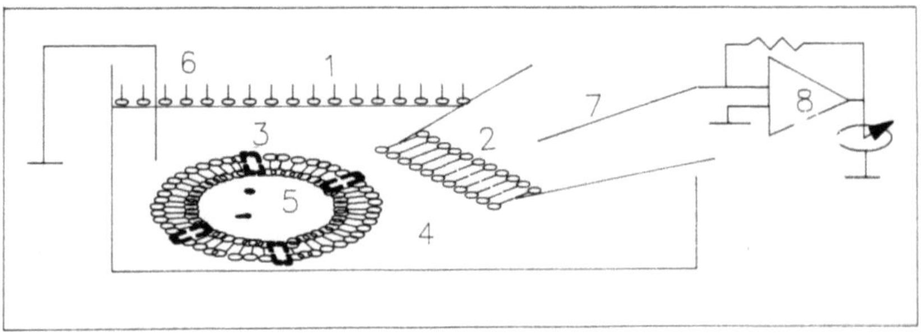

Fig. 2. Reconstitution of receptors into patch electrode bilayers. (1) monolayer. (2) bilayer. (3) ion channel. (4) vesicle. (5) internal content of vesicle. (6,7) Ag/AgCl electrodes. (8) patch-clamp.

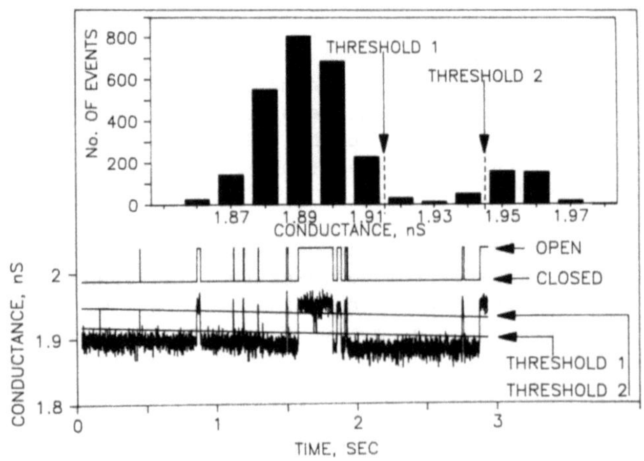

Fig. 3. Single channel conductances activated by 20 μM c-AMP and recorded at +50 mV and identification of single-channel transitions. (See text)

ing vesicles was monitored by the measurement of electric parameters. On rare occasions a transmembrane osmotic gradient of sucrose was used to promote fusion.

The second reconstituted system utilized very small bilayers formed on the tips of patch electrodes. This "tip-dip" method, which is shown in Fig. 2 was introduced by Wilmsen et al.[4] In our modification of the "tip-dip" method the bare bilayer was first formed by the successive transfer of two phospholipid monolayers upon the tip of a patch pipette, and then vesicles were added to the bulk solution bathing the *cis*-side of the preformed bare phospholipid membrane. We do not add lipid to vesicles and we do not expose membrane protein to air on the air-liquid interface.

Preparation of 1.5 μM membrane vesicles from rat olfactory epithelium has been described.[1]

Very small quantities of odorant solutions were injected by the Pneumatic PicoPump PV800 through the micropipette tip of about 1 μm diameter. Both patch-pipettes and micropipettes for injection were made from Glass 1BBL W/FIL 1.5 mm (World Precision Instruments, Inc.). The entire membrane system together with micromanipulator (Narishige MM-3) was contained within a grounded metal box which was placed on a GS-34 Vibration Isolated Lab Bench (Newport Corporation). The water solution in the trough was stirred and temperature was monitored to a precision of 0.4°C.

The electrical properties of the large planar bilayers were measured with a computer based system.[3] The Yale Mk V Patch-clamp system was utilized for the small membranes. All data recorded by the VCR system were analyzed off-line. Recorded single-channel events were subjected to computer analysis of amplitude and time distributions, autocorrelation functions and Fourier spectra.[5] Recorded signals were filtered at 2-5 kHz and sampled at 0.2, 1, 5, or 10 ms intervals. Records were reduced to series of data sets, each containing 7,000 data points. In order to analyze such recordings we used a simple two threshold algorithm[6] capable of identifying single-channel transitions in the presence of substantial levels of noise and drift (Fig. 3). The minimum detectable dwell time (~0.2 ms) was estimated by calibration of detection of brief events.

RESULTS AND DISCUSSIONS

Single channel behavior

Excitation by cAMP. The results of a typical experiment are shown in Fig. 3, which illustrates a small portion of the record. The bare BLM was first formed by the successive transfer of two phospholipid monolayers upon the tip of a patch pipette, and rat olfactory vesicles (final protein concentration about 3 μg/ml) were then added to the *cis*-side of the membrane. Approximately 5 min after that, cAMP (final concentration 20 μM) was added to the same side of membrane and a set of conductance fluctuations was activated (Fig. 3, lower record). Because analyses of these fluctuation events indicated that step heights were uniform in size, we operationally defined the minimal conductance increments that were observed as single unit steps. We attributed such discrete steps to the opening and closing of individual ion channels. The upper portion of Fig. 3 represents the amplitude histogram and the reconstructed record of membrane conductance consisting of the rectangular steps corresponding to open and closed states of the channel. Analysis of single channel fluctuations indicates the existence of an ion channel with a conductance of about 70 pS in 30 mM KCl, 30 mM NaCl, 2 mM $CaCl_2$, activated by cAMP without ATP or GTP. The mean open time was about 1 sec. Subsequent addition of ATP did not change the unitary amplitude of this

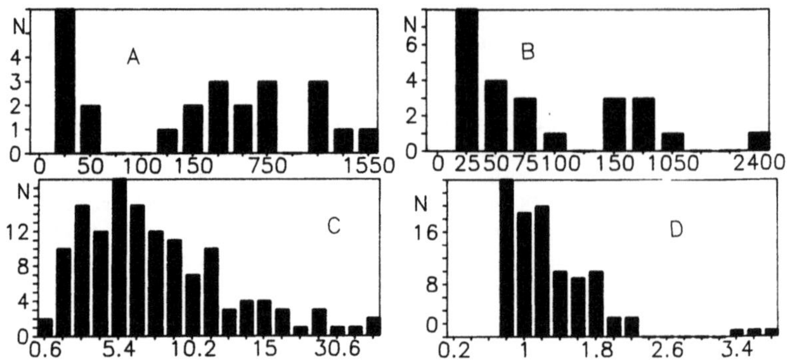

Fig. 4. Frequency distribution of a single channel life times. Time, ms. (A) and (B) - Open and closed time distributions after addition of 20 μM c-AMP. (C) and (D) - Open and closed time distributions after addition of 20 μM ATP. (Sampled at 0.2 ms).

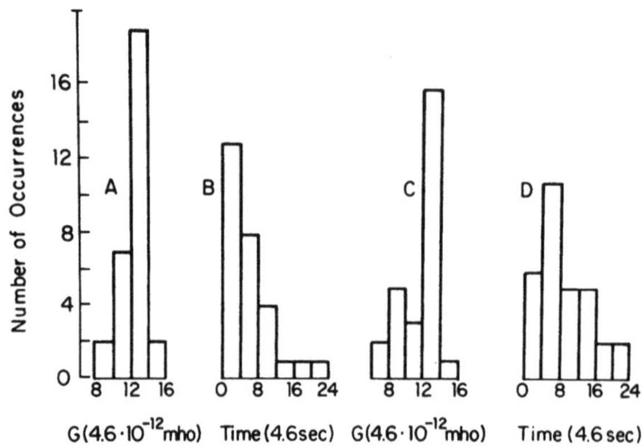

Fig. 5. The amplitude and open time distributions of the spontaneously activated (A) and (B) and the diethyl sulfide-activated (C) and (D) single-channel conductance fluctuations in BLMs which have been treated with rat olfactory homogenates.

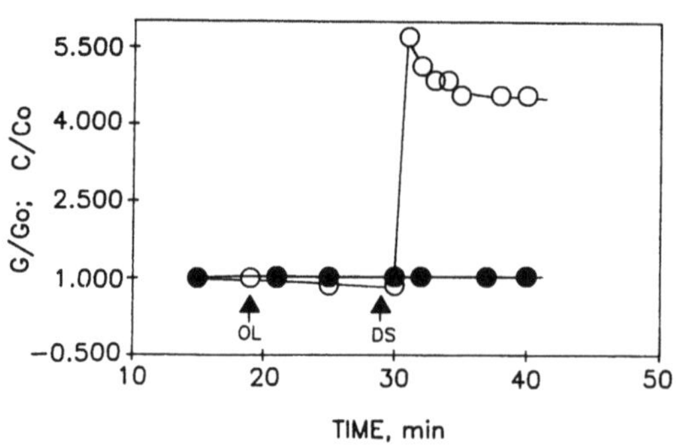

Fig. 6. Variation of the relative dc conductance (open circles) and the relative capacitance (filled circles) with time for a planar lipid bilayer. Addition of rat olfactory epithelial homogenate (3 μg/ml) is denoted by "OL"; the addition of odorant diethyl sulfide (4 nM) is denoted by "DS".

channel but caused a significant decrease of the mean open time to 6 msec (Fig. 4).[5] The histogram shows also a maximum in the time distribution of the open state (Fig. 4C). We believe that phosphorylation by ATP may take place, which may modulate the mean open time in our reconstituted system. This suggests that cAMP might regulate the channel activity in two ways: (a) directly and (b) via a protein kinase system. These data are consistent with the hypothesis that cAMP is a second messenger in the initial steps of olfaction and that a protein kinase may also be involved in ion channel modulation.[5]

Further analysis of autocorrelation functions and Fourier spectra of the sequences of dwell times indicated clustering of single-channel openings in the presence of cAMP and ATP. We suggested that a cyclic gating scheme may result in the correlation of successive dwell times, and the irreversible step included in this cycle may require an energy supply to maintain the steady state.[5]

Excitation by odorant. When olfactory vesicles were delivered to a BLM in the presence of ATP (10 μM) and GTP (10 μM) slow single-channel fluctuations of about 60 pS were observed with no cAMP or odorant added. These channels had a mean open time of about 30 s and their activity decayed spontaneously in 20 min. They reappeared after the addition of diethylsulfide (25 nM) and in the presence of this odorant displayed a mean open time of about 40 s. We analyzed statistically the temporal distribution of events. Figure 5B demonstrates the time distribution of the spontaneously activated channels. The histogram shows the simple logarithmic character of the distribution which can be explained by the independence of the individual events. In contrast to this picture, the time distribution of the fluctuations activated by diethyl sulfide is illustrated in Fig. 5D. The histogram shows a maximum in the time distribution of the open state similar to the maximum of open time distribution for the fluctuations excited by cAMP in the presence of ATP.

The differences in the temporal distributions of the presently observed single ion channels upon chemical activation under our model conditions could be physiologically significant. It has been shown[7] that the distribution of the number of events in the frog olfactory receptor cells is correlated with the concentration of the odorous stimuli, which is consistent with our interpretation of the present data.

Multiple-channel state

Odorant concentration dependence. The record from a typical experiment with a large BLM is illustrated in Fig. 6. It is seen that the addition of the olfactory vesicles alone results only in a small decrease in conductance; the capacitance is not perturbed within the limit of experimental precision. Upon the addition of diethyl sulfide (~4 nM) in the presence of ATP and GTP a large increase in conductance is evident within several minutes. After the overshoot in conductance, a stationary value is reached. The reaction of the membrane conductance to the odorant appeared to be dose-dependent with a threshold for chemical excitation of about 10^{-11} M. The stationary value of membrane conductance can be expressed by the number of open channels and conductivity of a single channel, obtained in the single-channel experiments. The dependence of the number of open channels per unit concentration of diethyl sulfide upon the number of open channels is shown in Fig. 7. The slope yields a binding site-odorant dissociation constant K_D of 1.2 nM. This value appears to be in reasonable accord with the K_D observed for the association of [^{35}S]-diethyl sulfide with olfactory homogenate, 2.2 nM.

Electric field dependence. Membrane conductivity can be estimated from the slopes of the linear parts of the experimental I-V characteristics and confirms the applicability of a single-barrier model for the channel conductivity:[8] $\ln\{i/[1 - \exp(-eV/kT)]\} = \ln i_o + aeV/kT$, where i_o is an exchange current, a is a coefficient of the barrier symmetry, V is the transmembrane voltage, and e, k, T have their usual meanings. The results were obtained with the individual experimental I-V curves which represents the conductivity over a broad variation in the number of incorporated channels. The experiments yield a mean value of $a = 0.31 \pm 0.11$ (S.D.) over the range of i_o from 1.8 - 128 pA. From the results we may conclude that the electrical properties of an individual channel determine the I-V characteristics of both single-channel and multiple channel membranes. Furthermore, variation of

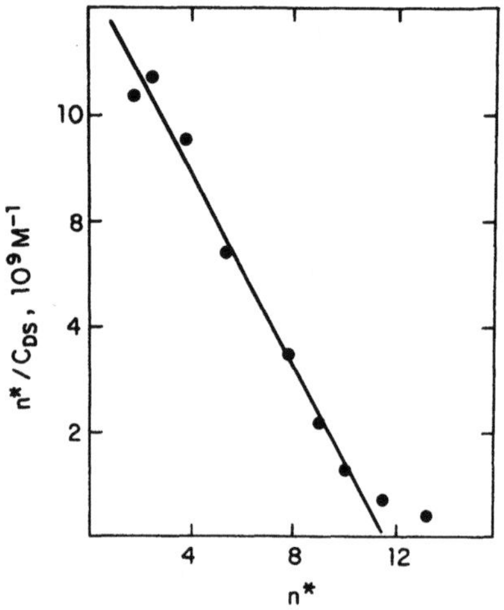

Fig. 7. The dependence of the number of open channels per unit concentration of diethyl sulfide upon the number of open channels. The circles represent experimental data. The line represents the fit of Langmuir isotherm with $K_D = 1.2\times10^{-9}$ M, $N_{max} = 11.5$.

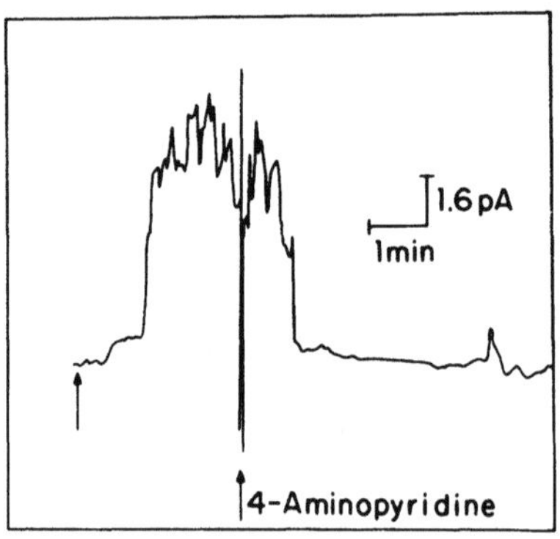

Fig. 8. The effect of 4-aminopyridine on the conductance of planar bilayer to which rat olfactory epithelium homogenate (5 μg/ml) was added (25 min prior to the far right arrow). 15 min prior to the far left arrow the membrane was clamped at 30 mV. The right arrow represent the addition of 4-aminopyridine (100 μl of 12.5 μm to give final concentration of 0.125 μM).

voltage (0-80 mV) does not influence the number of open channels. In this sense, the olfactory homogenate channels do not appear to be voltage-gated.

Ion selectivity and specificity. It has been shown that conductance of reconstituted membranes is directly proportional to potassium ion concentration and, in the absence of K^+, it is independent of sodium ion concentration below 0.08 M. The further evidence of potassium channels was obtained with 4-aminopyridine, a well-known specific blocker for K+ channels,[9] which has also been shown to inhibit the electrophysiological response of the olfactory epithelium. Figure 8 shows the result of adding 4-aminopyridine, when within a minute the odorant excited conductance was deactivated essentially to basal levels.

Second messenger. In our experiments the conductivity of the BLM treated with olfactory epithelial homogenates is increased in response to the odorant if the buffer solution used for homogenate preparation contains ATP and GTP, or ATP and GTP are added to the media bathing the membrane before the odorant addition.[10] If neither ATP nor GTP was added to the reconstituted system no conductivity increase was observed when diethyl sulfide was added. Experiments with the direct application of cyclic AMP suggest that cyclic AMP can mimic the effect of the odorant.[10] These data are also consistent with the hypothesis that cAMP is a second messenger in the initial steps of olfaction.

Olfactory sensor

Olfactory receptor based biosensor. It is well known that olfactory systems in living organisms provide excellent sensitivity, selectivity, and response dynamics.[13] Utilization of intact olfactory receptor structure as the transducer component of a biosensor will allow us to employ the unique characteristics of the olfactory system (Table 1). Several techniques appear practical for using the olfactory receptor to make chemical biosensors. One of the important questions is what structural level of reconstitution must be achieved to preserve the exceptional properties of the olfactory system and to accomplish a certain practical readout. For example, the specific binding olfactory receptor proteins (Table 1) can be immobilized on artificial membranes and coupled to any electric capacitance or charge sensitive device.[13,14] Because the estimated electric current response associated with specific binding of a single molecule appeared to be very small (Fig. 9A) compared with the 1/f noise, the biosensor based on this principle cannot be very sensitive. A different and, perhaps, better approach utilizes the reconstituted olfactory receptor subunits assembled in their membrane form together with the associated ion channels as the signal transducing portion of the sensor. In this case the specific binding of a single molecule can cause, in principle, an opening of the ion channel[1] (Fig. 9B), and can increase the current response by a factor of $\sim 10^3$. An additional amplification by a factor of $\sim 10^6$ can be achieved by reconstitution of the whole receptor, when the specific binding at the membrane surface receptor site causes the synthesis of the second messenger molecules[5,10] (cAMP or cGMP) at the inner layer of the membrane, and then these molecules in turn open millions of ion channels (Fig. 9C). Subsequent levels of reconstitution include the use of the whole olfactory receptor cell synapsed with the secondary neurons of the olfactory bulb.[15]

Bilayer reconstitution. The rationale behind bilayer or bilayer-like membranes which contain some biological material is that the binding of some small molecule to a protein will either cause a nonspecific change in the membrane which will result in a significant perturbation of its measurable physical properties, or alternatively, in the case of coupling to some putative ionophore, if such an assembly could be functionally transferred or purified and reconstituted, a change in ion currents due directly to the

Table 1. Quantitative characteristics of olfactory system

Characteristics	Value	Reference
Threshold	10^{-13} - 10^{-4} M	16,17
Amplification Mechanisms:		
Air-water partition	10 - 10^{3}	16,18
Area of receptive membrane	10^{2}	1,4
Biochemical	10^{3} - 10^{6}	16
Neuronal convergence	10^{2} - 10^{3}	16,20
Dynamic range	10^{-13} - 10^{-5} M	16
Log-log dose-response slope	0.2 - 0.4	16
Onset kinetics	200 - 300 msec	16,21
Adaptation fast and slow	1 sec and 1 min	16,22,23
Termination kinetics	200 - 300 msec	16,23
Density of receptor molecules	800 - 2,500 per μm^2	16,24
Concentration of receptor molecules	10^{-5} M (mucus)	24
Number of receptor molecules	10^{6} per cell	24
K_D for amino acids	10^{-5} - 10^{-10} M	25
Odorant vapor saturation	10^{-5} - 10^{-3} M	16
Number of neurons:		
Olfactory receptor cells	5×10^{7}	20
Mitral cells[a]	45,000	20
Tufted cell[b]	130,000	20
Number of glomeruli[c]	1,900	20
Number of specific anosmia[d] groups	43	37
Olfactory mucus:		
Surface area (salamander)	28 mm^2	25
Volume (salamander)	0.970 mm^3	25
Viscosity (canine tracheal)	2×10^{2} poise	26
Diffusion coefficient ("odorant")	2×10^{-5} cm^2/sec	27
Diffusion coefficient in air	5×10^{-2} cm^2/sec	12
Olfactory receptor cell:		
Resting potential	−35 - −65 mV	28
Spontaneous activity	0.3 - 2 impulses/sec	28
Excitatory activity	2 - 60 impulses/sec	28
Input resistance	165 - 600 mho	28
Membrane resistance	4×10^{3} Ω/cm^2	28
Single ion channel conductance	30 - 200 pS	1,39,40
Velocity of action potential	0.1 - 0.2 m/sec	28
Olfactory receptor proteins:		
Glycoprotein from cilia	95 kDa	29
Odorant-binding	18 kDa	38
Androstenone-binding	23 kDa	30
Anisole-binding	44 kDa	31
Pyrazine-binding	40 kDa	32
GTP-binding (α subunit)	42 kDa	34
Olfactory marker protein (OMP)	19 kDA	33
Possible second messenger	cAMP and/or cGMP	16
Adenylate cyclase activity (cilia)	25 nM cAMP/mg/min	35
Log-log odorant enhancement of adenylate cyclase activity	0.22	35
Carrier frequency (mitral cells)	50 - 80 Hz	36
Duration of carrier sweep	75 - 150 ms	36
Bandpass (mitral cells)	20 - 40 Hz	36

[a,b]Mitral and tufted cells are extrinsic neurons of olfactory bulb.
[c]Axon terminals of olfactory receptor neurons branch at the olfactory bulb in spherical concentrations of neuropil called glomeruli.
[d]Anosmia - a loss or absence of the sense of smell.

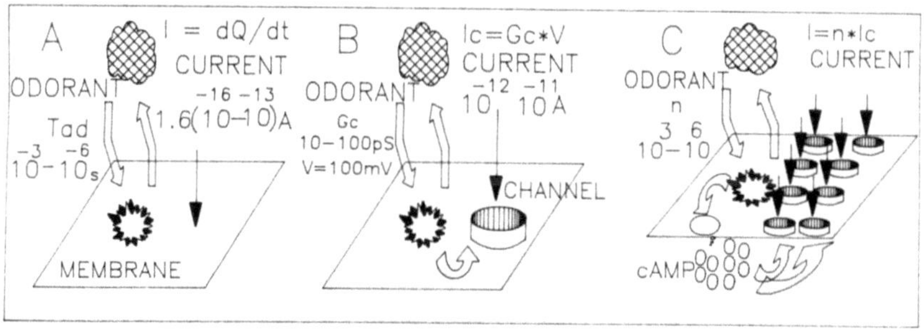

Fig. 9. Hypothetical mechanisms of initial events in olfactory transduction. (A) Momentary binding of odorant molecule to receptor site of receptor membrane. If the time constant of the adsorption (Tad) is 10^{-3}-10^{-6} s and the change of the interfacial charge Q is 1.6×10^{-9} C then the average charging current is $I = dQ/dt = 1.6\times10^{-16}$-$1.6\times10^{-13}$ A. (B) Binding of the odorant molecules causes the opening of a single ion channel associated with the receptor site. If the channel conductance (G_c) is 10-100 pS and membrane voltage (V) is 100 mV the current $I_c = 10^{-12}$-10^{-11} A. (C) Odorant as a primary transmitter activates the synthesis of internal second messenger. One odorant molecule can generate $n = 10^3$-10^6 molecules of second messenger amplifying the channel current I_c by a factor of n.

binding event could be measured. For example, using a membrane electrode, the binding of antigens to antibodies was found to produce an analytically useful response.[11] The hydrostatically closed BLM chamber (Fig. 1) with reconstituted chemosensitive membrane from olfactory receptor can be considered as a prototype of olfactory sensor.

Multiple sensor device. The rationale behind a multisensor assembly is simply that if a sufficient number of sensors which individually possess little chemical selectivity are combined such that the discriminate ability of each device is simultaneously polled, the discriminatory ability of the system as a whole is substantially augmented. A model of such a system would consist of an array of devices which individually would have little differences in electrical response to a variety of chemical agents. Such assemblies would be used to attempt to model response and to obtain a hopefully unique "signature" of a particular small organic molecule. Later, the chemosensitive device would include systems containing biological receptor macromolecules, and ultimately chemically-sensitive electronic device type substrates upon which monolayers containing receptor macromolecules can be deposited. We have constructed[12] a model of such an array, and have demonstrated that we can, with appropriate analog multiplexers, poll the low level currents from the devices in the array within time frames which are short (milliseconds) in comparison with the system response times.

CONCLUSION

We have found and characterized a chemosensitive K^+ channel from rat olfactory epithelium homogenates which can be functionally reconstituted into artificial planar lipid bimolecular membranes. Our data indicate that the steady-state conductance of bilayers modified with olfactory epithelial homogenates became sensitive to very low (subnanomolar) concentration of odorant in the presence of ATP and GTP. We hypothesize that chemosensitivity of the functionally reconstituted olfactory receptor is manifested as a change in the mean open time of single channel events in response to small (subnanomolar) concentrations of the odorants in the medium bathing the membrane under control of the activity of cyclic nucleotide-processing enzymes.

We have demonstrated a model system in which olfactory receptors can be functionally transferred into artificial planar lipid bilayer membranes. We have described a change of membrane steady-state conductance which quantitatively correlates with the odorant concentration. The detection limits of our model system are superior to those of many enzyme-based electrode systems. Our data are preliminary in nature, but on the basis of biochemical evidence our system could be sensitive to a variety of odorants.

We speculate that the detection limits of the olfactory sensor could approach the dissociation constant (e.g. $1/K_{affinity}$) for the membrane receptor. As these constants are found in many receptor systems in nanomolar and in some cases even in the picomolar concentration regimes, it is clear that in principle this type of electrochemical system could have considerable utility in a concentration regime unavailable to current enzyme-based systems. Additionally, this sensor could respond to a number of substances for which no other simple instrumental assay is presently available.

We expect the sensor based on the reconstituted olfactory system to be rapid, sensitive, and selective.

ACKNOWLEDGEMENTS

This work is supported by Grants: U.S. Army DAAG29-85-K-0109, DAAL03-86-G-0131, DAAL03-88-K-0077, NSF Grant BNS-8508495.

REFERENCES

1. V. Vodyanoy and R. B. Murphy, Single channel fluctuation in lipid bimolecular lipid membranes induced by rat olfactory epithelial homogenates, Science, 220:717 (1983).
2. V. Vodyanoy, and R. B. Murphy, Solvent-free lipid bimolecular membranes of large surface area, Biochem. Biophys. Acta, 687:189 (1982).
3. V. Vodyanoy, P. Halverson, and R. B. Murphy, Hydrostatic stabilization of solvent free lipid bimolecular membranes, J. Coll. Inter. Sci., 88:247 (1982).
4. U. Wilmsen, C. Methfessel, W. Hanke, and G. Boheim, in: "Physical Chemistry of Transmembrane Ion Motions," G. Troyanowsky, ed., p. 479, Elsevier, Amsterdam (1983).
5. V. Vodyanoy, Cyclic nucleotide-gated electrical activity in olfactory receptors, in: "Receptor and Transduction Mechanisms in Taste and Olfaction," Joseph G. Brand, and John H. Teeter, eds., p. 319, Marcel Dekker, New York (1989).
6. C. J. Kerry, K. S. Kits, R. L. Ramsey, M. S. P. Sansom, and N. R. Usherwood, Single channel kinetics of a glutamate receptor, Biophys. J., 51:137 (1987).
7. W. Van Dronglen, Unitary recordings of near threshold responses of receptor cells in the olfactory mucosa of the frog, J. Physiol., 277:424 (1978).
8. V. Vodyanoy, I. Vodyanoy, N. Fedorovich, Current-voltage characteristics

of bimolecular phospholipid membranes modified by iodine, Soviet Phys. Semicond., 5:604 (1971).
9. J. Z. Yeh, G. S. Oxford, C. H. Wu, and T. Narahashi, Dynamics of amino-yridine block of potassium channels in squid axon membrane, J. Gen. Physiol., 68:519 (1976).
10. V. Vodyanoy and I. Vodyanoy, ATP and GTP are essential for olfactory response, Neurosci. Lett., 73:253 (1987).
11. Y. Umezawa, M. Kataoka, M. Sugawara, H. Abe, K. Kojima, M. Takinami, H. Sazawa, and Y. Yasuda, Immunosensor systems using liposomes and planar lipid bilayer membranes for ion-channel model sensors, in: "Biosensors International Workshop," R. D. Schmid, ed., p. 127, VCH Publishers, New York (1987).
12. V. Vodyanoy, Olfactory sensor, in: "IEEE 1988 Engineering in Medicine and Biology," G. F. Harris, and C. Walker, eds., pp. 997-998, Core Communications, Arlington, VA (1988).
13. R. M. Cohen and J. Janata, Measurement of excess charge at polarized electrodes with field effect transistors, Electroanal. Chem., 151:41 (1983).
14. K. K. Likharev, Single-electron transistors: electrostatic analog of the dc squids, IEEE Trans. Magnet., 23(2):1142 (1987).
15. S. P. Fracek, Jr., G. W. Gross and R. Schafer, An *in vitro* neural network model of the olfactory system, Abstracts 17th Annual Meeting Soc. Neurosci., 13(2):1409 (1987).
16. D. Lancet, Vertebrate olfactory reception, Ann. Rev. Neurosci., 9:329 (1986).
17. F. Patte, M. Etcheto, and P. Laffort, Selected and standardized values of suprathreshold odor intensities for 110 substances, Chem. Senses Flavour., 1:238 (1975).
18. J. E. Amoore, R. G. Buttery, Partition coefficient and comparative olfactometry, Chem. Senses Flavour., 3:57 (1978).
19. T. S. Reese, Olfactory cilia in frog, J. Cell Biol., 25:209 (1965).
20. G. M. Shepherd, Synaptic organization of the mammalian olfactory bulb, Physiol. Rev., 52:864 (1972).
21. R. J. O'Connel, and R. J. Mozel, Quantitative stimulation of frog olfactory receptors, J. Neurophys., 32:51 (1969).
22. D. Ottoson, Analysis of the electrical activity of the olfactory epithelium, Acta physiol. scand. 35 suppl., 122:1 (1956).
23. D. Ottoson, The Electro-olfactogram, in: "Handbook of Sensory Physiology. IV. Chemical Senses. I. Olfaction," L. M. Beidler, ed., pp. 96-131, Springer-Verlag, Berlin (1971).
24. B. Ph. M. Menco, Qualitative and quantitative freeze-fracture studies on olfactory and nasal respiratory epithelial surfaces of frog, ox, rat and dog. II. Cell apices, cilia and microvilli, Cell Tissue Res., 211:5 (1980).
25. T. V. Getchell, F. L. Margolis, and M. L. Getchell, Perireceptor and receptor events in vertebrate olfaction, Prog. Neurobiol., 23:317 (1984).
26. M. King, Viscoelastic properties of airway mucus, Fed. Proc., 39:3080 (1980).
27. J. A. DeSimone, G. L. Heck and S. Price, Physicochemical aspects of transduction by chemoreceptor cell, in: "Perception of Behavioral Chemicals," D. Norris, ed., pp. 205-225, Elsevier, New York (1981).
28. T. V. Getchell, Functional properties of vertebrate olfactory receptor neurons, Physiol. Rev., 66:772 (1986).
29. Z. Chen and D. Lancet, Membrane proteins unique to vertebrate olfactory cilia: candidates for sensory receptor molecules, Proc. Natl. Acad. Sci. USA, 81:1859 (1984).
30. J. N. Gennings, D. B. Gower, and L. H. Bannister, Studies on the receptors to 5α-androst-16-en-3-one and 5α-androst-16-3n-3α-ol in sow nasal mucosa, Biochim. Biophys Acta, 496:547 (1977).

31. S. Price, Anisole binding protein from dog olfactory epithelium, Chem. Senses Flavour., 3:51 (1978).
32. J. Pevsner, R. R. Trifiletti, S. M. Strittmatter, and S. H. Snyder, Isolation and Characterization of an olfactory receptor protein for odorant pyrazines, Proc. Natl. Acad Sci. USA, 82:3059 (1985).
33. F. L. Margolis, Biochemical markers of the primary olfactory pathway: A model neural system, in: "Advances in Neurochemistry," Vol. 1, B. W. Agranoff, and M. H. Aprison, eds., pp. 193-246, Plenum Press, New York (1975).
34. U. Pace and D. Lancet, Olfactory GTP-binding proteins: signal-transducing polypeptide of vertebrate chemosensory neurons, Proc. Natl. Acad. Sci. USA, 83:4947 (1986).
35. Z. Chen, U. Pace, J. Heldman, A. Shapira, and D. Lancet, Isolated Frog olfactory cilia: a preparation of dendritic membranes from chemosensory neurons, J. Neurosci., 6(8):2146 (1986).
36. W. J. Freeman, A physiological hypothesis of perception, Perspect. Biol. Med., 561 (1981).
37. J. E. Amoore, Olfactory genetics and anosmia, Handb. Sensory Physiol., 4(1):245 (1971).
38. J. Pevsner, R. R. Reed, P. G. Feinstein, S. H. Snyder, Molecular cloning of odorant-binding protein: member of a ligand carrier family, Science, 241:336 (1988).
39. R. H. Anholt, Primary events in olfactory reception, Trends Biochem. Sci., 12(2):58 (1987).
40. R. A. Maue and V. E. Dionne, Patch-clamp studies of isolated mouse olfactory receptor neurons, J. Gen. Physiol., 90:95 (1987).

MONOLAYER MOLECULAR RECOGNITION SITES AS A BASIS FOR BIOSENSOR DEVELOPMENT

J.-H. Kim, D. M. Schufeldt, T. M. Cotton, and R. A. Uphaus

Department of Chemistry, University of Nebraska
Lincoln, NE 68588-0304

D. A. Rintoul

Division of Biology, Kansas State University, Manhattan, KS 66506

INTRODUCTION

All molecular recognition systems must possess some, at least, of the three important attributes essential for their functionality; these are: complementarity of size, shape and chemical affinity. Present efforts to develop molecular recognition systems by chemists have relied largely on production of compounds which operate on the basis of the functional complementarity of specific interactions between chemical entities. The more biological approaches (i.e. biomimetic enzyme systems) must stress, as well, the size- and shape- topospecificty of the system. In both cases, however, extensive and involved synthesis of the system is usually required. Described here are molecular recognition systems based on simple monolayer constructions, applicable to a wide range of molecules and capable of being prepared by simple procedures. These systems rely exclusively on the size/shape topospecificity of the target and template molecule, although future efforts will be directed quite probably to the problem of specific chemical group interactions between guest molecules and surrounding matrix material.

The preparation of shape-specific sites in skeletonized monolayers is based on a technique originally described by Sagiv[1-3]. His early studies were based exclusively on amphiphilic dyes in site/guest systems. The following procedure was evolved to prepare systems with an imprinted "memory" which recognized the specific molecule used in their original formation. A surface active dye, which could be called a "template" molecule, was dissolved in hexadecane, along with octadecyltrichlorosilane, and the solution exposed to a clean glass surface. Both species were sorbed onto the glass surface, the dye by physical adsorption and the silane by chemical reaction with the silicate surface. A closely packed monolayer formed on the glass surface in a time scale of minutes. The dye molecules were extracted with solvent, leaving a skeletonized monolayer containing sites which closely approximate the molecular shape of the originally adsorbed dye. Sagiv showed that the system could be reconstituted into its original form by exposure of the skeletonized monolayer to a concentrated solution of the original dye used as the template molecule, whereas dyes of larger size or differing shape were largely rejected and failed to be incorporated.

These early results indicate the feasibility of development of molecular recognition systems and biosensors for a wide range of small molecules, as well as the possibility of extending the application of recognition systems to include larger and more complex compounds.

RECENT EXTENSIONS OF THE SAGIV TECHNIQUE TO OTHER CLASSES OF MOLECULES

The use of the Sagiv technique as originally described was largely ignored until the recent intense efforts to produce functional molecular recognition systems. It has now been shown that this approach is applicable to a range of biologically important molecules[4] such as porphyrins and chlorophylls, as well as to amphiphilic dyes. In the early work of Sagiv, the detection of molecules reconstituted into template sites was by means of optical absorption spectroscopy only; the later studies employed surface enhanced resonance Raman scattering (SERRS) spectroscopy as the method of species detection. SERRS has been shown to have the sensitivity necessary to detect one mole-percent of a suitable chromophore in a monolayer.[5,6] In addition to high sensitivity, the method provides species-specific identification and, in some cases, can give indications of specific interactions and modes of orientation.

Subsequent studies[7] have provided quantitative measures of the densities of template molecules as a function of the SERRS intensity. In addition, estimates of site coverage were provided by underpotential deposition of lead ions into open template sites. This technique makes possible the deposition of a single monolayer of lead onto an electrode surface.

These results have been substantiated in independent studies by Tabushi et al.[8] and Yamamura et al.[9,10] Those studies involved electrochemical measurements on silanized tin oxide electrodes. Additionally, template recognition sites have been produced on silica gel supports by silanization techniques.[11]

It should be noted that the technique of template formation discussed here appears to be more versatile and facile than methods which rely on the production of topospecific sites in polymers using polymerization-inclusion methods. Examples of this approach are to be seen in the efforts of Wulff et al.[12] and Shea et al.[13]

Clearly, monolayer template systems based on skeletonized silanized monolayers offer multiple possibilities for biosensor development. Of considerable interest is the possibility of using larger, more complex biological molecules having the intrinsic capacity to respond to changes in acidity, specific small ions, drugs or toxins. Reported here is an example of the behavior of a large protein which was incorporated into a monolayer matrix. It acts as the functional element which could form the basis for a practical operating biosensor.

STRUCTURAL AND OPERATIONAL CHARACTERISTICS OF CHANNEL PROTEINS

The traditional fluid mosaic model of the cell membrane, that of a phospholipid bilayer containing proteins, cholesterol and other materials, did not explicitly provide for transport of materials into and out of the cell. Various mechanisms were proposed to circumvent this difficulty, which was eventually resolved by the isolation and characterization of channel protein systems. These are structures which use α-helix peptide subunits which spontaneously aggregate into oligomeric forms taking the shape of a hollow tube. The distribution of amino acid residues is such that the outside surface is very hydrophobic, to accommodate the system in the lipid bilayer of the membrane; whereas the amino acids making up the inside tube wall are hydrophilic or ionic, to providing an aqueous interior. Channel proteins are oriented perpendicular to the membrane plane, and are sufficiently long to reach from the cytoplasmic cell interior to the external periplasm, or to form a juncture with a channel from an adjacent cell. Because the ends of the structure lie in aqueous milieu, hydrophilic amino acid residues are present in high concentration at the tube edges. The channels thus provide a functional conduit (core diameters are 1-3 nm) for materials passing to or from the cell interior. The channel edges of two proteins located in different cells will readily link to provide a cell-to-cell passageway. This provision is especially important in nonvascular structures, as are present in the eye lens. An important group of these proteins, the ion-channel proteins, has been extensively studied.[14,15] A second group, the junction systems, function as transport conduits between adjacent cells.

Channel proteins are universally distributed in plant and animal tissue, often at very high densities. Amino acid composition has been highly conserved throughout biological evolution. In addition to their function as conduits of water soluble material, most, if not all kinds of these systems possess the important attribute of gating or regulation of flow of materials through the inside core. Thus, these systems act as molecular valves which are controlled by a variety of factors such as pH, membrane voltage gradients, specific cation (e.g. K, Na, Ca) concentrations, calmodulin (phosphodiesterase 3':5'-cyclic nucleotide activator), or other factors. The external edges of channels are the sites of specific groups and structures which respond to a variety of chemical signals, i.e. changes in pH, ionic content, etc. The mode of action of venoms and neurotoxins involves inactivation of the channel by tight and irreversible binding to such triggering end groups. Many disease states and degenerative conditions have been directly related to channel dysfunction. It is obvious that isolation and characterization of channel protein systems will be of great future importance, and it has been asserted that the continuing study of channel proteins will eventually reach an importance which will be comparable to that now provided by the current knowledge of DNA function.[16]

Unlike DNA, however, the channel protein systems presently afford an immediate opportunity for exploitation by incorporation into functional microelectronic and biosensor devices. These devices should have the following advantages: ease of fabrication, fast response time, high specific sensitivity and direct conversion of specific molecular events into electrical signals which can be stored or manipulated. Some preliminary reports support these suppositions. Michaelis et al.[17] have incorporated the glutamate receptor ion channel system into bilayer membranes to produce a sensing device for ion flux; preliminary indications are that one glutamate species may be capable of detection limits reaching the attomolar range. The number of monovalent ions passing through one glutamate channel was estimated at 3×10^7 sec^{-1}, providing an amplification of 10^4 for a single event which binds 1-4 glutamate molecules.

Lipid multilayers have been produced on glassy carbon and appear to have some formalistic resemblance to the operation of natural ion-channel protein systems.[18] Used as analytical electrodes, a significant amplification of the presence of ion analyte appears attainable.

A severe impediment to current applications of channel protein systems is the pronounced limitation on the amounts of materials available in pure form. However, the use of automated peptide synthesizers is making it practical to produce the required amounts of synthetic peptides of predesigned structure.[19] In addition, DeGrado, Lear and coworkers have initiated a program which systematically varies compositional parameters of synthetic peptide structures, to produce various artificial channel-forming systems.[20]

In the present exploratory studies of channel proteins as sensors, a protein was sought with the following characteristics: ready availability in finite amounts (1 mg) and in pure form; previous characterization by biochemical and biophysical techniques; a knowledge of the complete structure and the capacity to form stable monolayers at the air/water interface. This last requirement was not thought to be too severe, since intuition indicates that cylindrical structures with hydrophobic exterior walls and hydrophilic end groups are intrinsically amphiphatic; moreover, because channel proteins are invariably capable of incorporation into bilayer lipid membranes, it appears reasonable that they should form stable monolayers at air/water interfaces. Except for the arguable case of mellitin (membrane disrupting peptide in bee venom), which is know to form stable monolayers[21], it appears that there were few, if any, reports of channel proteins reconstituted into spread monolayers at air/water interfaces or as monolayers on solid supports.

The protein discussed here is MIP 26, the major intrinsic protein of the eye lens, a simple gap-junction system of 26 Kdalton weight. Although less intensively studied than the closely related rat liver gap-junction protein, it is now a well characterized entity and can be prepared in large amounts in very pure form. Moreover, its channel diameter can be varied by changes in pH or other parameters.[22]

A schematic representation of MIP 26 is shown in Fig. 1, based on the structure proposed by Gorin et al.[23] It is composed of a single long peptide of 240 amino acid residues with six highly hydrophobic regions which spontaneously form a rod-like structure with the hydrophobic α-helical regions arranged with their long axes parallel. Continuous lengths of hydrophilic residues interconnect each hydrophobic region, and are especially prevalent at the end of the structure which resides in the cell cytoplasm (including both the N- and C-termini). The actual channel system is a tetramer of such units, so arranged that the residues facing the inside core surface are polar and/or hydrophilic, whereas the residues presented on the external wall are quite hydrophobic. The diameter of the inside core has been estimated at around 1.4 nm, by fluorescent insertion probe techniques.[24] Assuming the usual value for the diameter of an α-helix, the complete, assembled tetramer should have a diameter of around 5 nm and a length of 8 nm.

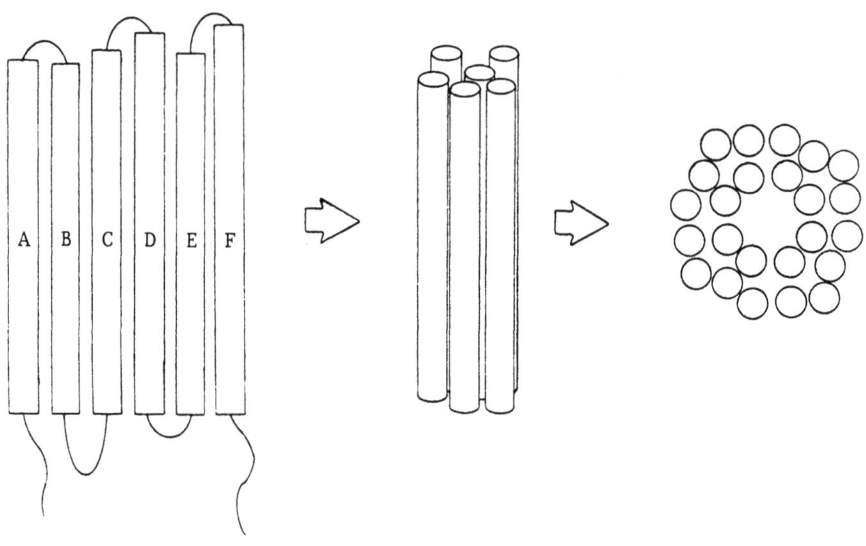

Fig. 1. Schematic of the Structure of MIP 26. Regions A-F are helical and hydrophobic.

The mode of channel opening and closing of MIP 26 is thought to be similar to that seen in the rat liver gap junction system. The latter is a closely related but distinct structure having much the same functions as MIP 26. It has been more intensively studied than MIP 26 and a detailed study of the open and closed forms suggests the following closure mechanism: The axes of subunits are parallel to the long axis of the protein but slightly tilted from the normal. Upon channel closure the tilt angle increases slightly and the subunits slide against each other into a configuration which decreases the channel diameter. In the case of MIP 26, channel diameter is evidently largest at pH 7, since maximum throughput of material is observed at this value. At pH 6 or 8 the channel diameter decreases appreciably.[25]

MONOLAYER CHARACTERISTICS OF THE MIP 26 CHANNEL PROTEIN

MIP 26 was prepared from fresh bovine eye lenses by published procedures;[26] after the final purification steps the protein was present in a nonaqueous phase consisting of 80 % chloroform and 20 % methanol. Chromatography indicated traces of phospholipid and cholesterol were also present. Monolayers were spread directly from the organic solvent system resulting from the preparative procedure. The monolayer trough and associated equipment were as previously described.[27] Appropriate amounts of the protein solution (ca. 0.05 ml) were spread on a subphase buffered with MES buffer (pH 6) or TES buffer (pH 7 or 8).

Surface isotherms of the system at two pH values are shown in Fig. 2. These isotherms were highly reproducible in several independent experiments. Expansion of the completely compressed film, followed by recompression, decreased the final area by several per cent, indicating loss of material to

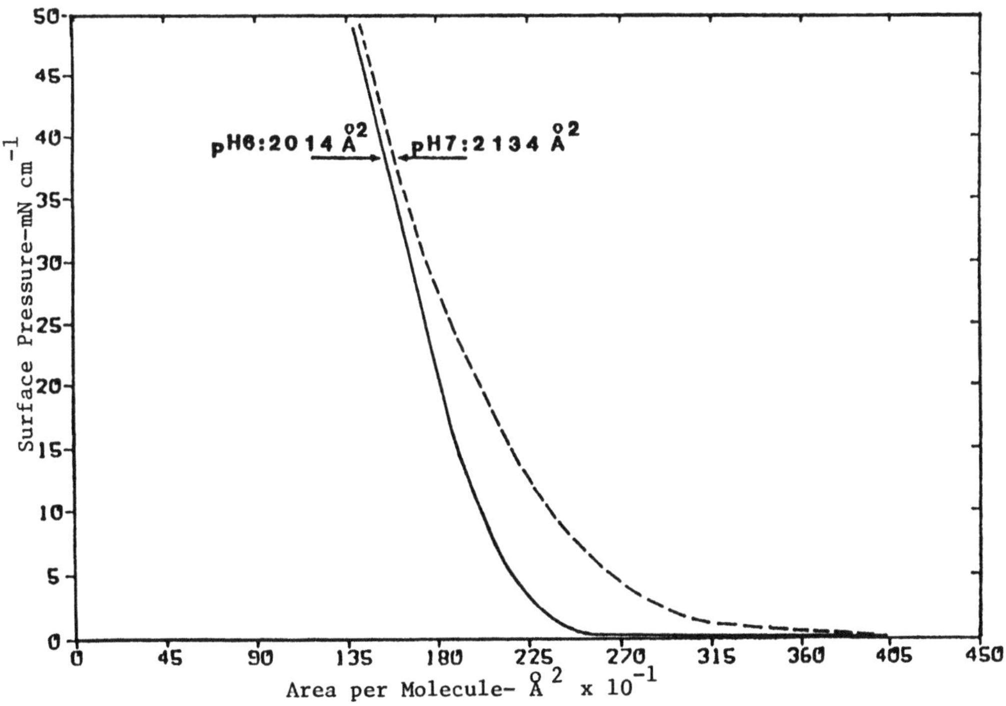

Fig. 2. Surface isotherms of MIP 26.

the subphase and/or a tendency of the monolayer to restructure into a bilayer system. Noteworthy is the great strength of the MIP 26 monolayer; no evidence was seen that the film collapsed at pressures of 50 mN cm^{-1} or more. Extrapolation of the steepest slopes of the curves in Fig. 1 indicated a somewhat smaller area/molecule for the system at pH 6. If the structure of MIP 26 in a spread monolayer were such that the long axis of the system were normal to the water surface, then the variation of molecular size at the different pH's can be attributable to a diameter decrease on the more acid subphase. This is consistent with the independent measurements of Gooden et al.,[25] who measured material flux through the system, reconstituted into vesicles, as a function of pH. Material throughput (i.e. channel diameter) was maximal at around pH 7.

More compelling evidence indicating a perpendicular orientation for the system is found by consideration of the observed areas per molecule. Assuming a tetrameric structure as indicated in Fig. 1, with an α-helix diameter of ca. 7 for each helix and a channel diameter of 14 Å, the experimentally determined molecular areas are consistent with a structure with perpendicular orientation.

The results indicate that a large protein unit such as MIP 26 can be treated much like traditional amphiphilic molecules such as fatty acids. In addition, it appears that MIP 26 could be used as a template molecule much as were the smaller compounds discussed in the Introduction, above. Accordingly, template systems were made up using MIP 26 and octadecyl mercaptan (OM) in methanol/chloroform. A silver electrode surface (to promote the Raman enhancement effect) was allowed to contact this solution for 30 min, after which the surface was washed briefly with methanol to remove excess and unadsorbed solutes. To test for the incorporation of a specific chromophore into the central channel, the monolayer was exposed to an aqueous solution of the dye 2-[4'-hydroxyphenylazo]-benzoic acid (HABA). This dye was chosen because it is insoluble in lipids, though water soluble and therefore probably capable of accommodation in the hydrophilic channel. In addition, resonance Raman and SERRS spectra of HABA-protein systems have been reported (cf. Ref. 24 and citations therein). The presence of HABA in the hydrophilic channel by detection of its spectra can be taken as evidence that the MIP 26 is oriented perpendicular to the silver surface, because end-to-end arrangements of the protein (with possible edge fusion of two units) would permit limited access from the aqueous phase to channels. The hydrophobic character of the external helix walls should promote little, if any, adsorption of dye by the system.

Results of the resonance Raman and SERRS study of this system are given in Fig. 3. All spectra were generated with an argon ion laser at 50 mW power, using the 457.9 nm line. Spectra are the integrated sum of 30 scans taken at a total time of 1 sec with a photodiode detector array. Spectra of Figs. 3A and 3B are the resonance Raman solution spectra of (a) HABA alone and (b) HABA with MIP 26, both in methylene chloride at room temperature. The spectra shown in Figs. 3C-3G are SERRS spectra at liquid nitrogen temperature. Spectrum of Fig. 3C is the SERRS of the system shown in Fig. 3B; Fig. 3D shows the spectrum of the HABA-MIP mixture with on OM present; Fig. 3E is a spectrum of the complete HABA-MIP-OM system; Fig. 3F the spectrum resulting from exposure of the surface to excess HABA, with attempt to wash off excess HABA solution; Fig. 3G the spectrum of the same system after washing and sonicating in methanol for 5 min. These results indicate that the HABA is strongly bound to the protein. Additionally, the normal resonance Raman (Fig. 3B) and the SERRS spectrum (Fig. 3G) differ, chiefly in the low frequency region (peaks at 1622-23 $cm^{-1}$and 1604-08 cm^{-1}) which can be interpreted as a specific HABA-protein interaction. A more detailed analysis of these spectra will be presented elsewhere.

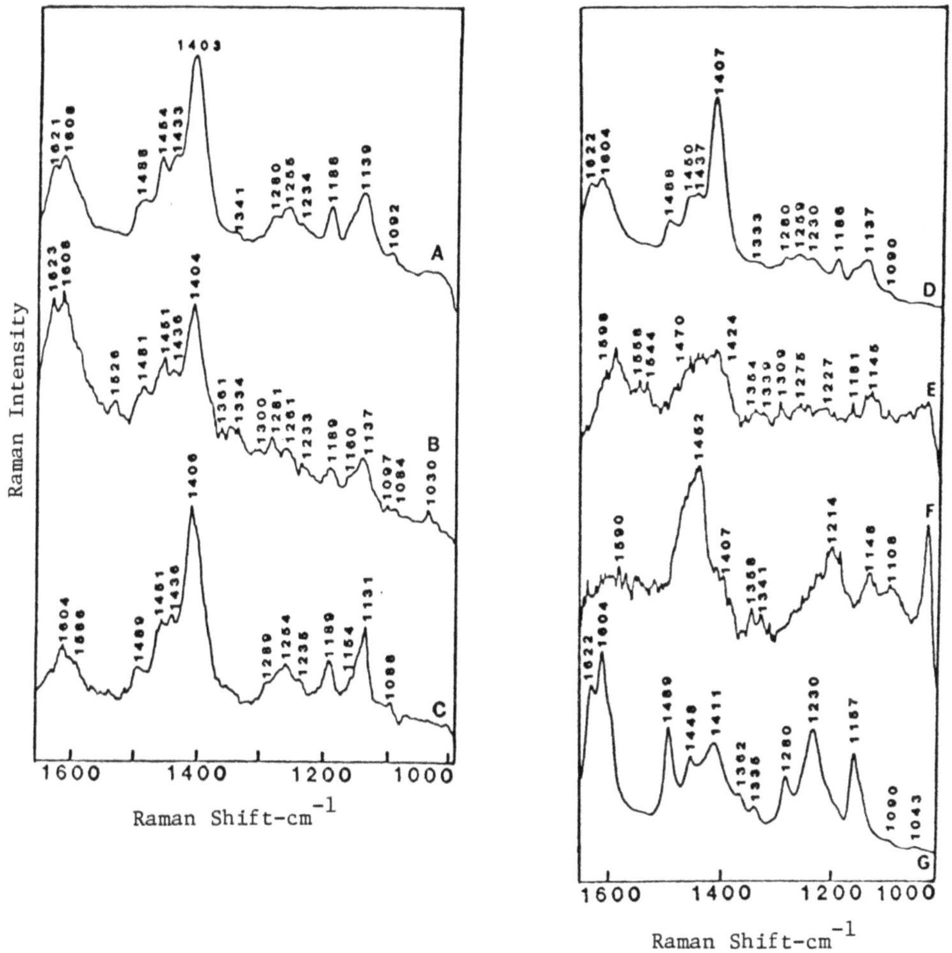

Fig. 3. Raman spectra of the MIP 26-OM-HABA systems.

The spectral results given above suggest, but do not prove, that the orientation of the MIP 26 system is perpendicular to the surface upon which it is adsorbed, and that the dye HABA is free to enter the channel from the exposed end of the protein. Results of electrochemical experiments support this assumption. If exposed channel ends are available at a solution interface and allow small ions to move freely through the channel to an electrode surface, then cyclic voltammmetry can be applied to demonstrate this. The underpotential deposition of lead ions onto an electrode surface was used successfully in a previous study of estimates of template site areas.[7] This technique was applied in the present instance to the MIP-OM template system. The system was exposed to an aqueous solution of HABA, as in the SERRS studies, with the objective of producing a HABA-blocked channel. The MIP 26-OM-HABA system was prepared on a silver electrode and briefly washed to remove excess surface HABA. The electrode was placed in a 1 mM solution of $Pb(NO)_2$ and cyclic voltammetric measurements initiated. The electrode potential was swept (50 mV sec^{-1}) from 0.00 to -0.40 volts vs. SCE, in order to produce a deposition of one monolayer of lead atoms on the electrode surface exposed at the ends of the MIP 26 channels. The system was purged continuously with nitrogen and stirred.

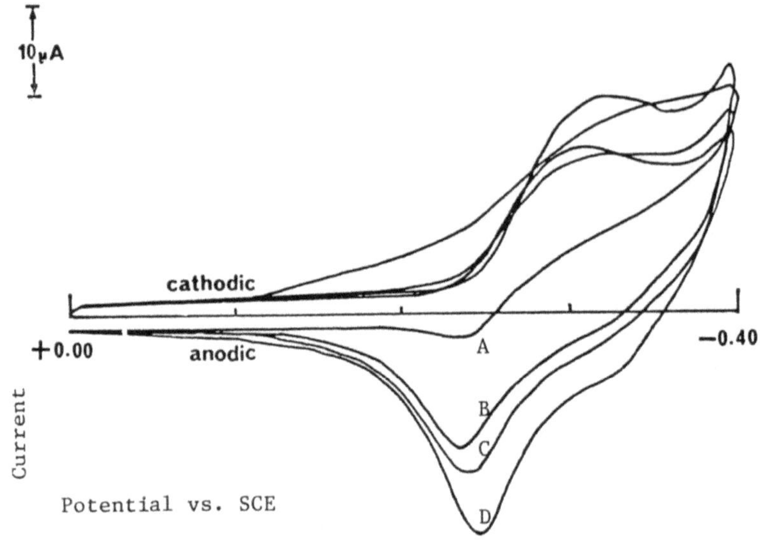

Fig. 4. Cyclic voltammetry of lead ions in MIP 26 Channels blocked with the dye HABA.

Results are shown in Fig. 4 for four redox cycles recorded at 4 min intervals. The areas under the curves for the anodic cycles can be compared as indicating the relative amounts of lead deposited during each cycle. Curve of Fig. 4A is the first, and Fig. 4D the final cycle of deposition. Curve of Fig. 4A indicates only a small ion flow present in the system, comparable to that seen in previous studies on completely covered OM or silylated surfaces. Curves of Figs. 4B-4D indicate successively larger ion currents, as presumably the HABA slowly diffuses out of its sites in the channels. A fifth cycle (not shown) was essentially identical to that shown in Fig. 4D, indicating that no further increase in channel transport is taking place.

These results are interpreted to indicate that the HABA is specifically occupying the hydrophilic channel of MIP-26, rather than present as a diffusely adsorbed layer on the hydrophobic exterior of the protein. In the latter situation the dye would be removed very rapidly by stirring action. The spectral results Fig. 3G indicate that the HABA remains in the system even after washing and sonication in methanol; thus, it would be expected that removal of HABA would be relatively slow and diffusion-limited, as inferred above from the underpotential experiments.

In summary, it has been shown that a typical channel protein has the potentiality for incorporation into a working biosensor device, for the following reasons: (1) It forms traditional monolayers on water, of high strength and stability; (2) It is probably favorably oriented to present its channel opening to the outside milieu; (3) it can be used as a template molecule on an electrode surface, much as other, smaller molecules have been used; (4) Its hydrophilic channel is a convenient cavity for the incorporation of a variety of small, water-soluble molecules; and (5) its capacity for control of material flow by modulation of pH or other factors indicates a potential use as a "molecular valve". Events taking place in the channel are readily manifest as electrical signals. The application of the monolayer template technique to other channel proteins, such as specific ion-channel systems, could produce a variety of species-specific sensor devices.

ACKNOWLEDGEMENTS

This study was supported by the DOE Chemical Sciences Division (DE-FG02-84ER-13261) and NIH (R01-EY05197-1A1).

REFERENCES

1. J. Sagiv, J. Am. Chem. Soc., 102:92 (1980).
2. J. Sagiv, Isr. J. Chem., 18:339 (1979).
3. J. Sagiv, Isr. J. Chem., 18:346 (1979).
4. J.-H. Kim, T. M. Cotton and R. A. Uphaus, Thin Solid Films, 160:389 (1988). Cf: Abstracts, Third International Conference on Langmuir-Blodgett Films, Götingen, FRG, July, 1987.
5. R. A. Uphaus, T. M. Cotton and D. Möbius, Thin Solid Films, 132:173 (1985).
6. T. M. Cotton, R. A. Uphaus and D. Möbius, Phys. Chem., 90:6071 (1986).
7. J.-H. Kim, T. M. Cotton and R. A. Uphaus, J. Phys. Chem., 92:5575 (1988).
8. I. Tabushi, K. Kurihara, K. Naka, K. Yamamura and H. Hatakeyama, Tetrahedron Lett., 28:4299 (1987).
9. K. Yamamura, H. Hatakeyama, K. Naka, I. Tabushi and K. Kurihara, J. Chem. Soc., Chem. Commun., 79 (1988).
10. K. Yamamura, H. Hatakeyama and I. Tabushi, Chem. Lett., 99 (1988).
11. Y.-T. Tao and Y.-H Ho, J. Chem. Soc., Chem. Commun., 417 (1988).
12. G. Wulff, ACS Symp. Ser., 308:186 (1986).
13. K. J. Shea and T. K. Dougherty, J. Am. Chem. Soc., 108:1091 (1986).
14. B. Hille, "Ionic Channels of Excitable Membranes," p. 225, Sinauer Associates, Sunderland, Massachusetts (1984).
15. C. Miller (Editor), "Ion Channel Reconstitution," Plenum Press, New York, (1986).
16. C. L. Schauf, Sci. Prog. Oxf., 71:459 (1987).
17. E. K. Michaelis, W. L. Chittenden, B. E. Johnson, N. Galton and C. Decedue, J. Neurochem., 42:397 (1984).
18. H. Sugawara, K. Kojima, H. Sazawa and Y. Umezawa, Anal. Chem., 59:2842 (1987).
19. G. Spach, Y. Merle and G. Molle, J. Chim. Phys., 82:719 (1985).
20. J. D. Lear, Z. R. Zimmerman and W. F. DeGrado, Science, 240:1177 (1988).
21. K. S. Birdi and V. S. Gevod, Coll. Polymer Sci., 265:257 (1987).
22. C. Peracchia and L. L. Peracchia, J. Cell. Biol., 87:719 (1980).
23. M. B. Gorin, B. Yancey, J. Cline, J.-P. Revel and J. Horwitz, Cell, 39:49 (1984).
24. I. Simpson, B. Rose and W. R. Loewenstein, Science, 195:294 (1977).
25. M. M. Gooden, L. J. Takemoto and D. A. Rintoul, Curr. Eye Res., 4:1107 (1985).
26. P. Russell, G. Robinson and J. Kinoshita, Exp. Eye Res., 32:511 (1981).
27. J.-H. Kim, T. M. Cotton and R. A. Uphaus, Thin Solid Films, 159:141 (1988).
28. F. Ni and T. M. Cotton, J. Raman Spectroscopy, in press.

Au/LANGMUIR-BLODGETT FILM/Zn Se TUNNEL INJECTION LIGHT-EMITTING STRUCTURES

N. G. Rambidi, A. N. Georgobiani, and P. A. Todua

All-Union Research Centre for
Surface and Vacuum Investigation
Moscow, U. S. S. R.

INTRODUCTION

During the years of rapid development of semiconductor optoelectronics and integrated optics, various active and passive device components were fabricated (radiation sources and receptors, modulators and light-switches, planar and fiber light-guides), which generated the premises for the realization of discrete and integrated elements of optical information processing systems.

Intensive work for perfecting the construction of optoelectronic devices and the technology of their fabrication is continuing now. In recent years utilization of organic materials and, first of all, utilization of surface-active substances, which can form mono- and multimolecular Langmuir-Blodgett (LB) films,[1-4] has begun to play a considerable role in this type of work. These films have a great advantage because they can be deposited onto semiconductor material surfaces at room temperature. The programmable components, the dielectric properties, the controllable thickness (from about 10 Å to 100 Å and more) ensure a wide range of utilization of LB films in various domains of solid state electronics and, especially, in micro- and optoelectronics.

The fabrication of electroluminescent light sources, radiation receptors, light-guides, which contain incorporated layers formed by the LB technique,[1,5-7] is continuing at present.

A special place in the electroluminescence excitation is occupied by processes connected with the carriers tunneling through an insulator layer. A thin insulator layer (usually an oxide) which separates the metal electrode from the semiconductor can ensure tunneling of electrons to the semiconductor as well as from the semiconductor (in the latter case a hole injection takes place). The minority carriers created by tunneling can recombine radiatively with the majority carriers directly as well as through radiative recombination centers. The best way to use such structures is to utilize them in the generation of short wavelength visible light sources, using wide band $A^{II}B^{VI}$ semiconductor materials which have a record high probability of radiative transitions. The operation efficiency of such an MIS structure is, first of all, determined by the uniformity of the properties of the thin (about 10 Å) insulating layer, which is destined to ensure efficient carriers tunneling. The Langmuir-Blodgett technique, which offers the possibility of forming ultrathin dielectric layers with desirable characteristics, can be an alternative solution to the well-known techniques for fabricating real injection light-emitting structures.

The aim of this work was to fabricate ZnSe tunnel injection structures using the LB technique, and to investigate their radiative and electrophysical properties.

Au/LB FILM/ZnSe STRUCTURES FABRICATION TECHNIQUE

To prepare Au/LB film/ZnSe structures we have used ZnSe single crystals with a resistivity $\rho \simeq 0.2\ \Omega$-cm. The ZnSe samples, grown as described above, had the form of flat wafers (10x10x1 mm^3), which was sliced perpendicular to the [111] axis. The samples were finished mechanically and then treated in a CrO_3 + HCl (20 g : 30 ml) polishing etchant for 10 minutes. The deposition of LB films of an organic surface-active material was performed on a Joyce-Loebl system.

The LB films consisted of layers of the Cd salt of the stearic acid $CdSt_2$ or $Cd[CH_3(CH_2)_{16}CO_2]_2$, and the Cd salt of the 10, 12-carbonic acid $Cd[CH_3(CH_2)_{15}-C{\equiv}C-C{\equiv}C-(CH_2)_8CO_2]_2$. To obtain the salt film on the liquid-gas interface the corresponding acid was not diluted on the pure water surface, as in Refs. 5 and 6, but a $CdCl_2$ solution ($2{\cdot}10^{-3}$ mol/l) was used as a sub-phase. In preparing the solution we used deionized water ($\rho \simeq 1.8{\cdot}10^7\ \Omega$-cm), purified on a Millipore system. The organic acid, which was first dissolved in chloroform (concentration ~10^{-3} mol/l) was injected on the liquid surface.

The surface tension serves as a quantitative characteristic of the monolayer's state. In our case the surface tension was registered by a Wilhelm sensor. The value of the surface tension for pure water was taken as a reference value.

Taking into account the ZnSe wettability the process of depositing the monolayers began when the sample in the vertical position crossed the interface going upward. After drying and "fixation" of a monolayer (the thickness of a $CdSt_2$ monolayer was ~25 Å; Ref. 8) the next layer was deposited when the substrate was going downward. Thus the next layer was oriented in the "tail" to "tail" direction, the subsequent movement of the substrate in the upward direction ensured the "head" to "head" direction, and so on. When deposition of an even number of layers was required, the substrate crossed the liquid-gas interface where surface-active materials had been previously removed.

The process of LB film deposition was performed at a constant value of surface tension as a result of a uniform decrease of the monolayer's area, which compensated the area taken off by the substrate. Stearic acid salt films were deposited at a surface tension value of 32 dyn/cm and for the diacetylenic acid 30 dyn/cm. The vertical speed, in both cases, was 3 mm/min for the first monolayer, and 10 mm/min for next layers. The metal contacts with thickness of 150-300 Å were deposited on the LB film with 3,5,6,9,12 monolayers of organic acids salts by evaporating gold under vacuum not worse than 10^{-6} Torr with a mean depositing rate of 5 Å/min. Every 2 minutes the evaporation process was interrupted for a time interval sufficient to prevent an unwanted overheating of the LB film. The ohmic contact to the ZnSe substrate was produced by the fusion of indium immediately before the LB film formation process.

For a reliable identification of the properties of the obtained Au/LB film/ZnSe structures, Schottky diodes were fabricated upon similar substrates by evaporating gold on a freshly polished ZnSe surface.

It must be noted that the results of investigation were the same, within the limits of the experimental errors, for both $CdSt_2$ layers and for layers of the Cd salt of the 10,12-carbonic acid. Because of this, we shall present mainly our results for the $CdSt_2$ structures. Similar experiments can be fully carried upon the structures containing the Cd salt of the 10,12-carbonic acid.

THE RADIATIVE PROPERTIES OF THE Au/LB FILM/ZnSe STRUCTURES

The electroluminescence of the Au/LB film/ZnSe structures, localized in the junction region, can be observed under forward bias; that is, when a positive potential (more than 1.7 v) is applied to the gold contact. Analogous conditions of excitation do not lead to the electroluminescence of the control samples - Au/ZnSe Schottky diodes.

The electroluminescence spectra taken at the room temperature and at 90K are presented on the Fig. 1 and 2, respectively. For comparison photoluminescence spectra excited by monochromatic pulse radiation (λ_{excit} = 337.1 nm, τ_{pulse} = 10 ns, P_{pulse} = 1.5 kW/cm^2) are presented in the same figures. In both excitation cases there was registered radiation which came through the semitransparent gold contact. It can be seen from the Fig. 1a that the electroluminescence spectrum of the structure at room temperature contains an intensive blue band with a maximum at 482 nm (halfwidth ~130 Å) and a wide yellow-orange band with a maximum in the 600 -- 610 nm region. A substantial difference exists between the electroluminescence spectra of the Au/$CdSt_2$ film/ZnSe structures and the analogous spectra of the Langmuir diodes Au/film phthalocyanine of silicon/ZnSe.[7] It should be noted that the radiation spectrum of the latter is a superposition of wide bands on maxima at 460 nm and 500 nm and is interpreted by the authors of Ref. 7 in terms of an inter-band recombination model and the recombination through an unknown defect center.

In our view this difference in the electroluminescence spectra is caused mainly by the difference in the used materials. In our case we used ZnSe single crystals, grown from the melt subsequently heat-treated in a Zn-melt, and the author of Ref. 7 had used an epitaxial ZnSe layer (the thickness of the layers ~3 μm) grown on GaAs:Si substrate by chemical metalloorganic vapor phase deposition. When the samples are cooled to liquid nitrogen temperature the intensity of the blue band is increasing (at the same line width) by about two orders of magnitude and the maximum is shifting toward the short wavelength region (λ_{max} = 465 nm). The intensity of the yellow-orange band increases by an order of magnitude, and the maximum is in the 600 -- 610 nm region. The authors of most papers dedicated to the investigation of the nature of the self-activated yellow-orange luminescence band are relating this band to D-A pairs, the acceptor level of which is formed by a background group III or VII donor impurity - Zn vacancy complex.[9] The nature of the electroluminescence blue band situated in the edge luminescence region is identified by a number of authors[10,11] with an electron radiative transition from the conduction band to a shallow acceptor level created by a background - impurity of the alkaline metals type which gives rise in ZnSe to Li_{Zn}, Na_{Zn} centers, etc. It is well known that the alkaline metals are the basic background impurities in ZnSe.[9]

In the laser mass-spectrograms of the studied ZnSe samples we have clearly identified the lines which are caused by the presence of Li, Na, K, Al. Assuming the validity of the self-activated luminescence model, it can be supposed that Al, which is an III group element, is the donor impurity that participates in the generation of centers responsible for the yellow-orange emission band.

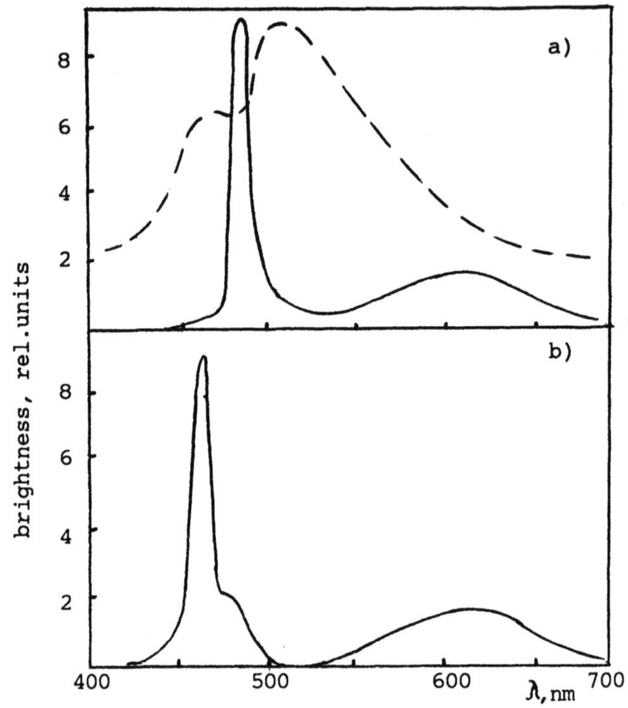

Fig. 1. (a) Electroluminescence spectra for I = 10 mA and (b) photoluminescence spectra; T = 300 K. The dotted curve is taken from Ref. 7.

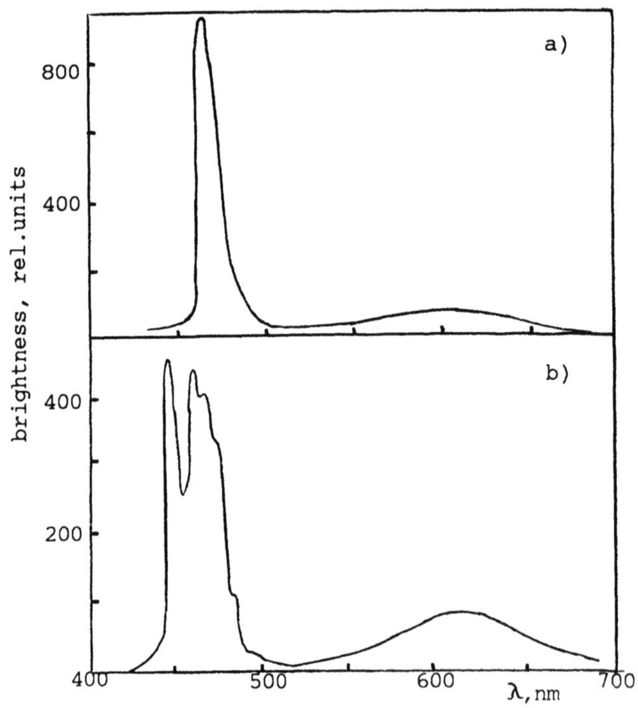

Fig. 2. (a) Electroluminescence spectra for I = 10 mA and (b) photoluminescence spectra; T = 90 K.

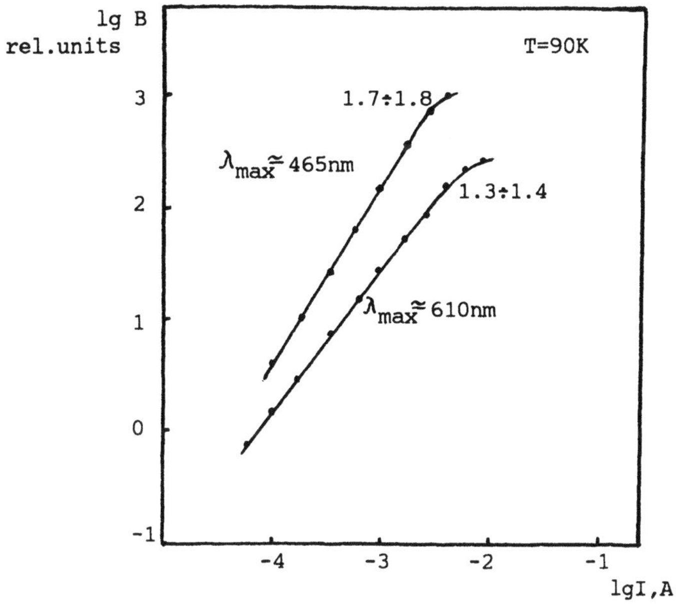

Fig. 3. Emission brightness vs. current dependence of the electroluminescence bands (T = 90 K).

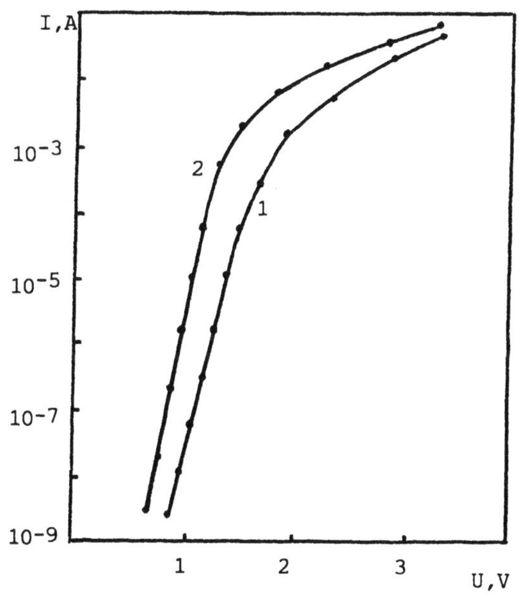

Fig. 4. The forward bias branch of the current-voltage curves for the Au/3 $CdSt_2$ monolayers /ZnSe structure.
T = 90 K (1), T = 300 K (2).

The activation energy, determined from temperature dependence of the electroluminescence blue band of Au/$CdSt_2$ film/ZnSe structures, is 115 ± 15 meV. This value correlates with the Li_{Zn} level activation energy (114 meV) in ZnSe determined in Ref. 11. This result together with the mass-spectrometrical measurements suggests the possibility of realization of the "conduction band - Li_{Zn} acceptor level" transition in forward biased Au/LB film/ZnSe structures.

When the electroluminescence is excited by 1 sec square wave voltage pulses with rise and decay times of 100 nsec on the intensity and emission oscillograms no visible lagging comparing with the voltage oscillograms was observed. Thus the relaxation time of the electroluminescence does not exceed 100 nsec, and is the same as in the case of Au/stearic acid film/ZnSe structures investigated in Ref. 5.

Figure 3 presents the intensity of emission vs. current dependence at T = 90 K for the blue and the self-activated electroluminescence bands, which can be approximated by $B \sim I^k$. The value of k, in the case of the blue and the yellow-orange bands is 1.7 -- 1.8 and 1.3 -- 1.4, respectively. These data correlate well with the data from Ref. 10; K = 1.66 and K = 1.33, respectively, where the above mentioned dependence for Me/SiO_2/ZnSe structures are investigated. In measuring the electroluminescence external quantum efficiency vs. the number of monolayers in the film no visible dependence on the insulating layer thickness was observed in the range of thicknesses used. The value of the electroluminescence quantum efficiency is $\eta_{ext} \simeq 10^{-5}$ photon per electron.

THE ELECTROPHYSICAL PROPERTIES OF THE STRUCTURES

The forward bias branches of current-voltage characteristics of Au/LB film/ZnSe structures having 3 monolayers of $CdSt_2$ measured at two fixed temperatures, T = 90 K and T = 300 K, are presented in Fig. 4. It is seen from the figure that the current-voltage characteristics are well approximated by the expression

$$I = I_s(T)\ \exp(\alpha U)$$

where α is a constant which has weak or no temperature dependence. At the same time the temperature dependence of the saturation current $I_s(T)$ (Fig. 5) can be written in the form:

$$I_s(T) = I'_s \exp(\beta T)$$

where the value of β in the 330 - 300 K temperature range is $1.6 \cdot 10^{-2}\ K^{-1}$. The value of α which gives us the curvature coefficient of current-voltage characteristic at a given temperature is 17.2 v^{-1} at T = 90 K, and is 20 v^{-1} at T = 300 K. Such a weak dependence of the current-voltage characteristics curvature coefficient speaks in favor of the tunnel injection mechanism of the current in the investigated structures. It can be seen from Fig. 4 that in the voltage region U $\gtrsim$ 2v the current begins to saturate owing to increasing contribution of the bulk resistance of the crystal. Fig. 6 presents a family of forward biased current-voltage characteristics of Au/LB film/ZnSe structures with various numbers of $CdSt_2$ monolayers. For comparison in the same figure, a current-voltage characteristic of a Schottky diode Au/ZnSe is also presented (curve 0). It is seen that the increase in the $CdSt_2$ insulating layer thickness results in the shifting of the characteristics toward higher voltages. From Fig. 7, where the forward bias voltage vs. the number of monolayers dependence is presented at various values of the current. The dependence can be expressed in the form

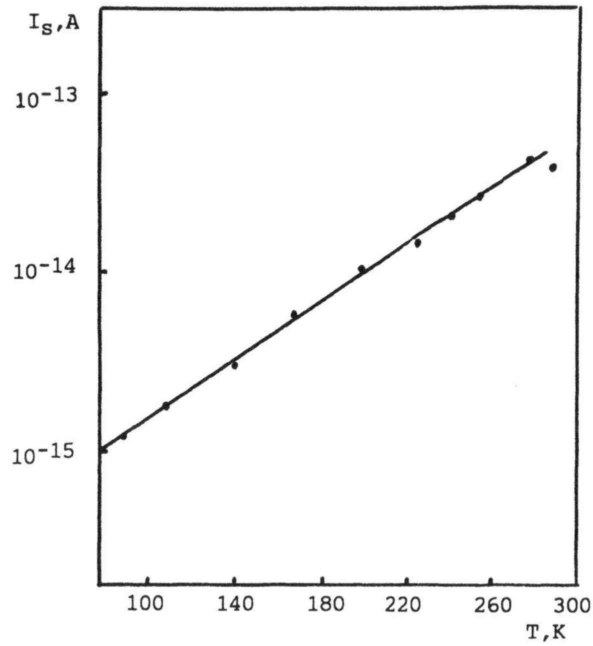

Fig. 5. Temperature dependence of the saturation current of the Au/3 $CdSt_2$ monolayers/ZnSe structure.

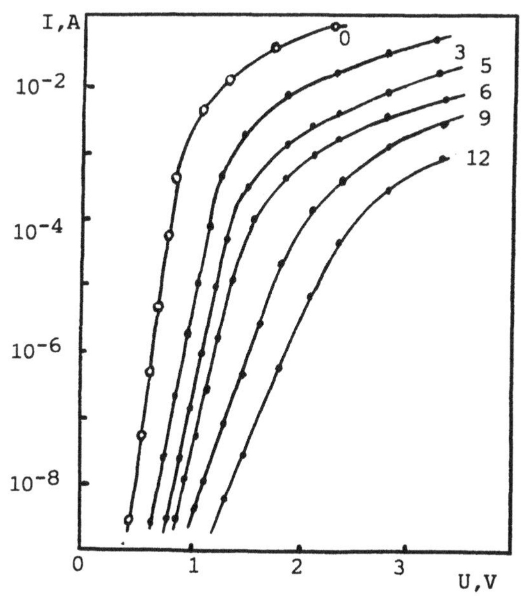

Fig. 6. The forward bias branches of the current-voltage characteristics of the Au/LB film/ZnSe structures with various number of $CdSt_2$ monolayers at T = 300 K.

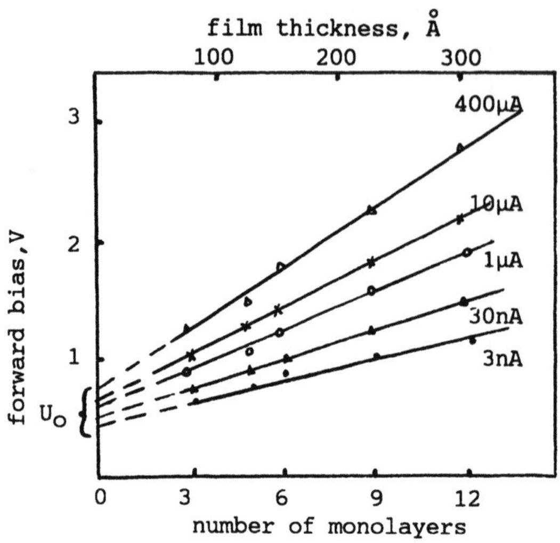

Fig. 7. Forward bias voltage vs. $CdSt_2$ film thickness for various fixed currents (T = 300 K).

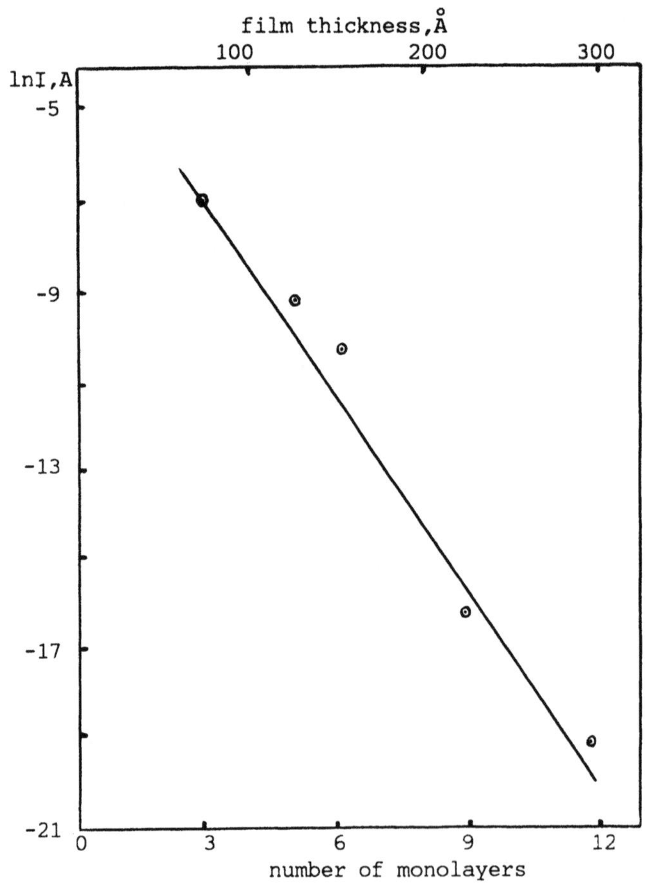

Fig. 8. Current vs. $CdSt_2$ film thickness for a given voltage (T = 300 K).

$$U(I) = U_o(I) + Ed$$

where U(I) is the applied forward voltage, $U_o(I)$ is the voltage on the depletion region of the crystal, E is the mean value of the electrical field intensity in the insulating film, and d is the film thickness.

It must be noted that the values of $U_o(I)$ obtained by extrapolation of the corresponding current straight lines to the zero value of the insulator thickness coincide with the corresponding values of the voltage and current for the Au/ZnSe Schottky diode, the current-voltage characteristics of which is shown in Fig. 6 (curve 0). Thus, the investigated structure can be represented equivalently as two diodes connected in series: a Schottky diode and a tunnel diode. For such MIS structures taking the form of a rectangular barrier with an effective height $q\phi_T$, which depends on the applied voltage, the authors of Ref. 14 obtained the relation

$$I \simeq A^{**}T^2 \exp(-\alpha_T\phi_T^{1/2}d) \exp(-q\phi_B/kT) [\exp(qU/kT)-1]$$

where the thermoelectrical emission current of the Schottky diode is multiplied by the factor $\exp(-\alpha_T\phi_T^{1/2}d)$, which represents the tunneling probability of the current carriers through the barrier with the width d, and A^{**} is the effective Richardson constant, ϕ_B is the Schottky barrier height. When the effective mass in the insulator is equal to the free electron mass, then $\alpha_T \simeq 1$, if ϕ_B and d are measured in volts and Å, respectively.[12]

From the above relation, if we apply the same voltage to structures having insulating layers of different thickness, then the corresponding values of the current will depend on thickness in the following way:

$$I = I_{sh}\exp(-bd)$$

where I_{sh} is the thermal emission current of a Schottky diode, and b is a constant which does not depend on the insulator thickness. Fig. 8 presents the current vs. insulating film thickness (the number of $CdSt_2$ monolayers) dependence for the Au/LB film/ZnSe structures which lead to the exponential dependence of the current on the thickness with the constant $b \simeq 5.3 \cdot 10^8\ m^{-1}$.

Using the results presented in Fig. 7 we can determine the voltage U and correspondingly the value of the field intensity in the insulating layers, and then plot them against the corresponding values of the current. From Fig. 9, where the dependence 1/E vs. I, determined as indicated above, is presented for the Au/LB film/ZnSe structures at room temperature. It is seen that this dependence is approximated by the expression

$$I = I_o\exp(-a/E)$$

where the value of a is $5.7 \cdot 10^6$ v/cm, which is of the same order of magnitude with the value of the field intensity 10^6 -- 10^7 v/cm needed for the development of tunnel conduction processes. This result is an important criterion for the good dielectric quality and the uniformity of the organic surface-active materials Langmuir films used in our work for fabrication of the insulating layers in the MIS structures.

Complementary information about the current mechanism in the studied structures can be extracted from the capacitance-voltage characteristics. Fig. 10 presents the high-frequency (f = 1 MHz) capacitance-voltage characteristics of Au/LB film/ZnSe structures with 3 and 5 $CdSt_2$ monolayers and the characteristics of a control Au/ZnSe diode. The form of this dependence differs considerably. For the Schottky diode the capacitance is decreasing when the reverse bias voltage is increasing. In the case of Au/LB film/ZnSe structures the capacitance-voltage curve has a typical ladder form which is

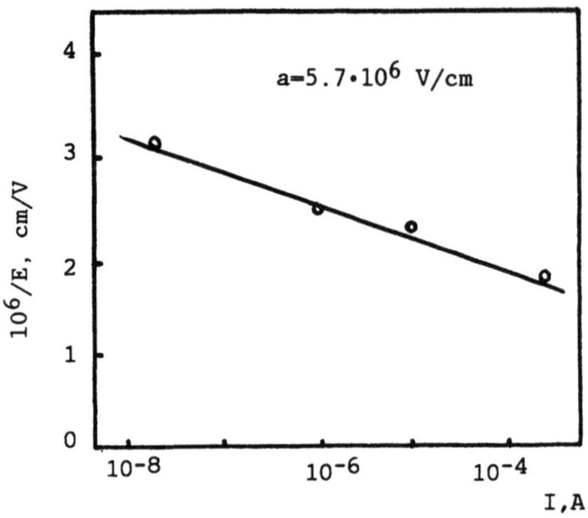

Fig. 9. I vs. 1/E characteristic of the Au/$CdSt_2$ film/ZnSe structures at T = 300 K.

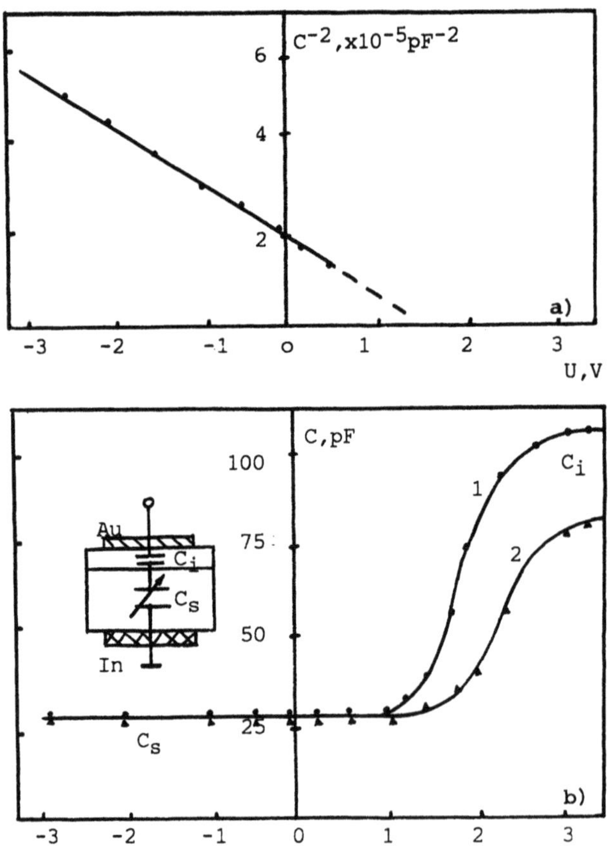

Fig. 10. Capacitance-voltage characteristics. (a) Schottky diode; (b) Au/LB film/ZnSe structures with 3 monolayers (1) and 5 monolayers (2) of $CdSt_2$.

characteristic of the tunnel structures of the MIS type. The total capacitance of such a structure can be represented as two capacitances connected in series (the insert in Fig. 10).

$$C = \frac{C_i C_s}{C_i + C_s}$$

where C_i is determined by the thickness of the insulating film and the electric permittivity and does not depend on the applied voltage, and C_s is determined by the electrophysical properties of the near-surface region of the semiconductor which depends on the applied voltage.

In the large positive bias region, when the accumulation of carriers takes place near the junction, the semiconductor capacitance becomes considerable and the resulting capacitance is determined mainly by the insulating film capacitance

$$C_{max} \simeq C_i = \frac{\epsilon_i \epsilon_o s}{d}$$

where ϵ_o is vacuum permittivity ($\epsilon_o = 8.85 \cdot 10^{-12}$ F/m); ϵ_i is the permittivity of $CdSt_2$ insulating film ($\epsilon_i = 2.71$; Ref. 14); d is the film thickness; s is the gold contact area ($s = 7 \cdot 10^{-4}$ cm^2).

Using this relation from the experimentally measured values of C_{max}, the insulating film thickness can be determined. Thus, in the case of 3 monolayers of $CdSt_2$, $C_{max} \simeq 100$ pF and the film thickness is $d \simeq 100$ Å. The obtained value of the thickness correlates well with the estimated value of the thickness of 3 monolayers 75 Å (the thickness of a $CdSt_2$ monolayer is ~25 Å; Ref. 8). Fig. 10 shows that the transition to 5 monolayers results in a corresponding decrease of C_{max}.

Under negative biases the semiconductor depletion region capacitance C_s begins to reveal its contribution, C_s becomes comparable and less than the insulating film capacitance C_i and virtually determines the resulting capacitance

$$C_{min} = \frac{\epsilon_i \epsilon_o s}{d + (\epsilon_i / \epsilon_s) W_m}$$

where ϵ_s is the semiconductor's permittivity (in the case a ZnSe $\epsilon_s = 8.66$; Ref. 15), W_m is the maximum thickness of the depletion layers.

From the value of $C_{min} = 25$ pF (Fig. 10) and supposing that $d \ll (\epsilon_i/\epsilon_s) W_m$, we obtain that $W_m = 0.2 \pm 0.05$ μm. On the other hand, the maximum thickness of the depletion layer is given by the relation[16]

$$W_m = \left[\frac{4 \epsilon_s kT \ln(N_d/n_i)}{q^2 N_d}\right]^{1/2}$$

where q is the electron charge, n_i is the concentration of the intrinsic carriers.

In case of ZnSe at T = 300 K, $n_i \simeq 8.8 \cdot 10^{-5}$ cm^{-3}. Then, substituting the experimentally determined value of W_m in this relation, we can find the concentration of the ionized donors to be $N_d \simeq 8 \cdot 10^{17}$ cm^{-3}.

From the Hall effect measurements performed at the room temperature we determined the mobility and the concentration which are, respectively, $\mu_e \simeq 140\ cm^2/v\text{-}sec$, and $n_e \simeq 2 \cdot 10^{17}\ cm^{-3}$. It can be seen that the value of n_e determined independently is in good agreement with the concentration of the ionized donor N_d which was obtained above.

Knowing the maximum thickness of the depletion region we can determine the charge upon the interface unit area to be[16]:

$$Q_s = qN_dW_m.$$

Substituting in this relation with $q = 1.6 \cdot 10^{-19}$ Q, $N_d = 8 \cdot 10^{17}\ cm^{-3}$, and $W_m \simeq 0.2\ \mu m$, we obtain $Q_s \simeq 2.6 \cdot 10^{-6}\ Q/cm^2$.

The observed shift in the Au/LB film/ZnSe structures capacitance-voltage characteristics toward the positive values of the voltage (Fig. 10), in our view, is explained by the fact that at U = 0 the flat-band state does not occur as it takes place in an ideal MIS structure. Apparently, as it is seen analyzing the current-voltage characteristics (Fig. 7), there exists a barrier at the insulator-semiconductor interface and correspondingly a depletion region. The depth of this region can be estimated as follows. As it was shown in Ref. 16 from the absolute shift U_f of the capacitance-voltage characteristics we can determine the surface density of the charge fixed in the insulator

$$Q_s = U_fC_i/S$$

substituting in this formula with $U_f \simeq 2v$, $C_i \simeq 100$ pF and $S = 7 \cdot 10^{-4}\ cm^{-2}$ we obtain $Q_f \simeq 3 \cdot 10^{-7} Q/cm^2$. Then the depletion region thickness for a zero bias (U = 0) is

$$W_o = \frac{Q_s}{qN_d}$$

equal to $W_o \simeq 2 \cdot 10^{-2}\ \mu m$.

CONCLUSIONS

The Au/LB film/ZnSe structure, in which the insulating layers consist of $CdSt_2$ and Cd salt of carbonic acid are MIS structures characterized by tunnel-injection current and electroluminescence excitation mechanisms. The high reproducibility of the electrophysical and emission properties of these structures underlines the desirability of utilization of the LB technique in generating short wavelength visible light-sources using wide-band $A^{III}B^{VI}$ semiconductor compounds.

ACKNOWLEDGMENTS

The authors are deeply grateful to S. I. Vedenejev, A. S. Kholmansky, E. F. Shestakova and B. T. Eltazarov for help in performing various stages of the work.

REFERENCES

1. First International Conference on Langmuir-Blodgett Films, Durham, 1982. - Thin Solid Films v. 99, N1-3 (1983).
2. L. M. Blinov, Uspechi Chimii v. 52, N8, p. 1263-1300 (1983).

3. N. G. Rambidi, W. M. Zamalin, J. M. Sandler, P. A. Todua, A. S. Kholmanski, "Molecular element base of perspective informational-logic devices", Moscow, VINITI, ser. Electronika, v. 22, 168 p. (1987).
4. G. L. Gaines, Jr. "Insoluble Monolayers at Liquid-Gas Interfaces," Interscience, New York (1966).
5. A. N. Georgobiani, N. G. Rambidi, P. A. Todua et al., Soviet Physics-Lebedev Institute Reports N9, p. 46-49 (1987).
6. A. N. Georgobiani, N. G. Rambidi, P. A. Todua et al., Soviet Physics-Lebedev Institute Reports N9, p. 42-45 (1987).
7. J. L. Hua, M. C. Petty, G. G. Roberts et al., Electron. Letters v. 23, N5, p. 231-232 (1987).
8. M. Pomerant, A. Segmuller, Thin Solid Films v. 68, p. 33-45 (1980).
9. D. D. Nedeoglo, A. V. Simashkevich, "Electrical and luminescence properties of ZnSe," Kishinev, Shtiinza, 150 p. (1984).
10. M. Yamagushi, A. Yamamoto, Jap. J. Appl. Phys. v. 16, N1, p. 77-84 (1977).
11. P. J. Dean, J. L. Merz, Phys. Rev. v. 178, N3, p. 1310-1318 (1969).
12. V. Kumar, W. E. Dahlke, Solid State Electron v. 20, p. 143-149 (1977).
13. W. Franz, "Dielektrischer Durchschlag," Berlin u.a., Springer-Verl., 155 p. (1956).
14. B. Mann, H. Kuhn, J. Appl. Phys. v. 42, N11, p. 4398-5003 (1971).
15. J. I. Pankove, "Optical Processes in Semiconductors," Printice-Hall, Englewood Cliffs, NJ (1971).
16. S. M. Sze, "Physics of Semiconductor Devices," 2nd ed., Wiley, London, (1981).

DESIGN OF A TRUE MOLECULAR ELECTRONIC DEVICE: THE ELECTRON TRANSFER SHIFT REGISTER MEMORY

David N. Beratan[a], José Nelson Onuchic[b], and J.J. Hopfield[c]

[a]Jet Propulsion Laboratory, California Institute of Technology, Pasadena, CA 91109; [b]Instituto de Física e Química de São Carlos, Universidade de São Paulo 13560, São Carlos, SP, Brazil; [c]Divisions of Chemistry and Biology, California Institute of Technology Pasadena, CA 91125 and AT&T Bell Laboratories Murray Hill, NJ 07974

INRODUCTION

An early and central goal in the field of "molecular electronics" was to create molecular size analogues of larger conventional electronic components [1-6]. The interest, as reflected in these proceedings, has shifted towards ensembles of molecules with novel switching properties (photochromism, channel controlling, etc.). This shift in emphasis was caused by the difficulty of implementing proposed molecular systems and the fact that ensembles of relatively simple chemical and biochemical systems can be reversibly switched between states, and these properties exploited.

The promise of truly molecular size computational elements, designed to possess prescribed microscopic properties, remains essentially unfulfilled because of the absence of particular and practical suggestions for structures and functions [7]. In this paper we present the design of a molecular size shift register memory element based on electron transfer reactions. The molecular elements could actually be made and interfaced with conventional microelectronic technology [8]. The memory elements could be constructed using known strategies of organic synthesis, and would provide real progress towards achieving devices with energy costs (per bit processed) approaching the energies of single molecule electronic or conformational energy transitions (a few eV or less) [1,9] and information storage densities much larger than possible with existing technologies. Most importantly, this is an example of a molecular size electronic device which could actually be made.

DESIGN QUESTIONS FOR A MOLECULAR ELECTRONIC DEVICE

The goal of molecular electronics is to provide molecular computational elements [2] which exploit single molecule rather than bulk properties of the material. The fact that single molecule properties are used allows design of the device from small up rather than from large down, and tuning of the device properties by making molecular changes.

The essential questions which must be addressed when designing a molecular electronic device which will function are: How is the energy to be supplied at a molecular level? How does the clock signal get to the molecular device? How are the molecular devices to be interconnected, and how are the connections to be chemically fabricated? How is communication between the molecular size features of a chip and the molecular device to be achieved? How are errors to be dealt with? The molecular shift register described below answers all of these design questions.

THE ELECTRON TRANSFER MOLECULAR SHIFT REGISTER

A shift register is a form of memory with the memory elements connected in a line. During each clock cycle, information is written at the left terminal cell and the contents of each cell are shifted one unit down the chain to the right. One bit of information is also read out at the right terminal cell.

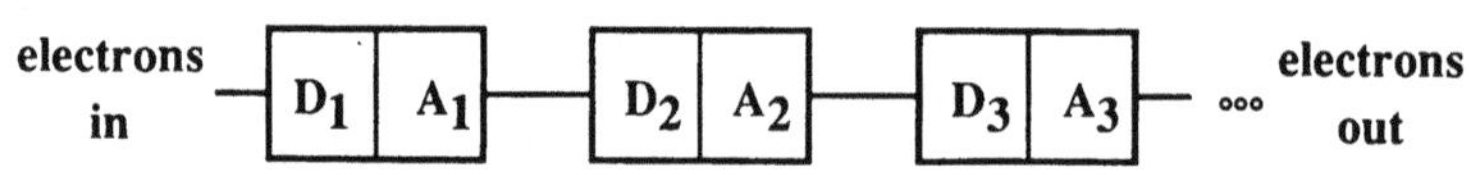

Figure 1. A schematic representation of the the cells in a shift register memory are shown.

Shift registers are used in conventional memory elements (magnetic bubble memories are shift registers) and information delay lines.

Perhaps the simplest molecular switching process is electron transfer. It involves moving a well localized electron from one species (D) to another (A) without breaking or forming any bonds. Electron transfer rates are sensitive functions of reaction free energy, vibronic coupling, and distance [10]. The rates of such reactions can be tuned over many orders of magnitude and the transfer may be either permanent or reversible.

A one-dimensional array of DA molecules would form a shift register memory element as shown in Figure 2. Information is written as a "1" (or "0") in each cell by the presence (or absence) of an electron on D. D has an accessible excited state from which the electron transfer to A occurs followed by the shift reaction. Information is written on the left side of the chain (one bit at a time) and advances down the chain one unit per clock cycle (light flash).

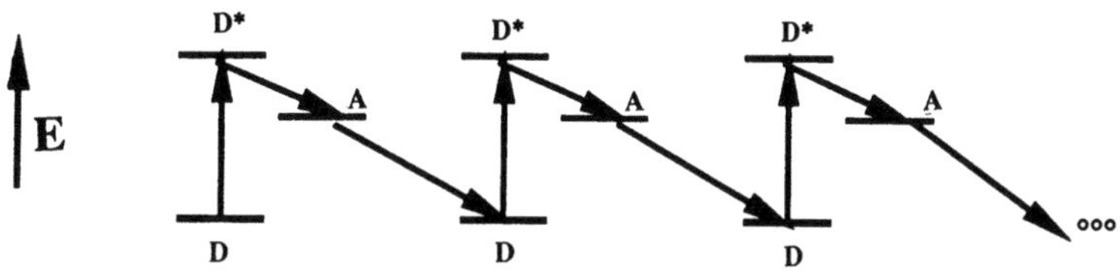

Figure 2. The donor and acceptor levels in a typical shift register polymer are shown. The donors have accessible ground and excitied states. Intermediate states are suppressed in this representation.

This scheme solves the design requirements: the energy is provided by light; the clock signal is provided by pulsing the light; the chemistry for linking donors and acceptors is known; interconnections with silicon may be provided by appropriate surface binding groups; errors can be dealt with to some degree with multiple chains or methods described below. Information in this implementation is written and read at electrodes bound to the first and last groups of the polymer.

Although the internal details of each unit will be more complicated than indicated in Figure 2, it shows the three characteristic time-scales of the system. These are the time for charge separation within a monomer unit (τ_C), the lifetime of the charge separated state in an isolated monomer (τ_r), and the time for the charge to be shifted one unit down the polymer (τ_S).

Based on known model compounds, typical monomer units with appropriate rates would be roughly 20 Å long. Interfacing with micron scale architecture, then, would require chains of about 600 repeat units (bits) or longer. Chains comprised of these units would be bound between two electrodes on a VLSI substrate [11] and immersed in a solution to maintain a Debye length of roughly 20 Å. To simplify detection of the arriving electrons, roughly 5000 identical chains would be needed between the electrodes [11]. Smaller numbers could be used but then the signal would require external amplification. The 5000 chain redundancy also provides a means of error correction. The redundancy allows averaging of the signal at the receiving electrode over all of the chains.

For half of the written electrons to arrive at the receiving electrode after 600 hops, the efficiency of transfer at each of the 600 steps must be greater than or equal to 99.9%. Biological photosynthetic systems suggest that such efficiency may be achievable in tailored systems [12]. Strategies which give each electron several chances to transfer are possible, in addition to simply optimizing the rate of forward reactions relative to recombination reactions. Other strategies to keep the electrons moving along parallel chains in synchrony (analogous to cooperativity in biological systems), or to provide some form of "molecular" charge amplification, are also under study. The 99.9% efficiency requirement can probably be satisfied in real systems by manipulating the duration of the excitation pulse. Pulses must be shorter than τ_S in duration to avoid multiple electron transfer steps down the chain during a flash. Pulses of length $\tau < \tau_S$ would provide $\sim\tau/\tau_C$ "chances" for the electron to transfer. Thus, if the donor is excited three times, and it has a 90% chance of electron transfer per excitation, the 99.9% efficiency constraint is satisfied. Polymers with more complicated repeat units and multiple wavelength excitation schemes might also be used to increase the efficiency [8].

POSSIBLE IMPLEMENTATIONS

We now give two examples of orbital energy level schemes which would provide proper shifting of information. In the first scheme, Figure 3a, bits are written as electrons which move from left to right in the register. Three sites per unit, rather than two, provide greater adjustability in the charge separation and recombination rates as well as enhanced efficiency. In the second example, Figure 3b, the bits are written as "holes" which move from left to right. Other implementations are also possible [8].

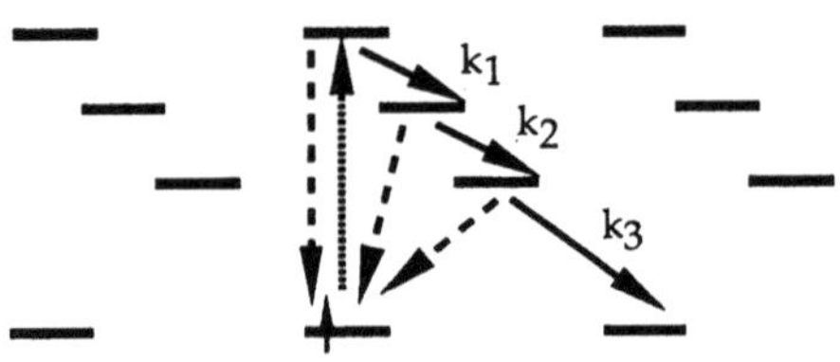

Figure 3a. In this scheme a donor, primary acceptor, and secondary acceptor are used in each repeat unit. Donor excitation initiates the electron transport. The string "010" is written.

Recently, we described two possible molecular implementations [8] of these schemes using electron donors and acceptors known from the photosynthetic model compound (Figure 4) or solar energy conversion literature [13]. These are meant as illustrations of the kinds of systems which might be of use. Modifications may be needed to produce a chain with the proper balance of forward and reverse rates. Since electron transfer reactions are relatively well understood from a theoretical point of view, rates can be tuned over many orders of magnitude by a combination of energetic, distance, and vibronic coupling effects.

Other strategies for building up the shift registers might include attaching chromophores to DNA [14], or building up Langmuir Blodgett films with the chromophores [15]. Also, relatively rigid organic polymers to which the chromophores would be attached might also be of use [16].

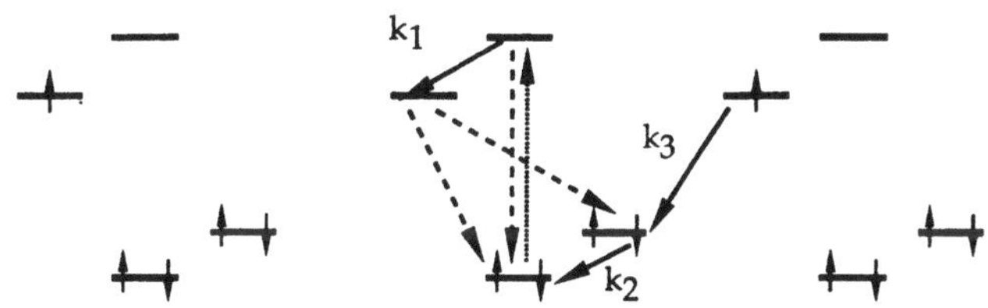

Figure 3b. In this scheme a donor, intermediate, and acceptor are used in each repeat unit. Intermediate excitation initiates the electron transport. The string "010" is written.

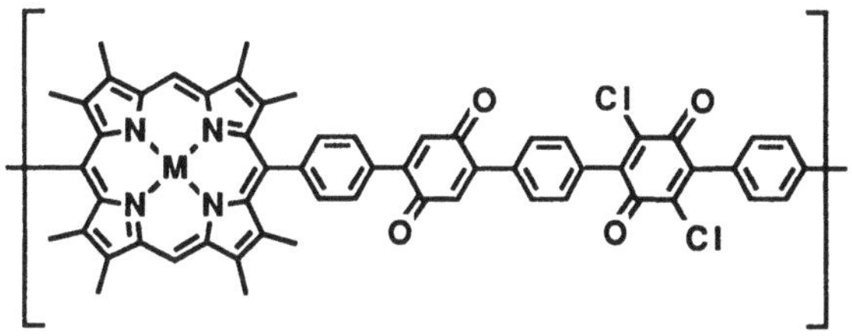

Figure 4. A possible molecular implementations of the scheme in Figure 3a for the shift register polymer is shown. Ref. 8 gives an implementation of the energy level scheme in Figure 3b.

PERSPECTIVES

One might have expected that moving from the micrometer to the nanometer size scale would cause uncertainty limitations on switching events and the loss of information purity in real systems. (On this size scale events need not be purely deterministic because of quantum effects). Also, there is the problem of coupling molecular size devices to macroscopic electronic components. For these reasons, one might have expected the need to work with ensembles of molecules and to move away from the truly molecular regime. However, if we can guarantee very high probabilities for the desired directionality of the reactions, as discussed above, there is no need for an ensemble of molecules to avoid the uncertainty limitations. The constraint of extremely high transfer probabilities may restrict the permissible values of τ_c, τ_r, and τ_s, and therefore the kinds of allowable structures, but the systems should be buildable.

The issues of charge amplification and error correction must be answered if one wants to decrease the size of the device further (beyond several thousand chains) and decrease its energy consumption (to "molecular" energies per bit processed, ~50 kT). This can be achieved by including a mechanism to correct for error propagation every few units of the chain, and by introducing some kind of molecular size charge amplifier, coupling the microscopic device to the macroscopic components.

In this initial description of a molecular shift register with 600 monomer groups, we need 99.9% efficiency for electron transfer between cells to guarantee that half of the electrons arrive at the receiving electrode at the proper time. Therefore, the errors propagate over the full length of the chains and "error correction" is only performed at the reading electrode. Including units to restore electrons to their proper places in the polymer would substantially improve the device performance. One way that this can be implemented is by building the system of roughly 10 polymer chains, and at a restoration step putting all corresponding cells of the system into the majority state of the cells. Such an implementation would also permit building an operational device with transfer efficiencies less than 99.9%.

The restoration described above may allow a device free of error limitations to be composed of a small number of chains. However, this does not solve the need for several thousand electrons at the receiving electrode so that the signal can be easily read. The way to eventually solve this problem is not by carrying 5000 electrons along the full length of the chains. Rather, the number of chains should be kept small and, near the chain ends, the charge should be amplified prior to arrival at the electrode. One way to achieve this amplification is through a molecular fork structure [8]. A molecular fork is conceived as a unit in the chain at which the arrival of a single electron results in the departure of two electrons and a hole moving away on separate branches. An assembly of such units would amplify one electron into 1000 in 10 branchings. Fewer than 10 polymer strands, each with such a branched tree structure at the end, would provide the needed number of electrons. Other charge amplification schemes are conceivable. For example, the arrival of a few electrons might induce a channel in a membrane to open releasing a large number of signaling species.

These two schemes, information restoration and fork amplification, can only be implemented if we have a reservoir of charge available. In some biological processes, such as photosynthesis and respiration, charge equilibrations are performed by sophisticated mechanisms [12]. When designing our fork, the simplest mechanism will probably involve electron transfer with species in solution. One way to couple the reservoir of charge to the device is by having some units of the fork with energy levels that may give (or receive) electrons to the solution. Light can then be used to inject (or remove) the electrons from these intermediate units to the rest of the fork at the appropriate times. One way to make these intermediate units more effective is by guaranteeing that they can exchange only two or more electrons with the solution. By doing this, no group in solution will be able to donate electrons to the shift register cells, so the further introduction of errors will be avoided. This may be achieved by using molecules that have stable redox states differing by two electrons, or by some cooperative mechanism of electron transfer. With this strategy there is no way that exchange of electrons between the device and the solution will effect the processing of information, which is a one-electron (or hole) transition.

ACKNOWLEDGMENTS

The research described in this paper was supported by the Office of Naval Research (contract N00014-K-0377), the Department of Energy's Energy Conversion and Utilization Technologies Division (ECUT) through an agreement with the National Aeronautics and Space Administration, and the Brazilian agencies CNPq and FINEP. The work was carried out, in part, at the Jet Propulsion Laboratory, California Institute of Technology. This manuscript was written during a visit by D.N.B to São Carlos, Brazil, supported in part by CNPq

REFERENCES

1. (a) F.L. Carter, Editor, Molecular Electronic Devices (Marcel Dekker, New York, 1982); (b) F.L. Carter, Editor, Molecular Electronic Devices II (Marcel Dekker, New York , 1987).

2. R.C. Haddon and A.A. Lamola, Proc. Natl. Acad. Sci. (U.S.A.) 82, 1874 (1985).

3. J.R. Milch, "Computers Based on Molecular Implementations of Cellular Automata", To appear in the proceedings of the Third International Symposium of Molecular Electronic Devices, North Holland, in press.

4. J. Van Brunt, J. Biotechnology 3, 209 (1985).

5. R.R. Birge, A.F. Lawrence, and L.A. Findsen, Proceedings of the 1986 International Congress on Technology and Technology Exchange, Pittsburgh, PA 1986.

6. A. Aviram, J. Am. Chem. Soc. 110, 5687 (1988).

7. Recently, interest in addressing single molecules has reemerged because of the invention of the scanning tunneling microscope. A. Aviram, C. Joachim, and M. Pomerantz, preprint, 1988; J.S. Foster, J.E. Frommer, and P.C. Arnett, Nature, 331, 324 (1988).

8. (a) J.J. Hopfield, J.N. Onuchic, and D.N. Beratan, Science, 241, 817 (1988); (b) J.J. Hopfield, J.N. Onuchic, and D.N. Beratan, submitted for publication, 1988.

9. Energies of the order of one eV are involved in typical molecular electronic and conformational transitions. This is the expected energy

cost for writing a bit of information at the molecular level.

10. (a) See volume 90, No. 16, J. Phys. Chem. (1986), for numerous reviews of current work in the field of electron transfer reactions; (b) J.N. Onuchic and D.N. Beratan, J. Am. Chem. Soc. 109, 6771 (1987); R. Austin, et. al., Editors., Protein Structure, Molecular and Electronic Reactivity (Springer-Verlag, New York, 1987); (c) J.N. Onuchic, D.N. Beratan, and J.J. Hopfield, J. Phys. Chem. 90, 3707 (1986).

11. (a) C. Mead and L. Conway, Introduction to VLSI Systems (Addison-Wesley, Reading, 1980); (b) B. Hoeneisen and C.A. Mead, Solid State Electronics 15, 819 (1972).

12. (a) W.W. Parson, B. Ke in Photosynthesis: Energy Conservation in Plants and Bacteria, Govindjee, ed. (Academic Press, New York, 1984) p. 331; (b) J.P. Allen, G. Feher, T.O. Yeates, H. Komiya, and D.C. Rees, Proc. Natl. Acad. Sci. (U.S.A.) 84, 5730 (1987); also 84, 6162 (1987); (c) J. Deisenhofer, O. Epp, K. Miki, R. Huber and H. Michel, Nature 318, 618 (1985).

13. (a) J. Olmsted III, S.F. McClanahan, E. Danielson, J.N. Younathan, and T.J. Meyer, J. Am. Chem. Soc. 109, 3297 (1987); (b) A.D. Joran, B.A. Leland, P.M. Felker, A.H. Zewail, J.J. Hopfield, and P.B. Dervan, Nature 327, 508 (1987); (c) S. Nishitani, N. Kurata, Y. Sakata, S. Misumi, A. Karen, T. Okada, N. Mataga, J. Am. Chem. Soc. 105, 7771 (1983); (d) P. Fromherz and B. Rieger, J. Am. Chem. Soc. 108, 5362 (1986).

14. (a) P.B. Dervan, private communication; (b) B.H. Robinson and N.C. Seeman, Protein Engineering, 1, 295 (1987); (c) J.K. Barton, C.V. Kumar, and N.J. Turro, J. Am. Chem. Soc. 108, 6391 (1986); (d) T.L. Netzel, preprints, 1988.

15. H. Kuhn, Pure and Appl. Chem. 53, 2105 (1985).

16. (a) P.W. Kenny and L.L. Miller, J. Chem. Soc., Chem. Commun. 84 (1988). (b) A.E. Stiegman, V.M. Miskowski, J.W. Perry, D.R. Coulter, J. Am. Chem. Soc. 109, 5884 (1987); (c) T.M. Swager and R.H. Grubbs, J. Am. Chem. Soc. 109, 894 (1987).

ULTRA HIGH SPEED BACTERIORHODOPSIN PHOTODETECTORS

G.W. Rayfield

Physics Department
University of Oregon
Eugene, OR 97403

ABSTRACT

Photodetectors can be made that contain bacteriorhodopsin (BR) as the photoactive material. These photodetectors consist of dried oriented purple membrane sheets of BR sandwiched between transparent electrodes. The oriented BR layer is made by electrodeposition from a solution of purple membrane in distilled water. The photoresponse is due to a very rapid light-induced polarization of the sample. Experiments indicate that the rapid polarization induces a photovoltage with a rise time in the picosecond time domain. An experiment can be devised to allow the ultrafast photoresponse of these BR based photodetectors to be measured.

INTRODUCTION

The objective of this research project is to develop an ultra high speed photodetector. We propose using a material of biological origin, bacteriorhodopsin, as the photosensitive material. We believe that these photodetectors will have picosecond or even subpicosecond response times. An electrooptic sampling system is being constructed to measure these extremely fast photoresponses.

Bacteriorhodopsin (BR) is a protein found in the membrane cell wall of <u>Halobacterium</u> halobium (Hb) when the bacteria is grown in low oxygen tension [1]. It is found in the form of a large two-dimensional crystalline sheet called "purple membrane" (PM). Each PM sheet is about one micron in diameter and 45 Angstroms thick. A single sheet contains about 10^5 BR molecules on a triangular lattice with three of the BR molecules--a trimer--on each lattice site. A single BR molecule, or monomer, consists of two main parts: 1) the opsin, and 2) a retinal chromophore. The opsin is a seven-alpha helical array of 248 amino acids and has a molecular weight of 26,000. Each of the seven linked alpha helices spans the membrane and is roughly perpendicular to it.

The retinal chromophore is responsible for visible-light (570 nm) absorption by BR and is located within a pocket of the opsin. It is bound via a Schiff base to a lysine in the amino acid sequence. The retinal chromophore is oriented with its long axis at 70° with respect to the normal of the PM sheet. Therefore, the electric dipole moment of the retinal lies almost in the plane of the membrane, and light incident along the normal to the PM sheet has the largest absorption cross section.

Within the cell wall, the BR molecules in the PM sheet function as a light driven proton pump, allowing the bacteria, under conditions of low oxygen tension, to replace oxidative phosphorylation with photophosphorylation. If a suspension of Hb cells is illuminated, the bathing solution becomes acidic, indicating that protons are moved from the interior of the cell to the exterior. PM sheets can be purified from a harvested cell culture and incorporated into vesicles. Illumination of these BR vesicles causes alkalization of the external bathing medium, indicating that protons are pumped inward, rather than outward as was the case for the cell suspension.

Flash illumination of an aqueous suspension of PM sheets shows a transient change in the visible absorption spectrum [1,2]. The optical absorption changes reflect a series of photointermediates, the photoreaction cycle, with characteristic rise and fall times that range from less than a picosecond to more than 10 milliseconds. A simplified series of the sequential photo intermediates in the photoreaction cycle is shown Figure 1. Flash illumination of the PM suspension also causes a transient release of protons to the bathing solution followed by a slow uptake of protons. The uptake and release of protons during the photoreaction cycle is also indicated in Figure 1.

The Schiff base that attaches the retinal to the opsin is unprotonated in the M intermediate. The protonation and deprotonation of the Schiff base roughly correlates with the uptake and release of protons from the PM suspension after flash illumination. Isomerization of the retinal chromophore also occurs immediately following flash illumination, so that the retinal chromophore is cis in the K, L, and M intermediates but trans in BR570.

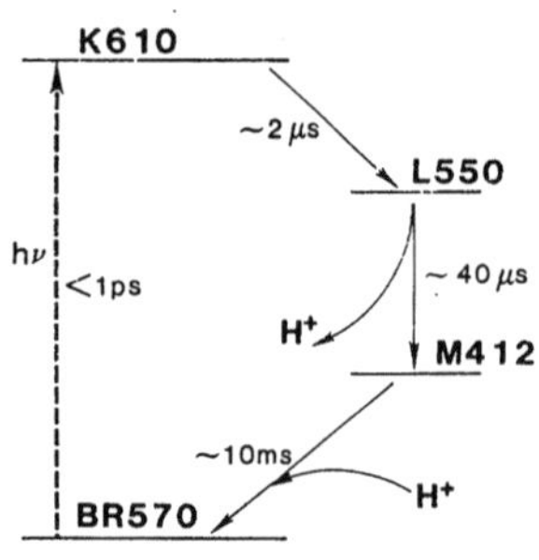

Figure 1. The Photoreaction Cycle of Bacteriorhodopsin is shown. The vertical arrow indicates the initial absorption of the photon.

PHOTOVOLTAGE MEASUREMENTS

The movement of charges within the membrane following a light flash can be detected electrically [3,4]. This is done by attaching the PM sheets in an oriented arrangement to either a lipid planar membrane (BLM) or a thin teflon septum separating two chambers filled with electrolyte. Flash illumination of the septum causes charge displacements within the membrane. These charge displacements cause a voltage (open circuit) to develop between two electrodes immersed in each electrolyte solution. A typical photovoltage response (normalized in magnitude) is shown in Figure 2. It was obtained using a very high input impedance voltage follower circuit as shown in Figure 2. A short-circuit measurement can be made by connecting a very low impedance transconductance amplifier between the electrodes. The short-circuit photoresponse reflects the capacitive charging currents in the membrane and is the derivative of Figure 2.

The time-dependent photovoltage following a light flash is plotted versus the log of time in Figure 2. Each charge displacement within the membrane contributes an exponential component to the photovoltage [3,4]. Only charge movements perpendicular to the membrane surface contribute to the photovoltage. Since photovoltage is plotted versus log time in Figure 2, the exponential components in the photovoltage manifests themselves as inflection points in the curve [4]. This graphical interpolation of the photovoltage data indicates that there are at least three components to the photovoltage. Two of the components have time constants not very different from the rise and fall time of the M intermediate. The initial negative photovoltage component is faster than all the photocycle intermediates except for K610, which suggests it is linked to that intermediate.

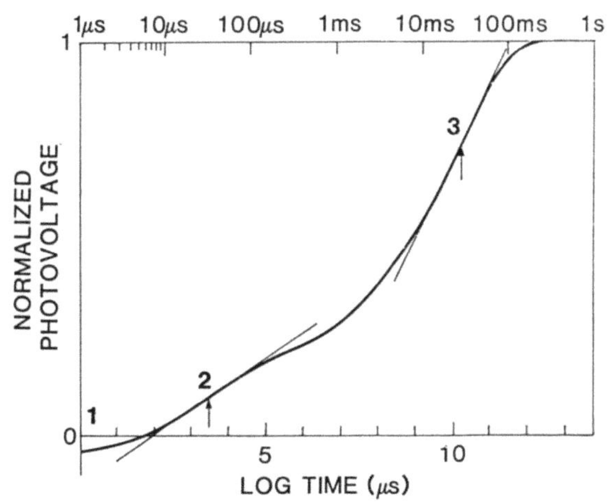

Figure 2. The photoresponse of oriented purple membrane sheets on a teflon septum separating two electrolytic solutions is shown. A 300 ns laser flash was used to illuminate the membrane.

ULTRAFAST EVENTS IN BACTERIORHODOPSIN

Femtosecond optical measurements [5] have been used to study the primary photoevents occurring immediately after the absorption of a photon by bacteriorhodopsin. Using these techniques the trans-cis isomerization of the retinal chromophore can be observed directly and was found to take place in about 500 femtoseconds. It has been proposed that the trans-cis isomerization of the retinal results in a movement of the protonated Schiff base away from a counter ion. This charge separation is a possible candidate for the early negative photovoltage component as well as the appearance of the K intermediate.

The appearance of the other intermediates (L,M, etc.) in the photoreaction cycle are thermally activated events and are, therefore, relatively slow. They can, in fact, be frozen out at liquid-nitrogen temperatures [2] so that the photocycle is aborted at the K intermediate. The time constants associated with the slower components (2 and 3 in Figure 2) of the photovoltage are also found to be temperature dependent [4]. However, no temperature dependence is observed for the initial negative transient in the photovoltage. These results also suggest that the initial negative transient in the photovoltage is associated with the appearance of the picosecond K intermediate in the photocycle.

FABRICATION OF BR PHOTODETECTORS

At a suitable pH, a PM sheet has both a net charge and an electric dipole moment. Taking advantage of this property, BR photodetectors were formed using the technique illustrated in Figure 3. First, the PM sample is washed in ultrapure distilled water and resuspended at about 4 mg/ml. A drop of this solution (30 microliters) is then placed on a transparent electrode (ITO glass) and covered with a second transparent electrode 1 mm away. The PM drop forms a bridge between the two electrodes so that a current flows between the electrodes when a voltage is applied. Usually 2 to 3 volts are applied for 2 or 3 minutes. When the transparent electrodes are separated, it is found that the negative-biased electrode has only clear water attached to it, while the positive electrode (anode) has all of the PM on its surface. This electrode is then dried under vacuum and covered with a counter transparent electrode similar to that used for the substrate. The active area of the resulting photodetector is about 0.25 cm^2.

TESTING THE PHOTODETECTORS

The photoresponse of the photodetector using dried PM is very similar to that obtained from oriented PM samples on a septum separating two electrolyte solutions. However, the rise times of the photovoltage components 2 and 3 in Figure 2 are slightly slower.

The measured response for the BR photodetectors is 0.35 mV per microjoule of incident light. A theoretical estimate of this value can be made by considering the voltage induced on two parallel capacitor plates when a charge between the

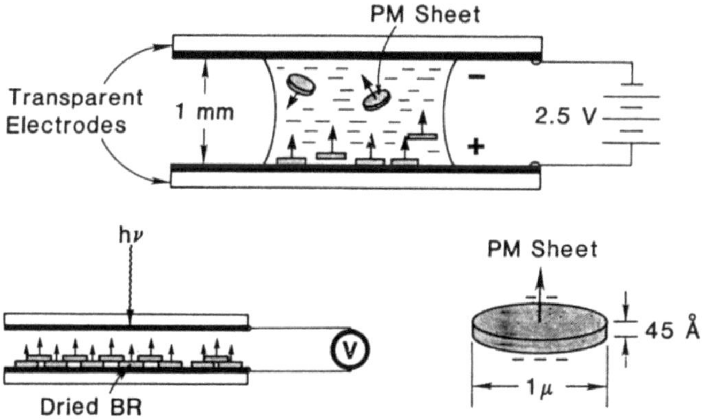

Figure 3. A diagram of the technique used to fabricate the BR photodetectors is shown.

plates is displaced [3]. A reasonable value for the displacement of the protonated Schiff base from its counter ion is about 2 angstroms, and the quantum efficiency for the formation of the K intermediate after the absorption of a photon is 2/3. Based on these values, the response of the detector should be approximately 350 mV per microjoule of incident light, or about three orders of magnitude larger than that measured for the current detectors.

The BR photodetectors should be functional from below liquid nitrogen temperatures to well above room temperature, since bacteriorhodopsin is stable and shows no thermal degradation at temperatures up to about 50 ^{o}C. We have stored the BR photodetectors for more than six months without any noticeable change in their performance.

An attempt to measure the rise time of the detector was made using a Q-switched Nd-YAG laser (Quanta Ray) and Tektronix 7904 oscilloscope with an S4 sampling head. The output from the sampling scope was recorded on a Nicolet digital oscilloscope. A nanosecond impedance converter (voltage follower) was also required due to the high output impedance of the detector. Figure 4, which shows a typical photoresponse, indicates a rise time of about 10ns, corresponding to the pulse width of the Q-switched Nd-YAG laser. These results with the Q-switched laser show that the rise time of the initial negative component in the photoresponse is certainly faster than all the photointermediates except K610. If the negative component in the photovoltage is indeed a subpicosecond event, then further efforts to measure the rise time of the photodetector using currently available electronics will be frustrated.

ELECTROOPTIC SAMPLING

In the past few years a technique called electrooptic sampling has been developed [6]. This technique is capable

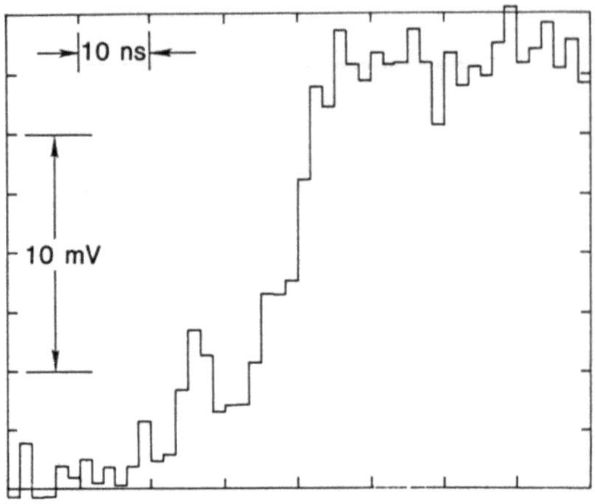

Figure 4. A photoresponse of the BR photodetector is shown. A 10 ns laser flash was used to illuminate the membrane. The photovoltage was recorded using a sampling oscilloscope.

of resolving subpicosecond photoinduced electrical transients. A diagram of an apparatus we are building to measure the rise time of the BR photodetectors is shown in Figure 5.

The electooptic sampling shown in Figure 5 uses a picosecond dye laser. A light pulse from the laser is split into two light pulses that traverse different optical paths. The length of one of the optical paths can be varied, and the light pulse along this path is used to excite the BR photodetector. The second light pulse passes through a polarizer and then through an electrooptic crystal ($LiNbO_3$) and finally through an analyzer and on to a "slow" photodetector. This slow photodetector measures the average incident light intensity; it does not resolve individual picosecond light pulses. The polarizer and analyzer form crossed polarizers.

Light Pulse I causes a photovoltage in the BR photodetector that is applied to electrodes on the electrooptic crystal. The electrooptic crystal responds to this impressed voltage by rotating the plane of polarization of the polarized light passing through it, resulting in an increased light transmission through the crossed polarizers. This increased light intensity through crossed polarizers is averaged over many light pulses by the slow photodetector. Since the optical path length of the first light pulse is variable, the temporal overlap of the voltage pulse and light pulse II in the electrooptic crystal can be adjusted. When they arrive simultaneously, the light intensity measured by the slow photodetector is a maximum. If the length of Light Path I is adjusted so that the voltage pulse arrives before or after Light Pulse II, then the light intensity at the slow photodetector is less than maximum. Thus variation of the optical path length of Light Pulse I permits the picosecond response time of the BR photodetector to be resolved.

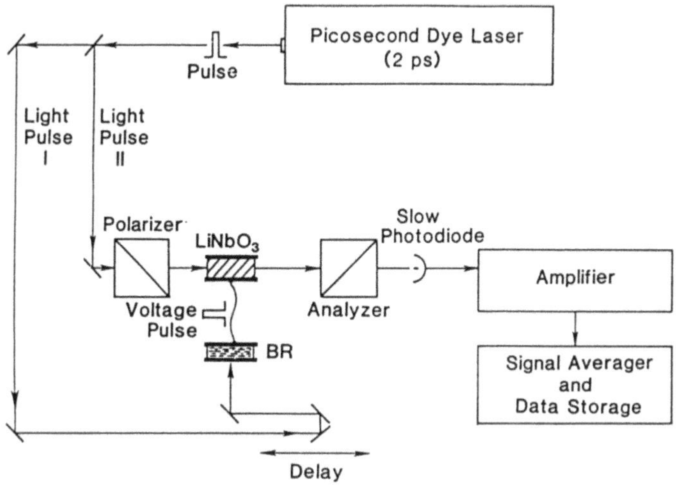

Figure 5. A schematic diagram of the picosecond electrooptic sampling apparatus is shown.

Usually, very high repetition rates are used for electrooptic sampling, which allows noise-reduction techniques to be used. The photoresponse characteristics of the BR photodetector require that a low repetition rate be used. Therefore, we are building a dye-amplified picosecond laser with a repetition rate of 10 Hz in order to obtain good "signal to noise" in the system.

SUMMARY

Photodetectors can be made, that use bacteriorhodopsin as the photoactive material. BR behaves as a photopolarizable material rather than as a photoconductor in these devices. The BR photodetectors indicates that BR can be used to make molecular electronic devices. Experimental evidence suggests that a subpicosecond response can be expected from these devices. An experimental apparatus has been designed that will be able to measure the expected subpicosecond photoinduced electrical signal from these devices.

REFERENCES

1. Stoeckenius, W. and R.A. Bogomolni, "Bacteriorhodopsin and Related Pigments of Halobacteria", Ann. Rev. Biochem., 52, 587-616 (1982).

2. Lozier, R.H. and W. Niederberger, "The Photochemical Cycle of Bacteriorhodopsin", Federation Proceedings, 36, 1805-1809 (1977).

3. Rayfield, G.W. "Events in Proton Pumping by Bacteriorhodoopsin", Biophys. J., 41, 109-117 (1983).

4. Rayfield, G.W. "Temperature Dependence of Photovoltages Generated by Bacteriorhodopsin", Biophys. J., 48, 111-115 (1985).

5. Mathies, R.A., C.H.B. Cruz, W.T. Pollard and C.V. Shank, "Direct Observation of the Femtosecond Excited-State cis-trans Isomerization in Bacteriorhodopsin", Science, 240, 777-779 (1988).

6. Valdmanis, J.A. and G. Mourou, "Subpicosecond Electrooptic Sampling: Principles and Applications", IEEE J. Quantum Electr., QE-22:1, 69-78 (1986).

OPTICAL RANDOM ACCESS MEMORY BASED ON BACTERIORHODOPSIN

Robert R. Birge, Chian-Fan Zhang and Albert F. Lawrence

Center for Molecular Electronics
Syracuse University
Syracuse, N.Y. 13244

INTRODUCTION

During the past five years, investigators have shown considerable interest in the potential use of light transducing proteins to perform optical switching functions.[1,2] There are significant advantages inherent in the use of biological molecules, either in their native form, or modified via organic synthesis or protein engineering, as active components in molecular electronic devices. These advantages derive in large part from the natural selection process, and the fact that nature has solved through trial and error problems of a similar nature to those encountered in harnessing organic molecules to carry out logic, switching and energy transducing functions. One example of this type of serendipity to be explored in the present paper is the use of the light harvesting protein, bacteriorhodopsin, as the photoactive element in an optically coupled random access memory.

THE PHOTOPHYSICAL PROPERTIES OF BACTERIORHODOPSIN

The protein, bacteriorhodopsin, is the light harvesting component in the purple membrane of the halophilic microorganism *Halobacterium halobium*.[3] This pigment system is grown by the bacterium under conditions of oxygen deprivation, when the normal respiration process of using oxygen to synthesize ATP becomes too inefficient to sustain growth. The role of the bacteriorhodopsin is to convert light energy into a hydrogen ion gradient that chemiosmotically drives the synthesis of ATP.[3,4]

The photocycle of bacteriorhodopsin is shown in Fig. 1. There are two binary photochemical reactions associated with the photocycle shown in Fig. 1 which can be utilized for optical switching and/or single bit data storage applications. At liquid nitrogen temperatures, the photocycle is "truncated" at the primary photoproduct, **K**, and light adapted bacteriorhodopsin participates in the following photoequilibrium:

$$\mathbf{bR}\ (\lambda_{max} = 576\text{nm}) \underset{\Phi_2 \sim 0.7}{\overset{\Phi_1 \sim 0.3}{\rightleftarrows}} \mathbf{K}\ (\lambda_{max} = 620\text{nm}) \quad (77\text{K}) \qquad (1)$$

where **bR** represents the light adapted form of the native protein, **K** represents the primary photoproduct, and Φ_1 and Φ_2 represent the forward and reverse photochemical quantum

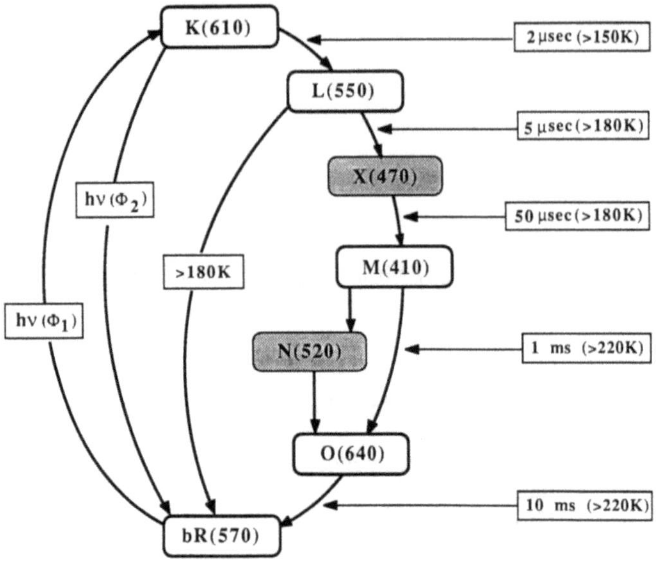

Figure 1. The photochemical cycle of light-adapted bacteriorhodopsin. Individual species are indicated within the rectangular polygons, and the numbers in parentheses following the labels indicate the approximate absorption maxima in nanometers extrapolated to ambient temperature. Optical random access memories operating at 77K limit the photoreaction to **bR** ⇔**K** interconversion. Optical RAM can also be operated at higher temperatures where the photocycle is truncated at the **M** intermediate (i.e. **bR** ⇔**M**).

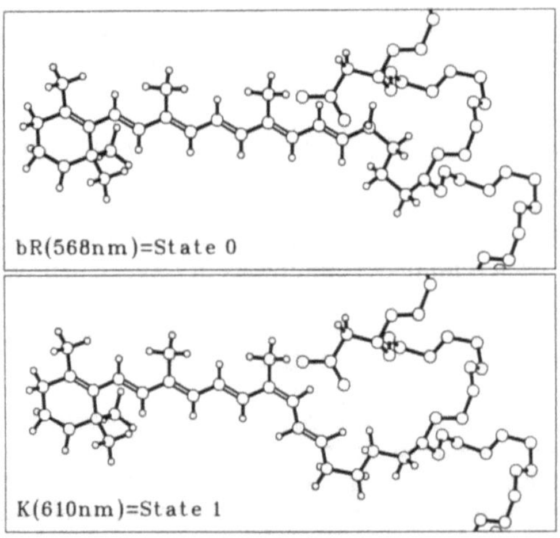

Figure 2. Our model of the primary event in light-adapted bacteriorhodopsin. The chromophore in **bR** is an all-*trans* retinyl protonated Schiff base attached to the protein α-helix via a lysine residue. The positive charge on the chromophore is stabilized by a negatively charged aspartic acid residue one α-helical turn away from the chromophore-lysine system. The primary event involves an all-*trans* to 13-*cis* photoisomerization to form **K**. Molecular dynamics calculations based on this model predict femtosecond photoisomerization rates (see Ref. 5). Standard photochromic data storage techniques require bit assignments based on photostationary states of **bR** and **K** (bit 0 → >97% **bR**; bit 1 → ~44% **K** & ~56% **bR**). The use of differential read data enhancement techniques permits the use of less differentiated concentration ratios to represent bit assignments (e.g. bit 0 = 84% **bR**: 16% **K**; bit 1 = 78% **bR**: 22% **K**).

yields, respectively. The absorption maxima are values measured at 77K. Our model[5] of the molecular transformations associated with the primary event is shown in Fig. 2. The formation times associated with both the forward and reverse photoreactions is less than 5 ps. Although the quantum yields have been the subject of intense study by a number of research groups, including our own, there remains disagreement in the literature.[5] The values shown above for the quantum yields are conservative estimates, and some researchers claim values for Φ_1 of 0.6-0.8 and values of Φ_2 approaching unity.[6] Although the above photoreaction is utilized for the present study, it should be noted that at higher temperatures (~200K), the photoequilibrium coupling **bR** and **M** can be used for moderate (μsec) speed optical switching and/or single bit data storage applications:

$$\mathbf{bR}\ (\lambda_{max} = 570\text{nm}) \underset{\Phi_3 \sim 0.7}{\overset{\Phi_1 \sim 0.3}{\rightleftarrows}} \mathbf{M}\ (\lambda_{max} = 412\text{nm}) \quad (200\text{K}) \qquad (2)$$

where **M** represents the blue-shifted intermediate that forms approximately 60 μs following excitation and is associated with deprotonation of the retinyl Schiff base. Surprisingly, the quantum yield of the reverse (**M** → **bR**) photoreaction, Φ_3, is observed to be identical to Φ_2 (**K** → **bR** photoreaction) within experimental error. This observation suggests that the back reaction may involve a rapid reprotonation of the excited state species prior to photochemistry:

$$\mathbf{M}\ (\mathcal{R}\text{=N-lysine}) \xrightarrow{h\nu} \mathrm{M}^{*} \xrightarrow{H^{+}} \mathrm{M}'^{*}\ (\mathcal{R}\text{=NH}^{+}\text{-lysine}) \longrightarrow \mathbf{bR} \qquad (3)$$

where $\mathcal{R}$ represents the $C_1..C_{15}$ portion of the retinyl polyene (see Fig. 2). If this interpretation is correct, it is likely that the **M** → **bR** photoreaction reaction takes place on a picosecond time scale. This prediction follows from the requirement that the reprotonation step must be extremely rapid, and once this occurs, the potential surface coupling the two species provides for direct ground state population of **bR**.

The intrinsic properties of the native protein make it an outstanding candidate for use in optical switching, optical RAM and optical disk applications.[1-10] These properties include: (1) long term stability of the protein to thermal and photochemical degradation, (2) picosecond photochemical reaction times (both the forward and reverse photoreactions produce stable products in less than 5 picoseconds at 77K), (3) high forward and reverse quantum yields permitting the use of low light levels for switching, (4) wavelength independent quantum yields, (5) a large shift in absorption spectrum accompanying photochemistry which permits accurate and reproducible assignment of state, (6) high two-photon cross sections for photoactivation permitting higher storage densities than is possible via one-photon excitation and (7) the ability to form thin films of bacteriorhodopsin with excellent optical properties by using the Langmuir-Blodgett technique. We will explore some of the above properties of bacteriorhodopsin in further detail below.

Photochemical and Thermal Stability. In general, proteins are not stable for extended periods of time at temperatures much higher than 35°C. Bacteriorhodopsin, however, has an enhanced stability due to the *in vivo* requirement that it operate at much higher temperatures in the native bacterium. *Halobacterium halobium* grows in salt marshes, and the ambient temperature often exceeds 40°C. While a detailed investigation of bacteriorhodopsin's thermal stability has not been carried out, investigations in our laboratory indicate that thin films prepared via the Langmuir-Blodgett technique are both photochemically and thermally stable under conditions encountered in an optical RAM as well as those required for long term shelf storage.

Photoreaction Kinetics. The **bR** ⇔ **K** photoreaction has been studied in detail via picosecond spectroscopy.[6] While there remains some disagreement in the literature concerning the nature of the intermediate, **J**, which couples these two species, as well as the effect of temperature and environment on the photoreaction kinetics, there is general agreement that the photoreaction occurs on a picosecond time scale. We will adopt, for the purposes of this discussion, a conservative photoreaction time of 5 ps. This photoreaction time is much shorter than the latencies associated with the electrooptical scanners, and it is unlikely that a system designed based on the **bR** ⇔**K** photoreaction will be able to take advantage of this protein's extremely rapid photochemistry during the next decade.

Photoreaction Quantum Yields. Bacteriorhodopsin is noted for its high photochemical efficiency. As noted above, the forward quantum yield, Φ_1, is observed to be 0.3 or larger, and the reverse quantum yield, Φ_2, is observed to be 0.6 or larger. The sum of the forward and reverse quantum yields is equal to, or greater than, unity. This exceptional efficiency, coupled with the high absorption cross-section (see below), combines to provide an optimal photostorage environment. Furthermore, the quantum yields are wavelength independent, which provides for flexibility in designing read/write algorithms.

Spectroscopic Properties. As shown in Figure 3, the formation of the primary photoproduct introduces a significant shift in the absorption maximum. While it is possible to generate almost pure (> 97%) **bR** by using red light in the wavelength region 630-700nm, it is not possible to generate comparable percentages of **K** regardless of excitation wavelength. This observation follows from the fact that the **K** intermediate has significant absorptivity throughout the 350 - 650 nm absorption region of **bR**. Experimental studies indicate that the highest population of **K** that can be formed is 44%, generated by using excitation in the region 500 - 540 nm (all experiments at 77K).[8] Accordingly, our optical RAM is based on the following photostationary state (PSS) assignments: State 0 ≡ > 97% **bR** (630 -635 nm PSS); State 1 ≡ 56% **bR** & 44% **K** (540 -545 nm PSS). The use of photostationary states to represent bit states does not preclude a very high signal/noise ratio for state assignment. As shown in Fig. 3, the photostationary state difference spectrum produces a large change in molar absorptivity. We will demonstrate below that the use of differential read techniques provides high speed read capability with high reliability.

Nonlinear Properties. Due to a large change in dipole moment upon excitation, the low-lying excited singlet states of bacteriorhodopsin have very large two-photon absorptivities.[7] Thus, photochemistry can be induced via two-photon excitation techniques, and this property of bacteriorhodopsin provides for increased storage densities. We have not taken advantage of this property of bacteriorhodopsin in our optical RAM because speed, rather than storage density, was the primary goal. Nevertheless, this capability should be noted by those investigators potentially interested in high density disk applications.

Protein Engineering Opportunities. An important advantage of using a protein system as a photoactive element derives from the potential ability to enhance selectively the spectroscopic and photochemical properties of the protein by using genetic engineering. Although the elegant studies by Khorana and coworkers[11] indicate the feasibility of modifying the photophysical properties of this protein, no concerted effort has been made to engineer new attributes to optimize the molecular electronic properties of bacteriorhodopsin. Collaborations with scientists at Genencor are currently underway to investigate the generation of modified proteins with red shifted absorption maxima better suited to excitation via commercially available laser diodes as well as the generation of modified proteins that have a binary photoreaction at ambient temperatures.

Preparation of Langmuir-Blodgett films. Bacteriorhodopsin is a membrane bound protein. Thus, it is not surprising to find that the Langmuir-Blodgett (L-B) technique[12] provides one of the most successful techniques investigated for preparing thin films of this protein with optimal optical properties. We typically prepare thin films containing 80% **bR** by weight in phospholipid bilayers on quartz slides (Fig. 4) or $\lambda/2$ flat or spherical mirrors. First two monolayers of the phospholipid are deposited on the surface to act as a buffer layer. Second, the bacteriorhodopsin is added in hexane solution to the phospholipid layer and the hexane is allowed to evaporate. Subsequent layers are put down recognizing that two sweeps are required for each monolayer of bacteriorhodopsin. Best results are obtained when 30 layers of **bR** are used (62 monolayers of phospholipid). More

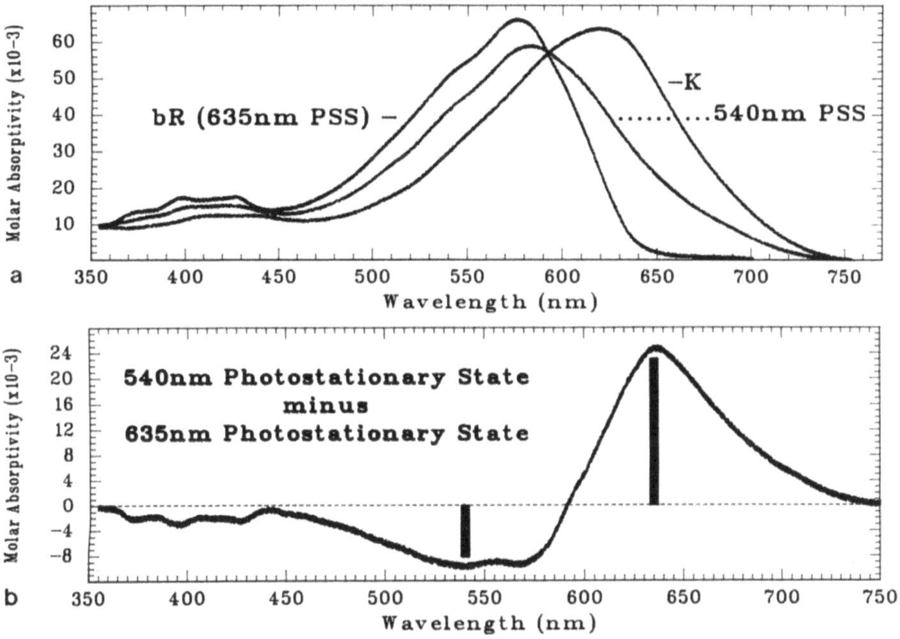

Figure 3. The optical random access memory is based on the photochemical conversion between two photostationary states. Green laser excitation in the region 540 - 545nm sets bit 1, and red laser excitation in the region 630 - 635nm sets bit 0. These wavelengths generate two photostationary states with absorption spectra shown in insert (a). Bits 0 and 1 can be assigned with high reliability by measuring the molar absorptivity difference for 540 - 545nm versus 630 - 635nm excitation (insert (**b**)). The differential absorptivity associated with maximal photochemical conversion (bit 0 = >97% **bR**; bit 1 = ~56% **bR** & ~44% **K**) is ~32,000 L M^{-1} (insert (**b**)). The differential absorptivity associated with 6% photochemical conversion (e.g. bit 0 = 84% **bR**: 16% **K**; bit 1 = 78% **bR**: 22% **K**) is ~4,500 L M^{-1}.

layers of **bR** are often required, however, to provide sufficient optical density. Optimizing the L-B procedures remains an active area of study. We are currently investigating the use of cross-linkable matrices to permit operation of the optical RAM at ambient temperatures. In that regard, a comparison of the properties of our L-B films with those observed for an ambient temperature film containing bacteriorhodopsin known as biochrome is useful.

Comparison with Biochrome. Approximately seven years ago, researchers at the Institute of Biological Physics of the USSR Academy of Sciences introduced a photochromic film called Biochrome (also known as Biochrom-BR and Biokrom).[9-10] The formula for preparing this film has, to our knowledge, never been published, but it is known that this material is based on a dehydrated suspension of bacteriorhodopsin in a transparent polymer matrix. It is also known that chemicals are added to enhance the lifetime of the **M** intermediate. The principal mechanistic difference between biochrome and our bR-based film is that biochrome operates at ambient temperature based on **bR**⇔M photochemistry (Eq. 2) and our film operates at liquid nitrogen temperature based on **bR**⇔K photochemistry (Eq. 1). The photosensitivity is similar (~5 x 10^{-4} J cm^{-2}) and the spatial resolution of both films is better than the diffraction limit. Biochrome displays a cyclicity of ~10^5, whereas our film displays a cyclicity better than ~10^7 based on extrapolation to 5% denaturation levels. We anticipate that the difference in cyclicity is due primarily to temperature rather than material, or that the difference is associated with the fact that limiting the photochromic reaction to **bR**⇔K, which maintains protonation of the retinyl chromophore, minimizes loss of photochromic activity due to side reactions. An important difference relevant to potential use of biochrome in optical RAM applications is

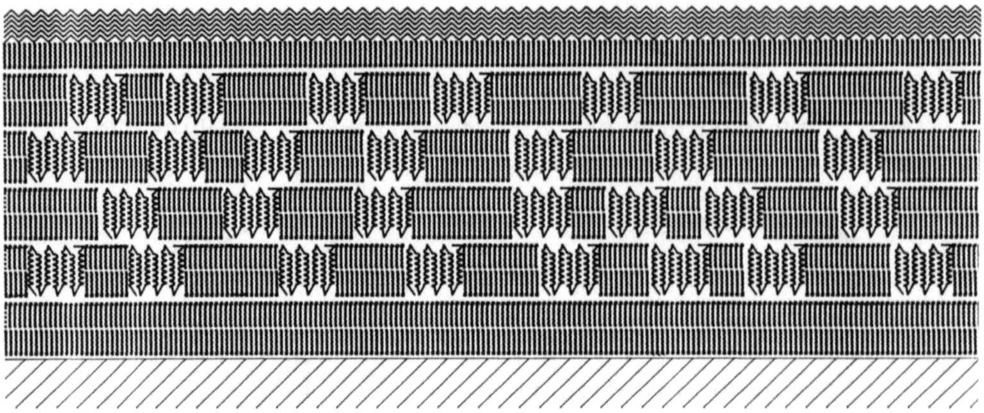

Figure 4. A schematic diagram of a Langmuir-Blodgett film containing light adapted bacteriorhodopsin. The thin film stabilizes the protein in a phospholipid bilayer containing from 30-100 layers of protein (and at least twice the number of monolayers of phospholipid binder). For convenience, only four protein layers are shown in the above diagram. Our experimental studies indicate that it is necessary to place at least one bilayer of lipid on the hydrophobic surface prior to adding the protein, probably because the protein extends partially into the upper and lower bilayer regions. A final coating of protein free lipid followed by a transparent cross-linked polymer is used to seal the coating in order to protect the thin film and prevent excessive loss of water content.

that biochrome can store data for a maximum of a few hours, after which time the **M** intermediate has thermally decayed back to **bR**. By operating our L-B based **bR** film at 77K, storage time is, for all intents and purposes, infinite. What remains to be explored is how biochrome film would behave at 77K. We anticipate that biochrome may have properties virtually identical to our film at liquid nitrogen temperatures, provided the polymer matrix can withstand liquid nitrogen temperatures without cracking.

DESIGN CRITERIA FOR bR-BASED PHOTOCHROMIC RAM

We have demonstrated that a thin film of bacteriorhodopsin provides an outstanding photochromic medium. Nevertheless, this materials aspect is but one component of the total engineering problem. Indeed, many researchers working on optical memories believe that photochromic media are intrinsically less reliable than alternatives such as magneto-optical and hole-burning technologies. To a large extent, these researchers base their opinions on the studies of Tomlinson.[13] This investigator concluded that the disadvantages of photochromic media are "likely to be overwhelming". Tomlinson notes two problems. First, photochromic storage media do not exhibit threshold response characteristics, and thus any attempt to read the state of the device will also disturb the state of the device. Thus, while it is possible to read the state of the device by using very low irradiation fluxes, each read operation will degrade the signal-to-noise ratio of the subsequent read process by photochemically modifying the photochromic material. Eventually, a "refresh" operation will be required to maintain data integrity. Second, in order to erase data at the same rate as data are written, the erase source must have a power output as much as an order of magnitude greater than that of the write source. For "typical" system parameters, Tomlinson estimates average erase powers of 22-44 mW, values which are roughly one order of magnitude larger than are realistic under normal design constraints.[13] It was the latter issue that dominated Tomlinson's lack of enthusiasm for photochromic storage media. While Tomlinson's analysis of the situation is correct for standard read-write optical formats, we have developed an alternative approach which has the potential of providing high reliability without requiring high laser powers (Fig. 5). When coupled with an optical scanner system to provide rapid random access to any data location (Fig. 6), our read-write technique offers both speed (~5 ns access time) and reliability. While a number of design issues remain to be resolved before our approach can be transformed into a commercially viable system, prototype studies indicate that our approach is feasible and that the component design requirements are realistic in terms of resolution and speed.

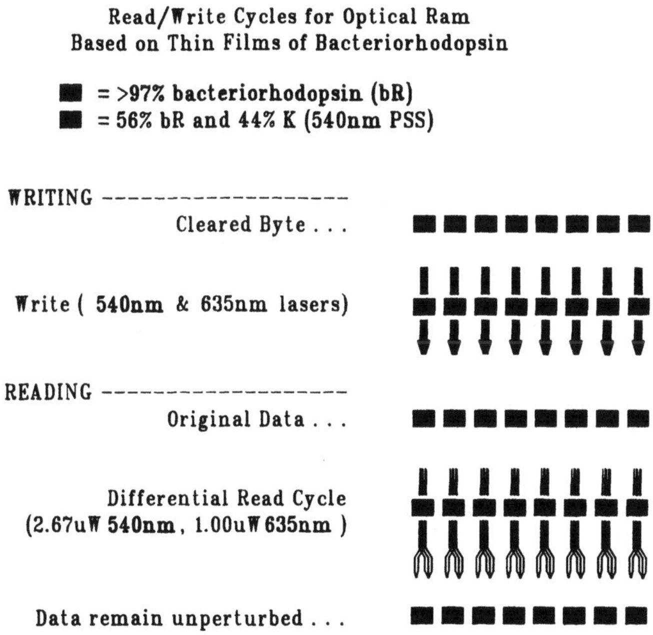

Figure 5. Our current read/write approach utilizes positive write techniques and differential read techniques to maximize speed and efficiency. The positive write technique sets bits to the desired state regardless of their original state by turning on the appropriate laser to drive the photochemistry in the correct direction. The differential read technique uses both lasers simultaneously to determine the differential absorptivity, and a high speed crowbar circuit to attenuate the green laser if the original state is observed to be bit 0 (>97% bR), or the red laser if the original state is observed to be bit 1. This technique permits rapid read cycles while simultaneously enhancing the delta concentration associated with the assignment of state based on the concentrations of **bR** and **K**. In the above example, bit designations are assigned based on the extremes in concentration, but the differential read algorithm allows bit assignments involving ~6% concentration differences (see text).

Differential Read Data Enhancement Techniques. There are two basic approaches to assigning the state of a photochromic system. The first, and more standard, approach relies on absolute assignment of state based on the use of a single laser source tuned so that it is absorbed by one of the states, but transmitted by the other. A reference absorption level is assumed, and if the laser beam is absorbed by a fraction equal to or greater than that level, the state of the bit is assigned accordingly. Such an approach requires that the photochromic bit be set with a relatively high conversion ratio, and thus bit 0 will be represented as >65% state 1 and bit 1 represented as >65% state 2. Lower conversion ratios will lead to bit errors due to uncertainties in the absolute absorptivity measurements. A corollary to this approach is that the laser power required to write or erase (but usually not both) must be high enough to generate significant conversion of the photochromic species within the irradiated volume. A quantitative analysis of this problem led Tomlinson[13] to conclude that under normal data transfer speeds, the erase powers exceeded realistic values (see discussion above).

An alternative approach uses two lasers to monitor the state of the device, and uses differential absorption and a pair of reference bits to assign the state of the bits that make up an individual word. The technique, which is diagrammatically outlined in Fig. 5, has rather severe hardware requirements, because it requires the use of dichromatic beam positioning scanners(Fig. 6) and ultra high speed crowbar electronics to control the laser intensity. Nevertheless, the advantages with respect to reliability and device speed are significant and the requirement for high laser intensities are obviated.

Our procedures define the state of a bit with reference to two standards. The first defines the differential absorptivity associated with a 0 bit, the second the differential absorptivity associated with a 1 bit. As we will demonstrate below, the 1 bit is represented by a ~6% increase in the concentration of **K** relative to **bR**. While this difference in concentration can be identified accurately by measuring the difference in absorbance for the red (633nm) and green (543nm) HeNe lines (e.g. Fig. 3), the assignment of state must be done by reference to standards which define the differential absorptivity levels for bit 0 and bit 1. The second aspect of our method requires that a crowbar circuit enhance the differential absorptivity difference during each read cycle. This is done by attenuating one laser prior to the other laser after carrying out the measurement and determining the state of the bit being read. The laser that is attenuated first is the one which is responsible for diminishing the concentration of that species which was enhanced during the write operation. In other words, if bit 0 is observed, the green laser is attenuated prior to the red laser. This read enhancement technique improves the signal/noise ratio of subsequent read operations by enhancing the differential absorptivity difference between the 0 and 1 bits.

Limitations of optical scanner technology dictate that we operate below the diffraction limits, and thus bits are separated by ~2 microns and the irradiated volume is ~ 10^{-12} cm^3. Approximately 10^7 protein molecules occupy this volume assuming a density of ~10^{19} molecules/cm^3. A one milliwatt laser operating at ~600nm provides a flux of ~3 x 10^6 photons/ns, and if we assume that this flux can be transferred to the sample without attenuation, and we assign an irradiation time of 4 ns, we provide a total excitation population of ~10^7 photons. Thus, the number of photons in the excitation pulse and the number of photochromic species within the irradiated volume of an individual bit are roughly equivalent. Maintaining a ratio of photons to irradiated species close to unity turns out to be optimal for DRDE techniques. The number of molecules excited within the irradiated volume, N*, is given by:[13,14]

$$N^* \cong N[1 - \exp(-Z_\lambda)] \tag{4}$$

where

$$Z_\lambda = \sigma_\lambda C L \tag{5}$$

and N is the number of photons in the laser pulse, σ_λ is the one-photon cross section of the photochromic material for absorption at wavelength λ, C is the concentration of the material in units of volume^{-1}, and L is the effective pathlength of the light trajectory through the photochromic material. We use the term "effective pathlength" because L will be twice the sample thickness when one is operating in reflection mode. The one-photon cross section is proportional to the molar absorptivity, ε_λ, of the sample at the appropriate wavelength,[14]

$$\sigma_\lambda = (3.8235 \times 10^{-21})\, \varepsilon_\lambda \tag{6}$$

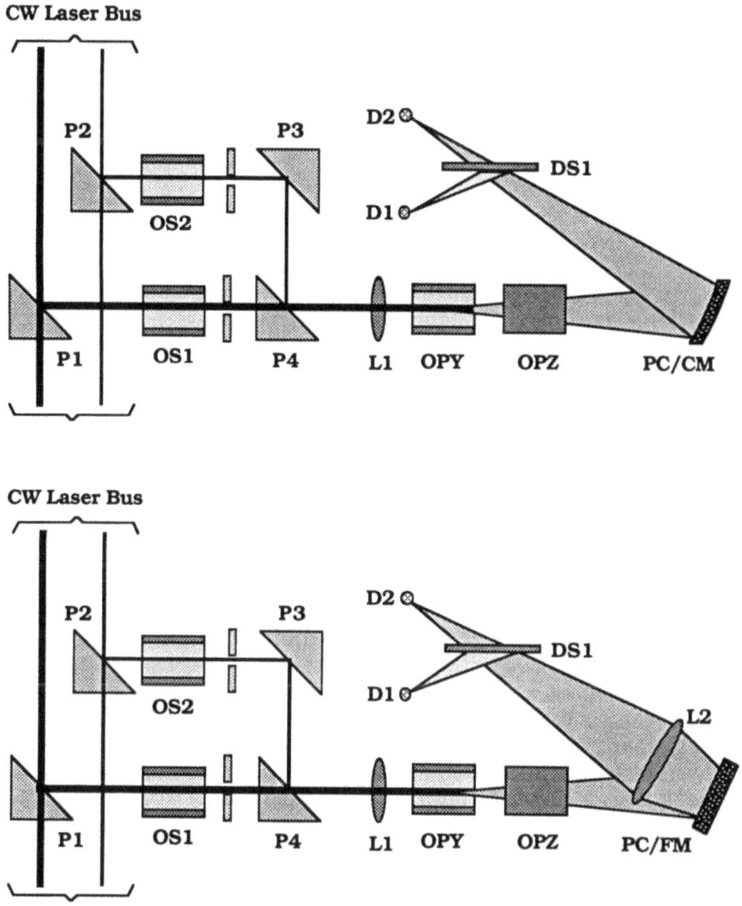

Figure 6. Schematic diagram of the principal components of a single memory plane of a bacteriorhodopsin based optical random access memory. Two different designs are shown. The design shown in the upper diagram stores information in reflection mode on a photochromic condensing mirror (PC/CM). The design shown in the lower diagram stores information in reflection mode on a photochromic flat mirror (PC/FM), and requires an additional component (lens L2) to redirect the scanned reflected light onto the laser diodes (D1 and D2). Laser excitation to drive the photochemistry is provided by continuous wave red (λ=633nm) and green (λ=543nm) helium-neon lasers, and the two beams travel down the vertical "CW Laser bus". Prisms 1 and 2 (P1 and P2) are coated so as to couple ~1 mW of the laser power into the optical shutters (OS1 and OS2). Thus, P1 and P2 require different reflective coatings depending upon the position of the memory plane assembly along the optical bus. The two beams are made colinear via P3 and P4. Lens L1, with a focal length roughly equal to the distance between L1 and the center of the photochromic surface (PC/CM or PC/FM), provides focussed irradiation (spot diameter of ~1.6 microns). The Langmuir-Blodgett film containing bacteriorhodopsin is deposited on either a condensing mirror (PC/CM) or a flat mirror (PC/FM), and the reflected laser light is directed onto the high speed photodiodes (D1 and D2). A dichroic beam splitter (DS1) directs the reflected green laser beam to D1 and the reflected red laser beam to D2. Scanning of the focussed laser beams across the photochromic surface is accomplished by using electro-optical scanners OPY and OPZ. Note that scanning in the plane of the page is shown to indicate the optical path as a function of this scanning operation. A memory word is represented by as many memory planes as there are bits in the word (a machine dependent number) plus two, which are used for reference bits (see text). Thus, a word is read in parallel so that access time is determined by the single bit access latency, which is determined primarily by the speed of the electro-optical scanners.

and based on the spectra shown in Fig. 3, we calculate a forward ($0 \rightarrow 1$) cross section of 1.9×10^{-16} cm^2 at 543 nm (pure **bR**) and a reverse ($1 \rightarrow 0$) cross section of 1.1×10^{-16} cm^2 at 633 nm (543 nm photostationary state). A typical sample thickness is ~5×10^{-5} cm ($L \cong 10^{-4}$ cm), and thus Z_{543}=0.18 and Z_{633}=0.10. The fraction of molecules converted from state 0 to state 1 during the excitation process is given approximately by

$$F_{1 \leftarrow 0} \cong [\Phi_1 N^*_{\mathbf{bR}} - \Phi_2 N^*_{\mathbf{K}}] / N_{iv} \tag{7}$$

where $N^*_{\mathbf{bR}}$ is the number of **bR** species excited within the irradiated volume, $N^*_{\mathbf{K}}$ is the number of **K** species excited within the irradiated volume, and N_{iv} is the number of molecules within the irradiated volume ($N_{iv} \cong 10^7$ molecules). A similar expression defines the fraction of molecules converted from state 1 to state 0 during the excitation process

$$F_{0 \leftarrow 1} \cong [\Phi_2 N^*_{\mathbf{K}} - \Phi_1 N^*_{\mathbf{bR}}] / N_{iv} \tag{8}$$

Eqs. 7 and 8 are reasonable approximations provided N, the number of photons in the pulse, is much smaller than N_{iv}, or provided N* is less than or equal to N_{iv} and Z < 0.3. The latter condition applies to the present situation. For the forward reaction ($\Phi_1 \cong 0.3$, $N^* \cong 2 \times 10^6$) we obtain a value for $F_{1 \leftarrow 0}$ of ~0.06, i.e. 6% of the **bR** molecules in the irradiated volume are converted to **K**. A similar value for the reverse reaction is obtained assuming one starts with the 543 nm photostationary state, because while $Z_{633} < Z_{543}$ the reverse quantum yield, Φ_2 is twice as large as Φ_1($F_{0 \leftarrow 1}$ is calculated to be equal ~0.07). These conversion fractions are smaller than would be required for normal read/write procedures using single frequency laser scanning. However, the use of differential (two laser) absorptivity measurement techniques increases the signal/noise ratio to 12 - 52 based on numerical simulations, where the range is assigned based on worst case (12) versus best case (52) assignments for the salient system parameters (such as noise imposed on the CW laser signals or the initial concentrations of **bR** and **K**). We consider a signal/noise ratio of at least 20 is required for satisfactory performance. We should note that our application of the term "signal/noise ratio" appears to be different from that used to assign the signal/noise ratio in other, related studies.[15] We base our numerical assignment on the ratio of the differential signal to the noise level that is present based on a worst case summation of all the noise sources, i.e., all of the noise sources are operating in the same direction to diminish the accuracy of the measurement. Thus, a signal/noise ratio of 20 provides satisfactory data reliability.

The exchange of identities problem. There is one complication associated with the DRDE technique that deserves mention. It would, in principle, be possible to carry out a sufficient number of read operations on a region of the optical RAM surface that the concentrations of the photoreactants representing bits 0 and 1 would become so different that a subsequent write operation at the laser powers discussed above would not "exchange identities". That is, changing a bit 0 to bit 1 and a bit 1 to bit 0 would yield a final pair of bits with nearly identical concentration ratios. Thus, a subsequent read operation on that section would be, at best, ambiguous. There are two methods of avoiding this situation. The first is to use the differential read technique to direct bits 0 and 1 toward discernible but always exhangeable concentration ratios (e.g. bit 0 = 84% **bR**: 16% **K**; bit 1 = 78% **bR**: 22% **K**). The alternative is to occasionally transfer the RAM data to disk memory, clear the entire surface by scanning multiple times with the red laser to reset all locations to bit 0 (a "global erase"), and then restore the data. The first approach is superior, because it permits continuous, uninterrupted operation. The second approach, however, is more easily accomplished within the controller firmware, and would decrease overall data through-put by less than 3%. Both methods are under investigation.

References

1. Communication with Submicron Structures. Perspectives in the Application of Biomolecules to Computer Technology. A.F. Lawrence and R.R. Birge, in *Nonlinear Electrodynamics in Biological Systems*, Plenum, New York, H.R. Adey and A.F. Lawrence, eds., pp. 207-218 (1984).

2. Current Research and Future Potential of the Application of Molecular Electronics to Computer Architecture, R.R. Birge, A.F. Lawrence and L.A. Findsen, in *Proceedings of the International Congress on Technology and Technology Exchange*, International Technology Institute, Pittsburgh, Pennsylvania, 1986, pp. 3-8.

3. Photophysics of Light Transduction in Rhodopsin and Bacteriorhodopsin. R.R. Birge, *Annual Review of Biophysics and Bioengineering* **10**, (1981), pp. 315-354.

4. Energy Storage in the Primary Step of the Photocycle of Bacteriorhodopsin. R.R. Birge and T.M. Cooper, *Biophys. J.* **42**, 61 (1983).

5. Molecular Dynamics of the Primary Photochemical Event in Bacteriorhodopsin. Theoretical Evidence for an Excited Singlet State Assignment for the **J** Intermediate. R.R. Birge, L.A. Findsen and B.M. Pierce, *J. Am. Chem. Soc.* **109**, 5041-5043 (1987).

6. Early Picosecond Events in the Photocycle of Bacteriorhodopsin. H.J. Polland, M.A. Franz, W. Zinth, W. Kaiser, E. Kolling and D. Oesterhelt, *Biophys. J.* **49**, 651-662 (1986).

7. Two-Photon Spectroscopy of Protein Bound Chromophores. R.R. Birge, *Accts. Chem. Research* **19**, 138-146 (1986).

8. A Spectroscopic, Photocalorimetric and Theoretical Investigation of the Quantum Efficiency of the Primary Event in Bacteriorhodopsin. R.R. Birge, T. M. Cooper, A. F. Lawrence, M. B. Masthay, C. Vasilakis, C.F. Zhang and R. Zidovetzki. *J. Am. Chem. Soc.* (in press).

9. Photosensitive Biological Complexes and Optical Recording of Information. Eds. G.R. Ivanitsky and N.N. Vsevolodov, Sci. Biol. Res. Center, Pushchino, 1985, pp. 1-209.

10. Biological Photosensory Complexes as Technical Information Carriers, N.N. Vsevolodov and G.R. Ivanitsky, Biofizika, **30**, 883 (1985).

11. Methods for Mutagenesis of the Bacterioopsin Gene, M.A. Gilles-Gonsalez, N.R. Hackett, S.J. Jones, H. G. Khorana, D.S. Lee, K.M. Lo and J.M. McCoy. *Methods Enzymol.*, **125**, 190-214 (1986).

12. An Applied Science Perspective of Langmuir-Blodgett Films. G.G. Roberts. *Advan. Phys.* **34**, 475-512 (1985).

13. Analyses of Bit-Oriented Optical Memories using Photochromic Media. W.J. Tomlinson. *Appl. Optics.* **23**, 4609-4612 (1984); *Appl. Optics.* **15**, 821-826 (1976).

14. One-Photon and Two-Photon Excitation Spectroscopy. R.R. Birge, in *Ultrasensitive Laser Spectroscopy*, Academic, New York, D.S. Kliger, ed., pp. 109-174 (1983).

15. Organic Photochromics for Spatial Light Modulation. C.J.G. Kirby and I. Bennion, *IEE Proc.*, **133**, Part J, 98-104 (1986).

ACTUAL POSSIBILITIES OF BACTERIORHODOPSIN APPLICATION IN OPTOELECTRONICS

N. N. Vsevolodov, A. B. Druzhko, and T. V. Djukova

Institute of Biological Physics
USSR Academy of Sciences
Pushchino, Moscow Region, 142292 USSR

INTRODUCTION

Recently several reports have appeared on the possibility of fabricating various technical micro- and macrodevices based on natural biological materials. Bacteriorhodopsin (BR), a retinal-protein complex of microbiological origin, is one of the most promising materials for such purposes today. A BR molecule can be regarded as a natural microelement transforming light into other types of energy or accumulating it.[1] Natural BR analogs have been found and artificial ones with different novel features have been synthesized.[2]

The light-dependent transport of protons resulting in the electrochemical membrane proton gradient is known to be the main function of BR in the purple membrane of *Halobacterium halobium*. Many physical characteristics of BR, such as photoelectric, piezoelectric, electric, electrochromic and photochromic, have been revealed.

The photochemical processes in BR can be considered to be a set of spectrally reversible processes. In view of all the features mentioned, a photochromic material called "Biochrom-BR" has been produced. It can be used in systems of constant and dynamic recording and optical information processing.[3] Recently it has been shown that biochromic films are suitable for holography (both usual and polarizable) in constant and dynamic regimes. The films can be used as light-sensitive materials in the elements of quantum electronics.[4]

1. PHOTOCHROMISM

The BR molecule can be considered as a photochromic element due to reversibility of its photoprocesses. The process of M412 photoproduct transformation back to the initial BR570 form can be slowed down by several orders of magnitude by the following treatments: dehydration of purple membrane (PM), addition of some chemical substances into the PM suspension, and embedding PM in solid transparent matrices.

Photochromic films can be produced by suspending PM in a polymeric solution, adding some chemical supplements controlling the photocycle, and spreading a thin layer of this mixture on any support. The films can be used for temporary or constant storage and processing of optical information.[5]

Eight years ago the photochromic reversible film called "Biochrom-BR" was developed at the Institute of Biological Physics of the USSR Academy of Sciences. Its optical characteristics are good enough to compete successfully with modern organic and inorganic photochroms. They are:

a) photosensitivity up to $5 \cdot 10^{-4}$ J/cm^2;
b) spatial resolution better than 1,000 mm^{-1};
c) cyclicity of more than 100,000;
d) information storage time from a fraction of a second to hours.

Such photochromic films can be used in holography to register dynamic amplitude and phase holograms. This film can be stored for more than ten years without any loss of its properties. Recently a material for constant recording of information has been produced. However, its photosensitivity is two orders of magnitude less than that of the reversible film "Biochrom-BR". The film production process is ecologically harmless.

Modern biochemical methods make it possible to synthesize many BR analogs with novel features. It can be achieved by substituting a natural chromophore by a modified one. Some BR analogs have already been used for production of photochromic films. Their spectral and kinetic characteristics appeared to be quite different from those of BR-films.[6]

2. REFRACTIVE INDEX CHANGE IN THE BR PHOTOCYCLE

Charge redistribution and protein conformation (tertiary structure) changes take place in the BR molecule in the course of photocycle. These changes are accumulated in the PM resulting in the alteration of the medium density and the refractive index. Media with a changing refractive index cause phase displacement of the transmitted light beam. Such media capable of changing the beam phase depending on the light intensity are widely used for hologram recording.

As shown experimentally, holograms on "Biochrom-BR" films are recorded only in the course of BR photocycle and live as long as the M412 photoproduct does.

The "Biochrom" capability of dynamic hologram recording can be used in vibrating object interferometry. One of the disadvantages of holographic interferometry is the lack of registering scheme adaptation to stray low-frequency oscillations. To avoid this, the photorefractive media scheme has been applied. Usually, photorefractive crystals of lithium niobath are used for this purpose. However, such crystals offer low sensitivity and, hence, a long recording time is required; it becomes too slow to register the medium oscillations.

Recently possibility of holographic processing of the phase modulation signal has been demonstrated experimentally. BR film is used as a dynamic photorefractive medium in a fiber-optic interferometer. This interferometer requires a light capacity which is two orders of magnitude less than was previously possible.[7]

3. PHOTOINDUCED ANISOTROPY

The BR molecule transition dipoles in each PM patch are known to be arranged practically in the plane of membrane but randomly, i.e. their arrangement has no stable direction within the plane. PMs in the film are also arranged in the same plane. Thus the film is initially isotropic with respect to the incident radiation of nonpolarized beam. However when the beam is linearly polarized and its wavelength is within the absorption range of the initial form, then there appears an optical axis in the film parallel to this polarization, i.e. optical anisotropy arises. At this moment, the medium becomes anisotropic when it interacts with any other light beam.

The effect described above allows us to record a polarizable hologram by two (orthogonal by polarization) beams.[4] The light intensity for polarizable recording in "Biochrom-BR" is by two orders of magnitude less than that for dyes used at present. In such a material the change of polarization type of the diffracted beam depends on the lattice recording type and the reconstructing beam polarization type.

Polarizable holography provides a possibility for development of parallel processing of large informational massives.

4. NONLINEAR PROPERTIES

As mentioned above, "Biochrom-BR" possesses a high energetic sensitivity, reaching in some samples as high as $5 \cdot 10^{-4}$ J/cm^2. In these cases there are no irreversible photochemical processes even at very high energy densities of incident irradiation. Nonlinear media with such characteristics can be used to observe the optical bistability and some other effects of self-influence. When "Biochrom-BR" is fixed in the arm of a three-mirror resonator of a powerful argon laser, optical bistability can be induced. This provides the possibility to control laser generation using low-powered light source with a switching rate being equal to 20 kHz.

In fact, the "Biochrom-BR" film can be considered a two-dimensional crystal (or a quasi-crystal) consisting of crystalline domains (PMs) with nonlinear properties. Our recent paper[8] described the effect of photochromic process in BR on nonlinear properties of the "Biochrom-BR" film, and in particular, on the generation of second harmonics (SH) of a laser pulse with radiation at the wavelength 1064 nm. Both reflected and transmitted SH generations were obtained in the 100 μm thick films. Upon illumination of the film in the region of harmonics generation with weak (0.1-100 mW/cm^2) irradiation of a tuned laser, the intensity of the SH depends both on the illumination intensity and the wavelength of the tuned laser. The SH intensity is maximal in the absence of illumination. When the intensity of illumination is increased in the region of absorption of the initial form (i.e. 570 nm), the transformation into M412 product occurs and the intensity of the SH decreases, whereas upon exposure to 412 nm light the intensity of SH increases. Commutation of the system of SH generation takes place. On the other hand, pumping irradiation at 1064 nm does not fall in the BR absorption band and has no effect on commutation. Thus in the course of reading out the SH information neither distortion nor erasure occurs.

5. PHOTOPOTENTIAL

Randomly arranged PMs can be oriented with the help of electrophoresis in multilayers. Total photopotential in this case reaches (in some samples) ten volts.[9]

6. ELECTROCHROMISM

The change of spectral properties in the BR photocycle is accompanied by charge displacement. One should expect that the charge displacement in the electric field induces changes of some spectral properties.[9] This effect (called electrochromic one) can be used as optical and electrooptical switches.

7. PIEZOEFFECT

This effect was observed by Kethis[10] in 1984, but its efficiency was small. It seems to be too early to consider its practical use.

CONCLUSION

It is quite probable that some BR properties mentioned above have already been applied in modern technologies. Photochromic materials such as "Biochrom-BR" can be used in optoelectronics. The photoelectric properties of BR may find their applications in biosensors (for example, in field-effect transistors already constructed in Japan).

It is important to attract the attention of scientists and engineers to the BR features mentioned above. Moreover, in addition to BR, there have been found some other rhodopsins (in mollusk, fish and insect eyes, etc.). The retinal-protein complexes are as inexhaustible as atoms, but less dangerous.

REFERENCES

1. S. P. Balashov and F. F. Litvin, "Photochemical transformations in bacteriorhodopsin", Moscow State Univ., Moscow (1985).
2. V. Balogh-Nair and K. Nakanishi, Synthetic analogs of retinal, bacteriorhodopsin and bovine rhodopsin, Meth. Enzymol. 88:496 (1982).
3. N. N. Vsevolodov, "Biochrom" films based on bacteriorhodopsin and some other light-sensitive biological complexes, in: collected articles "Light-sensitive biological complexes and optical information registration", Sci. Biol. Res. Center, Pushchino (1985).
4. N. N. Vsevolodov, G. R. Ivanitsky, M. S. Soskin and V. B. Taranenko, "Biochrom" films-reversible medium for optical recording, Avtometriya 2:41 (1986).
5. N. N. Vsevolodov and G. R. Ivanitsky, Biological photosensory complexes as technical information carriers, Biofizika 30(5):883 (1985).
6. A. B. Druzhko and S. K. Zharmukhamedov, "Biochrom" films based on bacteriorhodopsin analogs, in: collected articles "Light-sensitive biological complexes and optical information registration", Sci. Biol. Res. Center, Pushchino (1985).
7. Yu. O. Barmenkov, V. V. Zosimov, N. M. Kozhevnikov, O. I. Kotov, L. M. Lyamshev and V. M. Nikolaev, Recording of phase modulation signal of fibro-optical interferometer with the use of dynamic hologram in bacteriorhodopsin, Akust. Zh. 33(3):568 (1987).
8. O. A. Aktsypetrov, N. N. Akhmediev, N. N. Vsevolodov, D. A. Esikov and D. A. Shutov, Photochromism in nonlinear optics: photocontrolling generation of the second harmonics by bacteriorhodopsin molecules, Dokl. Akad. Nauk USSR 293:592 (1987).
9. A. V. Maximychev, E. P. Lukashev, A. A. Kononenko, L. N. Chekulaeva and S. F. Timashev, Photopotentials and bacteriorhodopsin photocycle regulation by electric field in highly oriented purple membrane films, Biolog. membrani 1:294 (1984).
10. B. P. Kethis, Piezoelectric mechanism of charge active transport in purple membrane from Halobacterium halobium, Biolog. membrani 1:1307 (1984).

TOWARDS THE MOLECULAR COMPUTER FACTORY

Michael Conrad

Dept. of Computer Science
Wayne State University
Detroit, Michigan 48202, USA

A NEW VENTURE

Scientists of diverse background have in recent years joined a new scientific adventure. Molecular computing, molecular and bioelectronics, biocomputing, biomolecular information processing, and biochip are some of the terms that have been used to describe this new direction. Another term is "molecular functional systems".

The term "functional" has a rather indefinite, tentative ring. Most investigators are concentrating on new molecular materials that exhibit novel physical effects. The idea is that these materials may some day be molded into molecule-sized tools that perform useful services. Information processing, energy transduction, and measurement are three examples.

The situation is very similar to biological evolution. All biological structures have indefinite, tentative functions. Teeth may be used for eating, as a weapon, for opening packages, or for getting elected to political office. It is this indefiniteness that allowed evolution to transform the swim bladder into the lung. Our course towards new molecular-functional devices is in this respect more closely linked to the self-design principles of biological evolution than to the other-design principles of conventional engineering and technology. The molecular biotechnology factory of the future should be an evolutionary factory, far different from the current technological infrastructure and requiring a radically different style of engineering thought.

A TAXONOMY OF FUNCTION

Despite our disclaimer, let us first consider some functions to which molecular biotechnology is being directed, without however attending to mechanism at this point. Our list can include:

- biosensor (signal detector)
- energy converter (light to chemical)
- bioreactor (chemical to chemical)
- biomotor (chemical to mechanical)
- signal generator (chemical or electrochemical)
- pattern processor (signal pattern to output control)

- molecular control element (milieu sensitive switch)
- morphogenerator (structural matrix or dissipative structure)
- biofilter (semipermeable membrane for milieu maintenance, pollution control)
- diagnostic/therapeutic (individualized medical or pharmaceutical agent)
- molecular repairman (for quality control or repair)
- molecular measurement instrument (detector, amplifier, recorder, display)
- molecular memory (storage/retrieval mechanism)
- molecular wire (for transfer of energy)
- communication link (between molecular devices or to conventional device)
- molecular computer/biocomputer

These are pretty much the functions we associate with cells and organisms. We cannot overemphasize that they are not as sharply demarcated as they first appear. A biosensor that detects photons by incorporating a light sensitive pigment, such as bacteriorhodopsin, in a membrane is extremely close to a photoenergy converter. With some modification it could also be transformed into a pattern recognizing unit[1] or an electrochemical signal generator.[2] A structural matrix built out of immunoglobin molecules might be used for signal generation or mechanical signal processing; built out of cytoskeletal components it might be used for mechanochemical signal processing or signal integration, or as a mechanochemical effector.[3,4] The semipermeable membrane that is so important for signal generation and energy transduction could serve as an ultrasensitive fractionator, especially if doped with suitable channel proteins (consider that some deep sea fish can concentrate helium). The difference between detection or fractionation and the process of measurement can be bridged by adding amplifying and memory elements. Bioreactors created by isolating or emulating second messenger or kinase-channel proteins provide downward (macroscopic to microscopic) mechanisms for setting the microphysical state of a system and upward (amplifying) mechanisms for recording measurement interactions. Bioreactors that incorporate reaction-diffusion processes with suitable nonlinearities can serve as spatial and temporal structure generators, signal pattern integrators, or as chaotic (pseudorandom) signal generators.

In this short note we cannot review the mechanistic details of proposed devices that perform the functions listed above. The biocomputer affords the greatest, though perhaps most distant, latitude for exploiting the various possibilities. We will therefore use the requirements for biocomputer development as the pincushion of our discussion.

A TAXONOMY OF ARCHITECTURES

It is convenient to think in terms of the following six architectural categories:

- conventional (von Neumann) architectures
- parallel (including neural) designs
- optical architectures (including integrated, fiber optic, and mirror-based open designs)
- analog or dynamics-based designs (including self-organizing dynamics)
- conformation-driven designs (that is, designs that utilize molecular shape, for example by transducing signals into a form that can control the activity of enzymes or structural proteins)
- hybrid designs (including those that incorporate sensors, measuring devices, or actuators)

The above list highlights the circumstance that both computer science and molecular biotechnology are a bit like art. They can take nature as a model or, at the other extreme, be guided by inventive imagination. In all cases, however, the possible designs fall into three general categories.

1. Simple units, complex networks. This is the principle of design in modern digital computers. A few types of simple, precisely specified components are knit together to perform a complex function. Such systems are *structurally programmable* since they can be made to perform desired actions with a simple user's manual (if run in a sequential mode).

2. Complex, highly integrated units. Examples are ratchets and musical instruments, such as the violin. The whole is molded from parts that are specifically tailored and adapted to one another. To design the unit it is necessary to perform experiments and calculations. A simple user's manual is insufficient. As a consequence the unit is structurally non-programmable, yet capable of performing a complex function.

3. Complex units, complex networks. Here complex, specifically adapted units are linked together into a larger unit. An automobile engine built from mutually adapted components is an example. A symphony orchestra is an example of sorts, in that the units are complex and can at least be specifically adapted to one another in their use. Component functions are defined by context, just as with the components (if any) of the more elemental units. Thus the system is nonprogrammable even though it has a marked network (connectionist) aspect.

Organizational types 1 and 3 both use the building block principle, the difference being that a few types of definitively specified building blocks are used in one case and an indefinite variety of functionally powerful building blocks are used in the other. Case 3 is the general case, since a technology that is capable of building networks of complex components would also be capable of building stand-alone complex units (case 2) as well as programmable networks of simple components (case 1).

Current computer technology, and electronic technology generally, places great emphasis on simple building blocks that can be linked together to perform desired functions; high and very high integration here refers to the closeness of the linkages and the methods for printing the whole network as a unit. The basic idea of connectionism is to build complex behavior out of simple building blocks. Thus today's parallel and so-called neurocomputer designs fall into organizational type 1 just as much as the conventional von Neumann architecture does. They are structurally programmable in that the units from which they are built could be connected to perform a desired function; but in practice they are not programmable because they are run in a parallel and not a sequential mode. Current optical designs also fall into the type 1 mode of organization.

Biosensors fall into the type 2 mode of organization since they often depend on the integration of a specially selected protein conformation into a matrix (such as a membrane) with dynamical properties.[5] Mechanical technologies attempt to use simple building blocks in so far as possible, but in fact often integrate molded materials into a complex unit in the mode 2 fashion.

Biology uses the building blocks principle with a vengeance (proteins and other macromolecules, subcellular organelles, cells, tissues and organs, organisms, social groups, populations). But it is the mode 3 type organization. The building blocks are complex, variegated, and context dependent in their function. The ultimate basis of this capability is the Darwinian-genetic mechanism of variation and selection. The genetic mechanism makes possible the generation and mass production of a wide variety of specially adapted forms at each level of organization. A variety of ontogenetic adaptive mechanisms are also available for tuning components.

EVOLUTIONARY ENGINEERING

Biotechnology, specifically gene technology, has opened up the possibility of using nature's genetic-evolutionary mechanism to generate and mass produce a wide variety of molecular building blocks. For the first time it is possible to construct a factory--we will call it a molecular computer factory--to produce on a massive scale components specifically tailored for particular functions and tailored to link to one another in specific ways (the case 3 mode of organization). A fortiori it would be possible to use such a factory to produce canonical components for programmable architectures (case 1) or highly integrated unitary aggregates (case 2).

The essential elements of the factory are illustrated in Fig. 1. It is really a technological infrastructure. It is also a radically different engineering paradigm than that prominent in any previous mass scale technology. The evolutionary methods of plant and animal breeding are mixed with the analytical methods of physical engineering. The new paradigm is implicit in a fundamental tradeoff principle. *A system cannot at the same time be programmable at the level of its physical structure, utilize the materials of which it is composed for the performance of function with high efficiency, and have high evolutionary adaptability.*[6,7] The tradeoff principle says that the engineer who tries to achieve all these commendable goals at once is wasting his time just as much as the engineer who tries to make a perpetual motion machine.

At the center of the factory is the device. It performs a function. This could be any of the functions listed earlier. The function is specified by an "evolutionary engineer". A main task for this engineer is to spell out the criteria for evaluating the device. If the device were created by modifying an extant organism the engineer might concentrate exclusively on the specification of function and the mechanism of evolution, ignoring completely how the system works. It would be a pure "artificial selection" system. But if it is composed by integrating macromolecular components an initial act of human creativity is necessary to produce a protodevice on which the selection scheme can operate. The design process thus divides into two broad stages. The first, that of conceptualizing a scheme for performing the function, emphasizes mechanism. The level of performance achieved is marginal. The second stage uses variation and selection (evolutionary programming) to transform the mechanism and tune it for excellent performance. Since the device is structurally nonprogammable a clear (machine-like) understanding of the mechanism is generally lost at this stage, though a general grasp of it can certainly be of heuristic value. If the biotechnologist insists on having the same kind of clear understanding of the structure-function relation as the digital computer engineer he will lock himself into the structurally programmable, inefficient, and nonevolvable domain of function.

The stage of initial conception may in part be inspired by the analysis of actual biological systems, or as already indicated, could be *de novo* inventive. Computer simulation can help as an aid for testing the value of the concept. Trial and error experimentation with plausible combinations of components might in some instances lead and in other instances follow the simulation studies. If the combination yields a primitive functional capability this could be amplified through the variation-selection procedure. The initial experiments and subsequent evolution require an initial trial set of components and subsequent variations on this trial set. Tools for creating the components include:

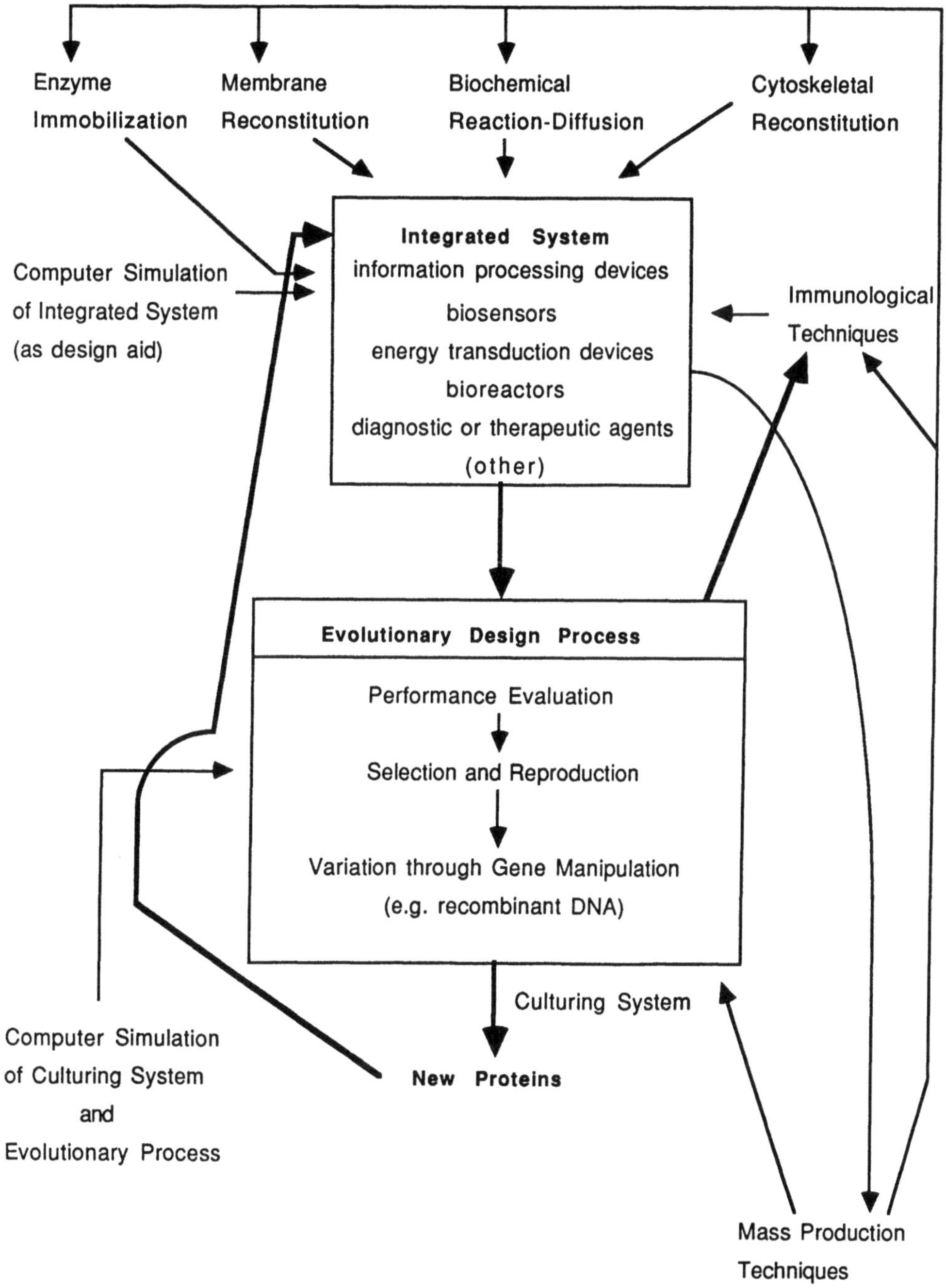

Fig. 1. Molecular computer development and production process.

- gene manipulation (e.g. recombinant DNA)
- protein engineering and self-assembly
- monoclonal antibodies and antigen-antibody networks
- immobilized enzymes
- membrane reconstitution (e.g. Langmuir-Blodgett films)
- cytoskeletal (e.g. microtubule) reconstitution
- bioreactors
- conventional analytic and synthetic methods

We cannot discuss these tools in detail in this fragmentary outline. But it is clear that the gene manipulation techniques are central, first because of their generative power and second, because of the ubiquitous structural and enzymatic importance of proteins. Variations upon which selection can act may also be introduced through antibody and membrane reconstitution techniques. Components (e.g. receptor-messenger-kinase systems) could be extracted, modified through gene manipulation techniques, and produced in quantity through biogenic culturing methods.

The important case of protein engineering illustrates the balance between evolutionary and analytical thinking. It is sometimes assumed that the ability to predict folded structure and function in at least an approximate manner is necessary for protein engineering. This is not feasible at this point in time, and workers in the field therefore generally proceed in the evolutionary manner indicated above. Many groups, following the analytical tradition, consider it important to perform structural analyses and to have access to extensive structural data bases to guide work. This may be of substantial heuristic value in many cases. But it is not necessary! The structure and function of proteins often change only gradually in response to single changes in amino acid sequence. This *gradualism* feature makes it possible to optimize protein function by exploring sequences of neighboring proteins in a step by step manner. This is the reason why natural evolution works.[8,9] Clearly a wide variety of structures still have to be explored; but the explosion of possibilities may be pruned by adding redundancies to the amino acid sequence that increase gradualism. This paradigm is alien to much conventional engineering, which seeks to eliminate mechanistically superfluous features. The successful evolutionary engineer will deliberately introduce such features in order to increase the efficacy of the search process. Some of the superfluous elements can be removed at a later stage. Trying to directly engineer the most efficient proteins by using structural data bases would in general involve a much greater explosion of possibilities.

The whole factory is thus in its essential structure an evolutionary bioreactor. The bioreactor may also be one of the device types at the center of the scheme. In this case the device at the center of Fig. 1 plays a second role as an agent for the highly specific synthesis of all sorts of variant materials. Bioreactors with bacteria as components can also serve this purpose. The scheme in short becomes entirely recursive (acts on itself).

The problem is that evolutionary systems tend to define their own goals. This is why it is important for the evolutionary engineer to specify with great care the criteria for selection. It is necessary to organize the culture system in the development stage so that evolutionary discovery and optimization are facilitated. This means using effective recombination and speciation strategies. But the opposite side of the coin is that it is necessary to prevent evolutionary change from occurring in the production phase. Here mutation and other genetic variation must be reduced to a minimum; repair mechanisms must be enhanced; the forces of selection must be stabilized. Our evolutionary factory is an ecosystem. The development of new components, whether the results of approved variation or the byproduct of extraneous variation, modifies the selecting environment. In order to stabilize the harvested product it is necessary to stabilize the environment.

Evolutionary and ecological theory, far removed from the floor of the ordinary factory and perhaps at first sight rather secondary to the physical, chemical, and preparative aspects of new materials, thus makes its pertinence known. Many fragmentary models for these processes have been developed. For operating the molecular computer factory a comprehensive predictive tool would be most helpful. Computer simulations that represent all the critical interactions in an evolutionary ecosystem have been developed that could serve this function.[10,11] The tradeoff principle implies that it is quite impossible for such models to keep pace with the real system. Here, as with the materials themselves, we cannot expect a precise predictive tool. What we need, and what we can expect, is computer aided cultivation (CAC), the biotechnological analog of computer aided design and manufacturing. CAC would allow the engineer to precisely specify and identify all the processes in the system, test assumptions about its principles of operation, and explore alternative evolution and production strategies.

Production would be too expensive and the production process too uncontrolled without a flexible machine line. Modern robotry is already capable of many of the manipulations associated with gene sequencing and manipulation, peptide synthesis, membrane reconstitution, mixing of chemicals and extraction of samples, culturing, and environmental control. A robot line can be reprogrammed for different regimes of operation, including alterations in component production (such as controlled alterations in the laying down of Langmuir-Blodgett films[12]), different regimes for integrating components, and for switching between culturing regimes directed to evolution of new products and those directed to production per se. Today's robotry has severe limitations whenever the context becomes ambiguous. Vision systems are slow, very fine control with variable materials is difficult, and quality control must be a human responsibility if the possible errors are not rigidly constrained. One of the goals of the molecular computer factory would be to develop molecular devices that increase the efficacy of robotry, and therefore that increase the efficacy of the factory.

As flexible mass production techniques become available the inventory of components will inevitably increase. Today's molecular biodevice engineer must go through the arduous process of mutually adapting each piece of a molecular jigsaw puzzle. Practically speaking this requires a large multidisciplinary group of the type that is barely visible on the scientific horizon; and it requires an intense and expensive effort that makes all but the most rudimentary devices look out of reach. The conventional electronics engineer, by contrast, can order the needed components off the shelf, or perhaps approach a specialty company to produce a component according to specification. The practicality of electronic, mechanical, and conventional materials technologies is based on a highly developed infrastructure of development and production. This is why the molecular computer factory should not be conceived as a unit housed in a single location. It is an infrastructure in which the increases in the scale of production should lead to increases in the inventory of components, leading to increases in the pace and sophistication of device development, in turn fueling the scale of production, and around again, in ever widening circles.

INFRASTRUCTURE FIRST

Elsewhere[6,13] we have described protodevices that could serve as starting points for the scheme illustrated in Fig. 1. We do not have the space to describe these in detail here; nor would it fit with our emphasis on a flexible concept of function to do so. Too tight a focus on a particular goal could mask significant potential uses of a device. Too strong a demand for a clear mechanistic rationale at the outset or even in the final stage is, according to the tradeoff principle, incompatible with the efficient coupling of the atoms and molecules in the device to its functional performance. The motto of the molecular computer factory should be "infrastructure first". It is not possible to fabricate a truly sophisticated prototype to demonstrate that building up the infrastructure is worthwhile. The growth of the infrastructure--which is more or less an inevitable development in any case--is a precondition for prototype fabrication.

The conventional digital computer followed a similar historical development. The infrastructure in Babbage's day was inadequate for the fabrication of his design. The difference is that the relation between the structure and (computational) function of Babbage's design could be understood with Cartesian clarity. This would only be the case for the least interesting of the devices that could be created and produced in the molecular computer factory.

This is because the devices of real interest have self-organizing dynamics. The elements are mixed, much like magnetized beads of varying shapes might be mixed. In the case of the beads the overall shape of the collection emerges from mutual interactions in accordance with the laws of physics. By selecting the shapes of the beads and appropriately ordering their addition to the aggregate it would in principle be possible to mold an overall shape (or mosaic) that performs a useful function. Such self-organizing dynamics are completely suppressed in digital computers; if they were not the computers would not be effectively programmable. As a consequence we can control the structure and hence the rules followed by the digital computer in precise detail. But it is at the expense of throwing away an enormous number of potential interactions that could contribute to the computational power of the system.[14]

Proteins, lipids, and nucleic acids are the magnetized beads of the molecular computer factory. The main feature of all these biopolymers is that the positions and motions of the atomic nuclei are of critical importance. This is in sharp contrast to digital switches, for which only the motions of the electrons need be considered. The first step of self-organization is folding of the amino acid sequence or in some cases the nucleotide sequence. Folded shape allows for specificity and changes in folded shape allow for control and memory. The second step is self-assembly of different shapes into a mosaic. This mosaic may disassemble and reassemble dynamically in response to changes in the milieu that affect the individual macromolecules. It may also provide a matrix for dynamic biochemical processes involving reaction and diffusion. This is another level of self-organization, since it may be concomitant to multiple steady states, limit cycles, or chaos. The molecular matrix may also provide a substrate for particle and quasi-particle exchanges that provide for a microphysical layer of self-organizing dynamics. Our hypothesis is that biological cells use all these layers of self-organizing dynamics to achieve their impressive capabilities for pattern recognition, internal control, and effector action. Networks of cells communicating via macroscopic signals, as in the brain, allow for yet a higher level of dynamic self-stabilization.

Let us consider, but only schematically, how self-assembly provides a powerful pattern processing capability. The sequence of events is: *input signal pattern ---> released macromolecules ---> self-assembled mosaic ---> readout adaptor ---> effector protein ---> output action*. The input signals activate receptors on a membrane, directly or indirectly causing molecular shapes to be released or altered. The important point is that the different input signals are represented by different molecular shapes. Each different input pattern will therefore give rise to a different self-assembled mosaic. The recognition sites of the readout adaptors may be engineered to respond to any of a large number of features of the self-assembled mosaic. Different features will represent different groupings of the input patterns. A second recognition site of the adaptor enzyme is engineered to recognize and trigger an effector protein. This controls the output of the module, for example, by triggering a membrane potential change.

Pattern recognition is a form of computing; by knitting together pattern recognizing modules of the above type in an appropriate way any computation can be duplicated. What in a conventional machine would be purely symbolic pattern processing is mediated by a *bona fide* free energy minimization process. The achievable efficiency is much higher, since a vastly greater fraction of the potential interactions among the particles in the system is directed to the computational function than could be the case in a system that achieves structural programmability by suppressing all possible emergent behaviors. The self-assembly mode of self-organization could be replaced by any of the other modes: particle and quasi-particle exchanges, coherent hydrogen bond dynamics, biochemical reaction-diffusion dynamics, or macroscopic signal integration involving a large number of modules.

It cannot be overemphasized that the efficiency advantage is not simply due to small size and speed of switching. The problem solving behavior is mediated by physical dynamics, not by the interlinkage of discrete switching processes that serve to map symbolic string processing operations in a one-to-one manner. The quantum nature of molecular systems is another factor that greatly enhances their computational power.[14] A conventional solid state switch is governed by a quantum mechanism; but this is completely irrelevant to the computational function of a digital computer, which is engineered to act as an absolutely classical device. In our self-assembly scheme the recognition capacity of the molecular pieces is all important to the computational capacity of the whole system. This capacity depends on the quantum nature of the electronic wave function and on electronic-conformational interactions in a way that could not be duplicated with a macroscopic analog.

Our protodevice scheme may be viewed as an advanced biosensor. It belongs to the conformation-driven and dynamics-based architectures, as would networks of such devices. Combined with a conventional machine it fits into the hybrid category. These are neuromolecular computers; if the internal dynamics of the modules are suppressed we have the special case of a conventional neural or parallel machine. If the machine is run in a sequential mode we have the main processing unit of a von Neumann machine. From this point of view computers built on conventional electronic technologies, whether standard or nonstandard in design, are a small cross-section of the possibilities.

One could also start with a conventional design as the protodevice. Molecular optoelectronic memory storage and retrieval devices are the nearest term possibility.[15,16] The development of new pigments could greatly contribute to optical computing. Optical realizations of so-called neural network computers might be achieved by fabricating modules in which input pigments each sensitive to a particular frequency are linked to an output pigment which emits in a frequency characteristic of its module. The linkage could be through a membrane potential change if the output pigment is voltage sensitive. Module 1 would be connected to module 2 if it contains an input pigment that responds to the frequency broadcast by module 2. This design can easily be elaborated on paper, though it is obviously rather artificial. Other starting points that are artificial from the standpoint of biology include attempts to self-assemble electron transfer proteins into conventional circuits or to self-assemble optoelectronically active polymers by attaching them to proteins.[17,18] Such prototypes would have to compete with conventional electronic technologies, whereas the self-organizing prototypes would be directed towards a totally unoccupied domain.

NEW SCIENCE, NEW TECHNOLOGY

But as stated more than once, we should be flexible about the product of the molecular computer factory. One main product could be new scientific knowledge. The cross-catalytic action of science and technology in this field may be expressed with the chicken-egg scheme:

Molecular functional technology <---> *Molecular functional science*

Science is probably the founding egg for most modern fields of technology. The scientific discoveries about electromagnetism, for example, provided the basis for electrical engineering. Solid state physics, as another example, provided the foundation for solid state electronics. But high energy physicists, to their dismay, might find that new engineering technology is the driving force behind the experimental part of their science. In the domain of molecular functional systems the connection between science and technology is much more dynamic. Advances in the science are as important for advancing the technology as conversely. The evolution process can always touch on new edges of nature, and as a consequence the evolutionary engineer cannot be confined to the known domains of physics and chemistry.

It is hardly necessary to list the possibilities here, but I shall mention a few of them since my perception is that the agenda of the field is in continual danger of being shrunk by the external pressure of a misplaced pragmatism. New measuring devices produced by the factory could make it possible to examine subtle dynamical processes in biomolecular systems that are currently out of reach. The factory should allow for the study of biopolymers to become sufficiently comparative to yield valuable perspectives on the mechanism of evolution and on the balance between accident and necessity in living systems. It should be possible to synthesize stunning organic structures that exhibit properties unknown to biological nature or to today's technology. The protodevice scheme used as our main example is analogous to a quantum measurement process. Macroscopic input is used to prepare the microstate of a system, the microstate develops in time, and finally it is amplified by molecular readout mechanisms to macroscopic proportions. The study of the material and design requirements of such devices may provide new understanding of the nature of quantum measurement and of the true physical basis of nonlinearity. The synthetic study of "proto-brain" computers could provide a more adequate concept of cognitive computation by elucidating the connection between computational capability and its material basis. These examples suggest that the analytic and synthetic study of molecular functional systems is pertinent to the deepest issues of science.

ACKNOWLEDGEMENTS

This paper is a condensed version of a larger (but incomplete) manuscript prepared in 1987 while the author was a Japanese Ministry of Education Special Visiting Professor at the Technological University of Nagaoka. The author acknowledges discussions with numerous colleagues in Japan that influenced his thinking and the Advanced Technology Institute in Tokyo for arranging a tour of an automated facility that illustrated future possibilities. Particular thanks are due to Professor Koichiro Matsuno for stimulating discussions of bioengineering and biotechnology, including discussions on the technical significance of "indefiniteness". Support by the U.S. National Science Foundation (Grant IRI 87-02600) and by the Digital Equipment Corporation is also acknowledged.

REFERENCES

1. M. Conrad and F. T. Hong, Molecular computer design and biological information processing: an electrochemical and membrane reconstitution approach to the synthesis of a cellular automaton, in: "Proceedings of the International Symposium on Future Electron Devices-- Bioelectronic and Molecular Electronic Devices," M. Aizawa, ed., pp. 89-94, Research and Development Association for Future Electron Devices, Tokyo (1985).
2. F. T. Hong, The bacteriorhodopsin model membrane system as a prototype molecular computing element, BioSystems, 19:223 (1986).
3. E. A. Liberman, S. V. Minina, O. L. Mjakotina, N. E. Shklovsky-Kordy, and M. Conrad, Neuron generator potentials evoked by intracellular injection of cyclic nucleotides, Brain Research, 338:33 (1985).
4. S. R. Hameroff, "Ultimate Computing," North Holland, New York (1987).
5. H. T. Tien, Ultrathin bilayer film: an experimental approach to biomolecular electronic devices, in: "Molecular Electronic Devices," F. L. Carter, R. E. Siatkowski, and H. Wohltjen, eds., pp. 209-226, North-Holland, Amsterdam (1988).
6. M. Conrad, On design principles for a molecular computer, Comm. ACM, 28:464 (1985).
7. M. Conrad, The price of programmability, in: "The Universal Turing Machine: A Fifty Year Survey," R. Herken, ed., pp. 285-307, Oxford and Verlag Kammerer & Unverzagt, Hamburg-Berlin (1988).
8. M. Conrad, Bootstrapping on the adaptive landscape, BioSystems, 11:167 (1979).
9. M. Conrad, "Adaptability," Plenum Press, New York (1983).
10. M. Conrad, Algorithmic specification as a technique for computing with informal biological models, BioSystems, 13:303 (1981).
11. M. M. Rizki and M. Conrad, EVOLVE III: A discrete events model of an evolutionary ecosystem, BioSystems, 18:121 (1985).
12. H. Kuhn, Molecular engineering--a begin and an endeavor, in: "Proceedings of the International Symposium on Future Electron Devices--Bioelectronic and Molecular Electronic Devices," M. Aizawa, ed., pp. 1-6, Research and Development Association for Future Electron Devices, Tokyo (1985).
13. M. Conrad, Molecular computer design: a synthetic approach to brain theory, in: "Real Brains, Artificial Minds," J. L. Casti and A. Karlqvist, eds., pp. 197-226, North-Holland, New York (1987).
14. M. Conrad, Quantum mechanics and molecular computing: mutual implications, Int. J. Quant. Chem., 15:287 (1988).
15. N. N. Vsevolodov and G. R. Ivanitskii, Biological light-sensitive complexes as technical information photocarriers, Biophysics, 30:962 (1985).
16. R. R. Birge, Photophysics of light transduction in rhodopsin and bacteriorhodopsin, Ann. Rev. Biophys. and Bioeng., 10:315 (1981).
17. R. I. Gilmanshin and P. I. Lazarev, Biotechnology as a source of materials for electronics, Biotekhnologiya, 3:421 (1987).
18. R. S. Potember, R. C. Hoffman, S. H. Kim, K. R. Speck and K. A. Stetyick, Molecular optical devices, J. Molec. Electronics., 4:5 (1988).

ULTRALOW FREQUENCY "OPTICAL" BIOCOMPUTERS: BIOPHYSICAL ARGUMENTS

D. Raković*+, Dj. Koruga+, D. Djaković+,
Ž. Martinović+, V. Desimirović+, and Dj. Minić*

*Faculty of Electrical Engineering, 11001 Belgrade, P.O. Box 816, Yugoslavia
+Molecular Machines Research Center, University of Belgrade, 27. Marta 80, 11000 Belgrade, Yugoslavia

INTRODUCTION

In the development of truly intelligent and creative computers, one of the most significant problems is the understanding of how the human brain operates, in the hope of imitating its processes.

Traditionally, the human brain has been regarded as a vast switching network consisting of tens of billions of individual processing neurons, communicating with one another through electrochemical impulses. The excitation of the nerve synapses requires the impulse fields of $\sim 10^5$ V/cm. Recent investigation of the new generation computers, the so-called "neural networks",[1] however, implies that the parallel interconnections of tens of thousands of neurons in the brain are essential, so that the collective rather than the individual neuronal activity is the basis of information processing in the brain.

However, in spite of the fact that the brain hardware has been somewhat properly simulated by parallel interconnections between the network subunits, such interconnections through the ultralow frequency (from 0 to 30 Hz) brainwaves were not taken into account. The electric fields in the brain tissue of the magnitude of $\sim 10^{-1}$ V/cm, characteristic of the electroencephalogram (EEG), have usually been considered to have no role in the excitation of brain neurons, and have been regarded merely as the noise of the "brain's motor". However, recent data strongly indicates a modulatory role for these fields, based on behavioral, neurophysiological, and neurochemical evidence.[2] Furthermore, it has been shown that even much weaker fields in the brain tissue ($\sim 10^{-7}$ V/cm) could produce different behavioral effects in animals and human beings. The influence of these considerably weaker fields, with characteristic windows in frequency (below 30 Hz) and intensity, has been associated with amplifying cooperative mechanisms, based on nonequilibrium resonant long-range interactions between biological macromolecules at the cell membrane.[2]

In accordance with the growing evidence that EEG patterns can be correlated with high-level cognitive processes in both animals and human beings, an alternative model of the brain, known as cooperative action,[3] has been proposed in the early 1970s. According to this model, neurons work together in large groups and interact with complex ultralow frequency electromagnetic fields that pervade the brain.

The role of electromagnetic fields of the brain in this process is still controversial; are these fields a reflection of a collective neuronal activity, transmitters of information, embodiment of consciousness itself, or all that together? In a recently proposed model,[4] it has been shown that the assumption that (sub)consciousness is associated with the electromagnetic component of the scanning brainwaves enables the explanation of striking time distortions in some psycho-physiological states, which cannot be explained by existing psychological[5] and biochemical[6] models (related to the amount and complexity of the processed and memorized stimuli, and the rate of biochemical metabolism affecting the "biological clocks", respectively).

In this paper we shall first present the main features of a somewhat improved model,[4] and then we shall discuss the influence of the proposed model on further development of brain-like biocomputers and molecular electronics.

MODEL

Let us consider a simplified model of the brain, as a pervasion of the brain neural network by ultralow frequency brainwaves, whose frequency spectrum contains four major frequency bands: delta (0.5 to 3.5 Hz), theta (3.5 to 8 Hz), alpha (8 to 13 Hz) and beta (13 to 30 Hz),[7] generated by independent factors.[8]. Brainwaves are basically ultralow frequency alternating ionic currents, generated by collective action of a large number of neurons in the brain.[9] Because of their alternating nature, these currents also generate electromagnetic fields of corresponding frequencies, according to the basic principles of electromagnetism; however, that field is difficult to irradiate because of its extremely low frequency as well as very small dimensions of its source,[10] but that very fact enables the localization of the field in the vicinity of the brain.

The brainwaves, essentially progressive in nature, scan across the brain surface, and we shall postulate in our model that (sub)consciousness is associated with the electromagnetic component of the scanning brainwaves (henceforth, brainwaves). Then, the "subjective" reference frame will be attached to the scanning brainwaves, and the "objective" one to the brain (i.e. laboratory). The relative velocity of the two reference frames will be equal to $v=c/(\epsilon_r\mu_r)^{1/2} \approx c/\epsilon_r^{1/2}$, where c donates the propagation velocity of the electromagnetic field in vacuum, μ_r (≈ 1) the ultralow frequency relative magnetic permittivity, and ϵ_r the ultralow frequency relative dielectric permittivity of the brain structure with corresponding reference brainwaves.[11] Most tissue water is in the free state, and has $\epsilon_r \approx 80$ at ultralow frequencies. On the other hand, the ultralow frequency relative dielectric permittivity of the cell membrane can have significantly higher values of up to $\sim 10^5$ (as a consequence of polarization of the volume ion density within the porous cell wall[12]), but can be reduced to the value of a lipid membrane medium of ~2 (in the absence of ions). For the extreme values of the membrane permittivity, the ultralow frequency electromagnetic waves, generated by the corresponding ion currents in brain tissue, will propagate along the cell membranes practically without energy dissipation in the neighboring tissue water, as the reflectivity (in the case when the relative refractive index of the two media is significantly higher or lower than unity) will be practically equal to unity.[13]

As the relative velocity v of the two reference frames can be comparable to the velocity of light in vacuum c, it is necessary to use relativistic relations between the physical variables measured in two reference frames.

So, the relation between the time intervals in the two reference frames is[14]

$$\Delta t = \frac{\Delta t_o}{(1-v^2/c^2)^{1/2}} = \frac{\Delta t_o}{(1-1/\epsilon_r)^{1/2}}, \tag{1}$$

where Δt_o denotes the proper time interval (which measures the clock that moves together with the moving frame), and Δt the time interval which measures the clock at rest.

The analogous relation holds true for frequencies measured in two reference frames, moving away from one another ($\alpha=\pi$)[15]

$$f = f_o \left. \frac{(1-v^2/c^2)^{1/2}}{1-(v/c)\cos\alpha} \right|_{\alpha=\pi} = f_o \frac{(1-1/\epsilon_r)^{1/2}}{1+1/\epsilon_r{}^{1/2}}, \tag{2}$$

where f_o denotes the proper frequency (which irradiates the moving source with respect to the moving frame), f the frequency that measures the observer at rest, and α the angle (in the resting frame) between the direction of the source movement and the direction from the observer to the source. The above relation is known as the relativistic Doppler effect.

The brain's ("objective") reference frame is attached to the Earth, and is practically inertial. On the other hand, the brainwaves' ("subjective") reference frame is approximately inertial only in the course of *stationary* physiological and emotional states (when ϵ_r =const.). Then both reference frames are inertial and move relatively to one another with the scanning velocity $v=c/\epsilon_r{}^{1/2}$ = const. According to the theory of relativity both frames are physically equivalent in the course of the scanning period, so that the observer from one reference frame can say that the other frame moves, and *vice versa*. The same may be true if one considers a situation in the course of many scanning periods with constant ϵ_r.

From the point of view of the inertial "subjective" observer, according to Eq. (1) there follows $\Delta t^{subj} = \Delta t_o^{obj}/(1-1/\epsilon_r)^{1/2} > \Delta t_o^{obj}$, which could account for a dilatation of the subjective time (Δt^{subj}), in comparison with the objective time that measures the Earth's clock (Δt_o^{obj}), in childhood[16] and under the influence of hypnosis,[17] meditation,[18] psychedelic drugs,[18] some psychopathological states[19] and near-death experiences![20] By using the extreme values of the membrane permittivity (from 2 to 10^5) it is obvious from Eq. 1 that Δt^{subj} can have the values in the interval from $(2)^{1/2}\Delta t_o^{obj}$ to Δt_o^{obj} (depending on the rate of the metabolic activity of ions in the cell membranes). However, that degree of the subjective time sense dilatation cannot account for such a striking time dilatation under the influence of hypnosis, meditation, psychedelic drugs, some psychopathological states, and near-death experiences. Such dilatations of the subjective time sense are possible, in the framework of our model, only for the values $\epsilon_r \gtrsim 1$, when according to Eq. 1 it follows $\Delta t^{subj} >> \Delta t_o^{obj}$. But, how can the values $\epsilon_r \gtrsim 1$ be reconciled with the known data for biological structures? A possible explanation is that ions play an important role in the acupuncture mechanisms,[21] forming around the acupuncture points at the lobe surface a layer of weakly ionized air, which has $\epsilon_r \gtrsim 1$. In such a case the striking subjective time distortions could be related to (sub)consciousness associated with the brainwaves localized in that weakly ionized layer.

What about the brainwave frequencies in those situations? From the viewpoint of the inertial "subjective" observer, according to Eq. 2 there follows $f^{subj}=f_o^{obj}(1-1/\epsilon_r)^{1/2}/(1+1/\epsilon_r^{1/2})<f_o^{obj}$.* This relativistic Doppler shift, observable from the "subjective" reference frame, can have very interesting consequences on the explanation of the mixing of conscious and subconscious contents. That is to say, the four main frequency bands, $f_{o\beta}^{obj}$ (13 to 30 Hz), $f_{o\alpha}^{obj}$ (8 to 13 Hz), $f_{o\theta}^{obj}$ (3.5 to 8 Hz) and $f_{o\delta}^{obj}$ (0.5 to 3.5 Hz) (the first two of them predominantly corresponding to conscious and the last two to subconscious states[22]) for $\epsilon_r \gtrsim 1$ start merging from the viewpoint of "subjective" reference frame: $f_\beta^{subj} \approx f_\alpha^{subj} \approx f_\theta^{subj} \approx f_\delta^{subj} \approx 0$ Hz. So, the low dielectric ($\epsilon_r \gtrsim 1$) weakly ionized layer in the vicinity of the lobe surface allows for conscious and subconscious brainwaves' contents from the whole "objective" spectrum (0 to 30 Hz) to be "subjectively" mixed in (sub)consciousness. Such situations are psychologically encountered in the dream phase (so called REM phase of sleep[23]), as well as under the influence of hypnosis, meditation, psychedelic drugs and some psychopathological states. Although the "objective" brainwave spectrum in such states does not differ significantly from the spectrum of the alert state,[24] the essential difference appears in the "subjective" brainwave spectrum due to propagation of brainwaves in the low dielectric weakly ionized layer with $\epsilon_r \gtrsim 1$ (when the whole "subjective" spectrum degenerates to ~0 Hz). For the sake of comparison, the brainwaves in the alert state are predominantly localized in the brain tissue with $\epsilon_r > 2$ (when the well differentiated "subjective" spectrum exists: $f_i^{subj} = f_{oi}^{obj} \cdot (1-1/\epsilon_r)^{1/2}/(1+1/\epsilon_r^{1/2})$, $i=\beta,\alpha,\theta,\delta$).

One could argue that the "subjective" observer attached to the scanning brainwaves could not register any brainwaves (frequencies) as it would be moving with the velocity of the wave front. However, it would be true only if the dielectric medium is homogeneous, with the same ϵ_r everywhere the brainwaves propagate. This is not necessarily true. So the "subjective" observer associated with the brainwaves which move through the part of medium with greater ϵ_r could register frequencies of faster brainwaves moving through the neighboring medium of lower ϵ_r. Consequently, to account for the mixing of conscious and subconscious contents as well as the striking time dilatation (when $\epsilon_r \gtrsim 1$ is necessary), the *structured* weakly ionized layer in the vicinity of the lobe surface is needed, further implying the acupuncture origin of that layer.

Although we previously assumed that the change of ϵ_r of the medium did not cause the change of the "objective" brainwaves' frequency (which is a reasonable physical assumption for linear media), it is not impossible that the change of ϵ_r itself can have activating or inhibiting influence on some of the structures in the brain, giving rise secondarily to the change of the "objective" spectrum. Such situations are encountered in sleep,[23] etc. However, details of such process are beyond the scope of our simplified model.

In *nonstationary* states (when $\epsilon_r \neq$ const.), the "subjective" brainwaves' reference frame is accelerated relative to the "objective" brain's reference frame, and according to the theory of relativity they are no more physically equivalent. Then the observer from the brain's (Earth's) reference frame could say that he is at rest, and that the brainwaves' reference frame is accelerated, but the reverse could not be said by the observer from the brainwaves' reference frame (because he would feel noninertial effects of the

*It should be pointed out that the above relation can be written only from the viewpoint of the "subjective" reference frame, as the source of brainwaves (of the frequency f_o^{obj}) is the brain itself, which is fixed relative to the "objective" reference frame. A reversed case is not possible, because a source of brainwaves fixed relative to the "subjective" reference frame does not exist.

acceleration of his reference frame). From Eq. 1 it is obvious that $\Delta t^{obj}=\Delta t_o^{subj}/(1-1/\epsilon_r)^{1/2}>\Delta t_o^{subj}$, which could be interpreted as a dilatation of the objective time, related to subjective anticipation of future events:[25] intuition, precognition, deep creative insights, etc. Obviously, most striking time distortions of that type are also possible for the values $\epsilon_r \geq 1$, when according to Eq. 1 follows $\Delta t^{obj}>>\Delta t_o^{subj}$, which could be related to (sub)consciousness associated with the brainwaves of the nonstationary degenerated "subjective" spectrum with $f_i^{subj} \approx 0$ Hz ($i=\beta,\alpha,\theta,\delta$), localized in the above mentioned weakly ionized layer at the lobe surface. Although the model predicts nonstationary changes of "subjective" brainwaves' frequencies during the change of ϵ_r, it could be expected that these changes of ϵ_r can have activating or inhibition influence on some of the structures in the brain, giving rise secondarily also to nonstationary changes of "objective" brainwaves' frequencies, which can be the cause for the appearance of a contingent negative variation (which appears in our brains every time we face some new situation), that plays a significant anticipative cognitive role.[26]

Is that the fully determined event which could be anticipated? From the viewpoint of quantum mechanics there are only possible futures,[27] so that one of them could be anticipated. Yet, according to the standard Copenhagen interpretation of quantum mechanics, the reduction of wave function and actualization of anticipating possibility would appear in such a process, which would be equivalent to determinism. However, according to the many-worlds interpretation of quantum mechanics,[28] one of the probable branches of development of our universe could be anticipated, by not affecting the others, leaving room for free will and the possibility for influence on the future. The feasibility that intuition and free will are compatible could be the reason why Nature has favored the survival of species operating at changeable "subjective" ultralow frequencies. On the contrary, the above evolutionist argument favors the many-worlds interpretation of quantum mechanics against the Copenhagen interpretation.

DISCUSSION

The unique ability of the proposed model to account for striking time dilatations in some psycho-physiological states, implies that (sub)consciousness is associated with the electromagnetic component of scanning brainwaves, and that the temporal aspects of the (sub)conscious functions of the brain are governed by relativistic (and quantum mechanical) laws, in spite of the apparently pure classical nature of the brain. This significantly supports the "cooperative action" model of the brain, which states that the interaction of the neurons, working together in large groups, with complex ultralow frequency electromagnetic fields pervading the brain, is the crucial one.

It should be added that under the influence of psychedelic drugs, besides a striking dilatation of the subjective time sense, there also exists a contraction of subjective space estimations, which is, for instance, manifested as enlarged handwriting.[6] This is consistent with the relativistic relation:[14] $\Delta l^{subj}=\Delta l_o^{obj}(1-1/\epsilon_r)^{1/2}$, where Δl_o^{obj} denotes the "objective" length measured by the Earth's meter, and Δl^{subj} the "subjective" sense of that length. According to the model, the subject enlarges handwriting Δl_o^{obj} up to the size when Δl^{subj} is experienced as in a normal state.

It is interesting to mention that the influence of tranquilizers is accompanied by reduced handwriting and contractions of subjective time sense.[6] These facts could be understood in the framework of the model, under the assumption that the normal subjective sense is correlated with some value $\epsilon_{r\alpha}$ (this value could be measured experimentally, by assuming that it acts resonantly on those brain structures responsible for generation of ordinary α-rhythm, which is supposed to determine the reference temporal rhythm[29]).

So, Eq. 1 for time intervals can be rewritten in the form

$$\Delta t^{subj} = \Delta t_{\alpha}^{subj} \frac{(1-1/\epsilon_{r\alpha})^{1/2}}{(1-1/\epsilon_{r})^{1/2}}, \tag{3}$$

from where it follows that $\Delta t^{subj} > \Delta t_{\alpha}^{subj}$ for $\epsilon_r < \epsilon_{r\alpha}$ (which corresponds to a dilatation of the subjective time sense), and that $\Delta t^{subj} < \Delta t_{\alpha}^{subj}$ for $\epsilon_r > \epsilon_{r\alpha}$ (which corresponds to a contraction of subjective time sense). The similar effects also appear in space relations.

In such a way, for quantitative testing of subjective time sense distortions versus relative permittivity ϵ_r, Eq. 3 should be used, where Δt_{α}^{subj} and $\epsilon_{r\alpha}$ would be the reference parameters determined previously in the ordinary α-state.

Finally, it should be pointed out that the proposed model has advantages in comparison with existing biochemical and psychological models. Biochemical models,[6] related to the rate of biochemical metabolism affecting the "biological clocks", cannot account for striking time dilatations, and also have some other weaknesses.[30] Psychological models,[5] related to the amount and complexity of the processed and memorized stimuli, cannot account for distortions of subjective space estimations, as well as anticipative abilities of consciousness. However, in the domain of overlapping of predictive abilities of the models, there presumably exists some superponning effects. For instance, the amount and complexity of the processed and memorized stimuli undoubtedly affect the subjective sense of duration of time intervals but the very mechanism of a relativistic dilatation of the subjective time presumably determines a "channel capacity" of information processing (which is the natural limitation of psychological models[31]). So, in the ordinary state all input information during 0.1 s (of objective time) is supposed[32] to be integrated and processed as a unit; however, in the states of a very distorted subjective time sense, during the same interval it could elapse much more subjective time ($\Delta t^{subj} \gg 0.1$ s), enabling us to be subjectively aware of a larger amount of quasi-simultaneous input information, i.e. parallel processing channels. Thus, the relativistic mechanism dilates the subjective time, and the psychological mechanism spans it.

The proposed simplified model could have significant influence on the development of brain-like biocomputers and molecular electronics, for several reasons:

1. The essential role in the temporal dilatations belongs both to the electromagnetic component of brainwaves and to the structured low dielectric ($\epsilon_r \gtrsim 1$) weakly ionized gaseous medium in the vicinity of the lobe surface, implying the significant additional role of ions in biophysical brain processes (the origin of that ionic layer should be sought in acupuncture mechanisms).

2. It enables us to be aware of the extremely fast information processing in low dielectric stationary states (ϵ_r=const.) and anticipative abilities in nonstationary states ($\epsilon_r \neq$const.), due to relativistic temporal dilatations.

3. Simultaneously, the structured low dielectric ionic medium enables the mixing of subconscious and conscious brainwaves' contents, due to relativistic Doppler mapping of the whole "objective" spectrum (0 to 30 Hz) on the degenerate "subjective" spectrum ($\approx$ 0 Hz).

4. The previous points 2 and 3 involve very important aspects of subconscious functioning, that have to be simulated by future brain-like biocomput-

ers. At the same time, the significant abilities of neural networks to simulate higher cognitive functions must not be neglected. For these reasons, the future biocomputer hardware should be made in the form of a biological neural network with an abundance of ions, and with the possible generation of ultralow frequency "brainwaves". This fact significantly determines the direction in the development of hardware for molecular electronics, set down in the foundations of future biocomputers.

5. The functional coupling of neural network and electromagnetic "brainwaves" component can be achieved through the ionic "brainwaves" component. The very mechanism of interaction of neural network and the ionic brainwaves' component can be described by Schrödinger wave holography,[33] while interaction of ionic and electromagnetic brainwaves' components can be described by the usual laws of electrodynamics.

6. It seems that our model implies that, besides the neural network level, the two other levels of the hardware of the brain are essentially important for the function of the brain: ionic and electromagnetic brainwaves' components. In some sense, the model suggests that the significant mode of the biocomputer operation could be the ultralow frequency "optical" network one.

Before artificial biological neural network production, with embedded ionic structure and induced "brainwaves", our model should be tested on yet existing solutions in living nature. That could be done in several ways:

(a) By monitoring the correlation of subjective time sense dilatations[34] and "channel capacity" of processing versus low dielectric ϵ_r values measured in the close vicinity of the lobe, under the influence of REM sleep phase, hypnosis, meditation, psychedelic drugs and some psychopathological states. Simultaneously, the degree of mixing of conscious and subconscious contents could be monitored. Also, the correlation of anticipative solving of conveniently designed problems versus variation of the above mentioned low dielectric ϵ_r values could be tested.

(b) To eliminate artifacts, similar experiments could also be done on animals, subjected to psychedelic drugs, convenient electrical stimuli, etc.

REFERENCES

1. J. J. Hopfield, Neural networks and physical systems with emergent collective computational abilities, Proc. Natl. Acad. Sci. USA 79:2554 (1982); J. J. Hopfield, Neurons with graded response have collective computational properties like those of two-state neurons, *ibid.* 81:3088 (1984); D. E. Rumelhart, J. L. McClelland, and the PDP Research Group, "Parallel Distributed Processing," MIT Press, Cambridge, MA (1986); G. North, Implementation and analysis, Nature 330:522 (1987).
2. W. R. Adey, Tissue interactions with nonionizing electromagnetic fields, Physiol. Rev. 61:435 (1981).
3. E. R. John, Switchboard versus statistical theories of learning and memory, Science 177:850 (1972); E. R. John, Y. Tang, A. B. Brill, R. Young, and K. Ono, Double-labeled metabolic maps of memory, *ibid.* 233:1167 (1986).
4. D. Raković, Biocomputers - the perspectives, Proc. XV Yug. Conf. Microelectr.: 143 (1987) (in Serbocroat); D. Raković, Dj. Koruga, D. Djaković, and Z. Martinović, Biocybernetical basis for data processing, Proc. II Int. Sem. Symp. Automat. Robot: 226 (1987); D. Raković, Dj. Koruga, D. Djaković, and Z. Martinović, Biocomputers: a biocybernet-

ical viewpoint, 1st Symp. Mol. Electr. Biocomp., Aug. 1987, Budapest (Hungary), commun.
5. R. E. Ornstein, "On the Experience of Time," Penguin Books, New York (1969).
6. R. Fischer, Biological time, in: "The Voices of Time," J. T. Fraser, ed., Braziller, New York (1969), and references therein.
7. I. Petersen, U. Sellden, and O. Eeg-Olofsson, The evolution of the EEG in normal children and adolescents from 1 to 21 years, in: "Handbook of EEG and Clinical Neurophysiology," Vol. 6, Part B: "The Evolution of the EEG from Birth to Adulthood," A. Remond, ed., Elsevier, Amsterdam (1975).
8. M. Defayolle and J. P. Dinard, Application de l'analyse factorielle a l'etude de la structure de l'EEG, Electroencephalogr. Clin. Neurophysiol. 36:319 (1974).
9. E. Basar, "EEG-Brain Dynamics: Relation Between EEG and Brain Evoked Potentials," Elsevier/North-Holland, Amsterdam (1980); O. Creutzfeldt and J. Houchin, Neuronal basis of EEG-waves, in: "Handbook of EEG and Clinical Neurophysiology," Vol.2, Part C: "The Neuronal Generation of the EEG," A. Remond, ed., Elsevier, Amsterdam (1974).
10. L. D. Landau and E. M. Lifshics, "Field Theory," 6th ed., Ch. 9, (in Russian), Nauka, Moscow (1973).
11. L. D. Landau and E. M. Lifshics, "Electrodynamics of Continuous Media," 2nd ed., Ch. 9, (in Russian), Nauka, Moscow (1982).
12. C. W. Einolf, Jr. and E. L. Carstensen, Low-frequency dielectric dispersion in suspensions of ion-exchange resins, J. Phys. Chem. 75:1091 (1971), and references therein.
13. M. Born and E. Wolf, "Principles of Optics," Ch. 1, Pergamon, Oxford (1964).
14. Chapter 1 in Ref. 10.
15. Chapter 6 in Ref. 10.
16. P. Fraisse, "The Psychology of Time," Eyre & Spottiswoode, London (1964).
17. M. H. Erickson and L. F. Cooper, "Time Distortion in Hypnosis," Williams & Wilkins, Baltimore (1959).
18. R. E. Ornstein, "The Psychology of Consciousness," Ch. 4, Freeman, San Francisco (1972).
19. K. Jaspers, "Algemeine Psychopathologie," Springer, Berlin (1953).
20. R. A. Moody, Jr., "Life after Life," Bantam Books, New York (1975).
21. Y. Omura, Current progress in acupuncture research, Acupunct. Elec. Therap. Res. Int. J. 5:1 (1980).
22. J. S. Meyer and N. Kok, Correlations of changes in mental state, EEG, cerebral blood flow and metabolism, in: "Handbook of EEG and Clinical Neurophysiology," Vol. 7, Part B: "Influence on the EEG of Certain Physiological States and Other Parameters," A. Remond, ed., Elsevier, Amsterdam (1974).
23. P. Passouant, Electro-clinical semeiology, in: "Handbook of EEG and Clinical Neurophysiology," Vol. 7, Part A: "EEG and Sleep," A. Remond, ed., Elsevier, Amsterdam (1975).
24. "Altered States of Consciousness," C. Tart, ed., Sect. 8, Anchor Books, New York (1972).
25. R. G. Jahn, The Persistent paradox of psychic phenomena: an engineering perspective, Proc. IEEE 70:136 (1982).
26. W. G. Walter, R. Cooper, V. J. Aldridge, W. C. McCallum, and A. L. Winter, Contingent negative variation: an electric sign of sensorimotor association and expectancy in the brain, Nature 203:380 (1964).
27. L. D. Landau and E. M. Lifshics, "Quantum Mechanics. Nonrelativistic Theory," 3rd ed., Ch.1 , (in Russian), Nauka, Moscow (1974).
28. "The Many-Worlds Interpretation of Quantum Mechanics," B.S. DeWitt and N. Graham, eds., Princeton Univ. Press, Princeton (1973).
29. A. J. Sanford, A periodic basis for perception and action, in: "Biological Rhythms and Human Performance," W. P. Colquhoun, ed., Academic, London and New York (1971).

30. For criticism of biochemical models, see Chapter 2 in Ref. 5.
31. Chapter 4 in Ref. 5.
32. C. T. White, Temporal numerosity and the psychological unit of duration, Psychol. Monogr. No. 575 (1963); J. M. Stroud, The fine structure of psychological time, in: "Interdisciplinary perspectives of time," R. Fischer, ed., Ann. NY Acad. Sci., Vol. 138 (1967).
33. R. Nobili, Schrödinger wave holography in brain cortex, Phys. Rev. A 32:3618 (1985).
34. R. T. W. L. Conroy and J. N. Mills, "Human Circadian Rhythms," Ch. 7, Churchill, London (1970).

MATHEMATICAL PROBLEMS ARISING IN MOLECULAR ELECTRONICS: GLOBAL GEOMETRY AND DYNAMICS OF THE DOUBLE-WELL POTENTIAL

Albert F. Lawrence and Robert. R. Birge

Center for Molecular Electronics, Department of Chemistry

Syracuse University, Syracuse, N. Y. 13244

INTRODUCTION

Device physics and molecular processes

In a demonstration that no physical limitations to computers arise from quantum mechanics and the uncertainty principle, Feynman [1] constructed Hamiltonians for spin systems which performed any desired sequence of logical operations. Although this work showed that there are no intrinsic reasons circuits could not be built at the molecular scale, numerous problems remain in constructing molecular analogs to the standard electronics building blocks: transmission lines, amplifiers, gates and switches. Numerous molecules or molecular systems have been proposed to serve these functions, but such questions such as interconnections, noise isolation, and immunity to thermal and quantum fluctuations remain unanswered. As with the engineering of computers in bulk matter, the engineering of computers in molecular systems requires attention to the detailed dynamics of the system. Part of the problem is at the foundations. Aside from numerical calculations, which are presently beyond the bounds of computer technology, and are likely to remain so for some time hence, the dynamics of large molecules are inaccessible to precise quantitative prediction. On the other hand, an analytical theory of molecular dynamics presents some serious technical challenges and conceptual difficulties. Rapid progress toward a science of information processing at the molecular level awaits the resolution of these fundamental difficulties.

INFORMATION PROCESSING IN BIOLOGICAL SYSTEMS

Molecular processes and biological activity

Biology provides many examples of molecular structures which through a precise control of the transport of mass, charge or energy act in an information processing capacity.

For example, numerous molecular systems in the cell membrane exhibit the transmission and control of signals through modulation of an effective potential. Often, the fundamental mechanics also include (i) a conformational switch from one metastable state to another, (ii) the motion of a charged particle via a combination of tunneling and escape processes along a periodic potential and (iii) the direct coupling of the conformational switch to the transport process [2,3,4,5]. A binding event or absorption of a photon of energy initiates this chain of events. In these membrane bound systems, movement in multiple potential well becomes a recurring central theme. In fact, information processing in biological systems is realized by motion in multiple wells potential. This is no accident. It is well known that the quantum mechanics of a two-level system is mathematically equivalent to the quantum mechanics of the spin system. Thus Feynman's computers can be built out of coupled two-level systems[1]. On the other hand, the mechanics of the two level system is itself a good model for motion in a double well potential when only the lowest energy level in each well is occupied [6]. Multiple potential well systems, specifically coupled two well systems, may be regarded as generalizations of spin chains. This observation makes the connection between information processing and quantum biology clearer.

Transduction and transport

Returning to the biological examples, modulation of an effective potential and coupling between conformational switches and transport in a multiple-well potential are characteristic features of a family of membrane-bound proteins (such as bacteriorhodopsin) related to the adrenergic receptors. This class includes rhodopsin, the muscarinic cholinergic receptors, other neurotransmitter receptors and peptide hormone receptors [3]. Adrenergic receptor proteins may be coupled to guanine nucleotide regulatory proteins (G proteins). A common feature of G protein coupled receptors is that agonist occupancy of the receptor leads to receptor activation of a G protein which in turn modulates the activity of an effector enzyme or ion channel [3]. Bacteriorhodopsin, the light-harvesting protein found in the purple membrane of *Halobacterium halobium,* exhibits all of the dynamical features exhibited by the adrenergic receptors [2]. As with the adrenergic receptors, the part of the protein which spans the membrane is composed of seven alpha-helices [3].

Fast photochemical events

The light-adapted form of the bacteriorhodopsin undergoes a photochemical cycle that is responsible for transporting protons from the inside (cytoplasm) to the outside (extracellular) of the membrane. Most workers believe the proton moves along a channel enclosed by the alpha helices through a sequence of tunneling events. Although the proton transport is interesting in its own right, we are mainly interested in the processes immediately following photon absorption. The primary event (bR_{570} -> K_{610}) is the only process of the cycle which requires energy input in the form of light [2]. This event involves an extremely rapid conformational change, less than 1.0 picosecond, in the retinyl chromophore. The major change is a *trans- cis* shift around the $C_{13} = C_{14}$ double bond. This rapid change is driven by a shift in electron charge along the polyene chain which, before photon absorption is stabilized by interaction with a negatively charged group on the protein substrate. The initial phototransduction, is in itself, a complex event, involving coupled motions of electrons and nuclei.

The retinyl chromophore exhibits multiple metastable configurations in both the excited and ground states. The primary event may involves a change in the shape of the conformational potential energy surface (due to the excitation of the electrons) followed by a conformational change and a nonradiative decay to the ground state of the electrons [4]. This is illustrated in Figure 1. Because the barrier to conformational change in the excited state is small, or negative, absorption of light leads to the exceedingly rapid cis-trans shift. In a sense, conformational transitions in the excited state act to effectively gate conformational changes in the ground state. A minor change in the potential for the excited state may act to change the relative probabilities of decay to the original conformational state and decay to the new conformational state [4], as is shown in the first part of Figure 1.

The dynamics of the initial phototransduction in bacteriorhodopsin exhibit a number of features which complicate the mathematical analysis. First, the rapidity of the conformational change eliminates the assumption of adiabaticity. The nuclear motion must be analyzed as occurring in a changing potential. Many of the standard analyses of tunneling in a double well potential assume weak coupling between the tunneling motion and the other degrees of freedom of the system. This is not the case with bacteriorhodopsin. Finally, as the potential wells become more shallow, states other than the lowest vibronic level in each well must be taken into account. Because of these difficulties, a rigorous mathematical analysis is not yet available. Nevertheless, attempts to solve various special cases of this problem have led to a number of interesting and useful mathematical developments [4]. At one time or another, works touching on several areas of pure mathematics have entered the picture. These include path integral techniques [7] asymptotic expansions [8,9,10] summation of series [11], singularity theory [12,13] operator algebras [14] and manifold theory [15].

Application of the double-well potential

The motion of a system in a double well potential serves as a model for switching. Birge and Lawrence [16] have shown how bacteriorhodopsin can be used as the basis of an optical memory. Through examination of the dynamics of the bacteriorhodopsin molecule we can also arrive at some understanding of the nature of amplification processes in molecular systems. In bacteriorhodopsin, we have seen that the switching is dependent on interactions with processes which are not specified in the abstract double-well potential

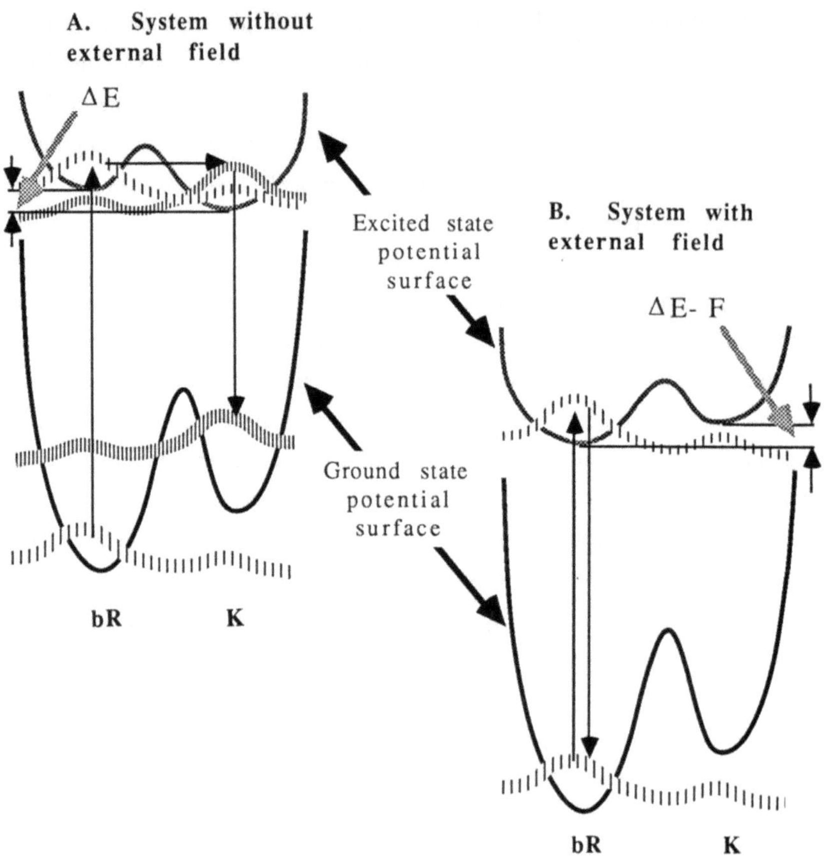

FIGURE 1
CONFORMATIONAL CHANGE VIA EXCITED STATE POTENTIAL

model. Outside forces are also required for amplification and gating. The molecular amplification of a relatively weak signal must be preceded by transformations which prepare the molecule in an excited state. A simple example of this is the electronic excitations which precede the conformational changes in bacteriorhodopsin. In the case of bacteriorhodopsin, for example, multiple potential wells (with respect to nuclear motion) are present in both the excited and ground states [4]. In order that the combined system act as an amplifier, the basic requirements are that the potential barriers to nuclear motion in the excited state are small, and the potential wells in the excited and ground states are properly aligned. Because a transition from the excited electronic state into the ground electronic state must end in one of several molecular configurations, a small ambient field may ensure that one of the molecular configurations available in the excited state is preferred over the others. This would shift the quantum yield of the entire excitation-transition-decay process in favor of the corresponding configuration in the ground state. In effect a small ambient field would act to gate a transition between two configurational states, as in Figure 1. Excited state configurational changes provide a quantum mechanical mechanism for biological transduction of electrical fields. A system of such molecules could, through the input of energy which maintains a substantial fraction of the population in an excited state, amplify small field changes to produce large changes in the relative populations in two conformational states.

PATH INTEGRAL TREATMENT OF MOLECULAR MOTIONS

Coupling between nuclear and electronic motions

A standard path-integral approach to the problem of coupled systems is given in the paper of Miller and George [17]. We will review their treatment of the general problem of electronic transitions in molecular systems. Recall that the molecular Hamiltonian is written

$$H = T_{q_e} + T_{q_n} + V(q_e, q_n), \tag{1}$$

where T_{q_e} is the electronic kinetic energy, T_{q_n} is the nuclear kinetic energy and $V(q_e, q_n)$ is the total potential energy of the the configuration. A rigorous treatment of the initial photon absorption in bacteriorhodopsin also requires additional terms representing the radiation field and its interaction with the electrons in the Hamiltonian. We ignore these terms in order to clarify the problems which arise in the treatment of coupled electronic and nuclear motions. The electronic Hamiltonian for fixed nuclear coordinates is given by:

$$h(q_e, q_n) = T_{q_e} + V(q_e, q_n) \tag{2}$$

The path integral representation for the propagator can be written as an iterated integral:

$$\langle q_{e,2} q_{n,2} | \exp[(-i\hat{H}(t_2 - t_1)/\hbar] | q_{e,1} q_{n,1} \rangle =$$

$$\int_{q_{e,1}}^{q_{e,2}} Dq_e \int_{q_{n,1}}^{q_{n,2}} Dq_n \exp[\, i\phi(q_e, q_n)/\hbar]. \tag{3}$$

The coordinates $q_{e,i}$ and $q_{n,i}$ represent the initial (i = 1) and final (i = 2) electronic and nuclear configurations, and ϕ is the classical action functional:

$$\phi(q_e, q_n) = \int_{t_1}^{t_2} dt\, [T_{q_e} + T_{q_n} - V(q_e, q_n)]. \tag{4}$$

We may integrate over the electronic variables first to obtain an expression relating the propagator for the molecule to the path-dependent electronic propagator, $<q_{e,2}| K[q_n(t)] |q_{e,1}>$

$$<q_{e,2}q_{n,2}| \exp[(- i\hat{H}(t_2 - t_1)/\hbar] |q_{e,1}q_{n,1}>$$

$$= \int_{q_{n,1}}^{q_{n,2}} Dq_n <q_{e,2}| K[q_n(t)] |q_{e,1}> \exp\left(\frac{i}{\hbar} \int_{t_1}^{t_2} dt\, T_{q_n} \right). \qquad (5)$$

The electronic propagator is given by

$$<q_{e,2}| K[q_n(t)] |q_{e,1}> =$$

$$\int_{q_{e,1}}^{q_{e,2}} Dq_e \exp\left(\frac{i}{\hbar} \int_{t_1}^{t_2} dt\, T_{q_e} - V(q_e,) \right). \qquad (6)$$

The electronic propagator is essentially a functional of the nuclear path. We actually need to evaluate the transition amplitudes related to the initial and final electronic states. If ϕ_1 and ϕ_2 are the wavefunctions for these states, then the reduced propagator is defined as

$$<2, q_{n,2}| \exp[(- i\hat{H}(t_2 - t_1)/\hbar] |1, q_{n,1}> =$$

$$\int dq_{e,1} \int dq_{e,2}\, \phi_2^* <q_{e,2}q_{n,2}| \exp[(- i\hat{H}(t_2 - t_1)/\hbar] |q_{e,1}q_{n,1}> \phi_1. \qquad (7)$$

The amplitude $K_{21}[q_n(t)]$ for the transition from state 1 to state 2 now appears in the formula

$$<2, q_{n,2}| \exp[(- i\hat{H}(t_2 - t_1)/\hbar] |1, q_{n,1}> =$$

$$\int_{q_{n,1}}^{q_{n,2}} Dq_n\, K_{12}[q_n(t)] \exp\left(\frac{i}{\hbar} \int_{t_1}^{t_2} dt\, T_{q_n} \right). \qquad (8)$$

The transition amplitude also depends upon the nuclear path. Given the path $q_n(t)$ we can compute $K_{21}[q_n(t)]$ directly. Because we are following the treatment of Miller and George [17], which is intended to account for the mechanics of molecular collisions, the electron wavefunctions ϕ_1 and ϕ_2 are defined for the nuclear configurations (or more properly, the nuclear wavefunctions) going into (1) and out from (2) the scattering event, but not necessarily for the nuclear configurations assumed during collision. The general problem of stability of states of finite particle systems has been reviewed recently by Sigal and Soffer [18]. (They also proved asymptotic completeness for short-range quantum-mechanical systems, i. e. that such a system decomposes into independently moving stable sub-systems as time goes to infinity.) Miller and George [17] assume that the wavefunctions ϕ_1 and ϕ_2 extend over the manifold of nuclear configurations so that the extended wavefunctions ψ_1 and ψ_2 are defined for the starting and ending configurations $q_n(t_1)$ and $q_n(t_2)$. This extension is also required for the case in which we are interested, the excitation of electrons by photon absorption, which results in a change in the potential seen by the nucleii. We also assume that the wavefunctions are orthogonal for each nuclear configuration. This is a reasonable assumption for the cases in which we are interested because we may choose ϕ_1 and ϕ_2 to be the ground and excited electron states, respectively. Finally, we also assume that we have chosen wavefunctions ψ_1 and ψ_2 which satisfy the Born-Oppenheimer approximation:

$$\hat{h}(q_e, q_n)\psi_i(q_e; q_n) = W_i(q_n)\psi_i(q_e; q_n). \tag{9}$$

Here, the $W_i(q_n)$ represent the adiabatic electronic eigenvalues which represent the potentials for nuclear motion. The time-dependent electronic wavefunction along the path $q_n(t)$ satisfies the time dependent Schrödinger equation:

$$i\hbar \frac{\partial}{\partial t}\Psi(q_e, t) = \hat{h}(q_e, q_n)(q_e, t). \tag{10}$$

Assuming that

$$\Psi(q_e, t) = c_1(t)y_1(q_e; q_n) + c_2(t)\psi_2(q_e; q_n) \tag{11}$$

then the Schrödinger equation and the assumptions on the ψ_i give the coupled equations:

$$i\hbar\dot{c}_1(t) = W_1 c_1(t) - i\hbar \left\langle \psi_1 \,\middle|\, \frac{\partial \psi_2}{\partial t} \right\rangle c_2(t) \tag{12a}$$

$$i\hbar\dot{c}_2(t) = W_2 c_2(t) - i\hbar \left\langle \psi_2 \middle| \frac{\partial \psi_1}{\partial t} \right\rangle c_1(t). \tag{12b}$$

The transition amplitudes are given as solutions of the above equations with initial conditions $c_1(t_1) = 1$, $c_2(t_1) = 0$; in particular,

$$K_{21}[q_n(t)] = c_2(t_2) \tag{13a}$$

$$K_{11}[q_n(t)] = c_1(t_2). \tag{13b}$$

These equations along with the path integral give an implicit solution for the problem of coupled motions in the Born-Oppenheimer approximation.

Hagedorn [19] has recently improved these results by showing how asymptotic solutions of the time dependent Schrödinger equation may be obtained without resorting to the path integral. In particular, he solves the equation

$$i\left(\frac{\partial \Psi}{\partial t}\right) = \hat{H}(\varepsilon)\Psi, \tag{14}$$

where

$$\hat{H}(\varepsilon) = -\sum_{j=1}^{K} \frac{\varepsilon^4}{2M_j} \Delta_j - \sum_{j=K+1}^{N} \frac{1}{2m_j} \Delta_j + \sum_{i<j} V_{ij}(q_i - q_j) \tag{15}$$

and the ratio of the masses of the electrons to the nuclei is given by the factor e^4. The q_i denote both electron and nuclear coordinates. These approximations are valid for times t on the order ε^{-2} and initial velocities of the heavy particles on the order e^2. These accuracies give the inherent limits of the dynamical version of the Born-Oppenheimer approximation. The speed of the initial photochemical event in bacteriorhodopsin is at these limits.

Before we leave the discussion of the coupling of the nuclear and electronic motions, we will indicate the differential geometry of the potential surface enters in a non-trivial way into multiple well tunneling. Following Miller and George [17], we can remove the phase advance in the coupled equations (12a) and (12b) by defining new coefficients $a_i(t)$:

$$a_i(t) = c_i(t) \exp\left(\frac{i}{\hbar} \int_{t_1}^{t} d\tau \, W_i(\tau) \right) \tag{16}$$

These coefficients satisfy the equations

$$\dot{a}_1(t) = a_2(t) \left\langle \psi_1 \middle| \frac{\partial \psi_2}{\partial t} \right\rangle \exp\left(\frac{-i}{\hbar} \int_{t_1}^{t} d\tau \, \Delta W(\tau) \right) \tag{17}$$

$$\dot{a}_2(t) = a_1(t) \left\langle \psi_2 \middle| \frac{\partial \psi_1}{\partial t} \right\rangle \exp\left(\frac{i}{\hbar} \int_{t_1}^{t} d\tau \, \Delta W(\tau) \right). \tag{18}$$

A first order perturbation solution gives

$$a_2(t_2) = \int_{t_1}^{t_2} dt \left\langle \psi_2 \middle| \frac{\partial y_1}{\partial t} \right\rangle \exp\left(\frac{i}{\hbar} \int_{t_1}^{t} d\tau \, \Delta W(\tau) \right). \tag{19}$$

The second factor on the right hand side of equation (19) is a trigonometric integral, which sometimes can be evaluated by the method of stationary phase. In this method, the integral is dominated by the contributions from regions where DW(t) = 0. This usually does not happen because the intersection between the surfaces W_1 and W_2 is generally of dimension f-2, where f is the number of parameters determining each surface separately. In this case we look for the times when the integrand is least rapidly varying, that is when

$$\frac{d}{dt} \Delta W(t) = 0. \tag{20}$$

Such points are given by times, t, where the time derivative of the path $q_n(t)$ vanishes and by the critical points of the surface $\Delta W(t)$. Because the former points can be regarded as constituting a subset of measure zero in the space of paths, the path integral is dominated by contributions from the critical points of the potential surface. The nature of these points and of the surface surrounding them is the concern of Morse theory [20]. The path integral may be expressed in terms of the geometric invariants of the critical points [10,12,13,21].

Asymptotic expansions

The evaluation of the path integral, as can be seen from the form of Equations (8) and (19) leads to calculation of trigonometric integrals such as:

$$I(\hbar, \alpha) = \int dq\, g(\mathbf{q}) \exp\left(\frac{i}{\hbar} f(\alpha; \mathbf{q}) \right) \tag{21}$$

where $\mathbf{q} = (q_1, q_2, \ldots, q_N)$. In this formulation, both f and g may depend on the parameters $\alpha = (\alpha_1, \alpha_2, \alpha_3, \ldots, \alpha_k)$ and f may be regarded as arising from a variable potential. Explicit derivations, resulting in trigonometric integrals applicable to a variety of physical systems, may be found in the book on path integrals by Schulman [22].

The usual technique [10] for finding approximations to Equation (21) is a generalization of Equation (20), that is finding points

$$\mathbf{q} = \mathbf{q}^{(i)}(\alpha), \quad i = 1, 2, \ldots, s$$

such that

$$\frac{\partial f(\alpha; \mathbf{q})}{\partial q_j} = 0, \quad j = 1, 2, \ldots, N. \tag{22}$$

These points correspond to the classical trajectories of the system, and are variously termed the stationary phase, saddle, singular or critical points. In the case where the classical trajectories coalesce, we have s critical points coincident for some value of the parameter vector, α_0. The problem is to find a uniform approximation for the integral near this value of α in the region which contains the s critical points. One usually proceeds by finding new integration variables $\mathbf{x}$ and parameters ζ, so that

$$\mathbf{x} = \mathbf{x}(\alpha; \mathbf{q}), \quad \mathbf{x} = (x_1, x_2, \ldots, x_n), \tag{23}$$

and

$$\zeta = \zeta(\alpha), \quad \zeta = (\zeta_1, \zeta_2, \ldots, \zeta_l) \tag{24}$$

where

$$f(\alpha; \mathbf{q}) = F(\zeta; \mathbf{x}) + \zeta_0. \tag{25}$$

We require that the map on the integration variables is one-to-one and uniformly analytic in a neighborhood which contains the s critical points and that the integral in Equation (21), rewritten in terms of F, **x**, and ζ is more tractable. As a matter of fact, if the number of parameters, α_i, is four or less, there are only seven canonical forms of F which arise as unfoldings of the singularity at α_0. These are given by Connor [10]. The trigonometric integral, in any case reduces to:

$$I(\hbar, \zeta) = \exp\left(\frac{i}{\hbar}\zeta_0\right)\int d\mathbf{x} G(\mathbf{x}) \exp\left(\frac{i}{\hbar}F(\zeta; \mathbf{x})\right), \tag{26}$$

where

$$G(\mathbf{x}) = g[\mathbf{q}(\mathbf{x})]\ J[\mathbf{q}(\mathbf{x})]. \tag{27}$$

The factor J(**q**) is the Jacobian of the transformation given by Equation (23).

In the case where there is only one integration variable, there are four canonical polynomial forms possible for F. These are given by

$$F_n(\zeta; x) = x^{n+2} + \sum_{i=1}^{n} \zeta_i x^i \tag{28}$$

If we write

$$G(x) = \sum_{k=0}^{n} q_k x^k + h(x) \frac{\partial F_n}{\partial x} \tag{29}$$

then the equation

$$I_1(\hbar, \zeta) = \exp\left(\frac{i}{\hbar}\zeta_0\right) \sum_{k=0}^{n} \hbar^{[k+1]/[n+2]} q_k \mathcal{U}_k(\xi_1, \xi_2, \ldots \xi_n) \tag{30}$$

gives the lowest order uniform approximation to $I(\hbar, \zeta)$. In particular

$$\mathcal{U}_k(\xi_1, \xi_2, \ldots \xi_n) =$$

$$\int x^k \exp[i(\xi_1 x + \xi_2 x^2 + \ldots + \xi_n x^n + x^{n+2})], \tag{31}$$

and

$$\xi_i = \hbar^{[i/(n+2)]-1}\zeta_i. \tag{32}$$

Albeverio and Höegh-Krohn [23] have given a rigorous derivation of formulas for the asymptotic approximation of solutions of the Schrödinger equation by trigonometric integrals in which the potential appears explicitly.

Some special cases of the double-well potential

If we assume that only the lowest state in each well is excited, then the quantum mechanics of a particle in a double-well potential specializes to the mechanics of the two-state system. Two-state models are very prevalent in quantum mechanics, as they are useful for the understanding a wide range of phenomena--ranging in scale from lasers and superconducting junctions through molecular reactions to subatomic particles. We can translate this picture into quantum-mechanical terms. To begin, we assume that the system is restricted to the ground state; that the potential barrier is large with respect to the ground state energy, $\hbar\omega_0$. A two state system without frictional forces is given by the Hamiltonian

$$\hat{H} = -\tfrac{1}{2}\hbar\Delta_0\sigma_x + \tfrac{1}{2}\varepsilon\sigma_z. \tag{33}$$

The phase space is two dimensional, and the operators σ_x, σ_y and σ_z are the Pauli matrices. If the parameter e is small, the eigenstates +1 and -1 for σ_z correspond to the particle localized in the right or left hand well respectively. The quantity $\frac{1}{2}\hbar\Delta_0$ is the matrix element for tunneling between the two wells and ε is the detuning parameter. This Hamiltonian may be diagonalized, with eigenvalues $(\pm\varepsilon^2 + (\hbar\Delta_0)^2)^{1/2}$. If the ratio $\hbar D_0/\varepsilon$ is large, the eigenstates are appreciably delocalized, otherwise, the eigenstates of the Hamiltonian operator are localized, i. e. they correspond to the eigenstates of σ_z and the particle is in one well or the other. This is in strong contrast with the classical picture of a switch. In order that we obtain some degree of localization with a small biasing force, we must include frictional forces [6].

The quantum-mechanical equivalent of frictional forces is a heat bath of harmonic oscillators, each coupled with the the system of interest. In the present case, the simplest model is the "spin-boson" Hamiltonian:

$$\hat{H}_{SB} = -\tfrac{1}{2}\hbar\Delta_0\sigma_x + \tfrac{1}{2}\varepsilon\sigma_z$$

$$+\sum_\alpha\left(\tfrac{1}{2}m_\alpha\omega_\alpha q^2_\alpha + p^2_\alpha/2m_\alpha\right) + \tfrac{1}{2}q_0\sigma_z\sum_\alpha C_\alpha q_\alpha. \tag{34}$$

In this formula $q_\alpha \equiv \hat{q}_\alpha$ and $p_\alpha \equiv \partial/\partial q_\alpha$ respectively. This system exhibits a rich variety of behaviors, depending on the tunneling matrix, the detuning parameter, the spectral content of the heat bath, and the coupling between the double-well system and the heat bath oscillators. We are primarily interested in the case where the double well potential governs a configurational change in a large protein molecule. The findings of Leggett, et. al.[5] cover many cases of biological interest: If there is no bias, ε, expectation function for the eigenvalue of σ_z, P(t), decays to zero, and for some finite value of the bias the wavefunction is eventually localized in the lower well. These results are dependent upon assumption that the process of tunneling does not disturb the environment too strongly.

An integrable double-well potential

Merzbacher [24] has shown that the wave equation for the double oscillator

$$i\hbar \frac{\partial}{\partial t}\psi(q) = -\left(\frac{\hbar^2}{2\mu}\frac{\partial^2}{\partial t^2} + V(q) \right)\psi(q) \tag{35}$$

where the potential given by

$$V(q) = \frac{1}{2}k(|q| - a)^2 \tag{36}$$

is integrable. Furthermore, this system gives good qualitative results for the zero temperature case of the symmetrical double well potential. This model potential is discontinuous at q = 0 and has minima at q = ±a. By setting

$$r = \left(\frac{2\mu\omega}{\hbar} \right)^{1\backslash 2} (q - a) \qquad q>0 \tag{37a}$$

$$r' = \left(\frac{2\mu\omega}{\hbar} \right)^{1\backslash 2} (q + a) \qquad q<0 \tag{37b}$$

we obtain stationary Schrödinger equations

$$\frac{d^2\psi}{ds^2} + \left(n + \frac{1}{2} - \frac{s^2}{4}\right)\psi = 0 \quad \text{for } s = r, r',\ E = \hbar\omega\left(v + \frac{1}{2}\right). \tag{38}$$

The solutions of Equations (38) are given in terms of parabolic cylinder functions D_v and $D_{v'}$ which are valid separately for the two intervals given in (37a) and (37b). The only problems in constructing a solution over the entire real line are: (i) eliminating the singularity at q = 0 and (ii) matching the two partial solutions at the common boundary of their domains.

These two conditions imply that both the slopes and the values of the functions should be matched at q = 0. In particular, the unnormalized wavefunctions ψ are given by two separate matching conditions (as equations for ν) in the odd and even cases:

$$D'_{\nu}\left[- \left(\sqrt{\frac{2\mu\omega}{\hbar}}\right) \; a \right] = 0, \tag{39}$$

for even ψ.

For odd ψ:

$$D_{\nu}\left[- \sqrt{\frac{2\mu\omega}{\hbar}} \; a \right] = 0. \tag{40}$$

The explicit unnormalized wavefunctions are given by

$$y(x) = \begin{cases} D_{\nu}\left[\sqrt{\frac{2\mu\omega}{\hbar}}(q - a)\right] & q \geq 0 \\ D_{\nu}\left[- \sqrt{\frac{2\mu\omega}{\hbar}}(q + a)\right] & q \leq 0. \end{cases} \tag{41}$$

Pairs of odd and even wavefunctions--ν_0 and ν_1 or ν_2 and ν_3, for example--have the same energy for large a (when the wells are widely separated), but the energies of the odd and even solutions to Equation (36) split as a approaches zero. Given that the separation parameter, a, is small, a state initially peaked at q = a will oscillate between the wells. The details are given by Merzbacher [24].

MOVING POTENTIALS

Time varying potentials and coherent states

Several special cases of the nonstationary quantum oscillator may be solved by means of the coherent state method. A quantum state is a coherent state if it is an eigenstate of an annhilation operator. Annhilation and creation operators for the states of harmonic oscillator generate a noncommutative Lie algebras of operators, the Heisenberg algebra. The Lie product in the Heisenberg algebra is the ordinary commutator of operators. Because of the connection between quantum states and operator algebras, the definition of a coherent state usually can be translated to an equivalent statement about the action of the dynamical group in the space of states. The group action gives the information necessary to find a solution of the time-dependent Schrödinger equation. Two cases solvable by coherent state methods are of particular relevance to the solution of the moving double well potential. These are the quantum oscillator acted upon by a variable external force and the quantum oscillator excited by a change in its parameters [14].

The forced harmonic oscillator

In this section we follow Perelomov [14] and simplify the notation by setting ħ and m = 1. The dynamics of the forced harmonic oscillator are given by the equation

$$i\frac{\partial}{\partial t}|\psi(t)\rangle = (\hat{H}_0 + \hat{H}_1)\,|\,\psi(t)\,\rangle, \tag{42}$$

where

$$\hat{H}_0 = \tfrac{1}{2}(p^2 + \omega^2 q^2) = \omega\,(a^\dagger a + \tfrac{1}{2}) \tag{43}$$

and

$$\hat{H}_1 = -f(t)\,q = -f(t)\,\frac{1}{\sqrt{2\omega}}(a + a^\dagger\,). \tag{44}$$

The annhilation and creation operators are given by a and $a^\dagger$ respectively. By employing a unitary transform

$$|\,\psi(t)\,\rangle = \exp(-i\,\hat{H}_1)\,|\,\tilde{\psi}\,\rangle \tag{45}$$

we can transform the Schrödinger equation

$$i\frac{\partial}{\partial t}|\,\tilde{\psi}(t)\,\rangle = \tilde{H}_1\,|\,\tilde{\psi}\,\rangle \tag{46}$$

where

$$\tilde{H}_1 = \exp(i\hat{H}_0 t)\,\hat{H}_1 \exp(-\,i\hat{H}_0 t) =$$
$$-\,f(t)\,\frac{1}{\sqrt{2\omega}}[a\;\exp(-\,i\omega t\,) + a^\dagger\;\exp(i\omega t\,)]. \tag{47}$$

We define functions $\beta(t)$ and $\bar{\beta}(t)$ via

$$\beta(t) = \frac{i}{\sqrt{2\omega}} f(t) \exp(i\omega t) \tag{48}$$

The Hamiltonian $\tilde{H}_1$ is a (time-varying) element of the Heisenberg algebra, so the evolution operator $\tilde{S}(t)$ defined by $| \tilde{\psi}(t) > = \tilde{S}(t) | \tilde{\psi}(0) >$ is given by the standard representation of the Heisenberg group [14]. In particular

$$\tilde{S}(t) = T(g(t)) = \exp[-i\phi(t)]\exp(\gamma(t)a^{\dagger} - \bar{\gamma}(t)a] \tag{49}$$

for some choice of ϕ and γ. If the initial state was a coherent state, it remains coherent for any moment of time. In this case

$$| \psi(t) > = \exp[-i\phi(t)] \, | \alpha(t) >. \tag{50}$$

The expectation value of the operator a is given by

$$< \tilde{\psi}(t) | a | \tilde{\psi}(t) > = \alpha(t). \tag{51}$$

The time evolution of α and ϕ are given by the equations [14]

$$\dot{\alpha} = \beta, \quad \dot{\phi} = \mathrm{Im}(\bar{\beta}\alpha). \tag{52}$$

This illustrates the application of coherent state methods to a time-dependent potential.

The forced parametric oscillator is a generalization of the forced harmonic oscillator:

$$i\frac{\partial}{\partial t} | \psi(t) > = \left[- \frac{1}{2} p^2 + \frac{1}{2}\omega(t)q^2 - f(t)q \right] | \psi(t) >. \tag{53}$$

The Hamiltonian operator in this equation is in the Lie algebra generated by the operators p, q, $(1/4\omega_0)p^2 + \omega_0 q^2$, and $(1/4)(pq + qp)$. These operators generate the Lie algebra of the inhomogeneous symplectic group ISp(2,**R**). The wavfunction $| \psi(t) >$ can be written:

$$| \psi(t) > = \exp[- i\phi(t)] \, | \zeta(t), \alpha(t) > \tag{54}$$

where $\zeta(t)$ and $\alpha(t)$ arise from representations of ISp(2,**R**). The coherent state $| \zeta, \alpha >$ is given by

$$| \zeta, \alpha > = N \exp[\tfrac{1}{2}\zeta(a^{\dagger})^2 + \alpha a^{\dagger} \; | 0 >, \quad N = (1 - |\zeta|^2)^{-1/2}\exp[- \tfrac{1}{2}|\alpha|^2]. \tag{55}$$

Differential equations for z and a can be derived by substituting Equation (55) into Equation (53).

A singular time dependent double-well potential

We can obtain a time-dependent double well potential by piecing together two separate time dependent potentials, each similiar to the potential in Equation (53). For the moment we assume that the singularity remains at $q = 0$. This gives a generalized version of Equation (35). The time evolution of the wavefunction can be solved in terms of coherent states over each separate subinterval. The problem is to piece together the coherent state solutions in each well so that they match at the singularity, in a manner analogous to Equation (38) and the following discussion. This enforces certain symmetry conditions on the solutions. Although we know that coherent state solutions can be constructed for general anharmonic potentials (see Eu [25]), the exact nature of the symmetry conditions required for an asymmetrical potential to be integrable have not, to the author's knowledge, been explicitly derived.

REFERENCES

1. R. P. Feynman, "Quantum Mechanical Computers," *Foundations of Physics* **16:** 507 (1986).
2. R. R. Birge, "The Photophysics of Light Transduction in Rhodopsin and Bacteriorhodopsin," *Ann. Rev. Biophys. Bioeng.*, **10**: 315 (1981).
3. B. W. Kobilka, T. S. Kobilka, K. Daniel, J. W. Regan, M. G. Caron and R. J. Lefkowitz *Science,* **240**: 1310 (1988).
4. R. R. Birge, L. A. Findsen and B. M. Pierce, "Molecular dynamics of the Primary Photochemical Event in Bacteriorhodopsin: Theoretical Evidence for an Excited State Assignment for the J Intermediate," *J. Am. Chem. Soc.,* **109**: 5041 (1987).
5. S. Pnevmatikos, "Soliton Dynamics of Hydrogen-Bonded Networks: A Mechanism for Proton Conductivity," *Phys. Rev. Lett.*, **60**: 1534 (1988).
6. A. J. Legett, S. Chakravarty, A. T. Dorsey, M. P. A. Fisher, A. Garg and W. Zwerger, "Dynamics of the Dissipative Two-State System," *Rev. Mod. Phys.*, **59**: 1 (1987).
7. C. DeWitt-Morette, A. Maheshwari, and B. Nelson, "Path Integration in Non-Relativistic Quantum Mechanics," *Physics Reports,* **50**: 255 (1979).
8. B. Simon, "Semiclassical Analysis of Low Lying Eigenvalues I. Non-Degenerate Minima: Asymptotic Expansions," *Ann. Inst. Henri Poincare,* **A38**: 295 (1983).
9. B. Simon, "Semiclassical Analysis of Low Lying Eigenvalues, II. Tunneling," *Annals of Mathematics,* **120**: 89 (1984).
10. J. N. L. Connor, "Catastrophes and Molecular Collisions." *Molecular Physics,* **31**: 33 (1976).
11. B. Simon, "Large Orders and Summability of Eigenvalue Perturbation Theory: A Mathematical Overview," *International Journal of Quantum Chemistry,* **21**: 3 (1982).
12. V. I. Arnold, "Remarks on the Stationary Phase Method and Coexter Numbers," *Uspekhi Mat. Nauk.*, **28**: 17 (1973).
13. J. J. Duistermaat, "Oscillatory Integrals, Lagrange Immersions and Unfolding of Singularities," *Communications on Pure and Applied Mathematics,* **27**: 207 (1974).
14. A. Perelomov, *Generalized Coherent States and Their Applications,* Springer-Verlag, Berlin (1986)

15. H. L. Cycon, R. G. Froese, W. Kirsch and B. Simon, *Schrödinger Operators*, Springer-Verlag, New York. (1987).
16. R. R. Birge and A. F. Lawrence, "Optical Random Access Memory Based on Bacteriorhodopsin," These Proceedings.
17. W. H. Miller and T. F. George, "Semiclassical Theory of Electronic Transitions in Low Energy Atomic and Molecular Collisions Involving Several Nuclear Degrees of Freedom," *J. Chem. Phys.*, **56**: 5637 (1972).
18. I. M. Sigal and A. Soffer, "The N-Particle Scattering Problem: Asymptotic Completeness for Short-Range Systems," *Annals of Mathematics,* **126**: 35 (1987).
19. G. A. Hagedorn, "High Order Corrections to the Time-Dependent Born-Oppenheimer Approximation I: Smooth Potentials," *Annals of Mathematics,* **124**: 571 (1986).
20. R. Bott, "Lectures on Morse Theory, Old and New," *Bulletin American Math. Soc.*, 7:2 331 (1982).
21. J. B. Keller and D. W. McLaughlin, "The Feynman Integral," *American Math. Monthly,* **82**: 457 (1975).
22. L. S. Schulman, *Techniques and Applications of Path Integration*, John Wiley and Sons, New York. (1981).
23. S. Albeverio and R. Höegh-Krohn, "Oscillatory Integrals and the Method of Stationary Phase in Infinitely Many Dimensions, with Applications to the Classical Limit of Quantum Mechanics I.," *Inventiones Math.,* **40**: 59 (1977).
24. E. Merzbacher, *Quantum Mechanics,* (2nd Edition), John Wiley & Sons, Inc., New York (1970).
25. B. C. Eu, "Quantum Theory of Large Amplitude Vibrational Motions in a One-Dimensional Morse Chain," *J. Chem. Phys.,* **73**: 2405 (1980).

NANOELECTRONICS AND SCANNING TUNNELING ENGINEERING

Conrad Schneiker and Stuart Hameroff
Advanced Biotechnology Laboratory
Department of Anesthesiology, College of Medicine

Mark Voelker, Jackson He, and Eustace Dereniak
Optical Sciences Center

Robert McCuskey
Department of Anatomy, College of Medicine
University of Arizona, Tucson, Arizona 85724

Dedicated to a pioneer of electronic NanoComputing,
Forrest L. Carter (29 April 1930——20 December 1987),
and also to the godfather of NanoTechnology,
Richard P. Feynman (11 May 1918——15 February 1988).

> It is difficult to suppress one's enthusiasm for the development of a viable *molecular technological base* when one recognizes the possible scientific, industrial and economic spin-off opportunities.
> —— *Forrest L. Carter, 1981*

> [Consider] the final questions as to whether, ultimately ... *we can arrange the atoms the way we want, the very atoms, all the way down!* ... Ultimately, *we can do chemical synthesis* .... When we get to the very, very small world —— say circuits of seven atoms —— we have a lot of new things that would happen that represent completely new opportunities for design. ... We can use, not just circuits, but some system involving the quantized energy levels, or the interactions of quantized spins, etc. ...
>
> In the year 2000, when they look back at this age, they will wonder why it was not until *the year 1960* that anybody began seriously to move in this direction.
> —— *Richard P. Feynman, 1960*

INTRODUCTION

Carter (1983) noted that *projected trends in microelectronic fabrication would intersect the molecular-nanometer ("nano"= 10^{-9}; nanometer = 10^{-9} meter) level around 2020 AD*. Why not simply charge ahead on this rather straightforward path to molecular scale devices and nanotechnology —— *which shows no sign of significantly slowing down and which is also extremely well funded and profitable* —— and then branch out to more sophisticated types of molecular electronics when new "industrial strength" infrastructures would then make it simpler, easier, and much more efficient to take such innovative steps? Considering the rate of progress this decade alone, *will current* *approaches to molecular devices be rendered obsolete* because they are *evolving too slowly* relative to integrated circuit microelectronics and nanoelectronics? (See Yamamura, Fujisawa and Namba, 1984; Haddon and Lamola, 1985; Bandyopadhyay, 1986; Gray and Campisi, 1986; Howard, Jackel and Skocpol, 1985; Kratschmer *et al.*, 1985; Whitehead, Isaacson and Wolfe, 1985.) Given the *enormous lead times and costs* for research, development, production learning curves, and gaining substantial market share, does the development of molecular electronic devices *for computers* make sense? Guided by such questions, we will suggest some hybrid possibilities below.

Carter's (1979, 1980, 1983, 1986) pioneering work on molecular electronic devices (MEDs) was preceded by other researchers concerned with fabricating nanostructures of various kinds; their work is summarized in Schneiker (1988). Feynman's (1960a, 1960b) proposals for molecular level nanofabrication of nanodevices and nanomachines are at last becoming well known; by way of brief review, we will mention some underpublicized work here. Pioneering efforts aimed at nanolithography and 3-d integration were made by Buck and Shoulders (1958) and Shoulders (1960, 1961, 1962, 1965). Ariadne (1966a, b, c, 1972) noted that: 1) information storage densities of 10^{18} words/gram could be achieved by writing on synthetic polymer molecules, 2) molecular bearings could be made by attaching reactive chemical groups at the center of molecular rotors to monomolecular filaments, 3) graphite sheets could be fashioned into geodesic spheres for use as ultralight bearings or, with appropriately structured openings, function as molecular sieves, and that, 4) various molecules might have useful electronic device properties and that they could use optical coupling for use as computing systems. Taniguchi (1974) noted that extrapolated trends in ultraprecision machining would reach the atomic, "*substance synthesizing*" level in a few decades; he coined the term *nanotechnology* to describe such possibilities. Taniguchi has since described some atomic bit machining processes (Taniguchi, 1985). Teague (1978) proposed the use of vacuum tunneling as a powerful tool for surface mapping using equipment designed by Young, Ward and Scire (1971, 1972); unfortunately they didn't receive sufficient support to develop what could have become the first scanning tunneling microscope. Teague (1986) has since suggested a new transconductance device based on tunneling. The scanning tunneling microscope (STM) utilizes computer controlled piezo actuators to scan an ultrasharp electrode to within a few angstroms of a conducting surface; an applied voltage leads to a tunneling current which provides a servo-feedback signal and permits surface mapping with atomic resolution. [Readers unfamiliar with STMs can find excellent reviews of the present state-of-the-art in Hansma and Tersoff (1987), Binnig and Rohrer (1987) and Hansma *et al.* (1988)]. STM co-inventors Binnig and Rohrer received the 1986 Nobel Prize in Physics, and STM capabilities for atomic level maneuvering and manipulation hold great promise for molecular devices. Franks' (1987) review paper on nanotechnology notes that STMs may be used for "scanning tunneling engineering". Perhaps the most promising example approaching scanning tunneling engineering was the recent manipulation of single molecules using an STM by Foster, Frommer and Arnett (1988), also heralded as "atomic-scale engineering" by Pethica (1988).

The development of a technologically and economically viable molecular electronics technology presents *extreme* technical challenges for the synthesis, characterization, testing, and system integration of molecular electronic devices (MEDs). Many of these problems have been discussed at length by Haddon and Lamola (1985) and Yates (1984). Our earlier papers have indicated how STM derivatives could be used to meet some of these challenges (Schneiker and Hameroff, 1988; Hameroff, 1987; Schneiker, 1988; Schneiker *et al.*, 1988); these papers and the references therein form the background for this discussion. A new version of a previously proposed "nanotechnology workstation" using two STMs combined with two optical microscopes and a micromanipulator for testing molecular electronic devices is shown in Figure 1.

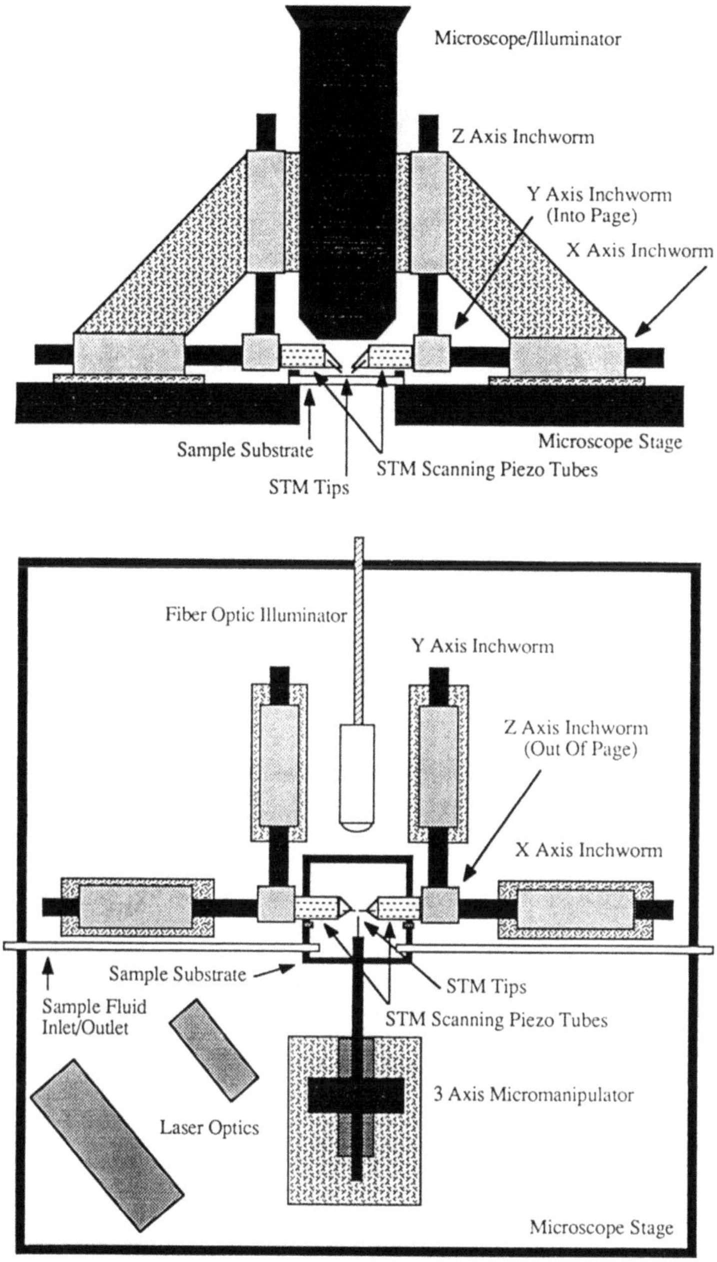

Fig. 1. Schematic top and front views of a dual tip STM. Microscope stage is for inverted fluorescence/phase contrast microscopes.

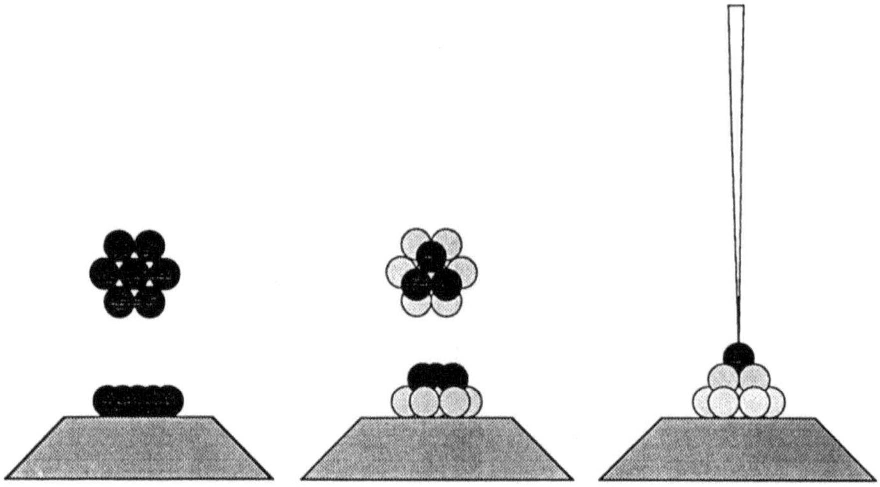

Fig. 2. Schematic examples of hexamer, trimer, and monoatomic pyramid STM/field emitter tips, with the extremely finely focussed electron/ion beams possible in the latter case indicated.

BEYOND THE SCANNING TUNNELING MICROSCOPE

> I would like to try and impress upon you the importance of improving the electron microscope by a hundred times. ... I put this out as a challenge: Is there no way to make the electron microscope more powerful? —— *Richard P. Feynman, 1960*

The answer to Feynman's question is "yes". First, the invention of the scanning tunneling microscope (STM) has led to important new types of microscopes including the atomic force microscope (Hansma *et al.*, 1988) and scanning near-field microscopy (Fischer, Durig and Pohl, 1988). These techniques take advantage of the *atomic resolution position control* of computer controlled piezo actuators. Second, the development of STM tips with atomically perfect pyramidal points terminating in a single atom (Fink, 1986, 1988) promises to revolutionize traditional scanning electron microscopy, improve atomic bit machining using ion beams, and have other major impacts on molecular scale technology. Indeed, Fink's invention of single atom tips may be at least as significant for science and technology as the invention of STMs —— furthermore, these tips are the ultimate example of nano-fabrication at the present time.

Fink's tips should stimulate major advances in low voltage scanning electron microscopy (LVSEM), which has several advantages over scanning electron microscopy (SEM) including: high secondary electron yield, minimized charging of insulators, reduced specimen damage, ability to image uncoated specimens, and potentially greatly improved resolution (Boyes, 1984; Joy, 1987; Shao, 1988). Fink's inventions are extremely important since they may help overcome problems involving electron beam sources and optics. Fink has demonstrated a very simple lens-less 15 eV SEM using an STM-like *mechanical* scanning system which has already demonstrated 30 Å resolution. A series of further important developments is sure to follow as *the surprising physical properties* of such fine beams are analyzed in more detail (Serena *et al.*, 1988).

MOLECULAR SCALE DEVICES

Research and development funding for presently successful and rapidly advancing integrated circuit electronics technology is going to vastly exceed that of any uncommercialized technology by orders of magnitude. Hence any products that build on this base using (bio)molecular technology should have economic and technological advantages over all (or mostly) MED systems. This multidisciplinary view is the reason we consider the *wider domain of nano-electronic devices*. These related technologies may be important bridges to "ultimate" *quantum mechanical* computing elements (using quantized spins, etc.,) suggested by Feynman (1960, 1986). At the systems level, evolution is almost always much more successful than revolution in the market place. Are there interesting possibilities in the region between these extremes which will ultimately get us to "ultimate" systems faster?

Vacuum NanoElectronic Devices

In his quest for extreme ultra-miniaturization of computing devices, Shoulders (1961) rejected the use of biological building blocks, even though biological "processes do work, and they can do so in a garbage can without supervision". He saw them as too limited environmentally and too difficult to control with available technology; instead, he sought to *directly* produce much simpler, more powerful, and much more rugged nanostructure arrays at video frequencies rates, which could in turn ultimately *aid in their own replication*. However, the progress in synthetic chemistry, macromolecular chemistry, and genetic engineering during the last 27 years may now make it feasible to incorporate *some* (bio)molecular fabrication processes into derivatives of Shoulders' original program.

The key elements of Shoulders' (1960) program evolved out of earlier efforts aimed at developing electron beam micromachining for (what is now called) nanolithography (Buck and Shoulders, 1958) and the development of submicron field emission triodes. These triodes depend on ultrasharp tip cathodes to facilitate tunneling of electrons into the surrounding vacuum at low voltages. Spindt, Holland and Stowell (1984) have continued the development of techniques for mass producing larger versions of such structures (see Figure 3). Fink (1986, 1988) has developed the means to make tungsten STM tips which terminate in perfect pyramid points ending in single atoms, and related points ending in trimers or hexamers. These tips are the ultimate in nanoscale field emitter cathodes, and Fink has demonstrated a 15 volt scanning field emission electron microscope using similar structures. The electron beams emanating from such tips are finely focussed, diverging less than a degree from the emission axis. (When operated in field ion microscope mode, extremely fine ion beams likewise result.) Binnig *et al.* (1985a, 1985b) suggested that STM tips could be used for sputtering and making nano-apertures in thin metal films. Feynman (1985) also (independently) suggested that STM tips could be used to make holes in thin metal films, and *further* noted that these apertures could serve as grids for nano-triodes using STM tips as cathodes. It may also be possible to utilize internal field emission and ballistic transport of electrons through very thin insulating and conducting layers; in this case an aperture-free grid could be used, and one or both vacuum regions might be filled with an appropriate material. Some subset of these techniques could be combined to produce 3-d integrated circuits as envisioned by Shoulders, but on a substantially smaller size scale with extremely high switching speeds. However, there are some mass production bottle necks which biological pattern formation/template processes might overcome. For example, self organizing protein assemblies like microtubules (which can be made to form 2-d arrays) or related structures (Hameroff, 1987) could serve as precise nanometer scale lattice templates.

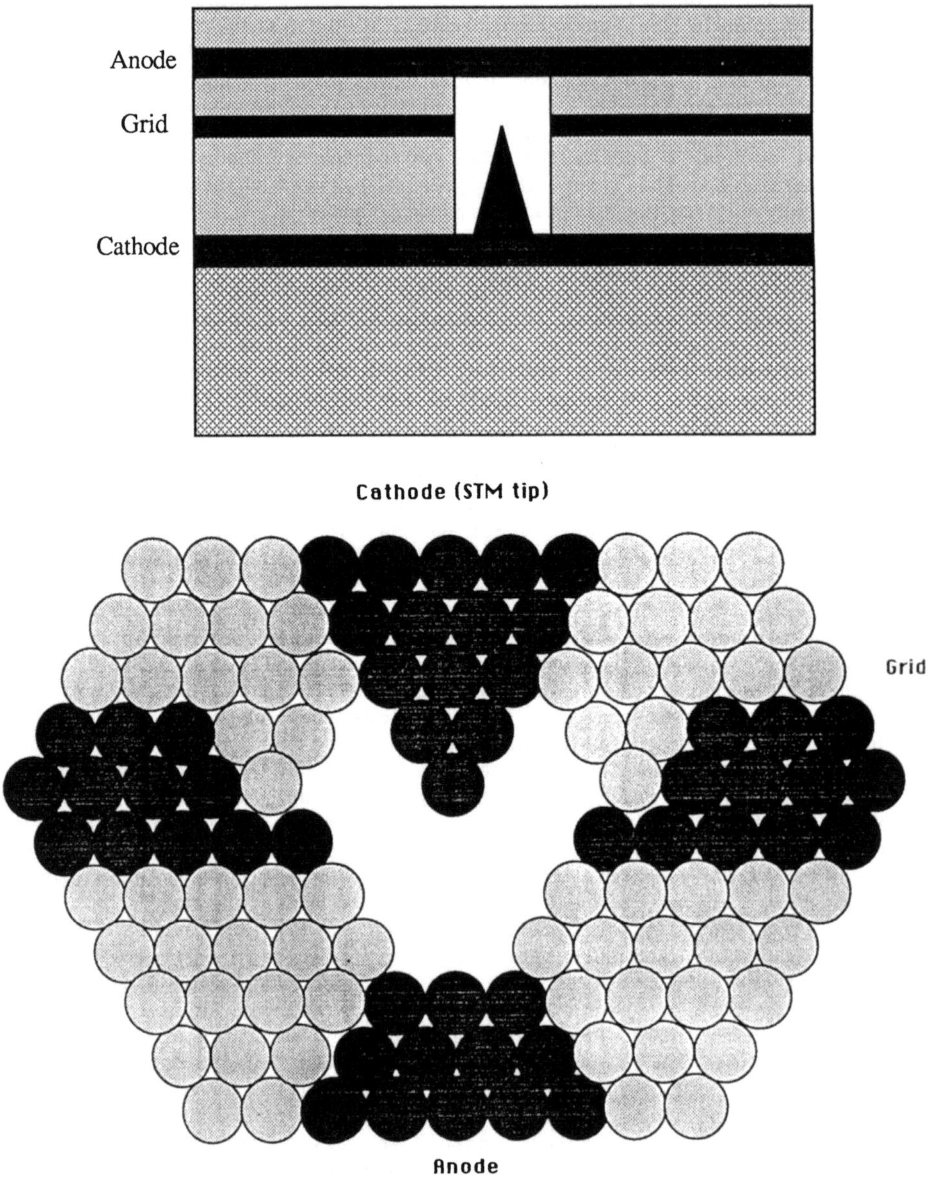

Fig. 3. Top: schematic example of an integrated circuit microtriode (cathode cone is approximately 1 μm tall). Bottom: speculative example of a nanotriode using a minimum scale monoatomic tip cathode (cathode cone is approximately 1 nm tall).

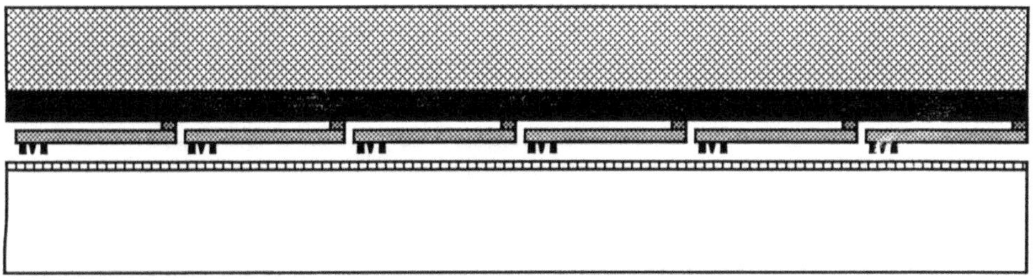

Fig. 4. Schematic cross section of a 100 × 100 planar array of tracking field emitters for use in: 1) high density memories; 2) large area, high resolution surface mapping; and 3) nanolithography.

Other aspects of Shoulders' program entailed using his field emitter structures with submicroscopic electrostatic electron beam lenses for lithography and microscopy so that the products of his technology could then be used in the production process itself. McCord and Pease (1986, 1987) have used STMs operated at 50 V in vacuum for nanolithography; however, they had to use a positively biased tip since the electric field otherwise caused secondary electrons to deflect back to the surface and lower the effective lithographic resolution and contrast. Fink's tips have the advantage of concentrating the potential drop in front of the tip; this results in a low positive field near the sample and does not drive secondary electrons back into the substrate. This makes this system a potential candidate for advanced nanolithography. Further, these tips can also generate exceedingly fine ion beams when operated in field ion microscope mode. Such beams might be used for ultraprecision ion milling, thus extending the scope of "atomic bit machining" processes discussed by Taniguchi (1985).

There are several advantages to using field emission processes: larger electron energies are available for initiating physical/chemical processes, the sensitivity of electrode-substrate gaps to vibration is reduced, larger electrode-substrate gaps permit much higher lateral scanning speeds to be used; and the larger voltages, currents, and gaps used place fewer constraints on the servo electronics. The main disadvantage is the need for clean high vacuum conditions for stable field emission and to prevent emitter degradation by contaminants. However, very small vacuum chambers may be used.

The first challenge is the mass production and placement of Fink's trimer emitter tips. The second challenge has to do with nano-lithography of a few critical regions of the nano-triodes where the statistical and graininess aspects of material deposition or removal would adversely affect device uniformity and operating margins.

We would like to be able to *mass produce* monoatomic, trimer, or even hexamer STM tips for use in multi-tip micro scanning electron microscopes, nanolithography, nanomachining, and switching devices. In addition we seek means to place such tips on existing structures in specific orientations. Further, in building devices, we require means of precisely spacing other parts in relation to such tips. Here is where use of monoclonal antibodies, Langmuir-Blodgett films, and other molecular assembly techniques could be used (Kuhn, 1983; Lehn, 1988; Carter, 1983, 1986).

There are some hybrid fabrication possibilities. At the level of pattern formation, conventional nanolithography seems best suited for determining long-range order in systems (data and address busses, power supply lines,

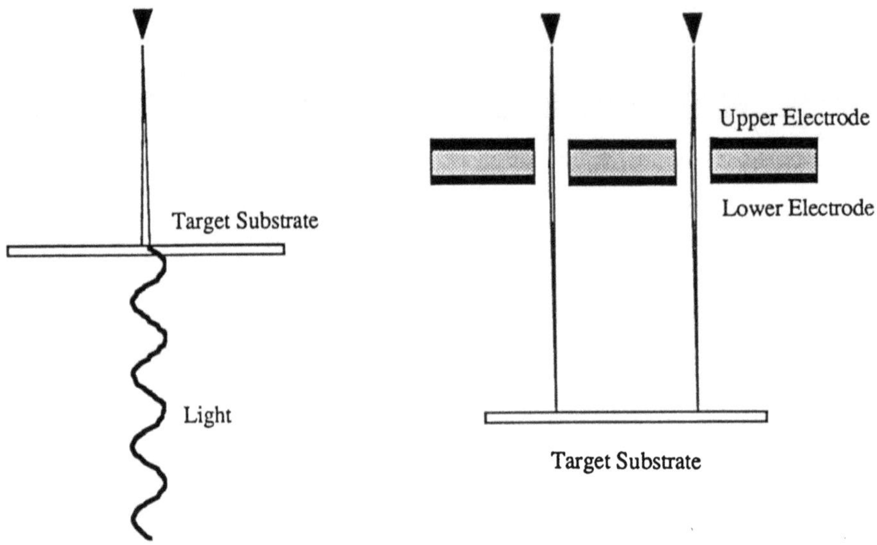

Fig. 5. Schematic cross section illustrating use of electron beam for nano-optical imaging or read out of thin film state.

etc.) whereas self-assembly processes seem best suited for establishing short-range order where extremely precise tolerances are required (within devices, between component layers, etc.).

How could trimer tips be formed? One possibility is biomolecular lithography, where the interstitial regions between molecules might be used as molds. Another possibility might be the use of monoclonal antibodies to select out ion milled nanoparticles which can then be detached from their supporting substrate.

<u>Single Electron Devices</u>

As we shrink atomic point source vacuum triodes (and related devices) toward their minimum physical dimensions, not only do we cross into the domain of potential 10^{-12} sec. operating speeds, we also reach the domain where the effects of single electron charging of cathodes, grids, and anodes become significant. Single electron charging phenomena has already been observed in STM junctions by Fulton and Doolan (1987). Likharev (1987) has done preliminary analysis of single electron transistors. He suggests that at the minimum size scale, his proposed configuration could be implemented using molecular structures. What happens if we use *vacuum* for the required insulating barriers? Indeed, what happens if we fabricate nanoscale versions of Mead's (1961) solid state triodes using *vacuum* for the tunneling barriers? The work on STM/ballistic-electron-emission microscopy (Kaiser and Bell, 1988) represents a possible step to obtaining the needed experimental data for progress in this direction. It's important to note that several important classes of electrical networks involving multiway feedback can function in (and sometimes utilize) the presence of shot noise, tunneling induced crosstalk, and other effects that would characterize the most compact and densely connected systems using such extremely fast and small devices.

Just as silicon micromechanical systems have been "spun off" of conventional VLSI fabrication technology, ultimately *nano-electro-mechanical* systems involving single electron charging may also likewise result from single electron systems. Shoulders (1965) proposed submicron electro-mechanical structures operating in the megahertz range and noted that the minimum size of his proposed systems were limited mainly by the resolution of the fabrication technology —— which now could (and should!) be extended all the way down into the one nanometer scale.

ELECTRON BEAM DATA STORAGE DEVICES

Electron beam storage systems for computers have been studied for over 20 years but were too expensive for widespread use. More recently, designs for silicon microSTMs and their application to storage devices have been patented (Pohl, Quate, 1986); substantial advances beyond these early developments are in the world's patent mills. Use of Fink's tips in such systems would increase potential speeds, extend the range of recording materials, simplify servo systems and reduce required dimensional tolerances.

SENSOR APPLICATIONS

There are many possibilities for applying STM technology to the development of microsensors. Young's (1966) ultramicrometer is an early illustration. Teague has proposed a device using a piezo modulated tunnel gap for a transconductance device. Many other possibilities are outlined in Schneiker (1988) and Schneiker *et al.* (1988). If single electron technology can be made to work and could be could be fabricated (say) on STM tips, it could become a great interfacing tool for molecular electronic devices. Use of Fink's tips in electron beam steering systems suggested by Shoulders (1965), but on a much smaller scale, may find application for high speed A/D converters and a host of other devices that mimic selected state-of-the-art microwave tube technology, etc.

THE FEYNMAN CONTEST FOR MICRO/NANO STMS

Feynman's original (1960) proposal for molecular and atomic scale machines included prizes and competition to "get kids interested in this field". He offered $1,000 to the first person to a) reduce the information on the page of a book to an area 1/25,000 smaller in linear scale to be read by an electron microscope, b) construct an electric motor which is only 1/64 inch cube. Both of these prizes have now been awarded.

Motivated by a desire to accelerate development of STM derived technology as a bridge to more powerful Feynman Machines, Hameroff (1987) and Schneiker (1985, 1986) have announced a series of micro-STM construction challenges and $1,000 prizes. As Feynman originally suggested in 1959:

> *... have some fun! Let's have a competition between laboratories.*

ACKNOWLEDGEMENTS

HDS Systems provided the computer drafted illustrations. Discussions with (and encouragement from) F. Carter, R. Feynman, P. Hansma, D. Pohl, and K. Shoulders were useful in helping CWS formulate some of the ideas presented above.

REFERENCES

Ariadne, 1966a,b,c, Daedalus, New Scientist, 21 July, 153; 3 Nov., 245, 15 Dec., 641.

Ariadne, 1972, Daedalus, New Scientist, 13 April, 104.
Bandyopadhyay, S., M. R. Melloch, S. Datta, B. Das, J. A. Cooper Jr. and M. S. Lundstrom, 1986, A Novel Quantum Interference Transistor (QUIT) with Extremely Low Power-Delay Product and Very High Transconductance, IEEE I.E.D.M., 86:76-79.
Becker, R. S., J. A. Golovchenko and B. S. Swartzentruber, 1987, Atomic-Scale Surface Modifications Using a Tunneling Microscope, Nature, 325:419-421.
Binnig, G., Ch. Gerber, H. Rohrer and E. Weibel, 1985a, Sputter Tip, IBM Tech. Discl. Bul., 27:4890.
Binnig, G., Ch. Gerber, H. Rohrer and E. Weibel, 1985b, Nano-Aperture, IBM Tech. Disclosure Bul., 27(8):4893.
Binnig, G. and H. Rohrer, 1987, Scanning Tunneling Microscopy —— From Birth to Adolescence, Rev. Mod. Phys., 59(3):615-625.
Bocko, M. F., K. A. Stephenson and R. H. Koch, 1988, Vacuum Tunneling Probe: A Nonreciprocal, Reduced-Back-Action Transducer, Phys. Rev. Lett., 61(6):726-729.
Boyes, E. D., 1984, High Resolution at Low Voltage: The SEM Philosopher's Stone? in: "Proceedings of the 42nd Annual Meeting of the Electron Microscopy Society of America," G. W. Bailey, ed., San Francisco Press, San Francisco, pp. 446-449.
Buck, D. A. and K. R. Shoulders, 1958, An Approach to Microminiature Printed Systems, in: "Proc. Eastern Joint Computer Conference," American Institute of Electrical Engineers, New York, pp. 55-59.
Carter, F. L., 1979, Problems and Prospects of Future Electroactive Polymers and 'Molecular' Electronic Devices, in: "the NRL Program on Electroactive Polymers, First Annual Report," L. D. Lockhart, Jr., ed., NRL Memorandum Report 3960, p. 121.
Carter, F. L., 1980, Further Considerations on 'Molecular' Electronic Devices, Second Annual Report, R. B. Fox, ed., NRL Memorandum Report 4335, 35.
Carter, F. L., 1983, Molecular Level Fabrication Techniques and Molecular Electronic Devices, J. Vac. Sci. Technol. B, 1(4):959-968.
Carter, F. L.,, 1986, Chemistry and Microstructures: Fabrication at the Molecular Size Level, Superlattices and Microstructures, 2(2):113-128.
Craighead, H. G., 1984, 10-nm Resolution Electron-Beam Lithography, J. Appl. Phys., 55(12):4430-4435.
Craighead, H. G. and P. M. Mankiewich, 1982, Ultra-Small Metal Particle Arrays Produced by High Resolution Electron-Beam Lithography, J. Appl. Phys., 53(11):7186-7188.
Crewe, A. V., 1976, Very Low Voltage Electron Microscopy, Ultramicroscopy, 1:267-269.
Csepregi, L., 1985, Micromechanics: A Silicon Microfabrication Technology, Microelectronic Engineering, 3:221-234.
De Brabander, M., R. Nuydens, G. Geuens, M. Moeremans, J. De Mey and C. Hopkins, 1988, Nanovid Ultramicroscopy: A New Non-Destructive Approach Providing New Insights in Subcellular Motility, in: "Microtubules and Microtubule Inhibitors 1985," M. De Brabander and J. De Mey, eds., Elsevier Science Publishers, North-Holland, pp. 187-196.
Deckman, H. W., B. Abeles, J. H. Dunsmuir and C. B. Roxlo, 1987, Microfabrication of Molecular Scale Microstructures, Appl. Phys. Lett., 50(9):504-506.
Dietrich, H. P., M. Lanz and D. F. Moore, 1984, Ion Beam Machining of Very Sharp Points, IBM Tech. Discl. Bul., 27:3039-3040.
Douglas, K. and N. A. Clark, 1986, Nanometer Molecular Lithography, Appl. Phys. Lett., 48(10):676-678.
Feynman, R. P., 1960a, There's Plenty of Room at the Bottom, Engrg. and Sci. (Cal. Inst. of Tech.), Feb., 22-36, and in: Miniaturization," H. D. Gilbert, ed., Reinhold, New York, 1961, pp. 282-296, and in: Schneiker, 1986b.

Feynman, R. P., 1960b, The Wonders that Await a Micro-Microscope, Saturday Review, 43, 2 April, 45-47.
Feynman, R. P., 1984, 1985, 1986, Personal communication.
Feynman, R. P., 1986, Quantum Mechanical Computers, Foundations of Physics, 16(6):507-531.
Fink, H.-W., 1986, Mono-Atomic Tips for Scanning Tunneling Microscopy, IBM J. Res. Develop., 30(5):460-465.
Fink, H.-W., 1988, Point Source for Ions and Electrons, Physica Scripta, 38:260-263.
Fischer, U. Ch., U. T. Durig and D. W. Pohl, 1988, Near-field Optical Scanning Microscopy in Reflection, Appl. Phys. Lett., 52(4):249-251.
Foster, J. S., J. E. Frommer and P. C. Arnett, 1988, Molecular Manipulation Using a Tunneling Microscope, Nature, 331:324-326.
Franks, A., 1987, Nanotechnology, J. Phys. E: Sci. Instrum., Dec., 1442-1451.
Fulton, T. A. and G. J. Dolan, 1987, Observations of Single-Electron Charging Effects in Small Tunnel Junctions, Phys. Rev. Lett., 59(1):109-112.
Gabriel, J. K., M. M. Mehragany and W. S. N. Trimmer, 1987, Micro Mechanical Components, in: "Modeling and Control of Robotic Manipulators and Manufacturing Process," Winter Annual Meeting of the American Society of Mechanical Engineers, December, pp. 397-401.
Gobrecht, J. and J. B. Pethica, 1986, The Potential of Mechanical Microlithography for Submicron Patterning, Microelectronic Engineering, 5:471-475.
Godin, T. J. and R. Haydock, 1986, Quantum Circuit Theory, Superlattices and Microstructures, 2(6):597-600.
Golvochenko, J. A., 1986, The Tunneling Microscope: A New Look at the Atomic World, Science, 232:48-53.
Gomer, R., 1986, Possible Mechanisms of Atom Transfer in Scanning Tunneling Microscopy, IBM J. Res. Develop., 30(4):428-430.
Gomez, J., L. Vazquez, A. M. Baro, N. Garcia, C. L. Perdriel, W. E. Triaca and A. J. Arvia, 1986, Surface Topography of (100)-type Electro-Faceted Platinum From Scanning Tunneling Microscopy and Electrochemistry, Nature, 323:612-614.
Gould, S., O. Marti, B. Drake, L. Hellemans, C. E. Bracker, P. K. Hansma, N. L. Keder, M. M. Eddy and G. D. Stucky, 1988, Molecular Resolution Images of Amino Acid Crystals with the Atomic Force Microscope, Nature, 332:332-334.
Granqvist, C. G. and R. A. Buhrman, 1976, Ultrafine Metal Particles, J. Appl. Phys., 47(5):2200-2219.
Gray, H. F. and G. J. Campisi, 1986, A Silicon Field Emitter Array Planar Vacuum FET Fabricated with Microfabrication Techniques, in: "Science and Technology of Microfabrication," Materials Research Society Symposia Proceedings, R. E. Howard, E. L. Hu, S. Namba and S. Pang, eds., pp. 25-30.
Haddon, R. C. and A. A. Lamola, 1985, The Molecular Electronic Device and the Biochip Computer: Present Status, Proc. Nat. Acad. Sci. USA, 82:1874-1878.
Hameroff, S. R., 1987, Ultimate Computing: Biomolecular Consciousness and Nanotechnology, Elsevier North-Holland, Amsterdam.
Hansma, P. K. and J. Tersoff, 1987, Scanning Tunneling Microscopy, J. Appl. Phys., 61(2):R1-R23.
Hansma, P. K., V. B. Elings, O. Marti and C. E. Bracker, 1988, Scanning Tunneling Microscopy and Atomic Force Microscopy: Application to Biology and Technology, Science, 242:209-216.
Howard, R. E., L. D. Jackel and W. J. Skocpol, 1985, Nanostructures: Fabrication and Applications, Microelectronic Engineering, 3:3-16.
Joy, D. C., 1987, Low Voltage Scanning Electron Microscopy, in: "Electron Microscopy and Analysis 1987," L. M. Brown, ed., Institute of Physics Conference Series, Bristol, pp. 175-180.
Kaiser, W. J. and L. D. Bell, 1988, Direct Investigation of Subsurface Inter-

face Electronic Structure by Ballistic-Electron-Emission Microscopy, Phys. Rev. Lett., 60(14):1406-1409.
Kaminsky, G., 1985, Micromachining of Silicon Mechanical Structures, J. Vac. Sci. Technol. B, 3(4):1015-1024.
Kiewit, D. A., 1973, Microtool Fabrication by Etch Pit Replication, Rev. Sci. Instrum., 44(12):1741-1742.
Kratschmer, E., and M. Isaacson, 1986, Nanostructure Fabrication in Metals, Insulators, and Semiconductors using Self-Developing Metal Inorganic Resist, J. Vac. Sci. Technol. B, 4(1):361-364.
Kuhn, H., 1983, Functionalized Monolayer Assembly Manipulation, Thin Solid Films, 99:1-16.
Lehn, J. M., 1988, Supramolecular Chemistry — Scope and Perspectives Molecules, Supermolecules, and Molecular Devices (Nobel Lecture), Angewandte Chemie, 27(1):89-112.
Likharev, K. K., 1987, Single-Electron Transistors: Electrostatic Analogs of the DC Squids, IEEE Transactions on Magnetics, 23(2):1142-1145.
Lin, C. W., F.-R. F. Fan and A. J. Bard, 1987, High Resolution Photoelectrochemical Etching of n-GaAs with the Scanning Electrochemical and Tunneling Microscope, J. Electrochem. Soc., 134:1038-1039.
McCord, M. A., and R. F. W. Pease, 1985, High Resolution, Low-Voltage Probes From a Field Emission Source Close to the Target Plane, J. Vac. Sci. Technol. B, 3(1):198-201.
McCord, M. A. and R. F. W. Pease, 1986, Lithography with the Scanning Tunneling Microscope, J. Vac. Sci. Technol. B, 4(1):86-88.
McCord, M. A. and R. F. W. Pease, 1987, Exposure of Calcium Fluoride Resist with the Scanning Tunneling Microscope, J. Vac. Sci. Technol. B, 5(1):430-433.
Mead, C. A., 1961, Operation of Tunnel-Emission Devices, J. Appl. Phys., 32(4):646-652.
Panitz, J. A., 1986, Immunologic Layer Formation on Metal Microelectrodes, J. Colloid and Interface Science, 111(2):516-528.
Pethica, J. B., 1988, Atomic-Scale Engineering, Nature, 331:301.
Pohl, D. W., 1986, Silicon MicroMechanical STM, IBM patent application.
Quate, C. F., 1986, Method and Means for Data Storage using Tunnel Current Data Readout, US Patent 4,575,822, 11 March.
Schneiker, C. W. and S. R. Hameroff, 1988, NanoTechnology Workstation Based on Scanning Tunneling/Optical Microscopy: Applications to Molecular Scale Devices, in: "Molecular Electronic Devices, Proceedings of the Third International Workshop on Molecular Electronic Devices, October, 1986," F. L. Carter, R. E. Siatkowski, and H. Wohltjen, Elsevier North-Holland, Amsterdam, pp. 69-90.
Schneiker, C. W., 1988, NanoTechnology with Feynman Machines: Scanning Tunneling Engineering and Artificial Life, in: "Artificial Life, Sante Fe Institute Studies in the Sciences of Complexity," C. Langton, ed., Addison-Wesley, pp. 443-500.
Schneiker, C. W., S. R. Hameroff, M. A. Voelker, J. He, E. Dereniak and R. McCuskey, 1988, Scanning Tunneling Engineering, J. Microscopy, 152(2):585-596.
Serena, P. A., L. Escapa, J. J. Saenz, N. Garcia and H. Rohrer, 1988, Coherent Electron Emission from Point Sources, J. Microscopy, 152(1):43-51.
Shoulders, K. R., 1960, On Microelectronic Components, Interconnections, and System Fabrication, Proc. Western Joint Computer Conference, 251-258.
Shoulders, K. R., 1961, Microelectronics Using Electron-Beam-Activated Machining Techniques, in: "Advances in Computers," Franz Alt, ed., Academic Press, New York, pp. 135-293.
Shoulders, K. R., 1962, On Microelectronic Components, Interconnections, and System Fabrication, in: "Aspects of the Theory of Artificial Intelligence," C. A. Muses, ed., Plenum Press, New York, pp. 217-235.
Shoulders, K. R., 1965, Toward Complex Systems, in: "Microelectronics and Large Systems," Spartan Books, pp. 97-128.
Shoulders, K. R., 1985, 1986, Personal Communication.

Simpson, M., P. Smith and G. A. Dederski, 1987, Atomic Layer Epitaxy, Surface Engineering, 3(4):343-348.

Sonnenfeld, R. and B. C. Schardt, 1986, Tunneling Microscopy in an Electrochemical Cell: Images of Ag Plating, Appl. Phys. Lett., 49(18):1172-1174.

Spindt, C. A., C. E. Holland and R. D. Stowell, 1984, Recent Progress in Low-Voltage Field-Emission Cathode Development, J. De Physique, Supplement, Dec., C9-269—C9-278.

Staufer, U., R. Wiesendanger, L. Eng, L. Rosenthaler, H. R. Hidber, H.-J. Güntherodt and N. Garcia, 1987, Nanometer Scale Structure Fabrication with the Scanning Tunneling Microscope, Appl. Phys. Lett., 51(4):244-246.

Stroscio, M. A., 1986a, Quantum-Based Electronic Devices, Superlattices and Microstructures, 2(1):45-47.

Taniguchi, N., 1974, On the Basic Concept of Nanotechnology, in: "Proc. Int. Conf. Prod. Eng. Tokyo," Part 2, JSPE, Tokyo, pp. 18-23.

Taniguchi, N., 1983, Current Status in, and Future Trends of, Ultraprecision Machining and Ultrafine Materials Processing, Ann. CIRP, 32:1-10.

Taniguchi, N., 1985, Atomic Bit Machining by Energy Beam Processes. Prec. Engr., 4:145-155.

Teague, E. C., 1978, Vacuum Tunneling, in: "Proc. NSF Workshop on Opportunities for Microstructure Science, Engineering and Technology, Nov. 19-22," Airlie House, Airlie, VA, p. 129.

Teague, E. C., 1986, Room Temperature Gold-Vacuum-Gold Tunneling Experiments, J. of Research of the National Bureau of Standards, 91(4):171-233.

Whitehead, M. Isaacson and E. Wolf, 1985, Nanometer Scale Metal Wire Fabrication, Microelectronic Engineering, 3:25-32.

Yamamura, Y., T. Fujisawa and S. Namba, 1984, Nanometer Structure Electronics, An Investigation of the Future of Microelectronics, in: "Proc. of the International Symposium on Nanometer Structure Electronics, April 16-18," Osaka University, Toyonaka, Japan.

Yates, F. E., 1984, Report on Conference on Chemically-Based Computer Designs, Crump Institute for Medical Engineering Report CIME TR/84/1, University of California, Los Angeles, CA.

Young, R. D., 1966, Field Emission Ultramicrometer, Rev. of Sci. Instruments, 37(3):275-278.

Young, R. D., J. Ward and F. Scire, 1971, Observation of Metal-Vacuum-Metal Tunneling, Field Emission, and the Transition Region, Phys. Rev. Lett., 27:922.

Young, R. D., J. Ward and F. Scire, 1972, The Topografiner: An Instrument for Measuring Surface Microtopography, Rev. Sci. Instrum., 43(7):999-1011.

Zingsheim, H. P., 1978, Molecular Engineering Using Nanometer Surface Microstructures, in: "Proceedings: NSF Workshop on Opportunities for Microstructure Science, Engineering and Technology in Cooperation with the NRC Panel on Thin Film Microstructure Science and Technology, November 19-22, 1978," Airlie House, Airlie, VA.

CONTRIBUTORS

P. L. Ahl
Naval Research Laboratory
Bio/Molecular Engineering Branch
Washington, DC 20375-5000, USA

Masuo Aizawa
Department of Bioengineering
Tokyo Institute of Technology
O-okayama, Meguro-ku
Tokyo 152, Japan

O. Albrecht
Molecular Electronics Corporation
4030 Spencer St., MS 108
Torrance, CA 90503, USA

David N. Beratan
Jet Propulsion Laboratory
California Institute of Technology
Pasadena, CA 91109, USA

Robert R. Birge
Center for Molecular Electronics
and Department of Chemistry
Syracuse University
Syracuse, NY 13244, USA

Martin Blank
Department of Physiology
and Cellular Biophysics
Columbia University
College of Physicians and Surgeons
630 West 168th Street
New York, NY 10032, USA

P. W. Bohn
Department of Chemistry
and Materials Research Laboratory
University of Illinois
at Urbana-Champaign
Urbana, IL 61801, USA

Vladimir N. Buravtsev
Research Computing Center
U.S.S.R. Academy of Sciences
Pushchino, Moscow Region 142292
U.S.S.R.

A. Cambiaso
Department of Biophysical
and Electronic Engineering
University of Genoa
Genoa, Italy

F. Caratozzolo
Department of Biophysical
and Electronic Engineering
University of Genoa
Genoa, Italy

S. Cenderelli
Department of Biophysical
and Electronic Engineering
University of Genoa
Genoa, Italy

G. Cerofolini
ENICHEM
Milan, Italy

A. Chiabrera
Department of Biophysical
and Electronic Engineering
University of Genoa
Genoa, Italy

R. Ciacarelli
Life Sciences Lab
Eastman Kodak Company
Rochester, NY 14650, USA

Michael Conrad
Department of Computer Science
Wayne State University
Detroit, Michigan 48202, USA

A. Corin
Life Sciences Lab
Eastman Kodak Company
Rochester, NY 14650, USA

T. M. Cotton
Department of Chemistry
University of Nebraska
Lincoln, NE 68588-0304, USA

Eustace Dereniak
Optical Sciences Center
University of Arizona
Tucson, Arizona 85724, USA

V. Desimirović
Molecular Machines Research Center
Faculty of Machine Engineering
University of Belgrade
27. Marta 80
11000 Belgrade, Yugoslavia

D. Djaković
Molecular Machines Research Center
Faculty of Machine Engineering
University of Belgrade
27. Marta 80
11000 Belgrade, Yugoslavia

T. V. Djukova
Institute of Biological Physics
U.S.S.R. Academy of Sciences
Pushchino, Moscow Region, 142292
U.S.S.R.

A. B. Druzhko
Institute of Biological Physics
U.S.S.R. Academy of Sciences
Pushchino, Moscow Region, 142292
U.S.S.R.

M. Esposito
Department of Biophysical
and Electronic Engineering
University of Genoa
Genoa, Italy

J. Falvo
Life Sciences Lab
Eastman Kodak Company
Rochester, NY 14650, USA

T. L. Fare
Naval Research Laboratory
Bio/Molecular Engineering Branch
Washington, DC 20375-5000, USA

D. T. Friesen
Bend Research, Inc.
Bend, OR 97701-8599, USA

A. N. Georgobiani
All-Union Research Centre for
Surface and Vacuum Investigation
Moscow, U.S.S.R.

T. Ginnai
Molecular Electronics Corporation
4030 Spencer St., MS 108
Torrance, CA 90503, USA

A. Gliozzi
Department of Physics
University of Genoa
Genoa, Italy

Girjesh Govil
Tata Institute
of Fundamental Research
Homi Bhabha Road
Bombay 400 005, India

M. Grattarola
Department of Biophysical
and Electronic Engineering
University of Genoa
Genoa, Italy

Oleg V. Gritsenko
Research Computing Center
U.S.S.R. Academy of Sciences
Pushchino, Moscow Region 142292
U.S.S.R.

Xiangming Gu
Institute of Biophysics
Academia Sinica
Beijing
People's Republic of China

Present Address:
National Synchrotron Radiation Lab
Institute of High Energy Physics
Academia Sinica
Beijing
People's Republic of China

Stuart R. Hameroff
Advanced Biotechnology Laboratory
Department of Anesthesiology
University of Arizona
Health Sciences Center
Tucson, Arizona 85724, USA

A. Harrington
Molecular Electronics Corporation
4030 Spencer St., MS 108
Torrance, CA 90503, USA

Jackson He
Optical Sciences Center
University of Arizona
Tucson, Arizona 85724, USA

C. A. Helm
Department of Chemical
and Nuclear Engineering
University of California,
Santa Barbara
Santa Barbara, CA 93106, USA

D. Holzschu
Life Sciences Lab
Eastman Kodak Company
Rochester, NY 14650, USA

Felix T. Hong
Department of Physiology
Wayne State University
Detroit, Michigan 48201, USA

J. J. Hopfield
Divisions of Chemistry and Biology
California Institute of Technology
Pasadena, CA 91125, USA
and
AT&T Bell Laboratories
Murray Hill, NJ 07974, USA

Kun-Sheng Hu
Institute of Biophysics
Academia Sinica
Beijing
People's Republic of China

Toshio Iijima
Electrotechnical Laboratory
Tsukuba City, Ibaraki 305
Japan

Guilford Jones, II
Department of Chemistry
Boston University
Boston, MA 02215, USA

A. D. Joran
Department of Chemistry
Case Western Reserve University
Cleveland, Ohio 44106, USA

Present Address:
Department of Chemistry
New York University
Washington Place
New York, NY 10003, USA

Yasujiro Kawabata
National Chemical Laboratory
for Industry
Higashi, Tsukuba, Ibaraki 305
Japan

J.-H. Kim
Department of Chemistry
University of Nebraska
Lincoln, NE 68588-0304, USA

Yonosuke Kobatake*
Faculty of Pharmaceutical Sciences
Hokkaido University
Sapporo, Japan
*deceased

V. Kochev
Membrane Biophysics Laboratory
Department of Physiology
Michigan State University
East Lansing, Michigan 48824, USA

Permanent Address:
Faculty of Physics
University of Sofia
Sofia, Bulgaria

Djuro Koruga
Molecular Machines Research Center
Faculty of Machine Engineering
University of Belgrade
27. Marta 80,
11000 Belgrade, Yugoslavia

Hans Kuhn
Max-Planck Institute
for Biophysical Chemistry
Göttingen, FRG

Present Address:
Ringoldswilstrasse 50
CH-3656 Tschingel ob Gunten
Switzerland

Albert F. Lawrence
Center for Molecular Electronics
and Department of Chemistry
Syracuse University
Syracuse, NY 13244, USA

Pavel I. Lazarev
Research Computing Center
U.S.S.R. Academy of Sciences
Pushchino, Moscow Region 142292
U.S.S.R.

Present Address:
Institute of Biophysics
U.S.S.R. Academy of Sciences
Pushchino, Moscow Region 142292
U.S.S.R.

F. S. Ligler
Naval Research Laboratory
Bio/Molecular Engineering Branch
Washington, DC 20375-5000, USA

D. Lorenz
Bend Research, Inc.
Bend, OR 97701-8599, USA

Rajesh Manchanda
Tata Institute
of Fundamental Research
Homi Bhabha Road
Bombay 400 005, India

D. Marr-Leisy
Molecular Electronics Corporation
4030 Spencer St., MS 108
Torrance, CA 90503, USA

Z. Martinović
Molecular Machines Research Center
Faculty of Machine Engineering
University of Belgrade
27. Marta 80
11000 Belgrade, Yugoslavia

Gen Matsumoto
Electrotechnical Laboratory
Tsukuba City, Ibaraki 305
Japan

Kenji Matsumoto
Faculty of Pharmaceutical Sciences
Hokkaido University
Sapporo, Japan

G. Mazzoni
Department of Physics
University of Genoa
Genoa, Italy

Robert McCuskey
Department of Anatomy
University of Arizona
College of Medicine
Tucson, Arizona 85724, USA

G. McLendon
Department of Chemistry
University of Rochester
Rochester, NY 14627, USA

L. Meda
SGS - Thompson
Agrate, Italy

D. R. Miller
Department of Chemistry
and Materials Research Laboratory
University of Illinois
at Urbana-Champaign
Urbana, IL 61801, USA

Dj. Minić
Faculty of Electrical Engineering
University of Belgrade
P.O. Box 816
11001 Belgrade, Yugoslavia

B. Nie
Department of Biophysical
and Electronic Engineering
University of Genoa
Genoa, Italy

Permanent address:
Department of Biomedical
Engineering
Sun Yat-sen University
of Medical Science
Guangzhou
People's Republic of China

José Nelson Onuchi
Instituto de Física
e Química de São Carlos
Universidade de São Paulo 13560
São Carlos, SP, Brazil

M. A. Ostrovsky
Institute of Chemical Physics
U.S.S.R. Academy of Sciences
Kosygin St. 4
Moscow, B-334, U.S.S.R.

A. Ottova
Membrane Biophysics Laboratory
Department of Physiology
Michigan State University
East Lansing, Michigan 48824, USA

Permanent Address:
Institute of Biotechnology
Slovak Technical University
Bratislava, Czechoslovakia

R. Paradiso
Department of Biophysical
and Electronic Engineering
University of Genoa
Genoa, Italy

K. Purdue
Department of Chemistry
University of Rochester
Rochester, NY 14627, USA

G. Parodi
Department of Biophysical
and Electronic Engineering
University of Genoa
Genoa, Italy

M. Parodi
Institute of Electrical
Engineering
University of Cagliari
Cagliari, Italy

Ratna S. Phadke
Tata Institute
of Fundamental Research
Homi Bhabha Road
Bombay 400 005, India

L. Powers
Center for Bio-catalysis Science and Technology
Utah State University
Logan, UT 84322-4630, USA

D. Raković
Faculty of Electrical Engineering
University of Belgrade
P.O. Box 816
11001 Belgrade, Yugoslavia
and
Molecular Machines Research Center
Faculty of Machine Engineering
University of Belgrade
27. Marta 80
11000 Belgrade, Yugoslavia

N. G. Rambidi
All-Union Research Centre for Surface and Vacuum Investigation
Moscow, U.S.S.R.

Present Address:
International Research Institute for Management Sciences
9 Prospekt 60-Pet Oktyabria
117312 Moscow, U.S.S.R.

Steen Rasmussen
Center for Nonlinear Studies and Theoretical Division
Los Alamos National Laboratories, MS-B258
Los Alamos, New Mexico 87545, USA

G. W. Rayfield
Physics Department
Eugene, OR 97403, USA
and
Bend Research, Inc.
Bend, OR 97701-8599, USA

D. A. Rintoul
Division of Biology
Kansas State University
Manhattan, KS 66506, USA

V. Rodov
Molecular Electronics Corporation
4030 Spencer St., MS 108
Torrance, CA 90503, USA

Alexander I. Rubinstein
Institute of Experimental Endocrinology and Hormone Chemistry
U.S.S.R. Academy of Medical Sciences
11 Dmitri Ulyanova St.
Moscow 117162, U.S.S.R.

Z. Salamon
Membrane Biophysics Laboratory
Department of Physiology
Michigan State University
East Lansing, Michigan 48824, USA

Permanent Address:
Institute of Physics
Technical University of Poznan
Poznan, Poland

Conrad Schneiker
Advanced Biotechnology Laboratory
Department of Anesthesiology
University of Arizona
College of Medicine
Tucson, Arizona 85724, USA

J. M. Schnur
Naval Research Laboratory
Bio/Molecular Engineering Branch
Washington, DC 20375-5000, USA

D. M. Schufeldt
Department of Chemistry
University of Nebraska
Lincoln, NE 68588-0304, USA

K. D. Seib
Naval Research Laboratory
Bio/Molecular Engineering Branch
Washington, DC 20375-5000, USA

F. Sherman
Department of Biochemistry
University of Rochester
Medical Center
Rochester, NY 14627, USA

Hitoshi Shichi
Eye Research Institute
Oakland University
Rochester, Michigan 48309, USA

Present Address:
Kresge Eye Institute
Wayne State University
Detroit, Michigan 48201, USA

Hiroaki Shinohara
Department of Bioengineering
Tokyo Institute of Technology
O-okayama, Meguro-ku
Tokyo 152, Japan

A. Singh
Naval Research Laboratory
Bio/Molecular Engineering Branch
Washington, DC 20375-5000, USA

Victor S. Sivozhelezov
Research Computing Center
U.S.S.R. Academy of Sciences
Pushchino, Moscow Region 142292
U.S.S.R.

J. W. Smuda
Naval Research Laboratory
Bio/Molecular Engineering Branch
Washington, DC 20375-5000, USA

S. Solmi
LAMEL, CNR
Bologna, Italy

Pill-Soon Song
Department of Chemistry
and Section of
Molecular Plant Biology,
School of Biological Sciences
University of Nebraska
Lincoln, NE 68588, USA

Walther Stoeckenius
Department of Biochemistry
and Biophysics, and
Cardiovascular Research Institute
University of California
San Francisco, CA 94143-0130, USA

Kazuhiro Takahashi
Electrotechnical Laboratory
Tsukuba, Ibaraki 305, Japan

Man-Qi Tan
Institute of Biophysics
Academia Sinica
Beijing
People's Republic of China

Hiroshi Tanino
Electrotechnical Laboratory
Tsukuba, Ibaraki 305, Japan

M. Tedesco
Department of Biophysical
and Electronic Engineering
University of Genoa
Genoa, Italy

H. T. Tien
Membrane Biophysics Laboratory
Department of Physiology
Michigan State University
East Lansing, Michigan 48824, USA

P. A. Todua
All-Union Research Centre for
Surface and Vacuum Investigation
Moscow, U.S.S.R.

Tian Yow Tsong
Department of Biochemistry
University of Minnesota
College of Biological Sciences
St. Paul, MN 55108, USA

Tetsuo Ueda
Faculty of Pharmaceutical Sciences
Hokkaido University
Sapporo, Japan

R. A. Uphaus
Department of Chemistry
University of Nebraska
Lincoln, NE 68588-0304, USA

Vitaly Vodyanoy
Department of Physiology
and Biophysics
University of California, Irvine
Irvine, California 92717, USA

Present Address:
Department of Physiology
and Pharmacology
Auburn University
Auburn, Alabama 36849, USA

Mark Voelker
Optical Sciences Center
University of Arizona
Tucson, Arizona 85724, USA

N. N. Vsevolodov
Institute of Biological Physics
U.S.S.R. Academy of Sciences
Pushchino, Moscow Region, 142292
U.S.S.R.

C. Wamser
Chemistry Department
Portland State University
Portland, OR 97207-0751, USA

Cathy Weiss
Department of Chemistry
Boston University
Boston, MA 02215, USA

Hank Wohltjen
Microsensor Systems, Inc.
Springfield, VA 22151, USA

So-ichi Yabuki
Department of Bioengineering
Tokyo Institute of Technology
O-okayama, Meguro-ku
Tokyo 152, Japan

Takafumi Yao
Electrotechnical Laboratory
Tsukuba, Ibaraki 305, Japan

Q. Zhang
Department of Chemistry
University of Rochester
Rochester, NY 14627, USA

Chian-Fan Zhang
Center for Molecular Electronics
and Department of Chemistry
Syracuse University
Syracuse, NY 13244, USA

M. Zviman
Membrane Biophysics Laboratory
Department of Physiology
Michigan State University
East Lansing, Michigan 48824, USA

INDEX

MIX
Papier aus verantwortungsvollen Quellen
Paper from responsible sources
FSC® C105338

If you have any concerns about our products,
you can contact us on
ProductSafety@springernature.com

In case Publisher is established outside the EU,
the EU authorized representative is:
Springer Nature Customer Service Center GmbH
Europaplatz 3, 69115 Heidelberg, Germany

Printed by Libri Plureos GmbH
in Hamburg, Germany